D0082743

PROTEIN
NMR
SPECTROSCOPY

PROTEIN
NMR
SPECTROSCOPY

PRINCIPLES AND PRACTICE

John Cavanagh

Wayne J. Fairbrother

Arthur G. Palmer III

Nicholas J. Skelton

ACADEMIC PRESS

San Diego   New York   Boston

London   Sydney   Tokyo   Toronto

This book is printed on acid-free paper. ∞

Academic Press, Inc.
A Division of Harcourt Brace & Company
525 B Street, Suite 1900, San Diego, California 92101-4495

United Kingdom Edition published by
Academic Press Limited
24-28 Oval Road, London NW1 7DX

Library of Congress Cataloging-in-Publication Data

Protein NMR spectroscopy : principles and practice / by John
  Cavanagh ... [et al.].
        p.   cm.
    ISBN 0-12-164490-1 (alk. paper)
    1. Proteins--Analysis--Laboratory manuals.   2. Nuclear magnetic
  resonance spectroscopy--Laboratory manuals.   I. Cavanagh, John, date.
    [DNLM: 1. Nuclear Magnetic Resonance--methods--laboratory manuals.
  2. Proteins--analysis--laboratory manuals.   QD 96.N8 P967 1995]
  QP551.P69726 1995
  574.19'245'072--dc20
  DNLM/DLC
  for Library of Congress   95-12822
                                                                   CIP

PRINTED IN THE UNITED STATES OF AMERICA
95 96 97 98 99 00 MM   9 8 7 6 5 4 3 2 1

CONTENTS

CHAPTER 3

EXPERIMENTAL ASPECTS OF NMR SPECTROSCOPY

3.1 NMR Instrumentation   95

CHAPTER 4

MULTIDIMENSIONAL NMR SPECTROSCOPY

CHAPTER 5

RELAXATION AND DYNAMIC PROCESSES

CHAPTER 6

EXPERIMENTAL ^{1}H NMR Methods

CHAPTER 7

HETERONUCLEAR NMR EXPERIMENTS

CHAPTER 8

SEQUENTIAL ASSIGNMENTS AND
STRUCTURE CALCULATIONS

PREFACE

Concomitant developments of modern molecular biology and multidimensional nuclear magnetic resonance (NMR) spectroscopy have increased explosively the use of NMR spectroscopy for generating structural and dynamical information on small- to medium-size biological macromolecules. Efficient molecular biological techniques for incorporation of the stable, NMR active, ^{13}C and ^{15}N isotopes into overexpressed proteins have resulted in dramatic advances in the design and implementation of multidimensional heteronuclear NMR spectroscopic techniques. Consequently, the maximum size protein amenable to complete structural investigation has increased from ~10 kDa using ^{1}H homonuclear NMR spectroscopy to ~30 kDa using ^{13}C and ^{15}N heteronuclear NMR spectroscopy and perhaps to ~40 or ~50 kDa using ^{13}C and ^{15}N heteronuclear NMR spectroscopy combined with fractional ^{2}H enrichment. Most recently, *in vitro* transcription techniques have expanded the application of ^{13}C and ^{15}N heteronuclear NMR spectroscopy to RNA molecules. Research programs for isotopically enriching DNA and carbohydrate molecules promise to further extend the reach of these powerful NMR techniques.

The maturation of the field of structural biology has made the study of structure–function relationships of biological macromolecules by NMR spectroscopy an integral part of diverse chemical and biological research efforts. As an indication of the success of the technique, NMR spectroscopy is being increasingly utilized by chemical and biological scientists not specifically trained as NMR spectroscopists. At the same time, a bewildering number of complex ^{13}C and ^{15}N heteronuclear NMR experiments that make increasingly sophisticated use of the quantum mechanics of nuclear spin systems have been developed (for example, compare the 2 ^{1}H radiofrequency pulses utilized in the COSY experiment with the 27 radiofrequency pulses applied at five different frequencies and four extended decoupling sequences utilized in the CBCA(CO)NH experiment). These developments occurred largely after the publication of the seminal texts *NMR of Proteins and Nucleic Acids* by K. Wüthrich in 1986 and *Principles of Nuclear Magnetic Resonance in One and Two Dimensions* by R. R. Ernst, G. Bodenhausen, and A. Wokaun in 1987.

In our view, a definite need exists for a graduate-level textbook that not only describes the practical aspects of state-of-the-art techniques in biomolecular NMR spectroscopy, but also presents the fundamental principles used to develop these techniques. Only a thorough understanding of the unifying principles of NMR spectroscopy empowers a student or researcher to evaluate, implement, and optimize new techniques that continue to emerge at a dizzying pace. In this spirit, *Protein NMR Spectroscopy: Principles and Practice* systematically explicates NMR spectroscopy from the basic theoretical and experimental principles, to powerful theoretical formulations of the quantum mechanics of nuclear spin systems, and ultimately to optimal experimental methods for biomolecular investigations. Although the text concentrates on applications of NMR spectroscopy to proteins, all of the theory and most of the experiments are equally relevant to nucleic acids, carbohydrates, and small organic molecules. The text focuses on the NMR spectroscopy of diamagnetic molecules (without unpaired electron spins); issues germane specifically to paramagnetic molecules (with unpaired electron spins) are discussed in other sources (see Suggested Reading). This text will serve a wide audience of students and researchers reflective of the variety of disciplines that employ NMR spectroscopy, including biochemistry, biology, chemistry, and physics.

Protein NMR Spectroscopy: Principles and Practice provides a comprehensive treatment of the principles and practice of biomolecular NMR spectroscopy. The theoretical basis of NMR spectroscopy is described in Chapters 1, 2, 4, and 5. Classical NMR spectroscopy of isolated spins is introduced through the Bloch equations in Chapter 1. The density matrix and product operator theoretical formalisms of NMR spectroscopy of coupled multi-spin systems are presented in Chapter 2. The major principles of multidimensional NMR spectroscopy, including frequency labeling of coherences, coherence transfer and mixing, and coherence pathway selection, are described in Chapter 4. The principles of nuclear spin relaxation and chemical exchange are developed by using the Bloch, Solomon, and semiclassical theoretical descriptions in Chapter 5. The experimental techniques used in modern multidimensional NMR spectroscopy of biological macromolecules in solution are described in Chapters 3, 6, and 7. Theoretical and practical aspects of experimental NMR spectroscopy, including data acquisition and data processing, are introduced in Chapter 3. Widely used spectroscopic techniques, such as spin decoupling, water suppression, composite pulses, selective pulses, and one-dimensional NMR spectroscopy, also are presented in Chapter 3. Multidimensional ^{1}H homonuclear NMR spectroscopy is described theoretically and illustrated with experimental examples in Chapter 6. Multidimensional ^{13}C/^{15}N heteronuclear NMR spectroscopy is described theoretically and illustrated with experimental examples in Chapter 7. Both Chapters 6 and 7 present the principal experimental techniques used to obtain resonance assignments, to measure internuclear distances, and to determine scalar coupling constants. Methods for the interpretation of NMR spectra, including resonance assignment strategies and protocols for structure calculations, are summarized in Chapter 8. These aspects of biomolecular NMR spectroscopy are evolving rapidly and detailed discussions could constitute an entire additional book. Consequently, Chapter 8 is intended to provide an overview of the subject and an entry into the primary literature.

In order to provide continuity and consistency throughout the text, a single protein, ubiquitin (76 amino acid residues, M_r 8565 Da), is used to demonstrate the experimental aspects of NMR spectroscopy. Unlabeled bovine ubiquitin was purchased from Sigma Chemical Company (product number U6253, St. Louis, MO). ^{15}N-labeled and ^{13}C/^{15}N-double-labeled human ubiquitin were purchased from VLI Research (Southeastern, PA). The human and bovine amino acid sequences are identical. NMR spectroscopy was performed using Bruker 500- and 600-MHz NMR spectrometers at a temperature of 300 K. Sample concentrations were 2.0 mM for unlabeled ubiquitin and 1.25 mM for labeled ubiquitin. Samples were prepared in aqueous (95% H_2O/5% D_2O or 100% D_2O) 50 mM potassium phosphate buffer at pH 5.8. NMR samples in 100% D_2O solutions were prepared from samples in 95% H_2O/5% D_2O by performing four cycles of lyophilizing and dissolving in D_2O (99.999 atom%) in the NMR tube.

A common lament of the scientist who wishes to understand a new discipline is "What books should I read?" We hope that *Protein NMR Spectroscopy: Principles and Practice* provides an answer for students and researchers with an interest in biomolecular NMR spectroscopy.

John Cavanagh
Wayne J. Fairbrother
Arthur G. Palmer III
Nicholas J. Skelton

ACKNOWLEDGMENTS

In writing this book, we have benefited greatly from helpful discussions with Mikael Akke, Jean Baum, Göran Carlström, Afshin Karimi, Lewis Kay, James Keeler, Johan Kördel, Ann McDermott, Daniel Raleigh, Mark Rance, and Jonathan Waltho.

We also thank Bruker Instruments, Inc., for providing Figures 3.2 and 3.3; Janet Cheetham and Duncan Smith (Amgen, Inc.) for providing data for Figures 7.41 and 7.44; and James Keeler for providing Figures 3.21, 3.22, and 4.3. [Figures 3.21 and 3.22 are reprinted from *Prog. NMR Spectrosc.* **19,** 47–129 (1987), with permission of Elsevier Science, Inc.] Figure 3.17 was prepared using the Azara program (generously provided by Wayne Boucher). Figures 6.44 and 7.25 were prepared using the Ztocsy program (generously provided by Mark Rance). Per Kraulis generously provided a database of ^1H and ^{13}C chemical shifts for use in Figures 8.2 and 8.7. Sarah Huntingdon (VLI Research, Inc.) provided assistance with the ^{15}N- and $^{13}C/^{15}N$-labeled ubiquitin samples. We are particularly indebted to James Keeler for permitting us to follow closely his lecture notes in preparing Section 4.3.

We thank Genentech, Inc., for access to the facilities used to acquire and process the NMR spectra of ubiquitin presented in this book. Walter Chazin, Mark Rance, and Peter Wright encouraged us to write this book during its initial planning stages. Lorraine Lica (Academic Press) provided continued assistance and encouragement throughout the project.

Finally, we are grateful particularly for the patience, support, and understanding of Patricia Bauer, Stacey Cavanagh, Jenni Heath, and Cindy Skelton throughout many evenings and weekends devoted to this project.

TABLE OF SYMBOLS

$1Q_{12\bar{3}}, 1Q_{1\bar{2}3},$ $1Q_{\bar{1}23}$ Three-spin single quantum operator, [6.20]

$3Q_x, 3Q_y$ Triple-quantum operator, [6.19]

$\mathbf{A}^\dagger$ Adjoint of a matrix or operator $\mathbf{A}$

$\langle A \rangle$ Expectation value of an operator $\mathbf{A}$; [2.17], [2.48]

$a_{mn}(t)$ Mixing coefficient for transfer of coherence from spin or site n to spin or site m; [4.18], [5.16], [5.138]

$A(\omega)$ Absorptive Lorentzian lineshape, [3.24]

$\mathbf{A}_k^q$ Tensor spin operator, [5.38]

$\mathbf{A}_{kp}^q$ Basis tensor operators of $\mathbf{A}_k^q$, [5.39]

$[\mathbf{A}, \mathbf{B}]$ Commutator of operators $\mathbf{A}$ and $\mathbf{B}$, [2.26]

$\mathbf{A} \otimes \mathbf{B}$ Direct product of matrices or operators $\mathbf{A}$ and $\mathbf{B}$, [2.133]

α Pulse rotation angle, [1.23]

$|\alpha\rangle$ Spin state with $m = \frac{1}{2}$

α_e Ernst angle, [3.115]

B_0 Static magnetic field strength

B_1 rf magnetic field amplitude

$\mathbf{B}_j$ jth basis operator

$\mathbf{B}_{rf}(t)$ rf magnetic field, [1.15]

$\mathbf{B}(t)$ Magnetic field vector

β Off-resonance phase shift of rf pulse, [3.62]

$|\beta\rangle$ Spin state with $m = -\frac{1}{2}$

$c_0(t)$ Function of physical constants and spatial variables for lattice–spin system coupling, [5.65]

$C_{00}^2(\tau)$ Orientational correlation function for rotation of an isotropic sphere, [5.71]

xix

$C(\tau)$	Orientational correlation function, [5.73]	$\mathscr{F}\{s(t)\}$	Fourier transformation of $s(t)$
d_{aN}	Sequential $^1\mathrm{H}^\alpha$–$^1\mathrm{H}^N$ NOE connectivity	$\mathbf{F}_z$	Nonselective z rotation, [4.31]
dB	Decibel [3.105]	G_z	z axis gradient strength
$d_{\beta N}$	Sequential $^1\mathrm{H}^\beta$–$^1\mathrm{H}^N$ NOE connectivity	γ	Gyromagnetic ratio
d_{NN}	Sequential $^1\mathrm{H}^N$–$^1\mathrm{H}^N$ NOE connectivity	Γ	Relaxation superoperator, [5.53]
$D(\omega)$	Dispersive Lorentzian lineshape, [3.25]	$\Gamma_1(t),$ $\Gamma_2(t)$	Intraresidue and inter-residue $^{15}\mathrm{N}$–$^{13}\mathrm{C}^\alpha$ coherence transfer functions, [7.75]
DQ_x, DQ_y	Double quantum operators; [2.242], [2.243]	Γ_{ij}	Rate constant for cross-relaxation between basis operators $\mathbf{B}_i$ and $\mathbf{B}_j$, [5.56]
δ	Chemical shift (ppm), [1.47]		
$\delta_{i,j}$	Kronecker delta function, [2.15]	$\Gamma_n(\tau)$	Refocused INEPT in-phase coherence transfer function for an $I_n S$ spin system, [7.13]
ΔB_0	Reduced static magnetic field strength, [1.20]		
$\Delta\nu_{\mathrm{FWHH}}$	Full-width-at-half-height linewidth in Hertz	$\mathscr{H}$	Hamiltonian operator
$\Delta\omega_{\mathrm{FWHH}}$	Full-width-at-half-height linewidth	$\mathscr{H}_0$	Time-independent Hamiltonian in absence of applied rf fields, [2.146]
$\Delta\sigma$	Nuclear shielding anisotropy, [1.45]	$\mathscr{H}_1(t)$	Stochastic Hamiltonian, [5.31]
Δt	Sampling interval	$\mathscr{H}_J$	Scalar coupling Hamiltonian, [2.146]
E	Energy; [1.4], [2.7]		
$\mathbf{E}$	Identity matrix or operator	$\mathscr{H}_{\mathrm{rf}}(t)$	rf magnetic field Hamiltonian, [2.87]
$\varepsilon, \varepsilon_I,$ $\varepsilon_{\mathrm{MQ}}$	PEP sensitivity enhancement factors, [7.40]–[7.42]	$h(t)$	Apodization function, [3.31]
η	NOE enhancement, [5.115]	$\mathscr{H}_z$	Zeeman Hamiltonian; [2.87], [2.146]
$\mathbf{F}^+$	Observation operator, [2.124]	$\hbar = h/2\pi$	Planck's constant divided by 2π
f_n	Nyquist frequency, [3.3]	i	$\sqrt{-1}$
$F_k^q(t)$	Random function of spatial lattice variables, [5.38]	I	Spin angular momentum quantum number
		$\mathbf{I}$	Angular momentum vector, [1.1]

I^+, I^-	Shift (raising and lowering) angular momentum operators	λ	Eigenvalue, [2.9], or exponential line-broadening constant, [3.39]
I_0, I^+, I^-	Spherical basis operators	λ_0	Exponential decay constant for damped oscillator, [3.22]
$I^\alpha, I^\beta, I^+, I^-$	Single element basis operators	λ_c	Decoupling scaling factor, [3.80]
I_x, I_y, I_z	Cartesian angular momentum operators	m	Magnetic quantum number
$I^+(rs), I^-(rs)$	Single transition shift operators	M_0	Equilibrium magnetization, [1.7]
I_z	z component of the angular momentum vector, [1.2]	$\mathbf{M}(t)$	Bulk magnetic moment or magnetization vector, [1.11]
$j(\omega)$	Spectral density function for isotropic phase in the high-temperature limit, [5.64]	$M^+(t)$	Complex magnetization, [1.37]
$J(\omega)$	Orientational spectral density function, [5.69]	$\boldsymbol{\mu}$	Magnetic moment vector, [1.3]
$j^q(\omega)$	Spectral density function, [5.50]	μ_0	Permittivity of free space
$j^q_{mn}(\omega)$	Cross spectral density function for the mth and nth relaxation interactions, [5.60]	ν_0	Resonance frequency in Hertz, [1.9]
$\mathbf{J}(t)$	Bulk angular momentum vector, [1.10]	$\boldsymbol{\omega}$	Angular velocity
$^nJ_{ij}$	n-bond scalar coupling constant between spins I_i and I_j	ω_0	Larmor resonance frequency, [1.9]
$\mathbf{K}$	Chemical rate matrix, [5.124]	ω_1	rf magnetic field strength
k_B	Boltzmann constant	ω_{NR}	Nonresonant frequency shift, [3.70]
k_{ij}	Microscopic rate constant for the ith and jth chemical species, [5.121]	ω_{rf}	rf magnetic field frequency
		Ω	Offset or chemical shift
		$\boldsymbol{\Omega}$	Diagonal matrix of chemical shift frequencies $d_{ij}\Omega_i$
$K_{IS} = |J_{IR} \pm J_{SR}|$	Double or zero quantum splitting for IRS three-spin system	$\Omega(t) = \{\theta(t), \phi(t)\}$	Polar angles
		p	Coherence order
		P	Power
		$\mathbf{P}$	Projection operator, [2.41]

$P_{\alpha\alpha}$, $P_{\alpha\beta}$, $P_{\beta\alpha}$, $P_{\beta\beta}$ — Spin state populations

$\mathscr{P}(\psi)$ — Probability density for lattice–spin system coupling, [2.46]

ϕ — Phase of coherence or rf magnetic field

ϕ_{NR} — Nonresonant phase shift, [3.69]

$\phi(z)$ — Spatially dependent phase, [4.38]

ψ — Basis function, eigenfunction, or stationary state function

ψ^* — Complex conjugate of ψ

$\Psi(t)$ — Wavefunction or state function

Q — Quality factor, [3.1]

$\mathbf{R}$ — Relaxation rate matrix; [5.12], [5.135]

R_1 — Spin–lattice or longitudinal relaxation rate constant, [1.24]

$R_{1\rho}$ — Rotating frame auto-relaxation rate constant

R_2 — Spin–spin or transverse relaxation rate constant, [1.26]

R_2^* — Inhomogeneously broadened transverse relaxation rate constant

R_{2MQ} — Relaxation rate constant for heteronuclear multiple quantum coherence, [7.27]

$\overline{R}_{2I}$ — Average relaxation rate constant for relaxation of transverse I spin coherence, [7.28]

$\overline{R}_{2S}$ — Average relaxation rate constant for relaxation of transverse S spin coherence, [7.26]

R_C — Cross rate constant, [5.19]

$R_\chi(\alpha)$ — Rotation matrix for rotation around the χ axis by an angle α, [1.34]

$R_I(\theta)$ — Relaxation rate constant for spin I in a tilted rotating frame [5.109]

r_{ij} — Distance between spins I_i and I_j

R_{inhom} — Inhomogeneous line broadening constant

$R_{IS}(\theta_I, \theta_S)$ — Cross-relaxation rate constant for spins I and S in a tilted rotating frame, [5.111]

R_L — Leakage rate constant, [5.20]

$r(t)$ — Rectangle function, [3.34]

ρ — Square root of the noise-power spectral density, [3.52]

ρ_I — Spin–lattice relaxation rate constant for spin I, [5.9]

$\mathscr{S}$ — Signal-to-noise ratio per unit time, [3.52]

S^2 — Order parameter, [5.77]

$S(\omega)$ — Frequency domain signal

$s(t)$ — Time domain signal

SW — Spectral width, [3.6]

σ — Isotropic nuclear shielding, [1.44]

$\sigma_\parallel = \sigma_{zz}$	Parallel component of nuclear shielding tensor	t_{max}	Maximum acquisition time for 1D NMR
$\sigma_\perp = (\sigma_{xx} + \sigma_{yy})/2$	Perpendicular component of nuclear shielding tensor	τ_{90}	90° pulse length
		τ_{180}	180° pulse length
σ^{eq}	Equilibrium spin density operator or matrix, [2.119]	τ_c	Rotational correlation time of a molecule, [1.41]
$\sigma_{IS}, \sigma_{IS}^{NOE}$	Cross-relaxation rate constant between spins I and S, [5.9]	τ_e	Effective correlation time for internal motions
σ_{IS}^{ROE}	Rotating frame cross-relaxation rate constant for spins I and S, [5.112]	τ_m	Mixing period
		τ_p	Pulse length
		τ_{RD}	Radiation damping time constant, [3.91]
$\sigma(t)$	Spin density matrix or operator, [2.47]	θ	Tilt angle, [1.21], or strong coupling parameter, [2.149]
$\sigma_{xx}, \sigma_{yy}, \sigma_{zz}$	Principle components of nuclear shielding tensor, [1.43]	$\mathrm{Tr}\{\mathbf{A}\}$	Trace of $\mathbf{A}$
		$\mathbf{U}$	Unitary operator, matrix, or propagator
$\mathbf{T}$	Inversion operator, [2.102]	$u(\omega)$	Imaginary part of complex frequency-domain spectrum [1.38]
T_1	Spin–lattice or longitudinal time constant	$v(\omega)$	Real part of complex frequency-domain spectrum, [1.38]
$t_1, t_2, \ldots$	Direct acquisition time or indirect evolution time	W_0, W_I, W_S, W_2	Transition frequencies, [5.91]
$t_{1,max}, t_{2,max}, \ldots$	Maximum value of $t_1, t_2, \ldots$	$Y_2^q[\Omega(t)]$	Modified spherical harmonic function
T_2	Spin-spin or transverse relaxation time constant	Z	Complex impedance, [3.2]
		ZQ_x, ZQ_y	Zero quantum operators; [2.246], [2.247]
T_c	Recycle delay	ζ_{NR}	Nonresonant rotation, [3.71]

CLASSICAL NMR SPECTROSCOPY

The explosive growth in the field of nuclear magnetic resonance (NMR) spectroscopy, particularly of large biomolecules, that continues today originated with the development of pulsed Fourier transform NMR spectroscopy by Ernst and Anderson (*1*) and the conception of multidimensional NMR spectroscopy by Jeener (*2*). NMR spectroscopy and X-ray crystallography currently are the only techniques capable of determining the three-dimensional structures of macromolecules at atomic resolution. In addition, NMR spectroscopy is a powerful technique for investigating time-dependent phenomena, including reaction kinetics and intramolecular dynamics of macromolecules. Historically, NMR spectroscopy of macromolecules was limited by the low inherent sensitivity of the technique and by the complexity (and consequent information content) of NMR spectra. The former limitation partially has been alleviated by the development of stronger magnets and more sensitive NMR spectrometers and by advances in techniques for sample preparation (both synthetic and biochemical). The latter limitation has been transmuted into a significant advantage by the phenomenal advances in the theoretical and experimental capabilities of NMR spectroscopy (and spectroscopists). The history of these developments has been reviewed by R. R. Ernst in his Nobel Laureate lecture (*3*). In light of subsequent developments, the conclusion of

Bloch's initial report of the observation of nuclear magnetic resonance in water seems prescient: "We have thought of various investigations in which this effect can be used fruitfully" (4).

1.1 Nuclear Magnetism

Nuclear magnetic resonances in bulk condensed phase were detected for the first time in 1946 by Bloch *et al.* (4) and Purcell *et al.* (5). Nuclear magnetism and NMR spectroscopy are manifestations of nuclear spin angular momentum. The theory of NMR spectroscopy is largely the quantum mechanics of nuclear spin angular momentum, an intrinsically quantum-mechanical property that does not have a classical analog. The spin angular momentum is characterized by the nuclear spin quantum number, I. Although NMR spectroscopy takes the nuclear spin as a given quantity, certain systematic features can be noted: (1) nuclei with odd mass numbers have half-integral spin quantum numbers, (2) nuclei with an even mass number and an even charge number have spin quantum numbers equal to zero, and (3) nuclei with an even mass number and an odd charge number have integral spin quantum numbers. Because the NMR phenomenon relies on the existence of nuclear spin, nuclei belonging to category 2 are NMR-inactive. Nuclei with spin quantum number greater than $\frac{1}{2}$ also possess electric quadrupole moments arising from nonspherical nuclear charge distributions. The lifetimes of the magnetic states for quadrupolar nuclei in solution normally are much shorter than the lifetimes for nuclei with $I = \frac{1}{2}$. NMR resonance lines for quadrupolar nuclei are correspondingly broad and can be more difficult to study. Relevant properties of nuclei commonly found in biomolecules are summarized in Table 1.1. For NMR spectroscopy of biomolecules, the most important nuclei with $I = \frac{1}{2}$ are ^{1}H, ^{13}C, ^{15}N and ^{31}P; the most important nucleus with $I = 1$ is the deuteron (^{2}H).

The nuclear spin angular momentum, $\mathbf{I}$, is a vector quantity with magnitude given by

$$|\mathbf{I}^2| = \mathbf{I} \cdot \mathbf{I} = \hbar^2 [I(I + 1)], \qquad [1.1]$$

in which I is the angular momentum quantum number and $\hbar$ is Planck's constant divided by 2π. Because of the restrictions of quantum mechanics, only one of the three Cartesian components of $\mathbf{I}$ can be specified simultaneously with $\mathbf{I}^2$. By convention, the value of the z component of $\mathbf{I}$ is specified by the equation

$$I_z = \hbar m, \qquad [1.2]$$

TABLE 1.1
Properties of Selected Nuclei[a]

Nucleus	I	γ (T · s)$^{-1}$	Natural abundance (%)
^{1}H	$\frac{1}{2}$	2.6752×10^8	99.98
^{2}H	1	4.107×10^7	0.02
^{13}C	$\frac{1}{2}$	6.728×10^7	1.11
^{14}N	1	1.934×10^7	99.64
^{15}N	$\frac{1}{2}$	-2.712×10^7	0.36
^{17}O	$\frac{5}{2}$	-3.628×10^7	0.04
^{19}F	$\frac{1}{2}$	2.5181×10^8	100.00
^{23}Na	$\frac{3}{2}$	7.080×10^7	100.00
^{31}P	$\frac{1}{2}$	1.0841×10^8	100.00
^{113}Cd	$\frac{1}{2}$	5.934×10^7	12.26

[a] The angular momentum quantum number, I, and the gyromagnetic ratio, γ, and natural isotopic abundance for nuclei of particular importance in biological NMR spectroscopy are shown.

in which the magnetic quantum number, $m = (-I, -I + 1, \ldots, I - 1, I)$. Thus, I_z has $2I + 1$ possible values. The orientation of the spin angular momentum vector in space is quantized, because the magnitude of the vector is constant and the z component has a set of discrete values. For an isolated spin in the absence of external fields, the quantum states corresponding to the $2I + 1$ values of m have the same energy and the spin angular momentum vector does not have a preferred orientation.

Nuclei that have nonzero spin angular momentum also possess nuclear magnetic moments. The nuclear magnetic moment, $\boldsymbol{\mu}$, is collinear with the vector representing the nuclear spin angular momentum vector and is defined by

$$\boldsymbol{\mu} = \gamma \mathbf{I}; \qquad \mu_z = \gamma I_z = \gamma \hbar m, \qquad [1.3]$$

in which the gyromagnetic ratio, γ, is a characteristic constant for a given nucleus (Table 1.1). Because angular momentum is a quantized property, so is the nuclear magnetic moment. The magnitude of γ, in part, determines the receptivity of a nucleus in NMR spectroscopy. In the presence of an external magnetic field, the spin states of the nucleus have energies given by

$$E = -\boldsymbol{\mu} \cdot \mathbf{B}, \qquad [1.4]$$

in which $\mathbf{B}$ is the magnetic field vector. Evidently, the minimum energy is obtained when the projection of $\boldsymbol{\mu}$ onto $\mathbf{B}$ is maximized. Since $|\mathbf{I}^2| > I_z^2$,

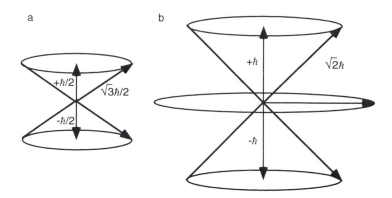

FIGURE 1.1 Angular momentum. The angular momentum vectors, **I**, and the allowed z components, I_z, for (a) a spin-$\frac{1}{2}$ particle and (b) a spin-1 particle are shown. The location of **I** on the surface of the cone of precession cannot be specified because of quantum-mechanical uncertainties in the I_x and I_y components.

μ cannot be collinear with **B** and the m spin states become quantized with energies proportional to their projection onto **B**. In a high-field NMR spectrometer, the static external magnetic field is directed along the z axis of the laboratory coordinate system. For this geometry, [1.4] reduces to

$$E_m = -\gamma I_z B_0 = -m\hbar\gamma B_0, \qquad [1.5]$$

in which B_0 is the static magnetic field strength. In the presence of a static magnetic field, the projections of the angular momentum of the nuclei onto the z axis of the laboratory frame results in $2I + 1$ equally spaced energy levels, which are known as the Zeeman levels. The quantization of I_z is illustrated by Fig. 1.1.

At equilibrium, the different energy states are unequally populated because lower-energy orientations of the magnetic dipole vector are more probable. The relative population of a state is given by the Boltzmann distribution:

$$\frac{N_m}{N} = \exp\left(\frac{-E_m}{k_B T}\right) \Big/ \sum_{m=-I}^{I} \exp\left(\frac{-E_m}{k_B T}\right)$$

$$= \exp\left(\frac{m\hbar\gamma B_0}{k_B T}\right) \Big/ \sum_{m=-I}^{I} \exp\left(\frac{m\hbar\gamma B_0}{k_B T}\right)$$

$$\approx \left(1 + \frac{m\hbar\gamma B_0}{k_{\rm B}T}\right)\Bigg/ \sum_{m=-I}^{I}\left(1 + \frac{m\hbar\gamma B_0}{k_{\rm B}T}\right)$$

$$\approx \left(1 + \frac{m\hbar\gamma B_0}{k_{\rm B}T}\right)\Bigg/(2I + 1), \qquad\qquad [1.6]$$

in which N_m is the number of nuclei in the mth state and N is the total number of spins, T is the absolute temperature, and $k_{\rm B}$ is the Boltzmann constant. At temperatures near 273 K, $m\hbar\gamma B_0/k_{\rm B}T \ll 1$ and the exponential functions in [1.6] can be expanded to first order using Taylor series. The populations of the states depend on both the nucleus type and the applied field strength. As the external field strength increases, the energy difference between the nuclear spin energy levels becomes larger and the population differences between the states increase. Of course, polarization of the spin system to generate a population difference between spin states does not occur instantaneously on application of the magnetic field; instead, the polarization, or magnetization, develops with a characteristic rate constant, called the spin–lattice relaxation rate constant (see Chapter 5).

The bulk magnetic moment, **M**, and the bulk angular momentum, **J**, of a macroscopic sample are given by the vector sum of the corresponding quantities for individual nuclei, $\boldsymbol{\mu}$ and **I**. At thermal equilibrium, the transverse components (e.g., the x or y components) of $\boldsymbol{\mu}$ and **I** for different nuclei in the sample are uncorrelated and sum to zero. The small population excess in the lower energy level gives rise to a bulk magnetization of the sample parallel (longitudinal) to the static magnetic field, $\mathbf{M} = M_0\mathbf{k}$, in which **k** is the unit vector in the z direction. Using [1.2], [1.3], and [1.6], M_0 is given by

$$M_0 = \gamma\hbar \sum_{m=-I}^{I} mN_m$$

$$= N\gamma\hbar \sum_{m=-I}^{I} m\exp(m\gamma\hbar B_0/k_{\rm B}T)\Bigg/\sum_{m=-I}^{I}\exp(m\gamma\hbar B_0/k_{\rm B}T)$$

$$\approx N\gamma\hbar \sum_{m=-I}^{I} m(1 + m\gamma\hbar B_0/k_{\rm B}T)\Bigg/\sum_{m=-I}^{I}(1 + m\gamma\hbar B_0/k_{\rm B}T)$$

$$\approx [N\gamma^2\hbar^2 B_0/\{k_{\rm B}T(2I + 1)\}] \sum_{m=-I}^{I} m^2$$

$$\approx N\gamma^2\hbar^2 B_0 I(I + 1)/(3k_{\rm B}T). \qquad\qquad [1.7]$$

By analogy with other areas of spectroscopy, transitions between Zeeman levels can be stimulated by applied electromagnetic radiation. The selection rule governing magnetic dipole transitions is $\Delta m = \pm 1$.

Thus, the photon energy required to excite a transition between the m and $m + 1$ Zeeman states is

$$\Delta E = \hbar \gamma B_0, \qquad [1.8]$$

which is seen to be directly proportional to the magnitude of the static magnetic field. By Planck's law, the frequency of the required electromagnetic radiation is given by

$$\omega_0 = \gamma B_0$$
$$\nu_0 = \omega_0/2\pi = \gamma B_0/2\pi \qquad [1.9]$$

in units of radians per second (rad/s) or hertz (Hz), respectively. The sensitivity of NMR spectroscopy depends on the population differences between Zeeman states. The population difference is on the order of 1 in 10^5 for ^1H spins in an 11.7-T (11.7-tesla) magnetic field; therefore, NMR is an insensitive spectroscopic technique compared to techniques such as visible or ultraviolet spectroscopy. This simple observation explains much of the impetus to construct more powerful magnets for use in NMR spectroscopy.

For the most part, this text is concerned with the NMR spectroscopy of spin $I = \frac{1}{2}$ (spin-$\frac{1}{2}$) nuclei. Only two nuclear spin states exist and only two energy levels are obtained by application of an external magnetic field. A single Zeeman transition between the energy levels exists. The spin state with $m = +\frac{1}{2}$ is referred to as the α state, and the state with $m = -\frac{1}{2}$ is referred to as the β state. If γ is positive, then the α state has lower energy than the β state.

1.2 The Bloch Equations

Bloch formulated a simple semiclassical vector model to describe the behavior of a sample of noninteracting spin-$\frac{1}{2}$ nuclei in a static magnetic field (6). The Bloch model will be outlined briefly in this section; many of the concepts and much of the terminology introduced persist throughout the text.

The evolution of the bulk magnetic moment (represented as a vector quantity) is central to the Bloch formalism. In the presence of a magnetic field, which may include components in addition to the static field, $\mathbf{M}(t)$ experiences a torque that is equal to the time derivative of the angular momentum:

$$\frac{d\mathbf{J}(t)}{dt} = \mathbf{M}(t) \times \mathbf{B}(t). \qquad [1.10]$$

Multiplying both sides by γ yields

$$\frac{d\mathbf{M}(t)}{dt} = \mathbf{M}(t) \times \gamma\mathbf{B}(t). \qquad [1.11]$$

The physical significance of this equation can be seen by using a frame of reference rotating with respect to the fixed axes with an angular velocity represented by the vector ω. Without loss of generality, the two coordinate systems are assumed to be superimposed initially. Vectors are represented identically in the two coordinate systems; however, time differentials are represented differently in the two coordinate systems. The equation of motion of $\mathbf{M}(t)$ in the laboratory frame and the rotating frame are related by (7)

$$\left(\frac{d\mathbf{M}(t)}{dt}\right)_{rot} = \left(\frac{d\mathbf{M}(t)}{dt}\right)_{lab} + \mathbf{M}(t) \times \omega$$

$$= \mathbf{M}(t) \times (\gamma\mathbf{B}(t) + \omega). \qquad [1.12]$$

The equation of motion for the magnetization in the rotating frame has the same form as in the laboratory frame provided that the field $B(t)$ is replaced by an effective field, $\mathbf{B}_{eff}$, given by

$$\mathbf{B}_{eff} = \mathbf{B}(t) + \frac{\omega}{\gamma}. \qquad [1.13]$$

For the choice $\omega = -\gamma\mathbf{B}(t)$, the effective field is zero, so that $\mathbf{M}(t)$ is time-independent in the rotating frame. As seen from the laboratory frame, $\mathbf{M}(t)$ precesses around $\mathbf{B}(t)$ with a frequency $\omega = -\gamma B$. For a static field of strength B_0, the precessional frequency, or the Larmor frequency, is given by

$$\omega_0 = -\gamma B_0. \qquad [1.14]$$

Thus, the bulk magnetization precesses around the main static field axis (defined as the z direction) at its Larmor frequency. The precessional frequency is identical to the frequency of electromagnetic radiation required to excite transitions between Zeeman levels. This identity is the reason that, within limits, a classical description of NMR spectroscopy is valid for systems of isolated spins.

Before proceeding further, the nomenclature used to refer to the strength of a magnetic field must be clarified. In NMR spectroscopy, the magnetic field strength B normally appears in the equation $\omega = -\gamma B$ that defines the precessional frequency of the nuclear magnetic moment. Conventionally, γB is referred to as the magnetic field strength measured in frequency units. Strictly speaking, the strength of the magnetic field is B, measured in gauss or tesla (10^4 G = 1 T); therefore, denoting γB as the magnetic field strength is incorrect (and has the obvious disadvantage of depending on the type of nucleus considered). That said, however, measuring magnetic field strength in frequency units (radians per second or hertz) is very convenient in many cases. Consequently, the terms γB and B both will be used to denote field strength in appropriate units throughout this text. For example, common usage refers to NMR spectrometers by the proton Larmor frequency of the magnet; thus, a spectrometer with a 11.7-T magnet is termed a 500-MHz spectrometer.

Precession of the bulk magnetic moment about the static magnetic field constitutes a time-varying magnetic field. By Faraday's law of induction, a time-varying magnetic field produces an induced electromotive force in a coil (of appropriate geometry) located in the vicinity of the bulk sample (8). Equation [1.11] suggests that precession of the bulk nuclear magnetization can be detected by such a mechanism. However, at thermal equilibrium, the bulk magnetization vector is collinear with the static field and no signal is produced in the coil. This text focuses on pulsed NMR experiments in which a short burst of radiofrequency (rf) electromagnetic radiation, typically of the order of several microseconds in length, perturbs the bulk magnetization from equilibrium. Such an rf burst is referred to as a pulse. After the rf field is turned off, the magnetization vector, $\mathbf{M}(t)$ will not, in general, be parallel to the static field. Consequently, the bulk magnetization will precess around the static field with an angular frequency $\omega_0 = -\gamma B_0$ and will generate a detectable signal in the coil.

The magnetic component of a rf field that is linearly polarized along the x axis of the laboratory frame is written as

$$\mathbf{B}_{\text{rf}}(t) = 2\,B_1 \cos(\omega_{\text{rf}}t + \phi)\mathbf{i}$$
$$= B_1 \{\cos(\omega_{\text{rf}}t + \phi)\mathbf{i} + \sin(\omega_{\text{rf}}t + \phi)\mathbf{j}\}$$
$$+ B_1 \{\cos(\omega_{\text{rf}}t + \phi)\mathbf{i} - \sin(\omega_{\text{rf}}t + \phi)\mathbf{j}\}, \qquad [1.15]$$

where B_1 is the amplitude of the applied field, ω_{rf} is the angular frequency of the rf field, often called the transmitter or carrier frequency; ϕ is the phase of the field; and $\mathbf{i}$ and $\mathbf{j}$ are unit vectors along the x and y axes, respectively. In the second equality, the rf field is decomposed into two circularly polarized fields rotating in opposite directions about the z axis. Only the field rotating in the same sense as the magnetic moment interacts significantly with the magnetic moment; the counterrotating, nonresonant field influences the spins to order $(B_1/2B_0)^2$, which is normally a very small number known as the Bloch–Siegert shift (but see Section 3.4.1). Thus, the nonresonant term can be ignored and the rf field written simply as

$$\mathbf{B}_{rf}(t) = B_1 \{\cos(\omega_{rf}t + \phi)\mathbf{i} + \sin(\omega_{rf}t + \phi)\mathbf{j}\}. \qquad [1.16]$$

The solution to equation [1.11] for a time-dependent rf field can be found by transforming to a rotating reference frame that makes the perturbing field time-independent. This is referred to as the rotating-frame transformation. The new frame is chosen to rotate at angular frequency ω_{rf} about the z axis. The equation of motion for the magnetization in the rotating frame, $\mathbf{M}^r(t)$, is given by

$$\frac{d\mathbf{M}^r(t)}{dt} = \mathbf{M}^r(t) \times \gamma\mathbf{B}^r, \qquad [1.17]$$

in which the effective field, $\mathbf{B}^r$, in the rotating frame is given by

$$\mathbf{B}^r = B_1 \cos\phi \; \mathbf{i}^r + B_1 \sin \phi \; \mathbf{j}^r + \frac{\Omega}{\gamma}\mathbf{k}^r, \qquad [1.18]$$

in which $\Omega = -\gamma B_0 - \omega_{rf} = \omega_0 - \omega_{rf}$ is known as the offset, and $\mathbf{i}^r$, $\mathbf{j}^r$, and $\mathbf{k}^r$ are unit vectors in the rotating frame. Equation [1.17] differs from [1.12] only because the quantities on both sides of the equality have been expressed in the rotating frame. The rf field is described by the amplitude B_1, and phase ϕ. In accordance with Ernst (9), the phase angle has been defined such that for an rf field of fixed phase x, $B_x = B_1$ and $B_y = 0$. The magnitude of the effective field is given by

$$B^r = \sqrt{(B_1)^2 + (\Delta B_0)^2} = B_1\sqrt{1 + (\tan \theta)^{-2}}, \qquad [1.19]$$

where ΔB_0 is known as the reduced static field and is equivalent to the z component of the effective field

$$\Delta B_0 = -\frac{\Omega}{\gamma}. \qquad [1.20]$$

The angle θ through which the effective field is tilted with respect to the z axis is defined by

$$\tan \theta = \frac{B_1}{\Delta B_0}. \qquad [1.21]$$

The direction of the effective field, as defined by θ and ϕ, depends on the strength of the rf field, $\mathbf{B}_{rf}(t)$; the difference between the transmitter and Larmor frequencies; and the phase of the rf field in the laboratory frame,

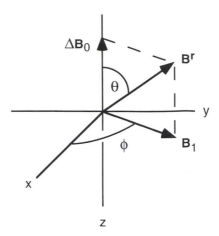

FIGURE 1.2 Orientations of $\Delta \mathbf{B}_0$, $\mathbf{B}_1$ and $\mathbf{B}^r$ in the rotating reference frame. Angles θ and ϕ are defined by [1.21] and [1.18].

as illustrated in Fig. 1.2. Frequently, $\mathbf{B}_{rf}(t)$ is referred to directly as the "B_1 field." In the rotating frame, on application of the B_1 field, $\mathbf{M}^r(t)$ precesses around the effective field $\mathbf{B}^r$ with an angular frequency ω^r

$$\omega^r = -\gamma B^r. \qquad [1.22]$$

If the rf field is turned on for a time period τ_p, called the pulse length, then the effective rotation angle α (or flip angle) is given by

$$\alpha = \omega^r \tau_p = -\gamma B^r \tau_p = -\gamma B_1 \tau_p \sqrt{1 + (\tan \theta)^{-2}} \qquad [1.23]$$

If the transmitter frequency, ω_{rf}, is equal to ω_0, then the irradiation is said to be applied on-resonance. In the on-resonance case, the offset term, Ω, equals zero, $B^r = B_1$, and, the effective field is collinear with the B_1 field in the rotating frame. These results have an important implication: the influence of the main static magnetic field, $\mathbf{B}_0$, has been removed. The bulk magnetization $\mathbf{M}^r(t)$ precesses around the axis defined by the B_1 field, with frequency $\omega^r = -\gamma B^r = -\gamma B_1 \equiv \omega_1$. Precession of the magnetization about the effective field in the rotating reference frame is illustrated in Fig. 1.3. As general practice in this text, the rotating frame will not be indicated explicitly, and unless otherwise stated, the rotating frame of reference will be assumed [i.e., $\mathbf{M}(t)$ will be written instead of $\mathbf{M}^r(t)$].

Following an rf pulse, the bulk magnetization precesses about the static magnetic field with a Larmor frequency ω_0. As described, the magnetization would continue to evolve freely in the transverse plane forever

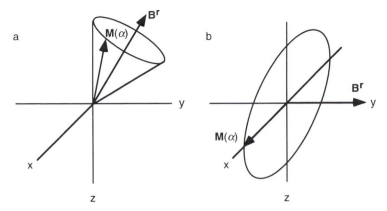

FIGURE 1.3 Effect of applied rf field. (a) In the presence of an applied rf field with y phase, the effective field, $\mathbf{B^r}$ is in the y–z plane in the rotating reference frame, and the magnetization vector precesses around $\mathbf{B^r}$. (b) If the rf field is applied on-resonance, then $\mathbf{B^r}$ is oriented along the y axis, and the magnetization vector rotates in the x–z plane orthogonal to $\mathbf{B^r}$.

following an initial pulse. Of course, this is not the case because eventually equilibrium must be reestablished. Bloch defined two processes to account for the observed decay of the NMR signal (6). These two relaxation mechanisms are responsible for the return of the bulk magnetization to the equilibrium state following some perturbation to the nuclear spin system. The first relaxation mechanism accounts for the return of the population difference across the Zeeman transition back to the Boltzmann equilibrium distribution, and is known as longitudinal or spin–lattice relaxation. Bloch assumed that spin–lattice relaxation is characterized by the first-order rate expression

$$\frac{dM_z(t)}{dt} = R_1[M_0 - M_z(t)], \qquad [1.24]$$

such that

$$M_z(t) = M_0 - [M_0 - M_z(0)] \exp(-R_1 t), \qquad [1.25]$$

in which R_1 is the spin–lattice relaxation rate constant (the spin–lattice relaxation time constant, $T_1 = 1/R_1$, is often encountered), and $M_z(0)$ is the value of the component of the magnetization along the z axis at $t = 0$. As shown, the z component or longitudinal magnetization returns to thermal equilibrium in an exponential fashion. A second relaxation process was introduced to account for the decay of the transverse magnetization in the x–y plane following a pulse. Transverse or spin–spin relaxation also is characterized by a first-order rate expression

$$\frac{dM_x(t)}{dt} = -R_2 M_x(t);$$

$$\frac{dM_y(t)}{dt} = -R_2 M_y(t) \qquad [1.26]$$

and

$$M_x(t) = M_x(0)\exp(-R_2 t);$$

$$M_y(t) = M_y(0)\exp(-R_2 t), \qquad [1.27]$$

in which R_2 is the spin–spin relaxation rate constant (the spin–spin relaxation time constant is $T_2 = 1/R_2$) and $M_x(0)$ and $M_y(0)$ are the values of the transverse magnetization at $t = 0$. The introduction of the concept of relaxation here is simply to assist in the initial description of the NMR phenomenon, and more detailed treatments of relaxation theory and processes will be presented in Chapter 5.

Combining [1.11], [1.24], and [1.26] yields the famous Bloch equations in the laboratory reference frame:

$$\frac{dM_z(t)}{dt} = \gamma(\mathbf{M}(t) \times \mathbf{B}(t))_z - R_1(M_z(t) - M_0)$$

$$= \gamma(M_x(t)B_y(t) - M_y(t)B_x(t)) - R_1(M_z(t) - M_0),$$

$$\frac{dM_x(t)}{dt} = \gamma(\mathbf{M}(t) \times \mathbf{B}(t))_x - R_2 M_x(t) \qquad [1.28]$$

$$= \gamma(M_y(t)B_z(t) - M_z(t)B_y(t)) - R_2 M_x(t),$$

$$\frac{dM_y(t)}{dt} = \gamma(\mathbf{M}(t) \times \mathbf{B}(t))_y - R_2 M_y(t)$$

$$= \gamma(M_z(t)B_x(t) - M_x(t)B_z(t)) - R_2 M_y(t),$$

describing the evolution of magnetization in a magnetic field. In the absence of an applied rf field, $B_x(t) = B_y(t) = 0$ and $B_z(t) = B_0$. Under these conditions, which are referred to as free-precession, the Bloch equations become

$$\frac{dM_z(t)}{dt} = -R_1(M_z(t) - M_0),$$

$$\frac{dM_x(t)}{dt} = \gamma M_y(t)B_0 - R_2 M_x(t),$$

$$\frac{dM_y(t)}{dt} = \gamma M_x(t)B_0 - R_2 M_y(t). \tag{1.29}$$

In the rotating reference frame, the Bloch equations are given by

$$\frac{dM_z(t)}{dt} = \gamma(M_x(t)B_y^r(t) - M_y(t)B_x^r(t)) - (M_z(t) - M_0)/T_1,$$

$$\frac{dM_x(t)}{dt} = -\Omega M_y(t) - M_z(t)\gamma B_y^r(t) - M_x(t)/T_2, \tag{1.30}$$

$$\frac{dM_y(t)}{dt} = \Omega M_x(t) + M_z(t)\gamma B_x^r(t) - M_y(t)/T_2,$$

in which $B_x^r(t) = B_1 \cos \phi$ and $B_y^r(t) = B_1 \sin \phi$.

In a common experimental situation in pulsed NMR spectroscopy, the B_1 field is applied for a time $\tau_p \ll 1/R_2$ or $1/R_1$. If neither B_1 nor ϕ is time-dependent, then the Bloch equations simplify to

$$\frac{dM_z(t)}{dt} = M_x(t)\gamma B_y^r - M_y(t)\gamma B_x^r,$$

$$\frac{dM_x(t)}{dt} = -\Omega M_y(t) - M_z(t)\gamma B_y^r, \tag{1.31}$$

$$\frac{dM_y(t)}{dt} = \Omega M_x(t) + M_z(t)\gamma B_x^r,$$

which can be written in matrix form as

$$\frac{d\mathbf{M}(t)}{dt} = \frac{d}{dt}\begin{bmatrix} M_x(t) \\ M_y(t) \\ M_z(t) \end{bmatrix} = \begin{bmatrix} 0 & -\Omega & -\gamma B_y^r \\ \Omega & 0 & \gamma B_x^r \\ \gamma B_y^r & -\gamma B_x^r & 0 \end{bmatrix} \mathbf{M}(t). \tag{1.32}$$

The solution to [1.32] can be represented as a series of rotations (9, 10):

$$\mathbf{M}(\tau_p) = \mathbf{R}_z(\phi)\mathbf{R}_y(\theta)\mathbf{R}_z(\alpha)\mathbf{R}_y(-\theta)\mathbf{R}_z(-\phi)\mathbf{M}(0), \tag{1.33}$$

in which the rotation matrices are

$$\mathbf{R}_x(\beta) = \begin{bmatrix} 1 & 0 & 0 \\ 0 & \cos \beta & -\sin \beta \\ 0 & \sin \beta & \cos \beta \end{bmatrix},$$

$$\mathbf{R}_y(\beta) = \begin{bmatrix} \cos\beta & 0 & -\sin\beta \\ 0 & 1 & 0 \\ \sin\beta & 0 & \cos\beta \end{bmatrix}, \qquad [1.34]$$

$$\mathbf{R}_z(\beta) = \begin{bmatrix} \cos\beta & -\sin\beta & 0 \\ \sin\beta & \cos\beta & 0 \\ 0 & 0 & 1 \end{bmatrix}.$$

In [1.34], the notation $\mathbf{R}_\chi(\beta)$ designates a rotation of angle β about the axis χ. Equation [1.33] will be used frequently to calculate the effect of rf pulses on isolated spins.

1.3 The One-Pulse NMR Experiment

Experimental aspects of NMR spectroscopy are described in detail in Chapter 3. In this section, a brief overview of a simple NMR experiment is presented. In the Bloch model, the maximum NMR signal is detected when the bulk magnetic moment is perpendicular (transverse) to the static magnetic field. As noted above, a rf pulse causes $\mathbf{M}(t)$ to precess about an axis defined by the direction of the effective magnetic field in the rotating frame; therefore, the properties of an rf pulse that cause rotation of $\mathbf{M}(t)$ from the z axis through an angle of 90° are particularly important in pulsed NMR spectroscopy.

An ideal pulse experiment that achieves a 90° rotation of $\mathbf{M}(t)$ will be considered. An on-resonance rf pulse of duration τ_p, strength B_1, and tilt angle $\theta = \pi/2$ is applied to the equilibrium magnetization state. If the rf pulse is applied along the y axis of the rotating frame (setting $\phi = \pi/2$ in [1.18]), then the magnetization following the pulse is given by ([1.33])

$$\mathbf{M}(\tau_p) = \mathbf{R}_y(\alpha)\mathbf{M}_0 = \mathbf{i}M_0\sin\alpha + \mathbf{k}M_0\cos\alpha = \begin{pmatrix} M_0\sin\alpha \\ 0 \\ M_0\cos\alpha \end{pmatrix} \quad [1.35]$$

where M_0 is the magnitude of the equilibrium magnetization and α is the rotation angle. The maximum transverse magnetization is generated for $\alpha = 90°$. The rf pulse used to achieve this state is conventionally called a 90° or $\pi/2$ pulse. A 90° pulse equalizes the populations of the α and β spin states. In a similar way, a 180° or π pulse generates no transverse magnetization; instead, the bulk magnetization is inverted from its original

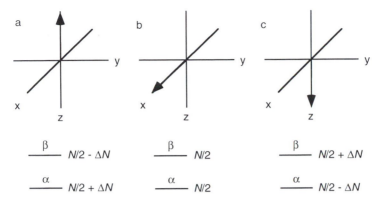

FIGURE 1.4 On-resonance pulses. The magnetization vectors and spin states α and β (a) for thermal equilibrium, (b) following a 90° pulse with y phase, and (c) following a 180° pulse are shown. The populations of each spin state are indicated. The total number of spins is N and $\Delta N = N\hbar\gamma B_0/(4k_B T)$.

state to yield $\mathbf{M}(\tau_p) = -M_0\mathbf{k}$. In the Bloch vector model, the bulk magnetization following a 180° pulse is aligned along the $-z$ axis. This corresponds to a population inversion between the α and β states, such that the β state now possesses excess population of nuclei. The populations of the Zeeman states and the net magnetization vectors following on-resonance pulses are illustrated in Fig. 1.4.

The magnetization precessing during the so-called acquisition period, t, generates the signal that is recorded by the NMR spectrometer. The signal is referred to as a free-induction decay (FID). The free-precession Bloch equations in the rotating frame [1.29] show that the FID can be described in terms of two components

$$M_x(t) = M_0 \sin\alpha \cos(\Omega t) \exp(-R_2 t),$$
$$M_y(t) = M_0 \sin\alpha \sin(\Omega t) \exp(-R_2 t),$$

[1.36]

which can be combined in complex notation as

$$M^+(t) = M_x(t) + iM_y(t) = M_0 \sin\alpha \exp(i\Omega t - R_2 t).$$

[1.37]

As a consequence of relaxation, the components of the bulk magnetization vector precessing in the transverse plane following a rf pulse are damped by the exponential factor $\exp(-R_2 t)$. In practice, both parts of the complex signal are detected simultaneously by the NMR spectrometer as $s^+(t) = \lambda M^+(t)$, with λ an experimental constant of proportionality. The complex time-domain signal can be Fourier-transformed to produce a complex

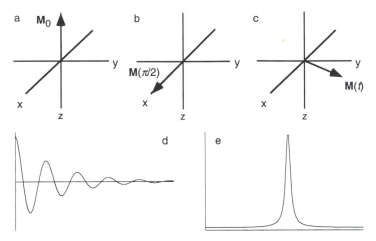

FIGURE 1.5 One-pulse NMR experiment. (a) The orientation along the z axis of the net magnetization at equilibrium, (b) the orientation along the x axis of the net magnetization following a 90° pulse with y phase, (c) the precessing magnetization in the x–y plane, (d) the FID recorded for the precessing magnetization during the acquisition period, and (e) the real component of the complex frequency domain NMR spectrum obtained by Fourier transformation of the FID.

frequency-domain spectrum

$$S(\omega) = \int_0^\infty s^+(t)\exp(-i\omega t)\,dt$$
$$= v(\omega) + iu(\omega), \qquad [1.38]$$

in which

$$v(\omega) = \lambda M_0 \frac{R_2}{R_2^2 + (\Omega - \omega)^2}, \qquad [1.39]$$

$$u(\omega) = \lambda M_0 \frac{\Omega - \omega}{R_2^2 + (\Omega - \omega)^2}. \qquad [1.40]$$

The function $v(\omega)$ represents an absorptive Lorentzian signal, and the function $u(\omega)$ represents a dispersive Lorentzian signal. The real part of the complex spectrum, $v(\omega)$, normally is displayed as the NMR spectrum. This simple one-pulse NMR experiment is illustrated schematically in Fig. 1.5.

1.4 Linewidth

The phenomenological linewidth is defined as the full-width at half-height of the resonance lineshape and is a primary factor affecting both resolution and signal-to-noise ratio of NMR spectra. For a Lorentzian lineshape [1.39], the homogeneous linewidth is given by $\Delta\nu_{FWHH} = R_2/\pi$ in hertz (or $\Delta\omega_{FWHH} = 2R_2$ in rad/s) and the inhomogeneous linewidth is $\Delta\nu_{FWHH} = R_2^*/\pi$, in which $R_2^*/\pi = R_2 + R_{inhom}$ and R_{inhom} represents the broadening of the resonance signal due to inhomogeneity of the magnetic field. In modern NMR spectrometers R_{inhom}/π is on the order of 1 Hz. As will be discussed in detail in Chapter 5, values of R_2 (and hence homogenous linewidths) are approximately proportional to the overall rotational correlation time of the protein and thus depend on molecular mass and shape of the molecule. As discussed in Section 6.1, observed linewidths significantly larger than expected based on the molecular mass of the protein imply that aggregation is increasing the apparent rotational correlation time or that chemical exchange effects (Section 5.6) contribute significantly to the inhomogeneous linewidth.

Given theoretical or experimental estimates of τ_c, the theoretical equations presented in Chapter 5 and Section 7.1.2.4 can be used to calculate approximate values of resonance linewidths. The resulting curves are shown in Fig. 1.6. The principal uncertainties in the calculation are due to the following factors: (1) anisotropic rotational diffusion of nonspherical molecules, (2) differential contributions from internal motions (particularly in loops or for side chains), (3) cross-correlation effects, (4) ^1H dipolar interactions with all nearby protons (which depend on detailed structures of the proteins), and (5) incomplete knowledge of fundamental parameters [such as chemical-shift anisotropies (CSAs)]. In light of these uncertainties, the results presented in Fig. 1.6 should be regarded as approximate guidelines. For example, ^1H (in an unlabeled sample), $^{13}C\alpha$, and ^{15}N linewidths in ubiquitin are ~6–9, ~7, and ~3 Hz, respectively. These values are consistent with values of 5, 6, and 2 Hz from Fig. 1.6.

The correlation time for Brownian rotational diffusion can be measured experimentally by using time-resolved fluorescence spectroscopy, light scattering, and NMR spin relaxation spectroscopy, or calculated by using a variety of hydrodynamic theories (that unfortunately require detailed information on the shape of the molecule) (11). In the absence of more accurate information, the simplest theoretical approach for approximately spherical globular proteins calculates the isotropic rotational correlation time from Stokes' law:

$$\tau_c = \frac{4\pi\eta_w r_H^3}{3k_B T},$$
[1.41]

in which η_w is the viscosity of the solvent, r_H is the effective hydrodynamic radius of the protein, k_B is Boltzmann's constant, and T is the temperature.

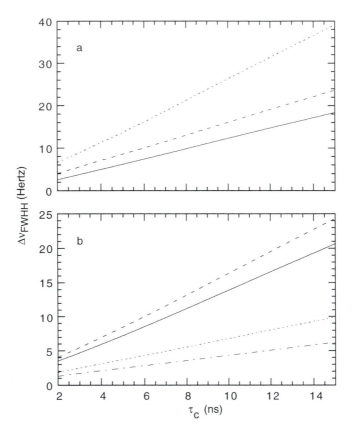

FIGURE 1.6 Protein resonance linewidths are shown as a function of rotational correlation time. (a) Linewidths for (—) ^{1}H spins, ($\cdots$) ^{1}H spins covalently bonded to ^{13}C, (---) ^{1}H spins covalently bonded to ^{15}N nuclei. (b) Heteronuclear linewidths for (—) proton-decoupled ^{13}C, (---) proton-coupled ^{13}C, (-·-·-) proton-decoupled ^{15}N, and ($\cdots$) proton-coupled ^{15}N spins. Calculations included dipolar relaxation of all spins, and CSA relaxation of ^{15}N spins. For ^{1}H–^{1}H dipolar interactions, $\Sigma_j r_{ij}^{-6} = 0.027$ Å^{-6} (21).

The hydrodynamic radius can be very roughly estimated from the molecular mass of the protein by assuming that the specific volume of the protein is $\overline{V} = 0.73$ cm^3/g and that a hydration layer of $r_w = 1.6$–3.2 Å (corresponding to one-half to one hydration shell) surrounds the protein:

$$r_H = [3\overline{V}M_r/(4\pi N_A)]^{1/3} + r_w. \qquad [1.42]$$

For the protein ubiquitin, $r_H = 16.5$ Å, and $\tau_c = 3.8$ ns at 300 K, compared with a value of 4.1 ns determined from NMR spectroscopy (*12*). Rotational correlation times in D_2O solution are approximately 25% greater than in H_2O solution because of the larger viscosity of D_2O.

1.5 Chemical Shift

A general feature of NMR spectroscopy is that the observed resonance frequencies depend on the local environments of individual nuclei and differ slightly from the frequencies predicted by [1.14]. The differences in resonance frequencies, referred to as chemical shifts, offer the possibility of distinguishing between nuclei in different chemical environments.

The phenomenon of chemical shift arises because motions of electrons induced by the external magnetic field generate secondary magnetic fields. The net magnetic field at the location of a nucleus depends on the static magnetic field and the secondary fields. The effect of the secondary fields, called nuclear shielding, can enhance or oppose the main field. In general, the electronic charge distribution in a molecule is anisotropic and the effects of shielding on a particular nucleus are described by the second-rank nuclear shielding tensor, represented by a 3×3 matrix. In the principal-coordinate system of the shielding tensor, the matrix representing the tensor is diagonal with principal components σ_{xx}, σ_{yy}, and σ_{zz}. The principal components of the shielding tensor have the following meaning: if the molecule is oriented such that the kth principal axis is oriented along the z axis of the static field, then the net magnetic field at the nucleus is given by

$$B = (1 - \sigma_{kk}) B_0. \qquad [1.43]$$

In isotropic liquid solution, collisions lead to rapid reorientation of the molecule, and consequently, of the shielding tensor. Under these circumstances, the effects of shielding on a particular nucleus can be accounted for by modifying [1.14] as

$$\omega = \gamma(1 - \sigma) B_0, \qquad [1.44]$$

in which σ is the average, isotropic, shielding constant for the nucleus:

$$\sigma = (\sigma_{xx} + \sigma_{yy} + \sigma_{zz})/3. \qquad [1.45]$$

The chemical-shift anisotropy (CSA) is defined conventionally as

$$\Delta\sigma = \sigma_{zz} - (\sigma_{xx} + \sigma_{yy})/2. \qquad [1.46]$$

Variations in σ due to different electronic environments cause variations in the resonance frequencies of the nuclei. In effect, each nucleus experiences its own local magnetic field. Fluctuations in the local magnetic field as the molecule rotates results in the CSA relaxation mechanism described in Section 5.4.4.

Resonance frequencies are directly proportional to the static field, B_0; consequently, the difference in chemical shift between two resonance signals measured in frequency units increases with B_0. In addition, the absolute value of the chemical shift of a resonance is difficult to determine because B_0 must be measured very accurately. In practice, chemical shifts are measured in parts per million (ppm or δ) relative to a reference resonance signal from a standard molecule:

$$\delta = \frac{\Omega - \Omega_{\text{ref}}}{\omega_0} \times 10^6 = (\sigma_{\text{ref}} - \sigma) \times 10^6, \qquad [1.47]$$

in which Ω and Ω_{ref} are the offset frequencies of the signal of interest and the reference signal, respectively. Chemical-shift differences measured in parts per million are independent of the static magnetic field strength. Referencing of NMR spectra is discussed in detail in Section 3.6.3.

Observed chemical shifts in proteins commonly are partitioned into the sum of two components: the so-called random-coil chemical shifts, δ_{rc}, and the conformation-dependent secondary chemical shifts, $\Delta\delta$. The random-coil chemical shift of a nucleus in an amino acid residue is the chemical shift that is observed in a conformationally disordered peptide (13–15). The secondary chemical shift contains the contributions from secondary and tertiary structure. This distinction is useful because secondary chemical shifts display characteristic patterns for secondary structural elements (16–18) and because semi-empirical theoretical treatments (19) are becoming increasingly accurate in predicting secondary chemical shifts. Distributions of chemical shifts observed in proteins are presented in Chapter 8.

1.6 Scalar Coupling and Limitations of the Bloch Equations

A brief treatment of a problem that will be discussed throughout this text will be used to illustrate the deficiencies of the Bloch theory. High-resolution NMR spectra of liquids reveal fine structure due to interactions between the nuclei. However, the splitting of the resonance signals into multiplets are not caused by direct dipolar interactions between magnetic dipole moments. Dipolar coupling, although extremely important in solids, is an anisotropic quantity that is averaged to zero to first order in isotropic solution (second-order effects are discussed in Chapter 5). Ramsey and Purcell suggested that the interaction was mediated by the electrons forming the chemical bonds between the nuclei (20). This interaction is known as spin–spin coupling or scalar coupling. The strength of the interaction is measured by the scalar coupling constant, $^nJ_{ab}$, in which n designates the number of covalent bonds separating the two nuclei, a and b. The magnitude of $^nJ_{ab}$ is usually expressed in hertz (Hz), and the most important scalar coupling interactions in proteins have $n = 1$ to 4. In the present text, n will be written explicitly only if the intended value of n is not clear from the context.

The scalar coupling modifies the energy levels of the system; hence, the NMR spectrum is modified. The prototypical example consists of two spin-$\frac{1}{2}$ nuclei (e.g., two ^1H spins or a ^1H spin and a ^{13}C spin). The two spins are designated I and S. The resonance frequencies are ω_I and ω_S, respectively,

$$\omega_I = -\gamma_I B_0 (1 - \sigma_I); \qquad \omega_S = -\gamma_S B_0 (1 - \sigma_S). \qquad [1.48]$$

The magnetic quantum numbers are m_I and m_S; each spin has two stationary states that correspond to the magnetic quantum numbers $1/2$ and $-1/2$. The complete two-spin system is described by four wavefunctions corresponding to all possible combinations of m_I and m_S

$$\Psi_1 = \Psi\left(\frac{1}{2}, \frac{1}{2}\right); \qquad \Psi_2 = \Psi\left(\frac{1}{2}, -\frac{1}{2}\right);$$

$$\Psi_3 = \Psi\left(-\frac{1}{2}, \frac{1}{2}\right); \qquad \Psi_4 = \Psi\left(-\frac{1}{2}, -\frac{1}{2}\right), \qquad [1.49]$$

where the first quantum number describes the state of the I spin and the second term describes the S spin. In the absence of scalar coupling between the spins, the energies of these four states are the sums of the energies for each spin. Because the β state has a higher energy than the α state, the energies are given by

$$E_1 = -\frac{1}{2}\hbar\omega_I - \frac{1}{2}\hbar\omega_S; \qquad E_2 = -\frac{1}{2}\hbar\omega_I + \frac{1}{2}\hbar\omega_S;$$

$$E_3 = \frac{1}{2}\hbar\omega_I - \frac{1}{2}\hbar\omega_S; \qquad E_4 = \frac{1}{2}\hbar\omega_I + \frac{1}{2}\hbar\omega_S. \qquad [1.50]$$

The total magnetic quantum number m for each energy level is the sum of the individual terms:

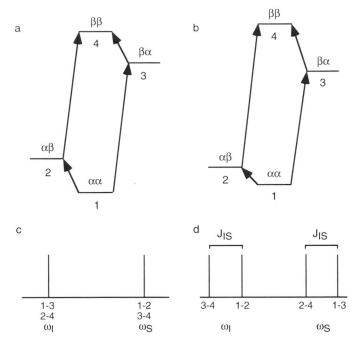

FIGURE 1.7 The energy levels for an AX spin system in the (a) absence and (b) presence of scalar coupling interactions between the spins. The allowed transitions are indicated between arrows. The energies of the four spin states are defined by (a) [1.50] and (b) [1.52].

$$m_1 = +\frac{1}{2} + \frac{1}{2} = +1; \qquad m_2 = +\frac{1}{2} - \frac{1}{2} = 0;$$

$$m_3 = -\frac{1}{2} + \frac{1}{2} = 0; \qquad m_4 = -\frac{1}{2} - \frac{1}{2} = -1.$$

[1.51]

The energy-level diagram for a two-spin system is shown in Fig. 1.7a. The observable transitions obey the selection rule $\Delta m = \pm 1$. Therefore, the allowed transitions occur between states 1 and 2, 3 and 4, 1 and 3, and 2 and 4 in Fig. 1.7; transitions between 2 and 3 or 1 and 4 are forbidden. The first two transitions involve a change in the spin state of the S spin whereas the latter two involve a change in the spin state of the I spin. Consequently, the NMR spectrum consists of one resonance line at ω_I, due to transitions 1–3 and 2–4, and one resonance line at ω_S, due to transitions 1–2 and 3–4.

Introducing the scalar coupling between I and S, with a value of J_{IS}, modifies the energy levels to

$$E_1 = -\frac{1}{2}\hbar\omega_I - \frac{1}{2}\hbar\omega_S + \frac{1}{2}\pi\hbar J_{IS}; \qquad E_2 = -\frac{1}{2}\hbar\omega_I + \frac{1}{2}\hbar\omega_S - \frac{1}{2}\pi\hbar J_{IS},$$

$$\qquad\qquad\qquad\qquad\qquad\qquad\qquad\qquad\qquad\qquad\qquad [1.52]$$

$$E_3 = \frac{1}{2}\hbar\omega_I - \frac{1}{2}\hbar\omega_S - \frac{1}{2}\pi\hbar J_{IS}; \qquad E_4 = \frac{1}{2}\hbar\omega_I + \frac{1}{2}\hbar\omega_S + \frac{1}{2}\pi\hbar J_{IS},$$

in which weak coupling has been assumed with $2\pi J_{IS} \ll |\omega_I - \omega_S|$. These expressions come from the equation (see Section 2.5.2)

$$E(m_I, m_S) = -m_I\omega_I - m_S\omega_S + 2\pi\, m_I\, m_S\, J_{IS}. \qquad [1.53]$$

The term in J_{IS} depends on the spin states of both nuclei, but the terms in ω_I and ω_S depend on the spin state of a single nucleus. The energy-level diagram for a scalar coupled two-spin system is shown in Fig. 1.7b. The resulting effect in the spectrum of the scalar coupled system is easily seen from the new values from the transition frequencies

$$\omega_{12} = \omega_S - \pi J_{IS}; \qquad \omega_{34} = \omega_S + \pi J_{IS};$$

$$\qquad\qquad\qquad\qquad\qquad\qquad\qquad [1.54]$$

$$\omega_{13} = \omega_I - \pi J_{IS}; \qquad \omega_{24} = \omega_I + \pi J_{IS}.$$

Now the spectrum consists of four lines: two centered around the transition frequency, ω_S, of the S spin but separated by $2\pi J_{IS}$ and two centered around the transition frequency of the I spin, ω_I, but separated by $2\pi J_{IS}$. A weakly coupled two-spin system is referred to as an AX spin system and a strongly coupled two-spin system is referred to as an AB spin system, in which A and X or A and B represent the pair of scalar coupled spins.

The Bloch vector model of NMR phenomena predicts that two resonance signals will be obtained for the two-spin system; in actuality, if the two spins share a nonzero scalar coupling interaction, four resonance signals are obtained. The basic Bloch model can be extended to describe the evolution of a scalar coupled system by treating each resonance line resulting from the scalar coupling interaction as an independent magnetization vector in the rotating frame. Although additional insights can be gained from using this approach, many problems still arise: (1) strong coupling effects that occur when $2\pi J_{IS} \approx |\omega_I - \omega_S|$ cannot be described, (2) the results of applying nonselective pulses to transverse magnetization in a homonuclear coupled system cannot be described without introducing additional ad hoc assumptions, and (3) transfer of magnetization via forbidden transitions when the spin system is not at equilibrium cannot be explained.

In principle, the Bloch picture is strictly only applicable to a system of noninteracting spin-$\frac{1}{2}$ nuclei. Despite these limitations, the Bloch model should not be abandoned completely. Many of the concepts and much of the terminology introduced by this model appear throughout the whole of NMR spectroscopy. Although the Bloch model is a valuable tool with which to visualize simple NMR experiments, more rigorous approaches are necessary to describe those NMR methods, particularly multidimensional NMR methods, that cannot be described classically. Much of the remaining theory of this text is devoted to developing methods of analysis that accurately predict the behavior of systems of two or more spins that interact via the scalar coupling or other interactions.

References

1. R. R. Ernst and W. A. Anderson, *Rev. Sci. Instrum.* **37**, 93–102 (1966).
2. J. Jeener. (Lecture, Ampère Summer School, Basko Polje, Yugoslavia, 1971).
3. R. R. Ernst, *Angew. Chem., Int. Eng. Ed.* **31**, 805–930 (1992).
4. F. Bloch, W. W. Hansen, and M. Packard, *Phys. Rev.* **69**, 127 (1946).
5. E. M. Purcell, H. C. Torrey, and R. V. Pound, *Phys. Rev.* **69**, 37–38 (1946).
6. F. Bloch, *Phys. Rev.* **70**, 460–474 (1946).
7. H. Goldstein, "Classical Mechanics," pp. 1–672. Addison-Wesley, Reading, Mass., 1980.
8. J. D. Jackson, "Classical Electrodynamics," pp. 1–848. Wiley, New York, 1975.
9. R. R. Ernst, G. Bodenhausen, and A. Wokaun, "Principles of Nuclear Magnetic Resonance in One and Two Dimensions," pp. 1–610. Clarendon Press, Oxford, 1987.
10. P. L. Corio, "Structure of High-Resolution NMR Spectra," pp. 1–548. Academic Press, New York, 1967.
11. R. C. Cantor and P. R. Schimmel, "Biophysical Chemistry," pp. 1–1371. Freeman, San Francisco, 1980.
12. D. M. Schneider, M. J. Dellwo, and A. J. Wand, *Biochemistry* **31**, 3645–3652 (1992).
13. R. Richarz and K. Wüthrich, *Biopolymers* **17**, 2133–2141 (1978).
14. A. Bundi and K. Wüthrich, *Biopolymers* **18**, 285–297 (1979).
15. D. Braun, G. Wider, and K. Wüthrich, *J. Am. Chem. Soc.* **116**, 8466–8469 (1994).
16. M. P. Williamson, *Biopolymers* **29**, 1423–1431 (1990).
17. S. Spera and A. Bax, *J. Am. Chem. Soc.* **113**, 5490–5492 (1991).
18. D. S. Wishart and B. D. Sykes, *Meth. Enzymol.* **239**, 363–392 (1994).
19. K. Ösapay and D. A. Case, *J. Am. Chem. Soc.* **113**, 9436–9444 (1991).
20. N. F. Ramsey and E. M. Purcell, *Phys. Rev.* **85**, 143–144 (1952).
21. A. G. Palmer, J. Cavanagh, P. E. Wright, and M. Rance, *J. Magn. Reson.* **93**, 151–170 (1991).

2

THEORETICAL DESCRIPTION OF NMR SPECTROSCOPY

A rigorous treatment of the dynamics of nuclear spin systems and NMR spectroscopy is afforded by the quantum-mechanical representation known as the *density matrix formalism* (*1,2*). The density matrix provides a complete description of the state of a spin system. Instead of following the evolution of only the bulk magnetization vector as in the Bloch model, the evolution of the density matrix provides complete information concerning a spin system at any particular point during an NMR experiment. The next few sections present a detailed overview of the development of the density matrix theory and its application in the simplest pulsed NMR experiment.

2.1 Postulates of Quantum Mechanics

A rather formal exposition of the mathematical concepts to be used through the remainder of the text is presented first. Commonly, in introductory quantum-mechanics texts (*3–5*), quantum-mechanical orbital angular momentum is introduced via the classical concepts of angular momentum. After establishing the relevant physics, the results are generalized to include the *intrinsic* angular momentum of electrons and nuclei. The intrinsic angular momentum does not have a classical analog; accordingly, orbital angular momentum will not be discussed in this text. Instead, the foundations of the theory of intrinsic angular momentum will be presented as postulates whose validity is established by comparison with experiment (NMR spectroscopy is a particularly powerful demonstration of the concepts).

2.1.1 THE SCHRÖDINGER EQUATION

The evolution in time of a quantum-mechanical system is governed by the Schrödinger equation:

$$i\hbar \frac{d\Psi(t)}{dt} = \mathcal{H}\Psi(t). \qquad [2.1]$$

The operator $\mathcal{H}$, termed the Hamiltonian of the system, incorporates the essential physics determining the evolution of the system. The Hamiltonian may be time-dependent or time-independent. Units in which $\hbar = 1$ will be assumed and factors of $\hbar$ will not be written explicitly; thus

$$i \frac{d\Psi(t)}{dt} = \mathcal{H}\Psi(t). \qquad [2.2]$$

When desired, necessary factors of $\hbar$ can be reintroduced by dimensional analysis; equivalently, all energies are measured in angular frequency units (rad/s). The solution of the Schrödinger equation is called the wavefunction for the system, $\Psi(t)$. The wavefunction contains *all* the knowable information about the state of the system and, consequently, is a function of the variables appropriate to the system of interest (e.g., spatial coordinates and spin coordinates). The probability density that the system is in the state described by $\Psi(t)$ at time t is given by

$$P(t) = \Psi^*(t)\Psi(t). \qquad [2.3]$$

If the wavefunction is known, then all the observable properties of that system can be deduced by performing the appropriate mathematical operations on the wavefunction. Wavefunctions generally will be assumed to be normalized such that

$$\int \Psi^*(t)\Psi(t)\, d\tau = 1, \qquad [2.4]$$

in which τ represents the generalized coordinates of the wavefunction (and may include sums over spin states). If necessary, wavefunctions can be normalized simply by defining

$$\Psi'(t) = \Psi(t) \bigg/ \left[\int \Psi^*(t)\Psi(t)\, d\tau \right]^{1/2}. \tag{2.5}$$

If $\mathcal{H}$ is time-independent, [2.2] can be solved by the method of separation of variables. Defining $\Psi(t) = \psi(\tau)\varphi(t)$, in which $\psi(\tau)$ contains the time-independent spatial and spin variables, and $\varphi(t)$ contains time-dependent terms:

$$i\frac{d\Psi(t)}{dt} = \mathcal{H}\Psi(t),$$

$$i\psi(\tau)\frac{d\varphi(t)}{dt} = \mathcal{H}\psi(\tau)\varphi(t),$$

$$i\int \psi^*(\tau)\psi(\tau)\, d\tau \frac{d\varphi(t)}{dt} = \varphi(t)\int \psi^*(\tau)\mathcal{H}\psi(\tau)\, d\tau, \tag{2.6}$$

$$\frac{d\varphi(t)}{dt} = -iE\varphi(t),$$

in which the energy of the system is defined by

$$E = \int \psi^*(\tau)\mathcal{H}\psi(\tau)\, d\tau. \tag{2.7}$$

Solving [2.6] yields

$$\Psi(t) = \psi(\tau)\exp(-iEt). \tag{2.8}$$

As shown by [2.8], if $\mathcal{H}$ is time-independent, then the time dependence of the wavefunctions is limited to a phase factor; this factor cancels when calculating probability densities using [2.3]. In this text, $\Psi(t)$ and $\psi(\tau)$ frequently are written as Ψ and ψ, respectively.

2.1.2 EIGENVALUE EQUATIONS

The purpose of quantum mechanics, at least insofar as it is applied to NMR spectroscopy, is to calculate the results expected from experiments. In the language of quantum mechanics, every physically observable quantity, A, has associated with it a Hermitian operator $\mathbf{A}$, that satisfies the eigenvalue equation:

$$\mathbf{A}\psi = \lambda\psi. \tag{2.9}$$

This equation defines a set of N eigenfunctions, ψ_i, and eigenvalues, λ_i, for $i = 1, 2, \ldots, N$ that satisfy in turn

$$\mathbf{A}\psi_i = \lambda_i\,\psi_i.$$ [2.10]

The adjoint of an operator satisfies the eigenvalue equation:

$$\psi^*\mathbf{A}^\dagger = \lambda^*\psi^*.$$ [2.11]

Hermitian operators are self-adjoint, $\mathbf{A} = \mathbf{A}^\dagger$, and satisfy the relationship

$$\int \psi^*\mathbf{A}\varphi\,d\tau = \left[\int \varphi^*\mathbf{A}\psi\,d\tau\right]^*.$$ [2.12]

Hermitian operators are important in quantum mechanics because, as demonstrated below, their eigenvalues are real numbers:

$$\mathbf{A}\psi = \mathbf{A}^\dagger\psi\,;$$

$$\psi^*\mathbf{A}\psi = \psi^*\mathbf{A}^\dagger\psi\,;$$

$$\lambda\,\psi^*\psi = \lambda^*\psi^*\psi\,;$$ [2.13]

$$\lambda\int \psi^*\psi\,d\tau = \lambda^*\int \psi^*\psi\,d\tau\,;$$

$$\lambda = \lambda^*.$$

Consequently, operators corresponding to observable quantities must be Hermitian. The eigenfunctions of a Hermitian operator form a complete orthonormal set. The orthonormality condition is

$$\int \psi_i^*\psi_j\,d\tau = \int \psi_j^*\psi_i\,d\tau = \delta_{i,j},$$ [2.14]

in which $\delta_{i,j}$ is the Kronecker delta with values

$$\delta_{i,j} = \begin{Bmatrix} 0 & \text{for} & i \neq j \\ 1 & \text{for} & i = j \end{Bmatrix}.$$ [2.15]

Unnormalized eigenfunctions can be normalized as in [2.5]; if necessary, the wavefunctions can be orthogonalized using a procedure known as the Gram–Schmidt process (3). A *complete* set of orthonormal functions, ψ, constitutes a set of basis functions for a vector space of dimension N, called the Hilbert space. Therefore, an arbitrary function defined in the vector space can be written as

$$\Psi = \sum_{n=1}^{N} c_n\psi_n,$$ [2.16]

in which the c_n are complex numbers and may depend on time.

The eigenvalue equation, [2.10], leads to the following interpretation of the relationship between an operator and its associated observable; the result of making a measurement of A on a system is one of the eigenvalues of $\mathbf{A}$. This statement illustrates the discrete nature of quantum mechanics; only a limited set of outcomes are possible for the measurement. In practice, however, the expectation value of $\mathbf{A}$ is measured experimentally. The expectation value means the average magnitude of a particular property obtained following a large number of measurements of that property carried out over an ensemble of identically prepared systems. The expectation value of some property, $\langle A \rangle$, is defined as the scalar product of Ψ and $\mathbf{A}\Psi$:

$$\langle A \rangle = \int \Psi^* \mathbf{A} \Psi \, d\tau. \qquad [2.17]$$

If the wavefunction for the system is an eigenfunction of the operator, $\Psi = \psi_n$, then

$$\langle A \rangle = \int \Psi^* \mathbf{A} \Psi \, d\tau = \int \psi_n^* \mathbf{A} \psi_n \, d\tau = \lambda_n \int \psi_n^* \psi_n \, d\tau = \lambda_n. \qquad [2.18]$$

This result shows that if Ψ is an eigenfunction of the operator $\mathbf{A}$, then measuring A for each member of the ensemble yields the identical result λ_n. In general, the wavefunction for the system will not be an eigenfunction of $\mathbf{A}$, and [2.16] is used to express [2.17] in terms of the eigenfunctions of $\mathbf{A}$. The derivation of $\langle A \rangle$ proceeds as follows:

$$\langle A \rangle = \int \Psi^* \mathbf{A} \Psi \, d\tau$$

$$= \int \left[\sum_{i=1}^{N} c_i \psi_i \right]^* \mathbf{A} \left[\sum_{j=1}^{N} c_j \psi_j \right] d\tau = \int \left[\sum_{i=1}^{N} c_i^* \psi_i^* \right] \mathbf{A} \left[\sum_{j=1}^{N} c_j \psi_j \right] d\tau$$

$$= \sum_{i=1}^{N} \sum_{j=1}^{N} c_i^* c_j \int \psi_i^* \mathbf{A} \psi_j \, d\tau = \sum_{i=1}^{N} \sum_{j=1}^{N} c_i^* c_j \lambda_j \int \psi_i^* \psi_j \, d\tau$$

$$= \sum_{j=1}^{N} c_j^* c_j \lambda_j = \sum_{j=1}^{N} c_j^2 \lambda_j. \qquad [2.19]$$

In obtaining [2.19], the orthonormality condition [2.14] has been used. The resulting equation for $\langle A \rangle$ has the following interpretation. When A is measured for a single member of the ensemble, the result obtained is one of the eigenvalues of $\mathbf{A}$; however, which eigenvalue is obtained cannot be specified in advance of the measurement. For the ensemble as a whole, the result λ_j is obtained in a proportion c_j^2; that is, c_j^2 is interpreted as the probability that the result λ_j is obtained in a single measurement.

Consequently, although the allowed values of A must be members of the discrete set of eigenvalues of **A**, the observable expectation value $\langle A \rangle$ can have any (continuous) value consistent with [2.19].

The time-independent Schrödinger equation is an eigenvalue equation for the Hamiltonian operator. Substitution of [2.8] into [2.2] yields

$$i\frac{d\Psi(t)}{dt} = \mathcal{H}\Psi(t);$$

$$i\frac{d\psi(\tau)\exp[-iEt]}{dt} = \mathcal{H}\psi(\tau)\exp[-iEt]; \qquad [2.20]$$

$$E\psi(\tau) = \mathcal{H}\psi(\tau).$$

The eigenvalues of the equation are the energies of the system, and the eigenfunctions are termed the *stationary states* of the system.

2.1.3 SIMULTANEOUS EIGENFUNCTIONS

Next, quantum mechanical restrictions on simultaneous measurement of different observable quantities are presented. Consider two operators, **A** and **B**, corresponding to observable properties A and B. The eigenfunctions and eigenvalues of **A** will be designated ψ and a; the eigenfunctions and eigenvalues of **B** will be designated φ and b. If the wavefunction for the system is ψ_1, then a measurement of A will yield a_1. Suppose that B is measured prior to the measurement of A. Prior to the measurement, the system is represented equivalently by the eigenfunction ψ_1 or by a superposition of the eigenfunctions of **B**, because

$$\psi_1 = \sum_{j=1}^{N} c_j \varphi_j. \qquad [2.21]$$

In this case, the measurement of B yields one of the eigenvalues of **B**—say, b_k with probability c_k^2. After the measurement of B, the wavefunction of the system is given by φ_k. In the language of quantum mechanics, the change of state (from a superposition of the N eigenfunctions of **B** to the single eigenfunction φ_k) on measuring B is called the collapse of the wavefunction. The state of the system is no longer represented by ψ_1. Instead, the system is represented as a superposition of the eigenfunctions of **A**:

$$\varphi_k = \sum_{i=1}^{N} d_i \psi_i. \qquad [2.22]$$

A subsequent measurement of A does not yield a_1; rather, one of the values a_i is obtained with probability d_i^2. In other words, the prior measurement of B prevents the precise outcome of the measurement of A to be specified in advance. This result is the basis for the famous Heisenberg uncertainty relationships (3). As the preceding discussion indicates, simultaneous measurements of A and B are possible only if **A** and **B** have the same eigenfunctions (but not necessarily the same eigenvalues). In such a case, the measurement of B does not alter the wavefunction; therefore, the same value of A is obtained prior to and subsequent to the measurement of B. This observation leads to the theorem that **A** and **B** have the same eigenfunctions if **AB** = **BA**. The proof of this statement is straightforward. If ψ represents the simultaneous eigenfunctions of **A** and **B**, then

$$\mathbf{AB}\psi_i = b_i\mathbf{A}\psi_i = b_i a_i \psi_i \qquad [2.23]$$

and

$$\mathbf{BA}\psi_i = a_i\mathbf{B}\psi_i = a_i b_i \psi_i = b_i a_i \psi_i. \qquad [2.24]$$

Thus

$$\mathbf{AB}\psi_i = \mathbf{BA}\psi_i, \qquad [2.25]$$

and since this equality is satisfied for all members of the set of eigenfunctions, ψ, the general theorem must hold.

Conventionally, the commutator of **A** and **B** is defined as

$$[\mathbf{A}, \mathbf{B}] = \mathbf{AB} - \mathbf{BA}. \qquad [2.26]$$

The earlier result can be restated: two operators have simultaneous eigenfunctions if their commutator vanishes.

2.1.4 EXPECTATION VALUE OF THE MAGNETIC MOMENT

As should now be clear, each operator for an observable quantity defines a set of basis vectors. Any complete orthonormal set can be used to expand an arbitrary wavefunction; consequently, a basis set can be chosen for computational convenience. In no case can the expectation value of an operator depend on the choice of the basis functions.

As an example of the ideas discussed above, the time-dependent expectation value of the magnetic moment of a single spin ($I = \frac{1}{2}$) will be calculated. Using [2.8] and [2.16], the wavefunction for the spin in the static magnetic field can be written as

$$\Psi = c_\alpha\psi_\alpha + c_\beta\psi_\beta = a\exp[-iE_\alpha t]\psi_\alpha + b\exp[-iE_\beta t]\psi_\beta, \qquad [2.27]$$

in which ψ_α and ψ_β are the stationary states and E_α and E_β are the energies of the states with $m = \frac{1}{2}$ and $m = -\frac{1}{2}$, respectively, and a and b are real numbers satisfying the normalization relation $a^2 + b^2 = 1$. Using [2.17] yields

$$\langle \mu_z \rangle = \int \Psi^* \mu_z \Psi \, d\tau = \gamma \int \Psi^* I_z \Psi \, d\tau$$

$$= \gamma \int (a \exp[iE_\alpha t]\psi_\alpha^* + b \exp[iE_\beta t]\psi_\beta^*)I_z$$

$$\times (a \exp[-iE_\alpha t]\psi_\alpha + b \exp[-iE_\beta t]\psi_\beta) \, d\tau$$

$$= \gamma a^2 \int \psi_\alpha^* I_z \psi_\alpha \, d\tau + \gamma ab \exp[i(E_\alpha - E_\beta)t] \int \psi_\alpha^* I_z \psi_\beta \, d\tau$$

$$+ \gamma ab \exp[-i(E_\alpha - E_\beta)t] \int \psi_\beta^* I_z \psi_\alpha \, d\tau + \gamma b^2 \int \psi_\beta^* I_z \psi_\beta \, d\tau$$

$$= \frac{\gamma}{2}(a^2 - b^2); \qquad\qquad [2.28]$$

$$\langle \mu_x \rangle = \int \Psi^* \mu_x \Psi \, d\tau = \gamma \int \Psi^* I_x \Psi \, d\tau$$

$$= \gamma a^2 \int \psi_\alpha^* I_x \psi_\alpha \, d\tau + \gamma ab \exp[i(E_\alpha - E_\beta)t] \int \psi_\alpha^* I_x \psi_\beta \, d\tau$$

$$+ \gamma ab \exp[-i(E_\alpha - E_\beta)t] \int \psi_\beta^* I_x \psi_\alpha \, d\tau + \gamma b^2 \int \psi_\beta^* I_x \psi_\beta \, d\tau$$

$$= \frac{\gamma ab}{2}(\exp[i(E_\alpha - E_\beta)t] + \exp[-i(E_\alpha - E_\beta)t])$$

$$= \gamma ab \cos[(E_\alpha - E_\beta)t] = \gamma ab \cos[\omega_0 t]; \qquad\qquad [2.29]$$

$$\langle \mu_y \rangle = \int \Psi^* \mu_y \Psi \, d\tau = \gamma \int \Psi^* I_y \Psi \, d\tau$$

$$= \gamma a^2 \int \psi_\alpha^* I_y \psi_\alpha \, d\tau + \gamma ab \exp[i(E_\alpha - E_\beta)t] \int \psi_\alpha^* I_y \psi_\beta \, d\tau$$

$$+ \gamma ab \exp[-i(E_\alpha - E_\beta)t] \int \psi_\beta^* I_y \psi_\alpha \, d\tau + \gamma b^2 \int \psi_\beta^* I_y \psi_\beta \, d\tau$$

$$= -i\frac{\gamma ab}{2}(\exp[i(E_\alpha - E_\beta)t] - \exp[-i(E_\alpha - E_\beta)t])$$

$$= \gamma ab \sin[(E_\alpha - E_\beta)t] = \gamma ab \sin[\omega_0 t], \qquad\qquad [2.30]$$

in which $\omega_0 = E_\alpha - E_\beta = -\gamma B_0$. These results utilize the following equations for the angular momentum operators (note that only the equations for I_z are eigenvalue equations):

$$I_z\psi_\alpha = \frac{1}{2}\psi_\alpha, \qquad I_z\psi_\beta = -\frac{1}{2}\psi_\beta;$$

$$I_x\psi_\alpha = \frac{1}{2}\psi_\beta, \qquad I_x\psi_\beta = \frac{1}{2}\psi_\alpha; \qquad [2.31]$$

$$I_y\psi_\alpha = \frac{i}{2}\psi_\beta, \qquad I_y\psi_\beta = -\frac{i}{2}\psi_\alpha,$$

together with the orthonormality of the wavefunctions. Equations [2.31] are derived from the Pauli spin matrices as shown in Section 2.2.5. The three equations [2.28]–[2.30] represent a vector of constant magnitude precessing about the z axis with an angular velocity ω_0. This result is identical to the predicted motion of the magnetic moment obtained from the Bloch model.

2.2 The Density Matrix

Calculations of scalar products and expectation values are frequent operations in quantum mechanics. Such calculations are facilitated by a formulation of quantum mechanics that focuses on the *density matrix* rather than the wavefunction for a system. Additionally, the symbolic manipulations required are simplified by using a notational system introduced into quantum mechanics by Dirac (6).

2.2.1 DIRAC NOTATION

The Dirac notation is a compact formalism for representing the scalar product. In this notation, a wavefunction, ψ, is represented by the *ket* function, $|\psi\rangle$, and the conjugate wavefunction, ψ^*, is represented by the *bra* function, $\langle\psi|$. In the Dirac notation, the scalar product of ψ and φ is written as the contraction of the bra $\langle\psi|$ and the ket $|\varphi\rangle$:

$$\langle\psi|\varphi\rangle \equiv \int \psi^*\varphi \, d\tau. \qquad [2.32]$$

Using the Dirac notation, an arbitrary wavefunction, Ψ, can be written as a superposition of a set of orthonormal time-independent kets, known as *eigenkets* or *basis kets*,

$$|\Psi\rangle = \sum_{n=1}^{N} c_n|n\rangle, \qquad [2.33]$$

where $|n\rangle$ are the basis kets (e.g., the α and β wavefunctions), c_n are complex numbers, and N is the dimensionality of the vector space. For example, the wavefunction for a system consisting of a single spin-$\frac{1}{2}$ nucleus can be described by the linear combination of the kets for the α and β states of that nucleus (which are the eigenfunctions of the angular momentum operator). The coefficients, c_n, can be regarded as amplitude factors that describe how much a particular basis ket contributes to the total wavefunction at any particular time. If the basis kets are time-independent, then any time dependence in Ψ is contained in the complex coefficients. Premultiplying [2.33] by the bra, $\langle m|$, and applying the orthogonality condition yields

$$c_m = \langle m|\Psi\rangle, \qquad [2.34]$$

so that

$$|\Psi\rangle = \sum_{n=1}^{N} c_n|n\rangle = \sum_{n=1}^{N} \langle n|\Psi\rangle|n\rangle = \sum_{n=1}^{N} |n\rangle\langle n|\Psi\rangle. \qquad [2.35]$$

The latter identity suggests that $|n\rangle\langle n|$ is an operator acting on Ψ such that

$$|n\rangle\langle n|\Psi\rangle = c_n|n\rangle. \qquad [2.36]$$

Because [2.35] must hold for arbitrary Ψ, the useful closure theorem is obtained immediately:

$$\sum_{n=1}^{N} |n\rangle\langle n| = \mathbf{E}, \qquad [2.37]$$

in which $\mathbf{E}$ is the identity operator. The operator $|n\rangle\langle n|$ is called a projection operator because it ''projects out'' from Ψ the component ket $|n\rangle$.

The expectation value of some property, $\langle A\rangle$, can be written in Dirac notation as

$$\langle A\rangle = \int \Psi^*\mathbf{A}\Psi \, d\tau = \langle\Psi|\mathbf{A}|\Psi\rangle. \qquad [2.38]$$

Now, using [2.33],

$$\langle A\rangle = \sum_{nm} c_m^* c_n \langle m|\mathbf{A}|n\rangle. \qquad [2.39]$$

In contrast to [2.19], the kets $|n\rangle$ are not necessarily the eigenfunctions of $\mathbf{A}$; therefore, the scalar products $\langle m|\mathbf{A}|n\rangle$ do not necessarily vanish for $m \neq n$. Equation [2.19] is a special case derived from [2.39] if the kets $|n\rangle$ are eigenfunctions of $\mathbf{A}$. For a given basis set, the terms $\langle m|\mathbf{A}|n\rangle$ are constants, and the value of the observable A for a particular state of the system is determined by the products of the coefficients $c_m^* c_n$. Both $\langle m|\mathbf{A}|n \rangle$ and $c_m^* c_n$ can be formed into matrices to yield

$$\langle A \rangle = \sum_{nm} c_m^* c_n A_{mn}, \qquad [2.40]$$

where $A_{mn} = \langle m|\mathbf{A}|n \rangle$ is the (mn)th element of the $N \times N$ matrix representation of the operator $\mathbf{A}$ in a given basis. Once the matrix of coefficients $c_m^* c_n$ is known, the expectation value of any observable can be calculated. The products $c_m^* c_n$ can be regarded as the elements of a matrix representation of an operator $\mathbf{P}$ defined by

$$P_{nm} = \langle n|\mathbf{P}|m \rangle = c_n c_m^*. \qquad [2.41]$$

Note that $\mathbf{P}$ can be explicitly written as a projection operator, $\mathbf{P} = |\Psi\rangle\langle\Psi|$. Substituting [2.41] into [2.40] yields

$$\langle A \rangle = \sum_{nm} c_n c_m^* \langle m|\mathbf{A}|n \rangle$$

$$= \sum_{nm} \langle n|\mathbf{P}|m \rangle\langle m|\mathbf{A}|n \rangle = \sum_{n} \langle n|\mathbf{PA}|n \rangle$$

$$= \sum_{nm} P_{nm} A_{mn} = \sum_{n} (PA)_{nn}$$

$$= \mathrm{Tr}\{\mathbf{PA}\}, \qquad [2.42]$$

where $\mathrm{Tr}\{\ \}$ is the *trace* of a matrix defined as the sum of the diagonal elements of the matrix. The equality on line 2 in [2.42] is a consequence of the closure theorem [2.37]; the equality on line 3 results from the definition of matrix multiplication of the matrix representations of the operators. Equation [2.42] states that the expectation value of some observable of a system, say, for example, the amount of x magnetization, is calculated as the trace of the product of $\mathbf{P}$ and $\mathbf{A}$. $\mathbf{P}$ is the operator that is defined by the coefficients $c_n c_m^*$ and so describes the state of the system at any particular point in time and $\mathbf{A}$ is the operator corresponding to the required observable. For the sake of completeness and formality, note that $\mathbf{P}$ is a Hermitian operator such that

$$\langle n|\mathbf{P}|m \rangle = \langle m|\mathbf{P}|n \rangle^*. \qquad [2.43]$$

The trace of a product of matrices is invariant to cyclic permutations of the matrices. Thus,

$$\mathrm{Tr}\{\mathbf{ABC}\} = \mathrm{Tr}\{\mathbf{CAB}\} = \mathrm{Tr}\{\mathbf{BCA}\}. \qquad [2.44]$$

A corollary of this theorem is that the trace of a commutator is zero:

$$\mathrm{Tr}\{[\mathbf{A}, \mathbf{B}]\} = \mathrm{Tr}\{\mathbf{AB} - \mathbf{BA}\} = \mathrm{Tr}\{\mathbf{AB}\} - \mathrm{Tr}\{\mathbf{BA}\} = 0. \qquad [2.45]$$

2.2.2 QUANTUM STATISTICAL MECHANICS

The analysis presented above is applicable to a system in a so-called pure state in which the entire system is described by the same wavefunction. The wavefunction for a macromolecule in an NMR solution is an enormously complicated function of the degrees of freedom of the molecule and includes contributions from the spin, rotational, vibrational, electronic, and translational properties of the molecule. Determining the complete wavefunction for the molecule is both infeasible and unnecessary because the properties of the nuclear spins are of primary interest in NMR spectroscopy. Accordingly, the system is divided into two components: the spin system and the surroundings (i.e., all other degrees of freedom). For historical reasons, the surroundings are termed the lattice. As a result of this division, the spin wavefunctions for different molecules in the NMR sample are no longer identical, but rather depend on the "hidden" lattice variables. Such a system is called a *mixed state*, and the effects of the lattice are incorporated by using statistical mechanics (2, 7). Each subensemble contained in the sample can be described by a wavefunction, Ψ, and a probability density, $\mathcal{P}(\Psi)$, that represents the contribution of the subensemble to the mixed state. The statistical value of the expectation value for a mixed state is then obtained by averaging over the probability distribution:

$$\overline{\langle A \rangle} = \int \mathcal{P}(\psi) \langle \Psi | \mathbf{A} | \Psi \rangle \, d\tau$$

$$= \sum_{nm} \int \mathcal{P}(\Psi) c_n c_m^* \, d\tau \, \langle m | \mathbf{A} | n \rangle$$

$$= \sum_{nm} \overline{c_n c_m^*} \, \langle m | \mathbf{A} | n \rangle. \qquad [2.46]$$

The matrix elements $c_n c_m^*$ will vary from system to system, but the matrix elements A_{mn} will not. A bar has been used to denote the statistical ensemble average in [2.46].

The ensemble average of coefficients, $\overline{c_n c_m^*}$, form a matrix that is referred to as the *density matrix*. The density matrix is the matrix representation of an operator σ referred to as the density operator, such that

$$\overline{c_n c_m^*} = \overline{\langle n | \mathbf{P} | m \rangle} = \langle n | \sigma | m \rangle = \sigma_{nm}. \qquad [2.47]$$

Since $\mathbf{P}$ is a Hermitian operator, so is σ. An expression similar to [2.42] for the expectation value of the property A in an ensemble of spins in a mixed state can be written as

$$\overline{\langle A \rangle} = \mathrm{Tr}\{\sigma \mathbf{A}\}. \qquad [2.48]$$

The bar will now be dropped for convenience, but an ensemble average is implied. To evaluate the expectation value of an observable, the matrix representation of the appropriate operator and, most importantly, the form of the density operator must be known. The time evolution of the system—say, as it passes through a particular sequence of rf pulses and delays—can be described by the time evolution of the density operator.

2.2.3 THE LIOUVILLE–VON NEUMANN EQUATION

A differential equation that describes the evolution in time of the density operator must be derived. The time-dependent Schrödinger equation, [2.2], can be written as

$$i \sum_n \frac{dc_n(t)}{dt} |n\rangle = \sum_n c_n(t) \mathcal{H} |n\rangle. \qquad [2.49]$$

Multiplying both sides by $\langle k|$ yields

$$i \sum_n \frac{dc_n(t)}{dt} \langle k|n\rangle = \sum_n c_n(t) \langle k|\mathcal{H}|n\rangle. \qquad [2.50]$$

The set of basis kets is orthonormal; therefore, $\langle k|n\rangle = 0$ unless $n = k$, and [2.50] reduces to

$$i \frac{dc_k(t)}{dt} = \sum_n c_n(t) \langle k|\mathcal{H}|n\rangle. \qquad [2.51]$$

Equation [2.51] can be used to find a differential equation for the matrix elements of the density operator:

$$\frac{d\langle k|\sigma|m\rangle}{dt} = \overline{\frac{dc_k c_m^*}{dt}} = c_k \overline{\frac{dc_m^*}{dt}} + \overline{\frac{dc_k}{dt}} c_m^*$$

$$= i \sum_n \overline{c_k c_n^*} \langle n|\mathcal{H}|m\rangle - i \sum_n \overline{c_n c_m^*} \langle k|\mathcal{H}|n\rangle$$

$$= i \sum_n \langle k|\sigma|n\rangle\langle n|\mathcal{H}|m\rangle - i \sum_n \langle k|\mathcal{H}|n\rangle\langle n|\sigma|m\rangle$$

$$= i[\langle k|\sigma\mathcal{H}|m\rangle - \langle k|\mathcal{H}\sigma|m\rangle], \qquad [2.52]$$

in which $\mathcal{H}$ is assumed to be identical for all members of the ensemble and the complex conjugate of [2.51] is written as

$$\frac{dc_k^*(t)}{dt} = \left[-i \sum_n c_n(t) \langle k|\mathcal{H}|n\rangle \right]^*$$

$$= i \sum_n c_n^*(t)\langle k|\mathcal{H}|n\rangle^*$$

$$= i \sum_n c_n^*(t)\langle n|\mathcal{H}|k\rangle. \tag{2.53}$$

The last line of [2.53] is obtained using the Hermitian property of $\mathcal{H}$ expressed in [2.12]. Equation [2.52] can be written in operator form as

$$\frac{d\sigma(t)}{dt} = i[\sigma(t), \mathcal{H}] = -i[\mathcal{H}, \sigma(t)]. \tag{2.54}$$

This is known as the *Liouville–von Neumann* equation and describes the time evolution of the density operator.

The solution to [2.54] is straightforward if the Hamiltonian is time-independent:

$$\sigma(t) = \exp(-i\mathcal{H}t)\, \sigma(0)\, \exp(i\mathcal{H}t). \tag{2.55}$$

Equation [2.55] can be shown to be a solution to [2.54] by simple differentiation:

$$\frac{d\sigma(t)}{dt} = -i\mathcal{H}\exp(-i\mathcal{H}t)\, \sigma(0)\, \exp(i\mathcal{H}t) + \exp(-i\mathcal{H}t)\, \sigma(0)\, i\mathcal{H}\exp(i\mathcal{H}t)$$

$$= i\,[\exp(-i\mathcal{H}t)\, \sigma(0)\, \mathcal{H}\exp(i\mathcal{H}t) - \mathcal{H}\exp(-i\mathcal{H}t)\, \sigma(0)\, \exp(i\mathcal{H}t)]$$

$$= i\,[\sigma(t)\mathcal{H} - \mathcal{H}\sigma(t)] = i\,[\sigma(t), \mathcal{H}]. \tag{2.56}$$

The exponential operator $\exp(\mathbf{A})$ is defined by its Taylor series expansion:

$$\exp(\mathbf{A}) = \sum_{k=0}^{\infty} \frac{1}{k!}\mathbf{A}^k = \mathbf{E} + \mathbf{A} + \frac{1}{2}\mathbf{AA} + \cdots, \tag{2.57}$$

in which $\mathbf{E}$ is the identity operator. The operators $\mathbf{A}$ and $\exp(\mathbf{A})$ necessarily commute. Furthermore, in the eigenbase of $\mathbf{A}$, the matrix representation of the exponential operator is

$$\langle m|\exp(\mathbf{A})|n\rangle = \langle m|\mathbf{E}|n\rangle + \langle m|\mathbf{A}|n\rangle + (\tfrac{1}{2})\langle m|\mathbf{AA}|n\rangle + \cdots$$

$$= \delta_{m,n}\,(1 + A_{mm} + (\tfrac{1}{2})\,A_{mm}^2 + \cdots)$$

$$= \delta_{m,n}\exp(A_{mm}) = \delta_{m,n}\exp(\lambda_m), \tag{2.58}$$

in which $\lambda_m = A_{mm}$ are the eigenvalues of $\mathbf{A}$. Thus, the exponential matrix is diagonal in the eigenbase of $\mathbf{A}$ and the diagonal elements are the exponentials of the eigenvalues of $\mathbf{A}$. An extremely important theorem concerning exponential operators states that $\exp(\mathbf{A} + \mathbf{B}) = \exp(\mathbf{A})\exp(\mathbf{B})$ if and only if $[\mathbf{A}, \mathbf{B}] = 0$ (3).

2.2.4 The Rotating-Frame Transformation

The solution to the Liouville–von Neumann equation is straightforward if the Hamiltonian is time-independent. A pulse sequence generally consists of two distinct parts: pulses and delays (during which no rf fields are present). For the present treatment, the time-dependent effects of the coupling between the spin system and the lattice will be neglected; these effects give rise to spin relaxation phenomena that will be discussed in Chapter 5. In this case, the Hamiltonian governing the delays will be time-independent; however, the rf field constituting the pulse remains a time-dependent perturbation. The simplest solution to this complication is to find a transformation that renders the pulse Hamiltonian time-independent and then apply [2.55]. The transformation that renders $\mathcal{H}$ time-independent is the quantum-mechanical equivalent of the rotating-frame transformation in the Bloch picture.

A *similarity transformation* applied to the laboratory-frame density operator σ generates a transformed density operator σ^r, such that

$$\sigma^r = \mathbf{U}\sigma\mathbf{U}^{-1}, \qquad [2.59]$$

in which $\mathbf{U}$ is a unitary operator. A unitary operator $\mathbf{U}$ is defined by the relationship $\mathbf{U}^{-1} = \mathbf{U}^\dagger$. The equation of motion for σ^r is described by

$$\frac{d\sigma^r(t)}{dt} = i\,[\sigma^r(t), \mathcal{H}_e], \qquad [2.60]$$

in which $\mathcal{H}_e$ is a transformed Hamiltonian. The form of $\mathcal{H}_e$ can be established as follows:

$$\frac{d\sigma^r}{dt} = \frac{d(\mathbf{U}\sigma\mathbf{U}^{-1})}{dt} = \mathbf{U}\frac{d\sigma}{dt}\mathbf{U}^{-1} + \frac{d\mathbf{U}}{dt}\sigma\mathbf{U}^{-1} + \mathbf{U}\sigma\frac{d\mathbf{U}^{-1}}{dt}$$

$$= i\mathbf{U}[\sigma, \mathcal{H}]\mathbf{U}^{-1} + \frac{d\mathbf{U}}{dt}\mathbf{U}^{-1}\mathbf{U}\sigma\mathbf{U}^{-1} + \mathbf{U}\sigma\mathbf{U}^{-1}\mathbf{U}\frac{d\mathbf{U}^{-1}}{dt}$$

$$= i\mathbf{U}[\sigma, \mathcal{H}]\mathbf{U}^{-1} + \frac{d\mathbf{U}}{dt}\mathbf{U}^{-1}\sigma^r + \sigma^r\mathbf{U}\frac{d\mathbf{U}^{-1}}{dt}. \qquad [2.61]$$

The common technique of inserting $\mathbf{E} = \mathbf{U}^{-1}\mathbf{U}$ has been utilized. To proceed, the following identities are established:

$$\frac{d\mathbf{E}}{dt} = \frac{d(\mathbf{U}\mathbf{U}^{-1})}{dt} = \frac{d\mathbf{U}}{dt}\mathbf{U}^{-1} + \mathbf{U}\frac{d\mathbf{U}^{-1}}{dt} = 0, \qquad [2.62]$$

which yields

$$\frac{d\mathbf{U}}{dt}\mathbf{U}^{-1} = -\mathbf{U}\frac{d\mathbf{U}^{-1}}{dt} \qquad [2.63]$$

and,

$$\mathbf{U}[\sigma, \mathcal{H}]\mathbf{U}^{-1} = \mathbf{U}(\sigma\mathcal{H} - \mathcal{H}\sigma)\mathbf{U}^{-1} = \mathbf{U}\sigma\mathbf{U}^{-1}\mathbf{U}\mathcal{H}\mathbf{U}^{-1} - \mathbf{U}\mathcal{H}\mathbf{U}^{-1}\mathbf{U}\sigma\mathbf{U}^{-1}$$

$$= [\sigma^r, \mathbf{U}\mathcal{H}\mathbf{U}^{-1}]. \qquad [2.64]$$

Substituting [2.63] and [2.64] into [2.61] yields

$$\frac{d\sigma^r}{dt} = i\mathbf{U}[\sigma, \mathcal{H}]\mathbf{U}^{-1} + \frac{d\mathbf{U}}{dt}\mathbf{U}^{-1}\sigma^r + \sigma^r\mathbf{U}\frac{d\mathbf{U}^{-1}}{dt}$$

$$= i[\sigma^r, \mathbf{U}\mathcal{H}\mathbf{U}^{-1}] - \mathbf{U}\frac{d\mathbf{U}^{-1}}{dt}\sigma^r + \sigma^r\mathbf{U}\frac{d\mathbf{U}^{-1}}{dt}$$

$$= i[\sigma^r, \mathbf{U}\mathcal{H}\mathbf{U}^{-1}] + \left[\sigma^r, \mathbf{U}\frac{d\mathbf{U}^{-1}}{dt}\right]$$

$$= i\left[\sigma^r, \mathbf{U}\mathcal{H}\mathbf{U}^{-1} - i\mathbf{U}\frac{d\mathbf{U}^{-1}}{dt}\right]. \qquad [2.65]$$

This system obeys [2.60] if the effective Hamiltonian, $\mathcal{H}_e$, is written as

$$\mathcal{H}_e = \mathbf{U}\mathcal{H}\mathbf{U}^{-1} - i\mathbf{U}\frac{d(\mathbf{U}^{-1})}{dt}. \qquad [2.66]$$

If a unitary transformation can be found that renders $\mathcal{H}_e$ time-independent, then the solution to [2.60] can be obtained by straightforward adaptation of [2.55]

$$\sigma^r(t) = \exp(-i\mathcal{H}_e t)\, \sigma^r(0)\, \exp(i\mathcal{H}_e t), \qquad [2.67]$$

in which $\mathbf{U} = \exp(-i\mathcal{H}_e t)$ is a unitary operator called the propagator. The general procedure for solving [2.54] is as follows. Find a unitary transformation that renders $\mathcal{H}$ time-independent; transform $\sigma(0)$ and $\mathcal{H}$ to $\sigma^r(0)$ and $\mathcal{H}_e$; solve [2.60] for $\sigma^r(t)$; and finally transform $\sigma^r(t)$ back to $\sigma(t)$. This procedure is diagrammed in Fig. 2.1.

2.2.5 MATRIX REPRESENTATIONS OF THE SPIN OPERATORS

To proceed, the matrix representation of the angular momentum operators that uses the $|\alpha\rangle$ and $|\beta\rangle$ states of the spin as basis functions must be presented. The Pauli spin matrices form a complete basis set for a single-spin-$\frac{1}{2}$ system (3):

$$I_x = \frac{1}{2}\begin{bmatrix} 0 & 1 \\ 1 & 0 \end{bmatrix}; \qquad I_y = \frac{1}{2}\begin{bmatrix} 0 & -i \\ i & 0 \end{bmatrix}; \qquad I_z = \frac{1}{2}\begin{bmatrix} 1 & 0 \\ 0 & -1 \end{bmatrix}. \qquad [2.68]$$

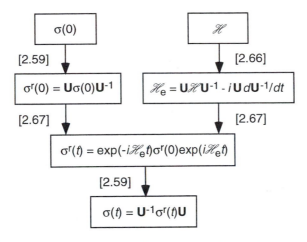

FIGURE 2.1 Flowchart illustrating the procedure for solving the Liouville–von Neumann equation.

Each of these operators is Hermitian. The spin operators satisfy the commutation relation

$$[I_x, I_y] = iI_z \qquad [2.69]$$

and any cyclic permutation of [2.69]. The kets are represented by the 2×1 column vectors:

$$|\alpha\rangle = \begin{bmatrix} 1 \\ 0 \end{bmatrix}; \qquad |\beta\rangle = \begin{bmatrix} 0 \\ 1 \end{bmatrix}, \qquad [2.70]$$

and the bras are represented by the 1×2 row vectors:

$$\langle\alpha| = [1 \quad 0]; \qquad \langle\beta| = [0 \quad 1]. \qquad [2.71]$$

Arbitrary kets and bras, expressed as a linear combination of the eigenkets or eigenbras, have the representations

$$|\Psi\rangle = c_\alpha|\alpha\rangle + c_\beta|\beta\rangle = c_\alpha \begin{bmatrix} 1 \\ 0 \end{bmatrix} + c_\beta \begin{bmatrix} 0 \\ 1 \end{bmatrix} = \begin{bmatrix} c_\alpha \\ c_\beta \end{bmatrix};$$

$$\langle\Psi| = c_\alpha^*\langle\alpha| + c_\beta^*\langle\beta| = c_\alpha^*[1 \quad 0] + c_\beta^*[0 \quad 1] = [c_\alpha^* \quad c_\beta^*]. \qquad [2.72]$$

Thus, the matrix representation of a ket is the column vector whose elements are the coefficients from the expansion in terms of basis kets. The results of operator manipulations can be expressed using matrix algebra. For example,

$$I_z|\alpha\rangle = \frac{1}{2}\begin{bmatrix} 1 & 0 \\ 0 & -1 \end{bmatrix}\begin{bmatrix} 1 \\ 0 \end{bmatrix} = \frac{1}{2}\begin{bmatrix} 1 \\ 0 \end{bmatrix} = \frac{1}{2}|\alpha\rangle,$$

$$I_z|\beta\rangle = \frac{1}{2}\begin{bmatrix} 1 & 0 \\ 0 & -1 \end{bmatrix}\begin{bmatrix} 0 \\ 1 \end{bmatrix} = \frac{1}{2}\begin{bmatrix} 0 \\ -1 \end{bmatrix} = -\frac{1}{2}|\beta\rangle;$$

$$I_x|\alpha\rangle = \frac{1}{2}\begin{bmatrix} 0 & 1 \\ 1 & 0 \end{bmatrix}\begin{bmatrix} 1 \\ 0 \end{bmatrix} = \frac{1}{2}\begin{bmatrix} 0 \\ 1 \end{bmatrix} = \frac{1}{2}|\beta\rangle,$$

$$I_x|\beta\rangle = \frac{1}{2}\begin{bmatrix} 0 & 1 \\ 1 & 0 \end{bmatrix}\begin{bmatrix} 0 \\ 1 \end{bmatrix} = \frac{1}{2}\begin{bmatrix} 1 \\ 0 \end{bmatrix} = \frac{1}{2}|\alpha\rangle; \qquad [2.73]$$

$$I_y|\alpha\rangle = \frac{1}{2}\begin{bmatrix} 0 & -i \\ i & 0 \end{bmatrix}\begin{bmatrix} 1 \\ 0 \end{bmatrix} = \frac{1}{2}\begin{bmatrix} 0 \\ i \end{bmatrix} = \frac{i}{2}|\beta\rangle,$$

$$I_y|\beta\rangle = \frac{1}{2}\begin{bmatrix} 0 & -i \\ i & 0 \end{bmatrix}\begin{bmatrix} 0 \\ 1 \end{bmatrix} = \frac{1}{2}\begin{bmatrix} -i \\ 0 \end{bmatrix} = -\frac{i}{2}|\alpha\rangle$$

express the results of the Cartesian spin operators acting on the $|\alpha\rangle$ and $|\beta\rangle$ kets. Similarly, the orthogonality relations are obtained as

$$\langle\alpha|\alpha\rangle = [1 \quad 0]\begin{bmatrix} 1 \\ 0 \end{bmatrix} = 1, \qquad \langle\alpha|\beta\rangle = [1 \quad 0]\begin{bmatrix} 0 \\ 1 \end{bmatrix} = 0;$$

$$\langle\beta|\alpha\rangle = [0 \quad 1]\begin{bmatrix} 1 \\ 0 \end{bmatrix} = 0, \qquad \langle\beta|\beta\rangle = [0 \quad 1]\begin{bmatrix} 0 \\ 1 \end{bmatrix} = 1. \qquad [2.74]$$

The matrix representations of operators and wavefunctions depends on the particular basis set employed. Matrix representations using different basis sets can be interconverted using unitary transformations. If Ψ' is the representation of a wavefunction in one (primed) basis set and Ψ is the representation in another (unprimed basis), then

$$|\Psi'\rangle = \mathbf{U}|\Psi\rangle, \qquad [2.75]$$

in which $\mathbf{U}$ is a unitary operator with matrix elements

$$U_{ij} = \langle i|\mathbf{U}|j\rangle = \langle i|j'\rangle. \qquad [2.76]$$

The representation of an operator in the two basis sets is then given by the similarity transformation:

$$\mathbf{A}' = \mathbf{U}\mathbf{A}\mathbf{U}^{-1}. \qquad [2.77]$$

Using these results, the expectation value of $\mathbf{A}'$ is

$$\langle\Psi'|\mathbf{A}'|\Psi'\rangle = \langle\Psi|\mathbf{U}^{-1}\mathbf{U}\mathbf{A}\mathbf{U}^{-1}\mathbf{U}|\Psi\rangle = \langle\Psi|\mathbf{A}|\Psi\rangle, \qquad [2.78]$$

which justifies earlier assertions that the results of calculating the expectation value of an operator do not depend on the choice of basis set.

In order to clarify these ideas, the transformation between a basis set consisting of the eigenfunctions of I_z and a basis set consisting of the eigenfunctions of I_x is presented. The eigenfunction equations for I_x are defined as

$$I_x|\varphi_1\rangle = \frac{1}{2}|\varphi_1\rangle; \qquad I_x|\varphi_2\rangle = -\frac{1}{2}|\varphi_2\rangle, \qquad [2.79]$$

in which φ_1 and φ_2 are the (as yet unspecified) eigenfunctions. An arbitrary wavefunction can be written as

$$\Psi = c_\alpha|\alpha\rangle + c_\beta|\beta\rangle \qquad [2.80]$$

in the basis functions of I_z and as

$$\Psi' = c_1|\varphi_1\rangle + c_2|\varphi_2\rangle \qquad [2.81]$$

in the basis functions of I_x. Application of [2.75] yields the matrix equation:

$$\Psi' = \mathbf{U}\Psi;$$

$$\begin{bmatrix} c_1 \\ c_2 \end{bmatrix} = \begin{bmatrix} U_{11} & U_{12} \\ U_{21} & U_{22} \end{bmatrix} \begin{bmatrix} c_\alpha \\ c_\beta \end{bmatrix}, \qquad [2.82]$$

which can be solved using the relationships in [2.79] to yield

$$\mathbf{U} = \frac{1}{\sqrt{2}}\begin{bmatrix} 1 & 1 \\ 1 & -1 \end{bmatrix}, \qquad [2.83]$$

from which the explicit relationships are obtained:

$$|\varphi_1\rangle = \frac{1}{\sqrt{2}}(|\alpha\rangle + |\beta\rangle); \qquad |\varphi_2\rangle = \frac{1}{\sqrt{2}}(|\alpha\rangle - |\beta\rangle), \qquad [2.84]$$

which can be used to verify the eigenfunction equations [2.79]. Using

[2.77], the operator, I_z for example, has a matrix representation in the basis set of the I_x eigenfunctions of

$$I'_z = \mathbf{U}I_z\mathbf{U}^{-1} = \frac{1}{4}\begin{bmatrix} 1 & 1 \\ 1 & -1 \end{bmatrix}\begin{bmatrix} 1 & 0 \\ 0 & -1 \end{bmatrix}\begin{bmatrix} 1 & 1 \\ 1 & -1 \end{bmatrix}$$

$$= \frac{1}{4}\begin{bmatrix} 1 & 1 \\ 1 & -1 \end{bmatrix}\begin{bmatrix} 1 & 1 \\ -1 & 1 \end{bmatrix} = \frac{1}{2}\begin{bmatrix} 0 & 1 \\ 1 & 0 \end{bmatrix}.$$ [2.85]

In a particularly important application of these ideas, the matrix representation of the Hamiltonian operator, $\mathcal{H}$, is calculated in some convenient basis (usually the product basis). The matrix $\mathbf{U}$ is then determined such that the new matrix representation of the operator, $\mathcal{H}'$, given by

$$\mathcal{H}' = \mathbf{U}\mathcal{H}\mathbf{U}^{-1},$$ [2.86]

is a diagonal matrix. The transformed basis functions given by [2.75] then represent the eigenfunctions of the Hamiltonian operator and the diagonal elements of $\mathcal{H}'$ are the energies associated with the stationary states of the system. An example of these calculations for a two-spin system is given in Section 2.3.

2.3 Pulses and Rotation Operators

The simple case of applying a rf pulse to a single spin-$\frac{1}{2}$ nucleus in a static field B_0 will be considered first. The pulse is a transverse rf field with magnitude $2B_1$ and angular frequency ω_{rf}. Remembering from the Bloch approach that this field can be decomposed into two counterrotating fields, only one of which has a noticeable effect on the spin, the Hamiltonian for the pulse is written as (8)

$$\mathcal{H} = -\boldsymbol{\mu} \cdot \mathbf{B}(t) = \mathcal{H}_z + \mathcal{H}_{rf}$$
$$= \omega_0 I_z + \omega_1 [I_x \cos(\omega_{rf}t + \phi) + I_y \sin(\omega_{rf}t + \phi)], \quad [2.87]$$

where I_α is the spin angular momentum operator along the axis α, $\omega_0 = -\gamma B_0$, and $\omega_1 = -\gamma B_1$. The first term after the second equality in [2.87] represents the influence of the static magnetic field, and the second term represents the rf pulse.

The choice of $\mathbf{U}$ that removes the time dependence from [2.87] is

$$\mathbf{U} = \exp(i\omega_{rf}I_z t).$$ [2.88]

Application of this unitary transformation, using [2.66], gives the effective Hamiltonian:

TABLE 2.1

Rotation Properties of Angular-Momentum Operators

u/v	x	y	z
x	I_x	$I_x \cos \theta - I_z \sin \theta$	$I_x \cos \theta + I_y \sin \theta$
y	$I_y \cos \theta + I_z \sin \theta$	I_y	$I_y \cos \theta - I_x \sin \theta$
z	$I_z \cos \theta - I_y \sin \theta$	$I_z \cos \theta + I_x \sin \theta$	I_z

The tabular entries (u, v) are the results of the similarity transformation $\exp(-i\theta I_v)I_u \exp(i\theta I_v)$.

$$\mathcal{H}_e = \omega_0 I_z + \omega_1 \exp(i\omega_{rf} I_z t)[I_x \cos(\omega_{rf} t + \phi)$$
$$+ I_y \sin(\omega_{rf} t + \phi)] \exp(-i\omega_{rf} I_z t)$$
$$- \exp(i\omega_{rf} I_z t)\omega_{rf} I_z \exp(-i\omega_{rf} I_z t). \qquad [2.89]$$

Using the rotation properties of the spin angular momentum operators presented in Table 2.1 (these properties will be derived below):

$$I_x \cos(\omega_{rf} t) + I_y \sin(\omega_{rf} t) = \exp(-i\omega_{rf} I_z t)I_x \exp(i\omega_{rf} I_z t);$$
$$-I_x \sin(\omega_{rf} t) + I_y \cos(\omega_{rf} t) = \exp(-i\omega_{rf} I_z t)I_y \exp(i\omega_{rf} I_z t), \qquad [2.90]$$

the second term in [2.89] is simplified to $\omega_1(I_x \cos \phi + I_y \sin \phi)$. The third term in [2.89] is simplified to $-\omega_{rf} I_z$ because an operator commutes with an exponential operator of itself. The effective Hamiltonian can be written as

$$\mathcal{H}_e = \omega_0 I_z + \omega_1(I_x \cos \phi + I_y \sin \phi) - \omega_{rf} I_z;$$
$$= (\omega_0 - \omega_{rf})I_z + \omega_1(I_x \cos \phi + I_y \sin \phi). \qquad [2.91]$$

This is now a *time-independent* effective Hamiltonian, and the solution in the form of [2.67] describes evolution of the density operator in the rotating frame. Note the strong similarity between [2.91] and [1.18].

If $\omega_0 - \omega_{rf} = 0$ and $\phi = 0$, then the Hamiltonian for an on-resonance x pulse becomes

$$\mathcal{H}_e = \omega_1 I_x, \qquad [2.92]$$

and, as follows from [2.67]

$$\sigma(\tau_p) = \exp(-i\mathcal{H}_e \tau_p)\sigma(0) \exp(i\mathcal{H}_e \tau_p)$$
$$= \exp(-i\omega_1 I_x \tau_p)\sigma(0) \exp(i\omega_1 I_x \tau_p)$$
$$= \exp(i\gamma B_1 I_x \tau_p)\sigma(0) \exp(-i\gamma B_1 I_x \tau_p). \qquad [2.93]$$

For simplicity, the superscript has been omitted from the rotating-frame density operator; in general, the context is sufficient to establish whether a rotating-frame or laboratory-frame density operator is intended. If $\alpha = -\gamma B_1 \tau_p$, is defined to be the flip angle of the pulse of length τ_p, then

$$\sigma(t) = \exp(-i\alpha I_x)\sigma(0)\exp(i\alpha I_x). \qquad [2.94]$$

The matrix representation of the exponential operators in [2.94] must be derived so that the effect on the density operator can be calculated. If the exponential rotation operators are defined as

$$\mathbf{R}_x(\alpha) = \exp(-i\alpha I_x) \quad \text{and} \quad \mathbf{R}_x^{-1}(\alpha) = \exp(i\alpha I_x), \qquad [2.95]$$

then [2.94] becomes

$$\sigma(t) = \mathbf{R}_x(\alpha)\sigma(0)\mathbf{R}_x^{-1}(\alpha). \qquad [2.96]$$

The rotation operators can be expanded as

$$\mathbf{R}_x^{-1}(\alpha) = \mathbf{E} + i\alpha I_x + \frac{1}{2!}(i\alpha I_x)^2 + \cdots. \qquad [2.97]$$

Using the Pauli spin matrices given in [2.68], the following relationships are easily derived:

$$I_x^2 = I_y^2 = I_z^2 = \frac{1}{4}\mathbf{E}; \qquad [2.98]$$

$$I_\eta^{2n} = \frac{1}{4^n}\mathbf{E}; \qquad [2.99]$$

$$I_\eta^{2n+1} = \frac{1}{4^n}I_\eta. \qquad [2.100]$$

Substituting the results contained in [2.98]–[2.100] into [2.97], and grouping together even and odd powers of iI_x yields

$$\mathbf{R}_x^{-1}(\alpha) = \mathbf{E}\left(1 - \frac{\alpha^2}{2!2^2} + \frac{\alpha^4}{4!2^4} + \cdots\right) + 2iI_x\left(\frac{\alpha}{2} - \frac{\alpha^3}{3!2^3} + \frac{\alpha^5}{5!2^5} + \cdots\right)$$

$$= \mathbf{E}\cos\frac{\alpha}{2} + 2iI_x\sin\frac{\alpha}{2}. \qquad [2.101]$$

Expanding I_x in terms of the raising and lowering operators, $I^+ = I_x + iI_y$ and $I^- = I_x - iI_y$, yields

$$2I_x = (I^+ + I^-) \equiv \mathbf{T}. \qquad [2.102]$$

T is known as the inversion operator and has the effect of changing the spin quantum number from $+\frac{1}{2}$ to $-\frac{1}{2}$ and vice versa. This leads to

$$\mathbf{R}_x^{-1}(\alpha) = \mathbf{E} \cos\frac{\alpha}{2} + i\mathbf{T} \sin\frac{\alpha}{2}. \qquad [2.103]$$

By similar reasoning,

$$\mathbf{R}_x(\alpha) = \exp(-i\alpha I_x) = \mathbf{E} \cos\frac{\alpha}{2} - i\mathbf{T} \sin\frac{\alpha}{2}. \qquad [2.104]$$

The rotation matrix corresponding to a pulse of flip angle, α, applied along the x axis can now be calculated. The elements of the matrix representations of the pulse rotation operators $\mathbf{R}_x^{-1}(\alpha)$ and $\mathbf{R}_x(\alpha)$ are constructed from the basis eigenfunctions using the expressions

$$R_x^{-1}(\alpha)_{rs} = \langle r|\mathbf{E} \cos\frac{\alpha}{2} + i\mathbf{T} \sin\frac{\alpha}{2}|s\rangle; \qquad [2.105]$$

$$R_x(\alpha)_{rs} = \langle r|\mathbf{E} \cos\frac{\alpha}{2} - i\mathbf{T} \sin\frac{\alpha}{2}|s\rangle. \qquad [2.106]$$

For example, if $\langle r| = \langle \alpha|$ and $|s\rangle = |\beta\rangle$, then matrix element $R_x^{-1}(\alpha)_{12}$ is

$$R_x^{-1}(\alpha)_{12} = \langle \alpha|\mathbf{E} \cos\frac{\alpha}{2} + i\mathbf{T} \sin\frac{\alpha}{2}|\beta\rangle = i \sin\frac{\alpha}{2}. \qquad [2.107]$$

The matrix resulting representations of the pulse operators are

$$\mathbf{R}_x^{-1}(\alpha) = \begin{bmatrix} c & is \\ is & c \end{bmatrix} \quad \text{and} \quad \mathbf{R}_x(\alpha) = \begin{bmatrix} c & -is \\ -is & c \end{bmatrix}, \qquad [2.108]$$

where $c = \cos(\alpha/2)$ and $s = \sin(\alpha/2)$.

Now the effect of an α pulse on I_z can be calculated as

$$\mathbf{R}_x(\alpha)I_z\mathbf{R}_x^{-1}(\alpha) = \frac{1}{2}\begin{bmatrix} c & -is \\ -is & c \end{bmatrix}\begin{bmatrix} 1 & 0 \\ 0 & -1 \end{bmatrix}\begin{bmatrix} c & is \\ is & c \end{bmatrix}$$

$$= \frac{1}{2}\begin{bmatrix} c & is \\ -is & -c \end{bmatrix}\begin{bmatrix} c & is \\ is & c \end{bmatrix}$$

$$= \frac{1}{2}\begin{bmatrix} c^2 - s^2 & 2ics \\ -2ics & s^2 - c^2 \end{bmatrix}$$

$$= \frac{1}{2}\begin{bmatrix} \cos\alpha & i\sin\alpha \\ -i\sin\alpha & -\cos\alpha \end{bmatrix}$$

$$= I_z \cos\alpha - I_y \sin\alpha, \qquad [2.109]$$

in which the last line is obtained by using [2.68]. If $\alpha = 180°$, the final matrix would be equal to

$$\mathbf{R}_x(\pi)I_z\mathbf{R}_x^{-1}(\pi) = \frac{1}{2}\begin{bmatrix} -1 & 0 \\ 0 & 1 \end{bmatrix} = -\frac{1}{2}\begin{bmatrix} 1 & 0 \\ 0 & -1 \end{bmatrix} = -I_z. \qquad [2.110]$$

If $\alpha = 90°$, then the final matrix becomes

$$\mathbf{R}_x(\pi/2)I_z\mathbf{R}_x^{-1}(\pi/2) = \frac{1}{2}\begin{bmatrix} 0 & i \\ -i & 0 \end{bmatrix} = -\frac{1}{2}\begin{bmatrix} 0 & -i \\ i & 0 \end{bmatrix} = -I_y. \qquad [2.111]$$

Simply, a 90° pulse applied with x phase to I_z magnetization generates $-I_y$ magnetization. These results are identical to the results obtained using the Bloch model.

Similar analysis for a pulse with y phase ($\phi = \pi/2$) generates a rotation matrix of the form

$$\mathbf{R}_y^{-1}(\alpha) = \begin{bmatrix} c & s \\ -s & c \end{bmatrix} \quad \text{and} \quad \mathbf{R}_y(\alpha) = \begin{bmatrix} c & -s \\ s & c \end{bmatrix}. \qquad [2.112]$$

Finally, a rotation about the z axis (which in practice is difficult to achieve experimentally with rf pulses) has the matrix representation:

$$\mathbf{R}_z^{-1}(\alpha) = \begin{bmatrix} c + is & 0 \\ 0 & c - is \end{bmatrix}$$

$$\mathbf{R}_z(\alpha) = \begin{bmatrix} c - is & 0 \\ 0 & c + is \end{bmatrix}. \qquad [2.113]$$

The rotation induced by the general Hamiltonian given by [2.91], which includes off-resonance effects and arbitrary pulse phases, can be written as

$$\mathbf{R}_\phi(\alpha) = \exp(-i\alpha\mathbf{n}\cdot\mathbf{I}) = \mathbf{E}\cos\frac{\alpha}{2} - 2i\,\mathbf{n}\cdot\mathbf{I}\sin\frac{\alpha}{2}, \qquad [2.114]$$

in which $\mathbf{n}$ is a unit vector along the direction of the effective field in the rotating frame $\mathbf{B}^r$, given by [1.18], α is given by [1.23]

$$\mathbf{n}\cdot\mathbf{I} = I_x\cos\phi\sin\theta + I_y\sin\phi\sin\theta + I_z\cos\theta, \qquad [2.115]$$

and θ is defined by [1.21]. Rather than derive a matrix representation of

[2.114], the following identity will be established:

$$\mathbf{R}_\phi(\alpha) = \mathbf{R}_z(\phi)\mathbf{R}_y(\theta)\mathbf{R}_z(\alpha)\mathbf{R}_y^{-1}(\theta)\mathbf{R}_z^{-1}(\phi). \qquad [2.116]$$

The proof of [2.116] depends on the following useful relationship:

$$\mathbf{U}f(\mathbf{A})\mathbf{U}^{-1} = f(\mathbf{U}\mathbf{A}\mathbf{U}^{-1}), \qquad [2.117]$$

in which $f(\mathbf{A})$ is an arbitrary function acting on the operator $\mathbf{A}$. Equation [2.117] can be verified by expanding $f(\mathbf{U}\mathbf{A}\mathbf{U}^{-1})$ as a Taylor series. Using [2.117],

$$
\begin{aligned}
\mathbf{R}_\phi(\alpha) &= \mathbf{R}_z(\phi)\mathbf{R}_y(\theta)\mathbf{R}_z(\alpha)\mathbf{R}_y^{-1}(\theta)\mathbf{R}_z^{-1}(\phi) \\
&= \mathbf{R}_z(\phi)\exp[-i\alpha\mathbf{R}_y(\theta)I_z\mathbf{R}_y^{-1}(\theta)]\mathbf{R}_z^{-1}(\phi) \\
&= \mathbf{R}_z(\phi)\exp[-i\alpha(I_z\cos\theta + I_x\sin\theta)]\mathbf{R}_z^{-1}(\phi) \\
&= \exp[-i\alpha\mathbf{R}_z(\phi)(I_z\cos\theta + I_x\sin\theta)\mathbf{R}_z^{-1}(\phi)] \\
&= \exp[-i\alpha(I_z\cos\theta + I_x\cos\phi\sin\theta + I_y\sin\phi\sin\theta)] \\
&= \exp[-i\alpha\mathbf{n}\cdot\mathbf{I}], \qquad [2.118]
\end{aligned}
$$

which completes the desired proof. Thus, the operator for rotation about an arbitrary angle can be represented as a series of rotations about the y and z axes. The five rotations used to represent $\mathbf{R}_\phi(\alpha)$ in [2.116] are not mutually independent; the rotation $\mathbf{R}_\phi(\alpha)$ can be reduced to three independent rotations using the Euler decomposition of the general three-dimensional rotation (8).

2.4 Quantum Mechanical NMR Spectroscopy

Theoretical analysis of an NMR experiment requires calculation of the signal observed following a sequence of rf pulses and delays. The overall procedure for performing the analysis is illustrated in Fig. 2.2. The initial state of the spin system is described by the *equilibrium density operator*. Evolution of the density operator through the sequence of pulses and delays is calculated using the Liouville–von Neumann equation [2.54] (Fig. 2.1). The Hamiltonian consists of the Zeeman, scalar coupling, and rf pulse terms that govern evolution of the density operator. The expectation value of the observed signal at the desired time is calculated using [2.48] as the trace of the product of the density operator and the *observation operator* corresponding to the observable magnetization. The equilibrium density operator and observation operator are described next.

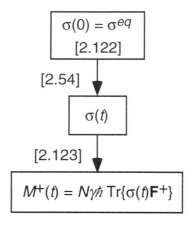

FIGURE 2.2 Flowchart illustrating the theoretical analysis of NMR experiments.

2.4.1 EQUILIBRIUM AND OBSERVATION OPERATORS

The lattice is assumed to always be in thermal equilibrium at a temperature T (equivalently, the lattice is assumed to have infinite heat capacity). At thermodynamic equilibrium, the nuclear spin states are assumed to be in thermal equilibrium with the lattice. Consequently, the values of $\mathscr{P}(\Psi)$ (see Section 2.2.2) are constrained such that the populations of the stationary states (given by the diagonal elements of the density matrix) have a Boltzmann distribution. Furthermore, the density matrix is diagonal at equilibrium because the members of the different subensembles described by $\mathscr{P}(\Psi)$ are uncorrelated. The form of the equilibrium density operator that satisfies these requirements is

$$\sigma^{\mathrm{eq}} = e^{-\mathscr{H}/k_\mathrm{B}T}/\mathrm{Tr}\{e^{-\mathscr{H}/k_\mathrm{B}T}\}. \qquad [2.119]$$

In the eigenbase of the Hamiltonian operator, the matrix elements of σ^{eq} are given by

$$\sigma_{mn}^{\mathrm{eq}} = \langle m|e^{-\mathscr{H}/k_\mathrm{B}T}|n\rangle \Big/ \sum_{i=1}^{N} \langle i|e^{-\mathscr{H}/k_\mathrm{B}T}|i\rangle$$

$$= \delta_{m,n} e^{-E_n/k_\mathrm{B}T} \Big/ \sum_{i=1}^{N} e^{-E_i/k_\mathrm{B}T}, \qquad [2.120]$$

which is a diagonal matrix whose elements are the required Boltzmann probabilities. In the high-temperature approximation, $E_n \ll k_B T$ and the equilibrium density operator can be approximated by

$$
\begin{aligned}
\sigma^{\text{eq}} &= e^{-\mathcal{H}/k_B T}/\text{Tr}\{e^{-\mathcal{H}/k_B T}\} \\
&\approx \{\mathbf{E} - \mathcal{H}/k_B T\}/\text{Tr}\{\mathbf{E} - \mathcal{H}/k_B T\} \\
&\approx \{\mathbf{E} - \mathcal{H}/k_B T\}/\text{Tr}\{\mathbf{E}\} \\
&\approx \mathbf{E}/N - \mathcal{H}/(N k_B T),
\end{aligned}
\qquad [2.121]
$$

in which N is the number of spins considered. The term $\mathbf{E}/N$ is a constant that does not affect the NMR experiment; accordingly, this term is seldom written explicitly, and the high-temperature approximation to the equilibrium density operator is simply written in terms of the Zeeman Hamiltonian as

$$
\sigma^{\text{eq}} = -\mathcal{H}/(N k_B T) = -\sum_{n=1}^{N} \frac{\omega_{0n}}{N k_B T} I_{nz}.
\qquad [2.122]
$$

By convention, the complex magnetization recorded during the acquisition period of an NMR experiment is given by (2):

$$
\mathbf{M}^+(t) = N \gamma \hbar \, \text{Tr}\{\sigma(t)\mathbf{F}^+\},
\qquad [2.123]
$$

in which N is the number of spins per unit volume

$$
\mathbf{F}^+ = \sum_{k=1}^{K} I_k^+ = \sum_{k=1}^{K} (I_{kx} + iI_{ky}),
\qquad [2.124]
$$

and K is the number of types of spins in the spin system (i.e., $K = 2$ for a two-spin system). The operator $\mathbf{F}^-$ could have been chosen as the observation operator equally well.

2.4.2 THE ONE-PULSE EXPERIMENT

The simplest NMR experiment consists of a single pulse followed by acquisition of the FID. For a single spin, [2.122] indicates that $\sigma^{\text{eq}} \propto I_z$. If a 90_x° pulse is applied to the equilibrium density operator, [2.111] indicates that the resulting density operator will be proportional to $-I_y$. The $-I_y$ magnetization will evolve during acquisition under the chemical-shift Hamiltonian (in the rotating frame) given as

$$
\mathcal{H}_{\text{eff}} = (\omega_0 - \omega_{\text{rf}})I_z = \Omega I_z.
\qquad [2.125]
$$

This is a time-independent Hamiltonian; therefore,

$$\sigma(t) = \exp(-i\mathcal{H}t)\sigma(0)\exp(i\mathcal{H}t)$$

$$= \exp(-i\Omega I_z t)\sigma(0)\exp(i\Omega I_z t)$$

$$= \mathbf{U}\sigma(0)\mathbf{U}^{-1}, \qquad\qquad [2.126]$$

and

$$\mathbf{U}^{-1} = \exp(i\Omega t I_z) = \begin{bmatrix} \exp(i\Omega t/2) & 0 \\ 0 & \exp(-i\Omega t/2) \end{bmatrix}. \qquad [2.127]$$

Performing the matrix manipulations for $\sigma(0) = -I_y$ yields

$$\sigma(t) = \frac{1}{2}\begin{bmatrix} 0 & i\exp(-i\Omega t) \\ -i\exp(i\Omega t) & 0 \end{bmatrix}$$

$$= \frac{1}{2}\begin{bmatrix} 0 & i[\cos(\Omega t) - i\sin(\Omega t)] \\ -i[\cos(\Omega t) + i\sin(\Omega t)] & 0 \end{bmatrix}$$

$$= \frac{1}{2}\begin{bmatrix} 0 & i\cos(\Omega t) + \sin(\Omega t) \\ -i\cos(\Omega t) + \sin(\Omega t) & 0 \end{bmatrix}$$

$$= \frac{1}{2}\begin{bmatrix} 0 & -\sin(\Omega t) \\ \sin(\Omega t) & 0 \end{bmatrix} + \frac{i}{2}\begin{bmatrix} 0 & \cos(\Omega t) \\ -\cos(\Omega t) & 0 \end{bmatrix}. \quad [2.128]$$

Using the Pauli spin matrices, [2.128] can be written as

$$\sigma(t) = \mathbf{U}\sigma(0)\mathbf{U}^{-1} = \exp(-i\Omega I_z)I_y\exp(i\Omega I_z)$$

$$= I_x\sin(\Omega t) - I_y\cos(\Omega t). \qquad\qquad [2.129]$$

Note that at $t = 0$, $\sigma(t) = -I_y$, and at $\Omega t = \pi/2$, $\sigma(t) = I_x$. Magnetization precesses at its chemical shift in the sense $-y \to x \to y \to -x$ in the rotating frame.

The preceding results yield the form of the detectable magnetization for a one-pulse sequence for a single isolated spin:

$$\mathbf{M}^+(t) = N\gamma\hbar\mathrm{Tr}\{(I_x\sin(\Omega t) - I_y\cos(\Omega t)(I_x + iI_y)\}$$

$$= N\gamma\hbar[\mathrm{Tr}\{I_x^2\}\sin(\Omega t) + i\mathrm{Tr}\{I_xI_y\}\sin(\Omega t)$$

$$- \mathrm{Tr}\{I_y I_x\} \cos(\Omega t) - i\mathrm{Tr}\{I_y^2\} \cos(\Omega t)]$$

$$= \frac{1}{4} N\gamma\hbar[\sin(\Omega t) - i\cos(\Omega t)]$$

$$= \frac{1}{4} N\gamma\hbar \exp[i(\Omega t - \pi/2)]. \qquad [2.130]$$

This signal has the form expected from the analysis of the same system using the Bloch model in the absence of relaxation (the factor $\exp[-i\pi/2]$ is a time-independent phase factor).

2.5 Quantum Mechanics of Multispin Systems

In this section, the use of the density operator approach to perform calculations on larger, scalar coupled spin systems will be illustrated; as discussed in Section 1.6, the Bloch model fails to properly account for the evolution of such spin systems. The problem is to establish the matrix representation of wavefunctions and operators in a two- (in general N-) spin system and derive an appropriate operator algebra. The central results will be derived using the direct product basis; transformations to other basis sets can be performed using similarity transformations as described above. Additional details can be found in the monograph by Corio (8).

2.5.1 DIRECT-PRODUCT SPACES

The wavefunctions in the *product basis* are given by the *direct products* of the wavefunctions for individual spins:

$$\Psi_m = |m_1\rangle \otimes |m_2\rangle \cdots \otimes |m_N\rangle \equiv \prod_{i=1}^{N} |m_i\rangle \equiv |m_1, m_2, ..., m_N\rangle, \qquad [2.131]$$

in which m_i takes on all possible values, yielding 2^N wavefunctions for spin-$\frac{1}{2}$ nuclei. The total magnetic quantum number associated with a wavefunction in the product basis is

$$m = \sum_{i=1}^{N} m_i. \qquad [2.132]$$

The direct product of two matrices is given by (illustrated for two 2×2 matrices)

$$\mathbf{A} \otimes \mathbf{B} = \begin{bmatrix} A_{11} & A_{12} \\ A_{21} & A_{22} \end{bmatrix} \otimes \begin{bmatrix} B_{11} & B_{12} \\ B_{21} & B_{22} \end{bmatrix} = \begin{bmatrix} A_{11}\mathbf{B} & A_{12}\mathbf{B} \\ A_{21}\mathbf{B} & A_{22}\mathbf{B} \end{bmatrix}$$

$$= \begin{bmatrix} A_{11}B_{11} & A_{11}B_{12} & A_{12}B_{11} & A_{12}B_{12} \\ A_{11}B_{21} & A_{11}B_{22} & A_{12}B_{21} & A_{12}B_{22} \\ A_{21}B_{11} & A_{21}B_{12} & A_{22}B_{11} & A_{22}B_{12} \\ A_{21}B_{21} & A_{11}B_{22} & A_{22}B_{21} & A_{22}B_{22} \end{bmatrix}. \qquad [2.133]$$

Thus, for example, the four wavefunctions in the product basis of a two-spin system are

$$\psi_1 = |\alpha\alpha\rangle = \begin{bmatrix} 1 \\ 0 \end{bmatrix} \otimes \begin{bmatrix} 1 \\ 0 \end{bmatrix} = \begin{bmatrix} 1 \\ 0 \\ 0 \\ 0 \end{bmatrix};$$

$$\psi_2 = |\alpha\beta\rangle = \begin{bmatrix} 1 \\ 0 \end{bmatrix} \otimes \begin{bmatrix} 0 \\ 1 \end{bmatrix} = \begin{bmatrix} 0 \\ 1 \\ 0 \\ 0 \end{bmatrix};$$

$$[2.134]$$

$$\psi_3 = |\beta\alpha\rangle = \begin{bmatrix} 0 \\ 1 \end{bmatrix} \otimes \begin{bmatrix} 1 \\ 0 \end{bmatrix} = \begin{bmatrix} 0 \\ 0 \\ 1 \\ 0 \end{bmatrix};$$

$$\psi_4 = |\beta\beta\rangle = \begin{bmatrix} 0 \\ 1 \end{bmatrix} \otimes \begin{bmatrix} 0 \\ 1 \end{bmatrix} = \begin{bmatrix} 0 \\ 0 \\ 0 \\ 1 \end{bmatrix}.$$

Next, consider the operator corresponding to the sum of the components I_z and S_z in a two-spin system. Clearly, the matrix representation of $I_z + S_z$ in a two-spin system must be a 4×4 matrix because the vector space is spanned by four wavefunctions; thus,

$$I_z + S_z \neq \frac{1}{2}\begin{bmatrix} 1 & 0 \\ 0 & -1 \end{bmatrix} + \frac{1}{2}\begin{bmatrix} 1 & 0 \\ 0 & -1 \end{bmatrix} = \begin{bmatrix} 1 & 0 \\ 0 & -1 \end{bmatrix}. \qquad [2.135]$$

A more formal analysis indicates that matrix representations of the operators in the two-spin system can be calculated from the direct product of the one-spin operators with the identity operator. The results for a two-spin system are

$$I_\eta^{(2\text{spin})} = I_\eta^{(1\text{spin})} \otimes E \quad \text{and} \quad S_\eta^{(2\text{spin})} = E \otimes S_\eta^{(1\text{spin})}, \quad [2.136]$$

where $\eta = x$, y, or z. In general, for a N-spin system, the representations of the angular momentum operators for the kth spin are given by

$$I_{\eta k}^{(N\text{spin})} = E_1 \otimes E_2 \otimes \cdots E_{k-1} \otimes I_{\eta k}^{(1\text{spin})} \otimes E_{k+1} \otimes \cdots E_N \quad [2.137]$$

Returning to the preceding example,

$$I_z^{(2\text{spin})} = I_z^{(1\text{spin})} \otimes E = \frac{1}{2} \begin{bmatrix} 1 & 0 \\ 0 & -1 \end{bmatrix} \otimes \begin{bmatrix} 1 & 0 \\ 0 & 1 \end{bmatrix}$$

$$= \frac{1}{2} \begin{bmatrix} 1 & 0 & 0 & 0 \\ 0 & 1 & 0 & 0 \\ 0 & 0 & -1 & 0 \\ 0 & 0 & 0 & -1 \end{bmatrix}; \quad [2.138]$$

$$S_z^{(2\text{spin})} = E \otimes S_z^{(1\text{spin})} = \begin{bmatrix} 1 & 0 \\ 0 & 1 \end{bmatrix} \otimes \frac{1}{2} \begin{bmatrix} 1 & 0 \\ 0 & -1 \end{bmatrix}$$

$$= \frac{1}{2} \begin{bmatrix} 1 & 0 & 0 & 0 \\ 0 & -1 & 0 & 0 \\ 0 & 0 & 1 & 0 \\ 0 & 0 & 0 & -1 \end{bmatrix}. \quad [2.139]$$

The combination of $I_z^{(2\text{spin})} + S_z^{(2\text{spin})}$ gives the correct matrix representation:

$$I_z^{(2\text{spin})} + S_z^{(2\text{spin})} = \begin{bmatrix} 1 & 0 & 0 & 0 \\ 0 & 0 & 0 & 0 \\ 0 & 0 & 0 & 0 \\ 0 & 0 & 0 & -1 \end{bmatrix}. \quad [2.140]$$

From now on, the (2spin) superscript will be implied. The fundamental rule of the operator algebra in direct-product spaces is (as illustrated for a two-spin system)

$$\mathbf{AB}|ij\rangle \equiv (\mathbf{A} \otimes \mathbf{B})(|i\rangle \otimes |j\rangle) = \mathbf{A}|i\rangle \otimes \mathbf{B}|j\rangle, \qquad [2.141]$$

in which $\mathbf{A}$ is an operator that acts on the i spin and $\mathbf{B}$ is an operator that acts on the j spin. Also note that

$$\mathbf{AB} \equiv \mathbf{A} \otimes \mathbf{B} = (\mathbf{A} \otimes \mathbf{E})(\mathbf{E} \otimes \mathbf{B}). \qquad [2.142]$$

Thus, for example,

$$I_z|\alpha\beta\rangle \equiv (I_z \otimes \mathbf{E})(|\alpha\rangle \otimes |\beta\rangle) = I_z|\alpha\rangle \otimes \mathbf{E}|\beta\rangle$$

$$= \frac{1}{2}|\alpha\rangle \otimes |\beta\rangle = \frac{1}{2}|\alpha\beta\rangle. \qquad [2.143]$$

As a second example,

$$I_z S_z|\alpha\beta\rangle \equiv (I_z \otimes S_z)(|\alpha\rangle \otimes |\beta\rangle) = I_z|\alpha\rangle \otimes S_z|\beta\rangle$$

$$= \frac{1}{2}|\alpha\rangle \otimes -\frac{1}{2}|\beta\rangle = -\frac{1}{4}|\alpha\beta\rangle, \qquad [2.144]$$

which can be written in matrix notation as

$$I_z S_z|\alpha\beta\rangle \equiv (I_z \otimes S_z)(|\alpha\rangle \otimes |\beta\rangle)$$

$$= \frac{1}{4}\left(\begin{bmatrix} 1 & 0 \\ 0 & -1 \end{bmatrix} \otimes \begin{bmatrix} 1 & 0 \\ 0 & -1 \end{bmatrix} \right) \begin{bmatrix} 0 \\ 1 \\ 0 \\ 0 \end{bmatrix}$$

$$= \frac{1}{4}\begin{bmatrix} 1 & 0 & 0 & 0 \\ 0 & -1 & 0 & 0 \\ 0 & 0 & -1 & 0 \\ 0 & 0 & 0 & 1 \end{bmatrix}\begin{bmatrix} 0 \\ 1 \\ 0 \\ 0 \end{bmatrix} = \frac{1}{4}\begin{bmatrix} 0 \\ -1 \\ 0 \\ 0 \end{bmatrix}$$

$$= -\frac{1}{4}|\alpha\beta\rangle. \qquad [2.145]$$

2.5.2 SCALAR COUPLING HAMILTONIAN

The free-precession laboratory-frame Hamiltonian for N scalar-coupled spins is

$$\mathcal{H} = \mathcal{H}_z + \mathcal{H}_J = \sum_{i=1}^{N} \omega_i I_{iz} + 2\pi \sum_{i=2}^{N} \sum_{j=1}^{i-1} J_{ij} \mathbf{I}_i \cdot \mathbf{I}_j, \qquad [2.146]$$

in which ω_i is the Larmor frequency of the ith spin and J_{ij} is the scalar coupling constant between the ith and jth spins. The eigenfunctions of this Hamiltonian are used as the basis functions for the construction of the matrix representation of the density operator. For completeness, the effects of *strong coupling* must be taken into account. The product wave-functions given by [2.131] are eigenfunctions of $\mathcal{H}$ only if $2\pi J_{ij}/|\omega_i - \omega_j| \ll 1$; this condition is known as the *weak coupling* regime. If the weak coupling condition does not hold, then the spins are said to be strongly coupled. In the strong coupling regime, the wavefunctions in the product basis with the same total magnetic quantum number become mixed and are no longer completely independent. A proper basis set is obtained by taking linear combinations of the subset of wavefunctions with the same value of m. Construction of wavefunctions for strongly coupled spin systems with $N > 2$ is facilitated by use of group theoretical methods (8). Strong coupling effects are particularly important in the analysis of coherence transfer in isotropic mixing experiments; group theoretical analyses are also important for treatment of identical spins (such as the three protons in a methyl group).

To illustrate these ideas, a scalar coupled two spin system, which was treated in the weak coupling limit in Section 1.6, will be analyzed. The two spins will be labeled I and S. The free precession Hamiltonian laboratory frame for the IS spin system is

$$\mathcal{H} = \omega_I I_z + \omega_S S_z + 2\pi J_{IS} \mathbf{I} \cdot \mathbf{S}, \qquad [2.147]$$

in which the scalar coupling constant is J_{IS}. A system of two coupled spins has the following four eigenfunctions:

$$\Psi_1 = |\alpha\alpha\rangle; \qquad\qquad \Psi_2 = \cos\theta \,|\alpha\beta\rangle + \sin\theta \,|\beta\alpha\rangle;$$
$$\Psi_3 = \cos\theta \,|\beta\alpha\rangle - \sin\theta \,|\alpha\beta\rangle; \qquad \Psi_4 = |\beta\beta\rangle, \qquad [2.148]$$

where θ is known as the strong coupling parameter and is defined as

$$\tan(2\theta) = \frac{2\pi J_{IS}}{\omega_I - \omega_S} \qquad [2.149]$$

for 2θ in the range 0 to π radians. If the spins are identical, then $\theta = \pi/4$ and the wavefunctions become

$$\Psi_1 = |\alpha\alpha\rangle; \qquad\qquad \Psi_2 = 2^{-1/2}(|\alpha\beta\rangle + |\beta\alpha\rangle);$$
$$\Psi_3 = 2^{-1/2}(|\beta\alpha\rangle - |\alpha\beta\rangle); \qquad \Psi_4 = |\beta\beta\rangle. \qquad [2.150]$$

The wavefunctions of [2.150] are symmetrical or antisymmetrical under the exchange of identical spins, as is required by the postulates of quantum mechanics (3). The energies of the four eigenstates are

$$E_1 = -\frac{1}{2}\omega_I - \frac{1}{2}\omega_S + \frac{1}{2}\pi J_{IS}; \qquad E_2 = D - \frac{1}{2}\pi J_{IS};$$

$$E_3 = -D - \frac{1}{2}\pi J_{IS}; \qquad\qquad E_4 = \frac{1}{2}\omega_I + \frac{1}{2}\omega_S + \frac{1}{2}\pi J_{IS},$$

[2.151]

where

$$D = \frac{1}{2}[(\omega_I - \omega_S)^2 + 4\pi^2 J_{IS}^2]^{1/2}.$$

[2.152]

In the strongly coupled spectrum, the energies of the stationary states and the positions of the resonance signals in the spectrum are altered, compared to the weakly coupled spin system (see [1.52]). Second, the intensities of the lines in the multiplet are no longer of equal intensity; specifically, the two outer lines reduce progressively in intensity as the strong coupling effect becomes more pronounced.

The results given in [2.148]–[2.152] are derived by diagonalizing the Hamiltonian matrix in the product basis; these results can be easily verified. For example, if Ψ_2 is an eigenfunction of $\mathcal{H}$, then

$$\mathcal{H}\Psi_2 = E_2\Psi_2$$

$$= (\omega_I I_z + \omega_S S_z + 2\pi J_{IS}\mathbf{I}\cdot\mathbf{S})(\cos\theta|\alpha\beta\rangle + \sin\theta|\beta\alpha\rangle)$$

$$= \frac{1}{2}\omega_I\cos\theta|\alpha\beta\rangle - \frac{1}{2}\omega_I\sin\theta|\beta\alpha\rangle - \frac{1}{2}\omega_S\cos\theta|\alpha\beta\rangle + \frac{1}{2}\omega_S\sin\theta|\beta\alpha\rangle$$

$$- \frac{1}{2}\pi J_{IS}\cos\theta|\alpha\beta\rangle - \frac{1}{2}\pi J_{IS}\sin\theta|\beta\alpha\rangle$$

$$+ \pi J_{IS}\cos\theta|\beta\alpha\rangle + \pi J_{IS}\sin\theta|\alpha\beta\rangle$$

$$= \frac{1}{2}(\omega_I\cos\theta - \omega_S\cos\theta - \pi J_{IS}\cos\theta + 2\pi J_{IS}\sin\theta)|\alpha\beta\rangle$$

$$+ \frac{1}{2}(-\omega_I\sin\theta + \omega_S\sin\theta - \pi J_{IS}\sin\theta + 2\pi J_{IS}\cos\theta)|\beta\alpha\rangle$$

$$= \frac{1}{2}(\omega_I - \omega_S - \pi J_{IS} + 2\pi J_{IS}\tan\theta)\cos\theta|\alpha\beta\rangle$$

$$+ \frac{1}{2}(-\omega_I + \omega_S - \pi J_{IS} + 2\pi J_{IS}/\tan\theta)\sin\theta|\beta\alpha\rangle.$$

[2.153]

In order for Ψ_2 to be an eigenfunction, the two terms in parentheses following the last equal sign must be identical. Thus

$$\omega_I - \omega_S - \pi J_{IS} + 2\pi J_{IS} \tan\theta = -\omega_I + \omega_S - \pi J_{IS} + 2\pi J_{IS}/\tan\theta;$$

$$2\tan\theta\,(\omega_I - \omega_S) - 2\pi J_{IS}(1 - \tan^2\theta) = 0;$$

$$\frac{2\tan\theta}{1 - \tan^2\theta} = \frac{2\pi J_{IS}}{(\omega_I - \omega_S)}; \qquad\qquad [2.154]$$

$$\tan(2\theta) = \frac{2\pi J_{IS}}{(\omega_I - \omega_S)},$$

which completes the demonstration because θ is defined according to [2.149]. By inspection,

$$E_2 = \tfrac{1}{2}(\omega_I - \omega_S - \pi J_{IS} + 2\pi J_{IS}\tan\theta), \qquad\qquad [2.155]$$

which is easily shown to be equal to [2.151] by solving [2.149] for $\tan\theta$. By comparison of [2.131] and [2.148], the transformation matrix $\mathbf{U}$ that converts the product basis into the strong coupling basis (and diagonalizes the Hamiltonian) is given by

$$\mathbf{U} = \begin{bmatrix} 1 & 0 & 0 & 0 \\ 0 & \cos\theta & \sin\theta & 0 \\ 0 & -\sin\theta & \cos\theta & 0 \\ 0 & 0 & 0 & 1 \end{bmatrix}. \qquad\qquad [2.156]$$

In the limit of weak scalar coupling, $\theta = 0$ and the wavefunctions of the two energy levels $|\alpha\beta\rangle$ and $|\beta\alpha\rangle$ are independent. The weak coupling Hamiltonian simplifies to

$$\mathcal{H} = \omega_I I_z + \omega_S S_z + 2\pi J_{IS} I_z S_z. \qquad\qquad [2.157]$$

To calculate evolution of the density operator under the weak coupling Hamiltonian, the effect of the operation

$$\sigma(t) = \exp[-i\alpha\,2I_z S_z]\,\sigma(0)\,\exp[i\alpha\,2I_z S_z] \qquad\qquad [2.158]$$

for $\alpha = \pi J_{IS} t$ must be calculated. The derivation is similar to the derivation of the rotation operators; thus

$$\exp[i\alpha\,2I_z S_z] = \mathbf{E} + i\alpha\,2I_z S_z + \frac{1}{2!}(i\alpha\,2I_z S_z)^2 + \cdots. \qquad\qquad [2.159]$$

Using the matrices given in Table 2.2, the following relationship is easily derived:

TABLE 2.2
Product Operators in the Cartesian Basis for a Two-Spin System

$$\frac{1}{2}\mathbf{E} = \frac{1}{2}\begin{bmatrix} 1 & 0 & 0 & 0 \\ 0 & 1 & 0 & 0 \\ 0 & 0 & 1 & 0 \\ 0 & 0 & 0 & 1 \end{bmatrix} \qquad I_z = \frac{1}{2}\begin{bmatrix} 1 & 0 & 0 & 0 \\ 0 & 1 & 0 & 0 \\ 0 & 0 & -1 & 0 \\ 0 & 0 & 0 & -1 \end{bmatrix} \qquad S_z = \frac{1}{2}\begin{bmatrix} 1 & 0 & 0 & 0 \\ 0 & -1 & 0 & 0 \\ 0 & 0 & 1 & 0 \\ 0 & 0 & 0 & -1 \end{bmatrix} \qquad 2I_zS_z = \frac{1}{2}\begin{bmatrix} 1 & 0 & 0 & 0 \\ 0 & -1 & 0 & 0 \\ 0 & 0 & -1 & 0 \\ 0 & 0 & 0 & 1 \end{bmatrix}$$

$$I_x = \frac{1}{2}\begin{bmatrix} 0 & 0 & 1 & 0 \\ 0 & 0 & 0 & 1 \\ 1 & 0 & 0 & 0 \\ 0 & 1 & 0 & 0 \end{bmatrix} \qquad I_y = \frac{1}{2}\begin{bmatrix} 0 & 0 & -i & 0 \\ 0 & 0 & 0 & -i \\ i & 0 & 0 & 0 \\ 0 & i & 0 & 0 \end{bmatrix} \qquad 2I_xS_z = \frac{1}{2}\begin{bmatrix} 0 & 0 & 1 & 0 \\ 0 & 0 & 0 & -1 \\ 1 & 0 & 0 & 0 \\ 0 & -1 & 0 & 0 \end{bmatrix} \qquad 2I_yS_z = \frac{1}{2}\begin{bmatrix} 0 & 0 & -i & 0 \\ 0 & 0 & 0 & i \\ i & 0 & 0 & 0 \\ 0 & -i & 0 & 0 \end{bmatrix}$$

$$S_x = \frac{1}{2}\begin{bmatrix} 0 & 1 & 0 & 0 \\ 1 & 0 & 0 & 0 \\ 0 & 0 & 0 & 1 \\ 0 & 0 & 1 & 0 \end{bmatrix} \qquad S_y = \frac{1}{2}\begin{bmatrix} 0 & -i & 0 & 0 \\ i & 0 & 0 & 0 \\ 0 & 0 & 0 & -i \\ 0 & 0 & i & 0 \end{bmatrix} \qquad 2I_zS_x = \frac{1}{2}\begin{bmatrix} 0 & 1 & 0 & 0 \\ 1 & 0 & 0 & 0 \\ 0 & 0 & 0 & -1 \\ 0 & 0 & -1 & 0 \end{bmatrix} \qquad 2I_zS_y = \frac{1}{2}\begin{bmatrix} 0 & -i & 0 & 0 \\ i & 0 & 0 & 0 \\ 0 & 0 & 0 & i \\ 0 & 0 & -i & 0 \end{bmatrix}$$

$$2I_xS_x = \frac{1}{2}\begin{bmatrix} 0 & 0 & 0 & 1 \\ 0 & 0 & 1 & 0 \\ 0 & 1 & 0 & 0 \\ 1 & 0 & 0 & 0 \end{bmatrix} \qquad 2I_yS_y = \frac{1}{2}\begin{bmatrix} 0 & 0 & 0 & -1 \\ 0 & 0 & 1 & 0 \\ 0 & 1 & 0 & 0 \\ -1 & 0 & 0 & 0 \end{bmatrix} \qquad 2I_xS_y = \frac{1}{2}\begin{bmatrix} 0 & 0 & 0 & -i \\ 0 & 0 & i & 0 \\ 0 & -i & 0 & 0 \\ i & 0 & 0 & 0 \end{bmatrix} \qquad 2I_yS_x = \frac{1}{2}\begin{bmatrix} 0 & 0 & 0 & -i \\ 0 & 0 & -i & 0 \\ 0 & i & 0 & 0 \\ i & 0 & 0 & 0 \end{bmatrix}$$

$$(I_z S_z)^{2n} = \frac{1}{4^n} \mathbf{E}. \qquad [2.160]$$

Substituting the results contained in [2.160] into [2.159], and grouping together even and odd powers of $iI_z S_z$ yields

$$\exp(i\alpha 2I_z S_z) = \mathbf{E}\left(1 - \frac{\alpha^2}{2!2^2} + \frac{\alpha^4}{4!2^4} + \cdots\right) + 4iI_z S_z\left(\frac{\alpha}{2} - \frac{\alpha^3}{3!2^3} + \frac{\alpha^5}{5!2^5} + \cdots\right)$$

$$= \mathbf{E}\cos\frac{\alpha}{2} + 4iI_z S_z \sin\frac{\alpha}{2}. \qquad [2.161]$$

Again using Table 2.2, the matrix representations of the operators become

$$\exp[i\alpha 2I_z S_z] = \begin{bmatrix} c + is & 0 & 0 & 0 \\ 0 & c - is & 0 & 0 \\ 0 & 0 & c - is & 0 \\ 0 & 0 & 0 & c + is \end{bmatrix};$$

$$\qquad [2.162]$$

$$\exp[-i\alpha 2I_z S_z] = \begin{bmatrix} c - is & 0 & 0 & 0 \\ 0 & c + is & 0 & 0 \\ 0 & 0 & c + is & 0 \\ 0 & 0 & 0 & c - is \end{bmatrix},$$

where $c = \cos(\alpha/2)$ and $s = \sin(\alpha/2)$.

2.5.3 ROTATIONS IN PRODUCT SPACES

For a system of N spins, the matrix representation of the pulse operator can be calculated from

$$\mathbf{R}_x^{-1}(\alpha) = \prod_{j=1}^{N} \mathbf{R}_{jx}^{-1}(\alpha) = \prod_{j=1}^{N}\left(\mathbf{E}\cos\frac{\alpha}{2} + i\sin\frac{\alpha}{2}\mathbf{T}_j\right). \qquad [2.163]$$

In [2.163], the effect of the scalar coupling term of the Hamiltonian has been ignored; this simplification requires that the length of the rf pulse, τ_p, satisfy $2\pi J_{ij}\tau_p \ll 1$. For a two-spin system,

$$\mathbf{R}_x^{-1}(\alpha) = \left(\mathbf{E}\cos\frac{\alpha}{2} + i\sin\frac{\alpha}{2}\mathbf{T}_1\right)\left(\mathbf{E}\cos\frac{\alpha}{2} + i\sin\frac{\alpha}{2}\mathbf{T}_2\right). \qquad [2.164]$$

The elements of the matrix representation of $\mathbf{R}$ may be constructed from the basis eigenfunctions using the expressions

$$R_x^{-1}(\alpha)_{rs} = \langle r | \prod_{j=1}^{N} \left(\mathbf{E} \cos \frac{\alpha}{2} + i \sin \frac{\alpha}{2} \mathbf{T}_j \right) | s \rangle. \qquad [2.165]$$

For example, the matrix element $R_x^{-1}(\alpha)_{12}$ can be calculated as follows:

$$R_x^{-1}(\alpha)_{12} = \langle \Psi_1 | \prod_{j=1}^{N} \left(\mathbf{E} \cos \frac{\alpha}{2} + i \sin \frac{\alpha}{2} \mathbf{T}_j \right) | \Psi_2 \rangle$$

$$= \langle \alpha\alpha | \prod_{j=1}^{N} \left(\mathbf{E} \cos \frac{\alpha}{2} + i \sin \frac{\alpha}{2} \mathbf{T}_j \right) (\cos\theta|\alpha\beta\rangle + \sin\theta|\beta\alpha\rangle)$$

$$= \langle \alpha\alpha | \left(\mathbf{E} \cos^2 \frac{\alpha}{2} + i \cos \frac{\alpha}{2} \sin \frac{\alpha}{2} \mathbf{T}_1 \right.$$

$$\left. + i \cos \frac{\alpha}{2} \sin \frac{\alpha}{2} \mathbf{T}_2 - \sin^2 \frac{\alpha}{2} \mathbf{T}_1 \mathbf{T}_2 \right)$$

$$\times (\cos\theta|\alpha\beta\rangle + \sin\theta|\beta\alpha\rangle)$$

$$= i \cos \frac{\alpha}{2} \sin \frac{\alpha}{2} \sin\theta + i \cos \frac{\alpha}{2} \sin \frac{\alpha}{2} \cos\theta$$

$$= i \cos \frac{\alpha}{2} \sin \frac{\alpha}{2} (\cos\theta + \sin\theta). \qquad [2.166]$$

This result is calculated using the property that the inversion operator $\mathbf{T}_j$ changes the spin state of spin j from α to β and vice versa. As another example, $R_x^{-1}(\alpha)_{14}$ is given by

$$R_x^{-1}(\alpha)_{14} = \langle \Psi_1 | \prod_{j=1}^{N} \left(\mathbf{E} \cos \frac{\alpha}{2} + i \sin \frac{\alpha}{2} \mathbf{T}_j \right) | \Psi_4 \rangle$$

$$= \langle \alpha\alpha | \prod_{j=1}^{N} \left(\mathbf{E} \cos \frac{\alpha}{2} + i \sin \frac{\alpha}{2} \mathbf{T}_j \right) | \beta\beta \rangle$$

$$= \langle \alpha\alpha | \left(\mathbf{E} \cos^2 \frac{\alpha}{2} + i \cos \frac{\alpha}{2} \sin \frac{\alpha}{2} \mathbf{T}_1 \right.$$

$$\left. + i \cos \frac{\alpha}{2} \sin \frac{\alpha}{2} \mathbf{T}_2 - \sin^2 \frac{\alpha}{2} \mathbf{T}_1 \mathbf{T}_2 \right) | \beta\beta \rangle$$

$$= -\sin^2 \frac{\alpha}{2}. \qquad [2.167]$$

Repeating these calculations for every element of the matrix representation of the pulse operator yields

$$\mathbf{R}_x^{-1}(\alpha) = \begin{bmatrix} c^2 & icsu & icsv & -s^2 \\ icsu & 1 - s^2u^2 & -s^2uv & icsu \\ icsv & -s^2uv & 1 - s^2v^2 & icsv \\ -s^2 & icsu & icsv & c^2 \end{bmatrix}, \qquad [2.168]$$

where $c = \cos(\alpha/2)$, $s = \sin(\alpha/2)$, $u = \cos\theta + \sin\theta$, and $v = \cos\theta - \sin\theta$. Since the rotation operators are unitary, $\mathbf{R}_x(\alpha)$ is the adjoint of $\mathbf{R}_x^{-1}(\alpha)$:

$$\mathbf{R}_x(\alpha) = \begin{bmatrix} c^2 & -icsu & -icsv & -s^2 \\ -icsu & 1 - s^2u^2 & -s^2uv & -icsu \\ -icsv & -s^2uv & 1 - s^2v^2 & -icsv \\ -s^2 & -icsu & -icsv & c^2 \end{bmatrix}. \qquad [2.169]$$

The same calculation can be performed using rotation matrices that concentrate on each spin in the two-spin system individually rather than both at the same time. This approach can be particularly useful in heteronuclear NMR experiments. The matrix representations of the rotation operators are obtained from the direct products of the single-spin rotation operators derived previously in [2.108], [2.112], and [2.113]. For example, for spin I,

$$\mathbf{R}_x(\alpha)[I] = \mathbf{R}_x(\alpha) \otimes \mathbf{E} = \begin{bmatrix} c & -is \\ -is & c \end{bmatrix} \otimes \begin{bmatrix} 1 & 0 \\ 0 & 1 \end{bmatrix}$$

$$= \begin{bmatrix} c & 0 & -is & 0 \\ 0 & c & 0 & -is \\ -is & 0 & c & 0 \\ 0 & -is & 0 & c \end{bmatrix}; \qquad [2.170]$$

$$\mathbf{R}_y(\alpha)[I] = \mathbf{R}_y(\alpha) \otimes \mathbf{E} = \begin{bmatrix} c & -s \\ s & c \end{bmatrix} \otimes \begin{bmatrix} 1 & 0 \\ 0 & 1 \end{bmatrix}$$

$$= \begin{bmatrix} c & 0 & -s & 0 \\ 0 & c & 0 & -s \\ s & 0 & c & 0 \\ 0 & s & 0 & c \end{bmatrix}, \qquad [2.171]$$

and for spin S,

$$\mathbf{R}_x(\alpha)[S] = \mathbf{E} \otimes \mathbf{R}_x(\alpha) = \begin{bmatrix} 1 & 0 \\ 0 & 1 \end{bmatrix} \otimes \begin{bmatrix} c & -is \\ -is & c \end{bmatrix}$$

$$= \begin{bmatrix} c & -is & 0 & 0 \\ -is & c & 0 & 0 \\ 0 & 0 & c & -is \\ 0 & 0 & -is & c \end{bmatrix}; \qquad [2.172]$$

$$\mathbf{R}_y(\alpha)[S] = \mathbf{E} \otimes \mathbf{R}_y(\alpha) = \begin{bmatrix} 1 & 0 \\ 0 & 1 \end{bmatrix} \otimes \begin{bmatrix} c & -s \\ s & c \end{bmatrix}$$

$$= \begin{bmatrix} c & -s & 0 & 0 \\ s & c & 0 & 0 \\ 0 & 0 & c & -s \\ 0 & 0 & s & c \end{bmatrix}. \qquad [2.173]$$

Using [2.163], the result $\mathbf{R}_x(\alpha) = \mathbf{R}_x(\alpha)[I] \, \mathbf{R}_x(\alpha)[S]$ is obtained by matrix multiplication and agrees with [2.169] with $\theta = 0$. These matrices would then be used in conjunction with those matrix representations of the two-spin product operators shown in Table 2.2.

2.5.4 ONE-PULSE EXPERIMENT FOR A TWO-SPIN SYSTEM

To compute the observable magnetization following a pulse and subsequent free precession, the evolution of the density operator beginning with the equilibrium matrix representation of the density operator for a two-spin system must be determined. Using [2.121], the initial density matrix is written as

$$\sigma(0) \approx \omega_I I_z + \omega_S S_z = \frac{1}{2} \begin{bmatrix} \omega_I + \omega_S & 0 & 0 & 0 \\ 0 & \omega_I - \omega_S & 0 & 0 \\ 0 & 0 & -\omega_I + \omega_S & 0 \\ 0 & 0 & 0 & -\omega_I - \omega_S \end{bmatrix},$$

$$[2.174]$$

in which a common divisor of $2k_BT$ has not been written for convenience and weak coupling has been assumed. A pulse α_x (with rotation angle α

and x phase) rotates an initial state of the density operator according to the now well-known general equation

$$\sigma(t) = \mathbf{R}_x(\alpha)\sigma(0)\mathbf{R}_x^{-1}(\alpha). \qquad [2.175]$$

For simplicity, weak scalar coupling and an on-resonance 90° pulse with x phase will be assumed. Using [2.174], [2.175], [2.168], and [2.169],

$$\sigma(t) = \mathbf{R}_x(\alpha)\sigma(0)\mathbf{R}_x^{-1}(\alpha)$$

$$= \frac{1}{2}\begin{bmatrix} 1 & -i & -i & -1 \\ -i & 1 & -1 & -i \\ -i & -1 & 1 & -i \\ -1 & -i & -i & 1 \end{bmatrix}$$

$$\times \begin{bmatrix} \omega_I + \omega_S & 0 & 0 & 0 \\ 0 & \omega_I - \omega_S & 0 & 0 \\ 0 & 0 & -\omega_I + \omega_S & 0 \\ 0 & 0 & 0 & -\omega_I - \omega_S \end{bmatrix}$$

$$\times \begin{bmatrix} 1 & i & i & -1 \\ i & 1 & -1 & i \\ i & -1 & 1 & i \\ -1 & i & i & 1 \end{bmatrix}$$

$$= \frac{1}{2}\begin{bmatrix} 0 & i\omega_S & i\omega_I & 0 \\ -i\omega_S & 0 & 0 & i\omega_I \\ -i\omega_I & 0 & 0 & i\omega_S \\ 0 & -i\omega_I & -i\omega_S & 0 \end{bmatrix} = -\omega_I I_y - \omega_S S_y, \qquad [2.176]$$

because [2.136] yields the results that

$$I_y = \frac{1}{2}\begin{bmatrix} 0 & 0 & -i & 0 \\ 0 & 0 & 0 & -i \\ i & 0 & 0 & 0 \\ 0 & i & 0 & 0 \end{bmatrix} \quad \text{and} \quad S_y = \frac{1}{2}\begin{bmatrix} 0 & -i & 0 & 0 \\ i & 0 & 0 & 0 \\ 0 & 0 & 0 & -i \\ 0 & 0 & i & 0 \end{bmatrix}. \qquad [2.177]$$

This is exactly the expected result: each term in the initial density operator transformed identically by the nonselective pulse. Following the pulse, density operator evolves under the free-precession Hamiltonian. Combining [2.127] with [2.162] yields the matrix representation of the exponential operator;

$$[i(\Omega_I I_z + \Omega_S S_z + 2\pi J_{IS} I_z S_z)t]$$

$$\begin{bmatrix} e^{i(\Omega_I + \Omega_S + \pi J_{IS})t/2} & 0 & 0 & 0 \\ 0 & e^{i(\Omega_I - \Omega_S - \pi J_{IS})t/2} & 0 & 0 \\ 0 & 0 & e^{i(-\Omega_I + \Omega_S - \pi J_{IS})t/2} & 0 \\ 0 & 0 & 0 & e^{i(-\Omega_I - \Omega_S + \pi J_{IS})t/2} \end{bmatrix}.$$

[2.178]

Performing the matrix multiplications yields

$$\sigma(t) = \exp[-i(\Omega_I I_z + \Omega_S S_z + 2\pi J_{IS} I_z S_z)t](-I_y - S_y)$$
$$\times \exp[i(\Omega_I I_z + \Omega_S S_z + 2\pi J_{IS} I_z S_z)t]$$

$$= \frac{i}{2} \begin{bmatrix} 0 & e^{-i(\Omega_S + \pi J_{IS})t} & e^{-i(\Omega_I + \pi J_{IS})t} & 0 \\ -e^{i(\Omega_S + \pi J_{IS})t} & 0 & 0 & e^{-i(\Omega_I - \pi J_{IS})t} \\ -e^{i(\Omega_I + \pi J_{IS})t} & 0 & 0 & e^{-i(\Omega_S - \pi J_{IS})t} \\ 0 & -e^{i(\Omega_I - \pi J_{IS})t} & -e^{i(\Omega_S - \pi J_{IS})t} & 0 \end{bmatrix}.$$

[2.179]

The observable signal is found by forming the product with operator $\mathbf{F}^+ \propto I^+ + S^+$:

$$\sigma(t)\mathbf{F}^+ =$$

$$\frac{i}{2} \begin{bmatrix} 0 & e^{-i(\Omega_S + \pi J_{IS})t} & e^{-i(\Omega_I + \pi J_{IS})t} & 0 \\ -e^{i(\Omega_S + \pi J_{IS})t} & 0 & 0 & e^{-i(\Omega_I - \pi J_{IS})t} \\ -e^{i(\Omega_I + \pi J_{IS})t} & 0 & 0 & e^{-i(\Omega_S - \pi J_{IS})t} \\ 0 & -e^{i(\Omega_I - \pi J_{IS})t} & -e^{i(\Omega_S - \pi J_{IS})t} & 0 \end{bmatrix} \begin{bmatrix} 0 & 1 & 1 & 0 \\ 0 & 0 & 0 & 1 \\ 0 & 0 & 0 & 1 \\ 0 & 0 & 0 & 0 \end{bmatrix}$$

$$= -\frac{i}{2} \begin{bmatrix} 0 & 0 & 0 & -e^{-i(\Omega_I + \pi J_{IS})t} - e^{-i(\Omega_S + \pi J_{IS})t} \\ 0 & e^{i(\Omega_S + \pi J_{IS})t} & e^{i(\Omega_S + \pi J_{IS})t} & 0 \\ 0 & e^{i(\Omega_I + \pi J_{IS})t} & e^{i(\Omega_I + \pi J_{IS})t} & 0 \\ 0 & 0 & 0 & e^{i(\Omega_I - \pi J_{IS})t} + e^{i(\Omega_S - \pi J_{IS})t} \end{bmatrix}.$$

[2.180]

The trace of this matrix is proportional to the observed complex magnetization:

$$\mathbf{M}^+(t) \propto e^{i(\Omega_I + \pi J_{IS})t} + e^{i(\Omega_I - \pi J_{IS})t} + e^{i(\Omega_S + \pi J_{IS})t} + e^{i(\Omega_S - \pi J_{IS})t}. \quad [2.181]$$

The spectrum consists of four signals arranged into two doublets. One doublet consists of the frequencies $\Omega_I \pm \pi J_{IS}$, and the other doublet consists of the frequencies $\Omega_S \pm \pi J_{IS}$.

2.6 Coherence

So far the density operator has been represented in terms of a Cartesian basis of the spin angular momentum operators I_x, I_y, and I_z. The density operator can also be expressed in a spherical operator basis, which provides additional insight into the density matrix theory. Operators in the Cartesian basis can be transformed to the spherical basis by taking linear combinations of I_x and I_y to yield the (normalized) raising and lowering operators:

$$I^+ = -\frac{1}{\sqrt{2}}(I_x + iI_y); \qquad I^- = \frac{1}{\sqrt{2}}(I_x - iI_y); \qquad I_z = I_z. \quad [2.182]$$

I_z is conserved in both representations and $(I^\pm)^\dagger \equiv -I^\mp$. The 16 basis operators for a system of two spin-$\frac{1}{2}$ nuclei are shown in Table 2.3. These operators are constructed simply from the direct products of the respective operators for each individual spin. For example,

$$I^- = \frac{1}{\sqrt{2}}\begin{bmatrix} 0 & 0 & 0 & 0 \\ 0 & 0 & 0 & 0 \\ 1 & 0 & 0 & 0 \\ 0 & 1 & 0 & 0 \end{bmatrix}; \qquad I^+S^+ = \begin{bmatrix} 0 & 0 & 0 & 1 \\ 0 & 0 & 0 & 0 \\ 0 & 0 & 0 & 0 \\ 0 & 0 & 0 & 0 \end{bmatrix};$$

$$I^+S^- = \begin{bmatrix} 0 & 0 & 0 & 0 \\ 0 & 0 & -1 & 0 \\ 0 & 0 & 0 & 0 \\ 0 & 0 & 0 & 0 \end{bmatrix}. \qquad [2.183]$$

For illustration, the matrix representation of I^- is written in [2.184] with the spin states of the system along the side and top of the matrix to indicate the spin states connected by each matrix element:

$$I^- = \frac{1}{\sqrt{2}} \begin{pmatrix} & \alpha\alpha & \alpha\beta & \beta\alpha & \beta\beta \\ 0 & 0 & 0 & 0 \\ 0 & 0 & 0 & 0 \\ 1 & 0 & 0 & 0 \\ 0 & 1 & 0 & 0 \end{pmatrix} \begin{matrix} \alpha\alpha \\ \alpha\beta \\ \beta\alpha \\ \beta\beta \end{matrix} . \qquad [2.184]$$

The only non-zero matrix elements present correspond to the transitions $\alpha\alpha \rightarrow \beta\alpha$ and $\alpha\beta \rightarrow \beta\beta$. The lowering operator, I^-, is associated with a change in the spin angular-momentum quantum number of $\Delta m = -1$ and a change in the state of the I spin from $\alpha(+\frac{1}{2}) \rightarrow \beta(-\frac{1}{2})$. In the case of the I^+S^+ operator, the only nonzero matrix element corresponds to $\Delta m = +2$ and a change in spin state from $|\alpha\alpha\rangle \rightarrow |\beta\beta\rangle$. In this instance, both the I spin and the S spin change state from α to β. Similarly for the I^+S^- operator, the nonzero matrix element corresponds to the transition $|\beta\alpha\rangle \rightarrow |\alpha\beta\rangle$ with $\Delta m = 0$. In this example, both spins change spin states in opposite senses.

The preceding examples illustrate the concept of *coherence,* which is one of the most fundamental aspects of NMR spectroscopy. As has been stated previously, a diagonal matrix element of the density operator, $\sigma_{nn} = \overline{c_n c_n^*}$, is a real, positive number that corresponds to the population of the state described by the basis function $|n\rangle$. Formally, an off-diagonal element of the density operator, σ_{nm}, represents coherence between eigenstates $|n\rangle$ and $|m\rangle$, in the sense that the time-dependent phase properties of the various members of the ensemble are correlated with respect to $|n\rangle$ and $|m\rangle$. Those matrix elements that denote $\Delta m = \pm 1$ are called *single-quantum coherence;* those that denote $\Delta m = \pm 2$ are called *double quantum coherence;* and not surprisingly, those that denote $\Delta m = 0$ are called *zero-quantum coherence.*

To make these ideas more concrete, consider the following example. The coefficients c_n for the two-level system for a spin $\frac{1}{2}$ can be written in polar notation in terms of an amplitude and a phase factor for the α and β states:

$$c_\alpha = |c_\alpha| \exp(i\phi_\alpha); \qquad [2.185]$$

$$c_\beta = |c_\beta| \exp(i\phi_\beta). \qquad [2.186]$$

TABLE 2.3

Product Operators in the Spherical Basis for a Two-Spin System

$$\mathbf{E} = \frac{1}{2}\begin{bmatrix} 1 & 0 & 0 & 0 \\ 0 & 1 & 0 & 0 \\ 0 & 0 & 1 & 0 \\ 0 & 0 & 0 & 1 \end{bmatrix} \qquad I_0 = \frac{1}{2}\begin{bmatrix} 1 & 0 & 0 & 0 \\ 0 & 1 & 0 & 0 \\ 0 & 0 & -1 & 0 \\ 0 & 0 & 0 & -1 \end{bmatrix} \qquad S_0 = \frac{1}{2}\begin{bmatrix} 1 & 0 & 0 & 0 \\ 0 & -1 & 0 & 0 \\ 0 & 0 & 1 & 0 \\ 0 & 0 & 0 & -1 \end{bmatrix} \qquad I_0 S_0 = \frac{1}{2}\begin{bmatrix} 1 & 0 & 0 & 0 \\ 0 & -1 & 0 & 0 \\ 0 & 0 & -1 & 0 \\ 0 & 0 & 0 & 1 \end{bmatrix}$$

$$I^{+} = \frac{1}{\sqrt{2}}\begin{bmatrix} 0 & -1 & 0 & 0 \\ 0 & 0 & 0 & 0 \\ 0 & 0 & 0 & -1 \\ 0 & 0 & 0 & 0 \end{bmatrix} \qquad I^{-} = \frac{1}{\sqrt{2}}\begin{bmatrix} 0 & 0 & 0 & 0 \\ 1 & 0 & 0 & 0 \\ 0 & 0 & 0 & 0 \\ 0 & 0 & 1 & 0 \end{bmatrix} \qquad I^{+}S_0 = \frac{1}{\sqrt{2}}\begin{bmatrix} 0 & -1 & 0 & 0 \\ 0 & 0 & 0 & 0 \\ 0 & 0 & 0 & 1 \\ 0 & 0 & 0 & 0 \end{bmatrix} \qquad I^{-}S_0 = \frac{1}{\sqrt{2}}\begin{bmatrix} 0 & 0 & 0 & 0 \\ 1 & 0 & 0 & 0 \\ 0 & 0 & 0 & 0 \\ 0 & 0 & -1 & 0 \end{bmatrix}$$

$$S^{+} = \frac{1}{\sqrt{2}}\begin{bmatrix} 0 & 0 & -1 & 0 \\ 0 & 0 & 0 & -1 \\ 0 & 0 & 0 & 0 \\ 0 & 0 & 0 & 0 \end{bmatrix} \qquad S^{-} = \frac{1}{\sqrt{2}}\begin{bmatrix} 0 & 0 & 0 & 0 \\ 0 & 0 & 0 & 0 \\ 1 & 0 & 0 & 0 \\ 0 & 1 & 0 & 0 \end{bmatrix} \qquad I_0 S^{+} = \frac{1}{\sqrt{2}}\begin{bmatrix} 0 & 0 & -1 & 0 \\ 0 & 0 & 0 & 1 \\ 0 & 0 & 0 & 0 \\ 0 & 0 & 0 & 0 \end{bmatrix} \qquad I_0 S^{-} = \frac{1}{\sqrt{2}}\begin{bmatrix} 0 & 0 & 0 & 0 \\ 0 & 0 & 0 & 0 \\ 1 & 0 & 0 & 0 \\ 0 & -1 & 0 & 0 \end{bmatrix}$$

$$I^{+}S^{+} = \begin{bmatrix} 0 & 0 & 0 & 1 \\ 0 & 0 & 0 & 0 \\ 0 & 0 & 0 & 0 \\ 0 & 0 & 0 & 0 \end{bmatrix} \qquad I^{-}S^{+} = \begin{bmatrix} 0 & 0 & 0 & 0 \\ 0 & 0 & 0 & 0 \\ 0 & -1 & 0 & 0 \\ 0 & 0 & 0 & 0 \end{bmatrix} \qquad I^{+}S^{-} = \begin{bmatrix} 0 & 0 & 0 & 0 \\ 0 & 0 & -1 & 0 \\ 0 & 0 & 0 & 0 \\ 0 & 0 & 0 & 0 \end{bmatrix} \qquad I^{-}S^{-} = \begin{bmatrix} 0 & 0 & 0 & 0 \\ 0 & 0 & 0 & 0 \\ 0 & 0 & 0 & 0 \\ 1 & 0 & 0 & 0 \end{bmatrix}$$

Any wavefunction can be expressed as

$$\Psi = c_\alpha|\alpha\rangle + c_\beta|\beta\rangle = |c_\alpha|\exp(i\phi_\alpha)|\alpha\rangle + |c_\beta|\exp(i\phi_\beta)|\beta\rangle. \quad [2.187]$$

Thus, for a pure state, the matrix elements of the projection operator are

$$\langle\alpha|\mathbf{P}|\alpha\rangle = |c_\alpha|^2, \qquad \langle\alpha|\mathbf{P}|\beta\rangle = |c_\alpha||c_\beta|\exp[i(\phi_\beta - \phi_\alpha)];$$
$$\langle\beta|\mathbf{P}|\beta\rangle = |c_\beta|^2, \qquad \langle\beta|\mathbf{P}|\alpha\rangle = |c_\alpha||c_\beta|\exp[i(\phi_\alpha - \phi_\beta)]. \qquad [2.188]$$

Because the state is pure, all members of the ensemble are identical and the terms $(\phi_\beta - \phi_\alpha)$ do not vary between members of the ensemble. For a mixed, macroscopic state, however,

$$\langle\alpha|\sigma|\alpha\rangle = \overline{|c_\alpha|^2}, \qquad \langle\alpha|\sigma|\beta\rangle = \overline{|c_\alpha||c_\beta|} \cdot \overline{\exp[i(\phi_\beta - \phi_\alpha)]};$$
$$\langle\beta|\sigma|\beta\rangle = \overline{|c_\beta|^2}, \qquad \langle\beta|\sigma|\alpha\rangle = \overline{|c_\alpha||c_\beta|} \cdot \overline{\exp[i(\phi_\alpha - \phi_\beta)]}. \qquad [2.189]$$

If no relationship exists between the macroscopic phase properties of the α state (across the ensemble) and the phase properties of the β state (across the ensemble), then $(\phi_\alpha - \phi_\beta)$ takes on all values in the range 0–2π, and $\overline{\exp[i(\phi_\beta - \phi_\alpha)]} = \overline{\exp[i(\phi_\alpha - \phi_\beta)]} = 0$. In this case, $\langle\alpha|\sigma|\beta\rangle = \langle\beta|\sigma|\alpha\rangle = 0$, and there is *no coherence between the two states*. Therefore, as has been stated previously, the equilibrium density matrix is diagonal.

The application of a rf pulse induces exchange of population (i.e., transitions) between stationary states for which $\Delta m = \pm 1$ and causes perturbations of the equilibrium population distribution. In the case of a spin-$\frac{1}{2}$ nucleus, a rf pulse that redistributes the population across the $\alpha \leftrightarrow \beta$ transition creates a phase relationship across that transition such that $\overline{\exp[i(\phi_\alpha - \phi_\beta)]} \neq 0$ (averaged over the ensemble). The density operator following the pulse is said to represent a coherent superposition between the two states; more commonly, this phenomenon is referred to simply as *coherence*. Coherence describes correlation of quantum-mechanical phase relationships among a number of systems (separate nuclei) that persists even after the rf field is removed. Coherence is a phenomenon *associated* with an NMR transition but is not a transition; coherence does *not* change the populations of the spin states. Nonzero off-diagonal elements of the density matrix denote the existence of coherent superposition states.

Both spherical and Cartesian basis operators are useful for describing NMR spectroscopy. The Cartesian operators are a convenient basis for describing the effects of rf pulses on the density operator, and the spherical operators are a convenient basis for describing the evolution of coherence

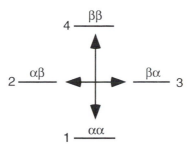

FIGURE 2.3 Multiple-quantum transitions for *IS* spin system. Shown are the zero-quantum flip-flop transitions between states $|\alpha\beta\rangle$ and $|\beta\alpha\rangle$ and the double-quantum flip-flip transitions between states $|\alpha\alpha\rangle$ and $|\beta\beta\rangle$.

in a NMR experiment. Only two eigenstates, $|\alpha\rangle$ and $|\beta\rangle$, exist for a single spin-$\frac{1}{2}$ nucleus; consequently, coherences associated with the $|\alpha\rangle \leftrightarrow |\beta\rangle$ transitions with $\Delta m = \pm1$ are conveniently represented by the raising and lowering operators I^+ and I^-. Four eigenstates exist for a two-spin system. Figure 2.3 illustrates the appearance of double- and zero-quantum coherence where eigenstates are connected in which $\Delta m = \pm2$ and $\Delta m = 0$, respectively. Double-quantum coherence is associated with transitions in which the spin states can change from $\alpha\alpha \leftrightarrow \beta\beta$. The change in eigenstate is identical for both the spins involved, and this is often called a "flip-flip" transition. On the other hand, zero-quantum coherence is associated with transitions in which the spin states change $\alpha\beta \leftrightarrow \beta\alpha$, i.e., in the opposite sense to each other, and are often called "flip-flop" terms.

The two-spin case will be seen to be the most commonly encountered as far as this text is concerned; however, spin systems consisting of three or more scalar coupled spins are evidently important and display additional features. Some of the salient features of larger spin systems will be briefly discussed using a weakly coupled three-spin system as an exemplar. In the two-spin case, each $\Delta m = \pm1$ transition involves the spin state of one nucleus changing whereas the spin state of the other nucleus remains constant. The spectrum can be conveniently labeled with the spin states of the coupled spins as shown in Fig. 2.3. Similar arguments apply to the three spin system, although the appearance of the spectrum is a little more complex. The three spins are denoted I, R, S and have three scalar coupling constants: J_{IS}, J_{IR}, and J_{SR}. The wavefunctions for the scalar coupled three-spin system are denoted $|m_I, m_R, m_S\rangle$ in the product basis, and the energies of the *eight* levels can be calculated by generalizing [1.53] or

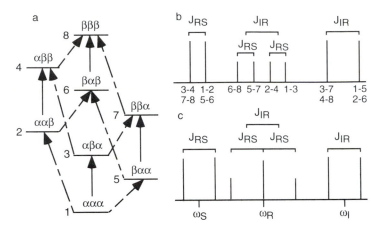

FIGURE 2.4 Spin states and spectrum for a three-spin IRS system. (a) The eight spin states and the allowed single-quantum transitions between states are shown. (- - -) Single-quantum transitions of the I spin; (—) single-quantum transitions of the R spin; (– – –) single-quantum transitions of the S spin. (b) A schematic spectrum for an IRS spin system is shown for the special case that $J_{IR} < J_{RS}$ and $J_{IS} = 0$. (c) A schematic spectrum for an IRS spin system is shown for the special case that $J_{IS} = J_{RS}$ and $J_{IS} = 0$.

direct application of [2.146] and [2.7]. The energy-level diagram for a three-spin system is shown in Fig. 2.4 along with a schematic spectrum. The appearance of the spectrum can change depending on the exact relative chemical shifts of the I, R, and S spins and on the relative sizes of J_{IS}, J_{IR}, and J_{SR}; however, Figure 2.4 is sufficient for illustrative purposes. As in the two spin case, each line of a specific multiplet (in this case the multiplets are *quartets*) can be associated with different spin states of the coupled spins. Looking at Fig. 2.4, consider the two transitions 1–2 and 2–4. Because they share a common energy level (2 in this case), these two transitions are referred to as *connected* transitions. The spin state of one of the three spins remains unchanged across connected transitions (e.g., the I spin state is $|\alpha\rangle$ for the connected transitions 1–2 and 2–4). The values of m_i for the stationary states are $m_1 = +\frac{3}{2}$, $m_2 = +\frac{1}{2}$ and $m_4 = -\frac{1}{2}$. The states represented by m_1 and m_4 are at opposite ends of the transition pathway under discussion and differ in their value of m by 2. In this case, the two connected transitions are said to be *progressively* connected. On the other hand, levels 6 and 7 in the connected transitions 5–6 and 5–7 *do not differ in their values of m* (e.g., $m_6 = m_7 = -\frac{1}{2}$). In this instance, the transitions are said to be *regressively* connected.

2.7 Product Operator Formalism

Although the density matrix theory provides a rigorous description of the evolution of a nuclear spin system, the requisite matrix calculations quickly become cumbersome as the number of spins and eigenstates increases unless implemented numerically on a computer. Unfortunately as well, the density matrix formalism provides little qualitative insight into NMR experiments. The design of new experiments and the optimization of existing experiments is facilitated if the spectroscopist has an intuitive feel for the evolution of the important components of the density operator at each point in the experiment.

Most elementary quantum mechanics is concerned with finding wavefunctions that are solutions to the Schrödinger equation. The Hamiltonian is an operator, and as has been stated previously, physically observable quantities such as energy, position, angular momentum are represented in quantum mechanics by operators. Since the aim of the theoretical analysis of NMR spectroscopy is prediction of the outcome of experiments, concentration on the operators themselves rather than the solutions to the Schrödinger equation proves to be a powerful approach. As an illustration, the analysis of the one-pulse experiment in Section 2.4.2 indicates that the equilibrium density operator can be expressed in terms of the Cartesian I_z spin operator. This operator is partially converted into the I_y operator by a pulse with x phase and rotation angle α; subsequent evolution under the Zeeman Hamiltonian converts the I_y operator into a linear combination of I_x and I_y spin operators. In this case, the evolution of the density operator is represented by the interconversion of single-spin operators. For heteronuclear NMR, and with the continued development of stronger magnets, more frequently in proton NMR, high-resolution NMR spectra are predominantly first order, and for the most part, spin systems are weakly coupled. In the weak coupling regime, a simplified formalism, referred to as the *product operator formalism*, that treats each weakly coupled system independently can be used to analyze evolution of the density operator (9–11). The product operator formalism retains much of the rigor of the full density matrix treatment while offering considerable insight into complex NMR experiments.

2.7.1 Operator Spaces

In general, an arbitrary density operator can be represented as a linear combination of a complete set of orthogonal basis operators, $\mathbf{B}_k$:

$$\sigma(t) = \sum_{k=1}^{K} b_k(t)\mathbf{B}_k, \qquad\qquad [2.190]$$

in which $b_k(t)$ are complex coefficients and K is the dimensionality of the Liouville operator space spanned by the basis operators. For a system of N spin-$\frac{1}{2}$ nuclei, $K = 4^N$. *Liouville operator space*, and its attendant operator algebra, can be regarded as an elaboration of the ideas of the Hilbert vector space and vector algebra (*2, 12*). The orthogonality condition is

$$\text{Tr}\{\mathbf{B}_j^\dagger \mathbf{B}_k\} = \langle \mathbf{B}_j | \mathbf{B}_k \rangle = \delta_{jk} \langle \mathbf{B}_k | \mathbf{B}_k \rangle. \qquad [2.191]$$

Unnormalized basis operators, $\mathbf{B}_k$, can be normalized using

$$\mathbf{B}_k' = \mathbf{B}_k / \langle \mathbf{B}_k | \mathbf{B}_k \rangle^{1/2}. \qquad [2.192]$$

The expectation value of an operator $\mathbf{A}$ can be written, by substitution of [2.190] into [2.48], as

$$\langle A(t) \rangle = \text{Tr}\{\sigma(t)\mathbf{A}\} = \text{Tr}\left\{ \sum_{k=1}^{K} b_k(t)\mathbf{B}_k \mathbf{A} \right\} = \sum_{k=1}^{K} b_k(t)\,\text{Tr}\{\mathbf{B}_k \mathbf{A}\}. \qquad [2.193]$$

Note that $\text{Tr}\{\mathbf{AB}\}$ used in [2.193] and $\text{Tr}\{\mathbf{A}^\dagger\mathbf{B}\}$ used in [2.191] in general are not equal unless $\mathbf{A}$ is a Hermitian operator. The time evolution of the density operator can be expressed, by substitution of [2.190] into [2.54] and [2.55], as

$$\frac{d\sigma(t)}{dt} = i[\sigma(t), \mathcal{H}] = i \sum_{k=1}^{K} b_k(t)[\mathbf{B}_k, \mathcal{H}]; \qquad [2.194]$$

$$\sigma(t) = \exp[-i\mathcal{H}t]\sigma(0)\exp[i\mathcal{H}t]$$

$$= \sum_{k=1}^{K} b_k(0)\exp[-i\mathcal{H}t]\mathbf{B}_k\exp[i\mathcal{H}t]. \qquad [2.195]$$

The usefulness of [2.193]–[2.195] is that the evolution of the density operator and expectation values can be calculated from a limited number of trace operations $\text{Tr}\{\mathbf{B}_k \mathbf{A}\}$ and transformation rules for $\exp[-i\mathcal{H}t]\mathbf{B}_k\exp[i\mathcal{H}t]$.

A transformation of the density operator is formally described as a rotation of an initial density operator σ^1 to a new operator σ^2 under the effect of a particular Hamiltonian, $\mathcal{H}$:

$$\sigma^2 = \exp(-i\mathcal{H}t)\,\sigma^1\,\exp(i\mathcal{H}t). \qquad [2.196]$$

The form of the $\mathcal{H}$ varies according to whether a pulse or a period of free precession is causing the rotation. The beauty of this formalism is that by expressing $\mathcal{H}$ and σ in terms of the angular-momentum operators, existing solutions to the above equation can be applied. These solutions, which were derived in Section 2.3, can be applied as a recipe by using a simple set of rules, which are presented below. The notation to be employed has the form

$$\sigma^1 \xrightarrow{\mathscr{H}t} \sigma^2, \qquad\qquad [2.197]$$

which represents the formal expression of [2.196]. The Hamiltonians consist of one or more of three things: (1) effect of spins under the influence of a rf pulse, (2) the development of chemical shift, and (3) evolution of scalar coupling. Most importantly, in the weak coupling regime, the chemical-shift and scalar coupling interactions commute with each other. Note that throughout this analysis relaxation of the spins back to equilibrium is not considered.

2.7.2 BASIS OPERATORS

The choice of basis operators is determined by the problem in hand at any specific time. Particularly useful are the angular-momentum operators I_x, I_y and I_z, which represent the x, y and z components of the spin angular momentum of the system. For a *single spin*, the state of a magnetization vector can be specified by the amounts of x, y, and z magnetization. In the same way the quantum-mechanical state of the system can be described by specifying the magnitudes of the operators that are present at any time. Formally, the state of the system is specified by the density operator, and the density operator is expressed as a linear combination of operators. In most cases, Cartesian basis operators, $E/2$, I_x, I_y, and I_z, will be employed, although other basis sets, such as the single element $(I^\alpha, I^\beta, I^+, I^-)$ and spherical basis operators $(E/2, I_0, I^+, I^-)$, are also useful.

The Cartesian and single element basis sets are related by

$$I_z = \frac{1}{2}(I^\alpha - I^\beta); \qquad I_x = \frac{1}{2}(I^+ + I^-);$$

$$\qquad\qquad [2.198]$$

$$I_y = \frac{1}{2i}(I^+ - I^-); \qquad \frac{1}{2}\mathbf{E} = \frac{1}{2}(I^\alpha + I^\beta).$$

Levitt notes that the three Cartesian operators form a three-dimensional space in which the density operator, represented by a vector $\mathbf{I}$, rotates at $|\omega|$ rad/s about a vector ω, with Cartesian components

$$\omega_x = \omega_1 \cos\phi; \qquad \omega_y = \omega_1 \sin\phi; \qquad \omega_z = \Omega, \qquad [2.199]$$

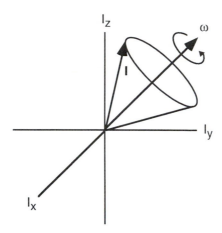

FIGURE 2.5 Geometric representations of rotations in an operator space. Precession of the angular-momentum operator about the effective field direction in angular-momentum operator space.

where Ω is the resonance offset in rad/s, ϕ is the phase of an applied rf pulse, $\omega_1 = -\gamma B_1$, and B_1 is the rf field strength (13). This geometric interpretation of the Cartesian operator space is illustrated in Fig. 2.5. The identity operator $\mathbf{E}$ is independent of rotation.

Single-transition-shift operators can be defined in terms of the Cartesian components or as products of kets and bras:

$$I^+(rs) = I_x(rs) + iI_y(rs) = |r\rangle\langle s|;$$
$$I^-(rs) = I_x(rs) - iI_y(rs) = |s\rangle\langle r|.$$ [2.200]

As noted by Ernst et al. (2), the indices are ordered such that $M_r > M_s$ to ensure that the raising and lowering operators increase and decrease the magnetic quantum numbers, respectively

$$I^+(rs)\,|s\rangle = |r\rangle; \qquad I^-(rs)\,|r\rangle = |s\rangle.$$ [2.201]

For the one-spin case, the eigenstates are $|\alpha\rangle$ and $|\beta\rangle$, and the density operator can be expanded in terms of four basis operators I^α, I^β, I^+, and I^-. The Dirac notation defines the matrix elements of the operators

$$I^\alpha = |\alpha\rangle\langle\alpha|; \qquad I^+ = |\alpha\rangle\langle\beta|;$$
$$I^\beta = |\beta\rangle\langle\beta|; \qquad I^- = |\beta\rangle\langle\alpha|,$$ [2.202]

such that, for example, $I^-|\alpha\rangle = |\beta\rangle\langle\alpha|\alpha\rangle = |\beta\rangle$ because the eigenstate $|\alpha\rangle$ is associated with $m = +\frac{1}{2}$ and $|\beta\rangle$ is associated with $m = -\frac{1}{2}$. Similarly $I^+|\beta\rangle = |\alpha\rangle\langle\beta|\beta\rangle = |\alpha\rangle$.

The potential of the product operator approach becomes evident in the case of two weakly scalar coupled spins. Each pair of spins has four eigenstates, $|\alpha\alpha\rangle$, $|\alpha\beta\rangle$, $|\beta\alpha\rangle$, and $|\beta\beta\rangle$, where the first symbol in each represents the state of the I spin and the second symbol represents the S spin. The single-element operator basis contains four so-called population terms

$$|\alpha\alpha\rangle\langle\alpha\alpha| = I^{\alpha}S^{\alpha}; \qquad |\alpha\beta\rangle\langle\alpha\beta| = I^{\alpha}S^{\beta};$$

$$|\beta\alpha\rangle\langle\beta\alpha| = I^{\beta}S^{\alpha}; \qquad |\beta\beta\rangle\langle\beta\beta| = I^{\beta}S^{\beta}.$$

[2.203]

The basis contains eight single-quantum transitions associated with the two spins (remembering the definitions of [2.202]):

$$|\alpha\alpha\rangle\langle\alpha\beta| = I^{\alpha}S^{+}; \qquad |\alpha\beta\rangle\langle\alpha\alpha| = I^{\alpha}S^{-};$$

$$|\beta\alpha\rangle\langle\beta\beta| = I^{\beta}S^{+}; \qquad |\beta\beta\rangle\langle\beta\alpha| = I^{\beta}S^{-};$$

$$|\alpha\alpha\rangle\langle\beta\alpha| = I^{+}S^{\alpha}; \qquad |\beta\alpha\rangle\langle\alpha\alpha| = I^{-}S^{\alpha};$$

$$|\alpha\beta\rangle\langle\beta\beta| = I^{+}S^{\beta}; \qquad |\beta\beta\rangle\langle\alpha\beta| = I^{-}S^{\beta}.$$

[2.204]

Note in these cases one spin remains "untouched" and the transition involves only the change in spin state of the other spin. These operators are given the name *single-quantum transitions*, and describe the *single-quantum coherence* associated with the transition. The coherence between eigenstates may be due to a transition in which both spins change their spin states simultaneously. These coherences can be classified as *double-quantum coherence* in which both spins change spin states in the same sense,

$$|\alpha\alpha\rangle\langle\beta\beta| = I^{+}S^{+}; \qquad |\beta\beta\rangle\langle\alpha\alpha| = I^{-}S^{-},$$

[2.205]

or *zero-quantum coherence* terms in which both spins change spin states in the opposite sense,

$$|\alpha\beta\rangle\langle\beta\alpha| = I^{+}S^{-}; \qquad |\beta\alpha\rangle\langle\alpha\beta| = I^{-}S^{+}.$$

[2.206]

Each of these product operators has a reasonably simple interpretation in terms of energy levels and transitions shown in Fig. 2.6. Note that for N scalar coupled spins-$\frac{1}{2}$, a full operator set contains 4^{N} members.

Product operators in the Cartesian basis will be used most often in this text because the Cartesian basis affords the simplest treatment of pulses and delays in a pulse sequence. For the two spin case, 16-Cartesian product operator terms are required:

$$\frac{1}{2}E \qquad I_{x} \qquad I_{y} \qquad I_{z} \qquad S_{x} \qquad S_{y} \qquad S_{z}$$

$$2I_{x}S_{z} \qquad 2I_{y}S_{z} \qquad 2I_{z}S_{z} \qquad 2I_{z}S_{x} \qquad 2I_{z}S_{y}$$

[2.207]

$$2I_{x}S_{x} \qquad 2I_{y}S_{y} \qquad 2I_{x}S_{y} \qquad 2I_{y}S_{x}$$

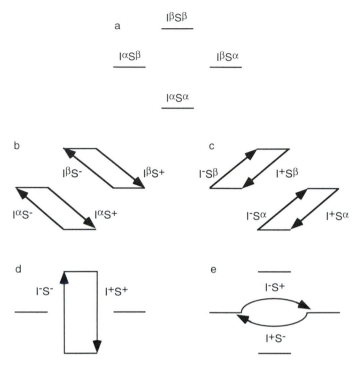

FIGURE 2.6 Single-transition basis operators for IS spin system. (a) Populations, (b, c) single-quantum transitions, (d) double-quantum transitions, and (e) zero-quantum transitions.

The relationship between the Cartesian and single-element operators [2.198] can be seen as, for example,

$$I_z = 2(I_z)\left(\frac{1}{2}\mathbf{E}\right) = \frac{1}{2}(I^\alpha - I^\beta)(S^\alpha + S^\beta);$$

$$S_x = 2\left(\frac{1}{2}\mathbf{E}\right)(S_x) = \frac{1}{2}(I^\alpha + I^\beta)(S^+ + S^-); \qquad [2.208]$$

$$2I_zS_y = 2(I_z)(S_y) = \frac{1}{2i}(I^\alpha - I^\beta)(S^+ - S^-)$$

The operators have physical interpretations; for example,

$$\mathbf{E} = I^\alpha S^\alpha + I^\alpha S^\beta + I^\beta S^\alpha + I^\beta S^\beta \qquad [2.209]$$

denotes equal populations of all energy levels, and

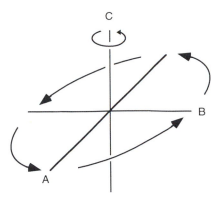

FIGURE 2.7 Operator rotations. The rotations induced by an operator **C** acting on an operator **A** are illustrated. The operators satisfy the commutation relationship [2.211], and the rotations are represented mathematically by [2.212].

$$I_z = \frac{1}{2}(I^\alpha S^\alpha + I^\alpha S^\beta - I^\beta S^\alpha - I^\beta S^\beta) \qquad [2.210]$$

denotes equal polarization across the two single-quantum transitions of the I spin.

2.7.3 EVOLUTION IN THE PRODUCT OPERATOR FORMALISM

The goal of the product operator formalism is to derive the evolution of a spin system through a particular pulse sequence as conveniently as possible. Effects of pulses and delays in terms of Cartesian product operators are extremely simple since each factor of the product is rotated independently. Rotation operator equations similar to [2.129] can be derived by the matrix derivations established above; however, this approach is rather laborious. Instead, the rules for transformations of product operators can be established using the following useful theorem. If three operators satisfy the commutation relationship (and its cyclic permutations)

$$[\mathbf{A}, \mathbf{B}] = i\mathbf{C}, \qquad [2.211]$$

then

$$\exp(-i\theta\mathbf{C})\,\mathbf{A}\,\exp(i\theta\mathbf{C}) = \mathbf{A}\cos\theta + \mathbf{B}\sin\theta. \qquad [2.212]$$

Equation [2.212] can be verified by differentiating $\exp(-i\theta\mathbf{C})\,\mathbf{A}\,\exp(i\theta\mathbf{C})$ twice with respect to θ, applying the commutation relations and solving the resulting harmonic differential equation. The evolution indicated by [2.212] can be illustrated succinctly by Fig. 2.7.

2.7.3.1 Free Precession   During periods of free precession, the effects of chemical-shift evolution and scalar coupling evolution must be considered. For a spin I, the Hamiltonian has the form $\mathcal{H} = \Omega_I I_z$, where Ω_I is the offset of spin I, and the rotation during a delay, t, can be given

$$I_z \xrightarrow{\;\Omega_I I_z t\;} I_z; \qquad\qquad\qquad [2.213]$$

$$I_x \xrightarrow{\;\Omega_I I_z t\;} I_x \cos(\Omega_I t) + I_y \sin(\Omega_I t); \qquad\qquad [2.214]$$

$$I_y \xrightarrow{\;\Omega_I I_z t\;} I_y \cos(\Omega_I t) - I_x \sin(\Omega_I t). \qquad\qquad [2.215]$$

For a two-spin system, I and S, the form of the Hamiltonian for evolution under a scalar coupling of J_{IS} is $\mathcal{H} = 2\pi J_{IS} I_z S_z$. The magnetization evolves as

$$I_z \xrightarrow{\;2\pi J_{IS} I_z S_z t\;} I_z; \qquad\qquad\qquad [2.216]$$

$$I_x \xrightarrow{\;2\pi J_{IS} I_z S_z t\;} I_x \cos(\pi J_{IS} t) + 2 I_y S_z \sin(\pi J_{IS} t); \qquad [2.217]$$

$$I_y \xrightarrow{\;2\pi J_{IS} I_z S_z t\;} I_y \cos(\pi J_{IS} t) - 2 I_x S_z \sin(\pi J_{IS} t). \qquad [2.218]$$

The last two equations, [2.217] and [2.218], demonstrate that in-phase magnetization evolves into antiphase magnetization under the influence of the scalar coupling interaction. The analogous evolution of the two-spin operators, $2 I_\eta S_z$, is given by

$$2 I_z S_z \xrightarrow{\;2\pi J_{IS} I_z S_z t\;} 2 I_z S_z; \qquad\qquad\qquad [2.219]$$

$$2 I_y S_z \xrightarrow{\;2\pi J_{IS} I_z S_z t\;} 2 I_y S_z \cos(\pi J_{IS} t) - I_x \sin(\pi J_{IS} t); \qquad [2.220]$$

$$2 I_x S_z \xrightarrow{\;2\pi J_{IS} I_z S_z t\;} 2 I_x S_z \cos(\pi J_{IS} t) + I_y \sin(\pi J_{IS} t). \qquad [2.221]$$

Evolution of the S_η and $2 I_z S_\eta$ operators is obtained by exchanging the I and S labels in [2.213]–[2.221].

2.7.3.2 Pulses Radiofrequency pulses applied down a specific axis induce rotations in a plane orthogonal to that axis. The Hamiltonian expression describing the pulses can be written as $\mathcal{H}t = \alpha I_x$ or αI_y, for an x pulse or y pulse, respectively, and α is the flip angle of the pulse. Pulses of arbitrary phase or that include the effects of resonance offset can be obtained using composite rotations as in [2.116]. The transformations for a pulse of phase $\pm x$ are given by

$$I_z \xrightarrow{\ \alpha I_{\pm x}\ } I_z \cos \alpha \mp I_y \sin \alpha; \qquad [2.222]$$

$$I_y \xrightarrow{\ \alpha I_{\pm x}\ } I_y \cos \alpha \pm I_z \sin \alpha; \qquad [2.223]$$

$$I_x \xrightarrow{\ \alpha I_{\pm x}\ } I_x, \qquad [2.224]$$

and, for a pulse of phase $\pm y$,

$$I_z \xrightarrow{\ \alpha I_{\pm y}\ } I_z \cos \alpha \pm I_x \sin \alpha; \qquad [2.225]$$

$$I_y \xrightarrow{\ \alpha I_{\pm y}\ } I_y; \qquad [2.226]$$

$$I_x \xrightarrow{\ \alpha I_{\pm y}\ } I_x \cos \alpha \mp I_z \sin \alpha. \qquad [2.227]$$

These transformations of the product operators are illustrated geometrically in Fig. 2.8.

2.7.3.3 Practical Points The rules presented above enable description of a wide variety of pulsed NMR experiments. Before examining some very useful specific examples, some formalities and practical points will be presented.

During a period of free precession for the two-spin system I and S, the evolution of the density operator is represented as

$$\sigma^1 \xrightarrow{\ \Omega_I I_z t + \Omega_S S_z t + 2\pi J_{IS} I_z S_z t\ } \sigma^4. \qquad [2.228]$$

Because each term in Eq. [2.228] commutes with the others, the one evolution period can be divided into a series of rotations or a *cascade:*

$$\sigma^1 \xrightarrow{\ \Omega_I I_z t\ } \sigma^2 \xrightarrow{\ \Omega_S S_z t\ } \sigma^3 \xrightarrow{\ 2\pi J_{IS} I_z S_z t\ } \sigma^4. \qquad [2.229]$$

The order in which the rotations due to shift and coupling evolution are applied is unimportant. Likewise, the effect of a nonselective pulse applied to the I and S spins is written as

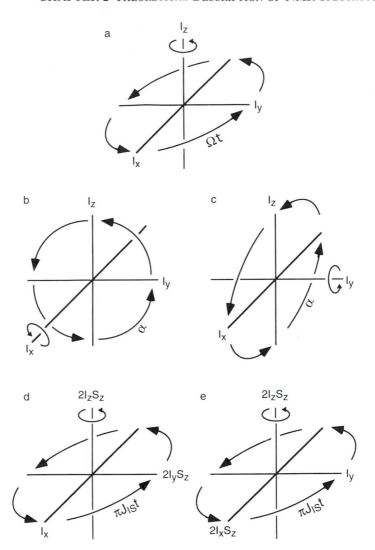

FIGURE 2.8 Transformations of product operators. The product operator transformations given in [2.213]–[2.227] are represented geometrically. (a) Transformations under the chemical-shift Hamiltonian; (b) rotations induced by a pulse of x phase, (c) rotations induced by a pulse of y phase; (d, e) transformations under the scalar coupling Hamiltonian.

$$\sigma^1 \xrightarrow{\;\alpha(I_x + S_x)\;} \sigma^3 . \qquad\qquad [2.230]$$

Because I and S operators commute, a nonselective pulse can be represented by a pulse on I first followed by a pulse on S (or by a pulse acting on S followed by a pulse acting on I):

$$\sigma^1 \xrightarrow{\;\alpha I_x\;} \sigma^2 \xrightarrow{\;\alpha S_x\;} \sigma^3 . \qquad\qquad [2.231]$$

The effect of a pulse applied selectively to the S spin of a product term such as $2I_x S_z$ is obtained using the rule that rotations affect only operators of the same spin. In other words, the I_x part of the product operator remains untouched by the pulse to the S spin and the S_z term is rotated normally. The result obtained is

$$2I_x S_z \xrightarrow{\;\alpha S_x\;} 2I_x S_z \cos\alpha - 2I_x S_y \sin\alpha . \qquad\qquad [2.232]$$

Operators are unaffected by rotations involving the same operator because an operator and the exponential of an operator commute. For example,

$$I_x \xrightarrow{\;\alpha I_x\;} I_x ; \qquad\qquad [2.233]$$

$$2I_y S_z \xrightarrow{\;\Omega_S S_z t\;} 2I_y S_z . \qquad\qquad [2.234]$$

2.7.4 SINGLE-QUANTUM COHERENCE AND OBSERVABLE OPERATORS

The single-quantum coherence term I_x can be expressed using [2.198] and [2.208] as

$$I_x = \frac{1}{2}(I^+ S^\alpha + I^- S^\alpha) + \frac{1}{2}(I^+ S^\beta + I^- S^\beta). \qquad\qquad [2.235]$$

This operator, involving a transverse Cartesian component, results from the sum of the single-quantum transitions of the I spin. Evolution under the free-precession Hamiltonian yields

$$\exp[-i\mathcal{H}t] I_x \exp[i\mathcal{H}t]$$

$$= \frac{1}{2}(I^+ S^\alpha \exp[-i(\Omega_I + \pi J)t] + I^- S^\alpha \exp[-i(-\Omega_I + \pi J)t])$$

$$+ \frac{1}{2}(I^+ S^\beta \exp[-i(\Omega_I - \pi J)t] + I^- S^\beta \exp[-i(-\Omega_I - \pi J)t]). \qquad [2.236]$$

The trace of this result with the observation operator yields

$$\text{Tr}\{\exp[-i\mathcal{H}t]I_x\exp[i\mathcal{H}t]\,\mathbf{F}^+\}$$

$$= \frac{1}{2}\exp[i(\Omega_I - \pi J)t]) + \frac{1}{2}\exp[i(\Omega_I + \pi J)t]).$$

[2.237]

Both terms contained in the operator are *positive*, indicating an *in-phase* component of the x magnetization. The frequencies of the two components of the in-phase signal are separated by the scalar coupling between the two spins. *An operator with a single transverse Cartesian component is observable.* Another example of a single-quantum coherence operator is

$$2I_xS_z = \frac{1}{2}(I^+S^\alpha + I^-S^\alpha) - \frac{1}{2}(I^+S^\beta + I^-S^\beta).$$

[2.238]

Evolution under the free-precession Hamiltonian yields

$$\exp[-i\mathcal{H}t]2I_xS_z\exp[i\mathcal{H}t]$$

$$= \frac{1}{2}(I^+S^\alpha\exp[-i(\Omega_I + \pi J)t] + I^-S^\alpha\exp[-i(-\Omega_I + \pi J)t])$$

$$- \frac{1}{2}(I^+S^\beta\exp[-i(\Omega_I - \pi J)t] + I^-S^\beta\exp[-i(-\Omega_I - \pi J)t]).$$

[2.239]

The trace of this result with the observation operator yields

$$\text{Tr}\{\exp[-i\mathcal{H}t]I_x\exp[i\mathcal{H}t]\,\mathbf{F}^+\} = \frac{1}{2}\exp[i(\Omega_I - \pi J)t]) - \frac{1}{2}\exp[i(\Omega_I + \pi J)t]).$$

[2.240]

In this case, the contributions from the S spin are of opposite sign, indicating an *antiphase* x component of the magnetization on the I spin, in which the two components of the signal have opposite sign. Formally, the antiphase terms are not directly observable in the sense that at a particular instant, a term such as $2I_xS_z$ does not contribute to the observed x magnetization. However, antiphase operators will evolve under the influence of the scalar coupling interaction, provided that I and S have a nonzero scalar coupling constant, into an in-phase operator that is detectable. Analogous terms for the I spin product operator involving y components are similar except that the phase of the magnetization is shifted by 90°.

For a system of N spins, operators containing *one* x or y component are observable and generate in-phase resonance signals. Operators of the form $I_{1x}I_{iz}I_{jz} \cdots I_{kz}$ are observable and generate antiphase resonance signals if J_{1i}, J_{1j}, ..., J_{1k} are all greater than zero. Operators containing an I_x factor and operators containing an I_y factor generate resonance signals that differ in phase by 90°. These simple principles can be used to identify observable terms and the form of the resulting spectral features without explicitly calculating the evolution of the density operator during the acquisition period.

2.7.5 MULTIPLE-QUANTUM COHERENCE

Multiple-quantum coherence states are represented by product operators in which both spins have transverse components. For example,

$$2I_xS_y = \frac{1}{2i}(I^+S^+ - I^-S^-) - \frac{1}{2i}(I^+S^- - I^-S^+). \qquad [2.241]$$

The first term on the right-hand side, $(I^+S^+ - I^-S^-)$, is pure double-quantum coherence ($|\Delta m| = 2$), whereas the second term, $(I^+S^- - I^-S^+)$, is pure zero-quantum coherence ($|\Delta m| = 0$). The multiple-quantum coherence term $2I_xS_y$ is a superposition of both double- and zero-quantum coherence. Multiple-quantum coherences can be prepared by suitable combinations of pulses and free-precession periods. Such terms have more than one transverse operator component and are not observable directly; however, multiple-quantum coherences possess some unique properties of considerable utility.

Multiple-quantum coherences can be expressed conveniently in terms of Cartesian and/or shift operators. Pure double-quantum coherence is represented by suitable combinations of bilinear product operators:

$$\frac{1}{2}(I^+S^+ + I^-S^-) = \frac{1}{2}(2I_xS_x - 2I_yS_y) = DQ_x; \qquad [2.242]$$

$$\frac{1}{2i}(I^+S^+ - I^-S^-) = \frac{1}{2}(2I_xS_y + 2I_yS_x) = DQ_y. \qquad [2.243]$$

Pure double-quantum coherence precesses at the *sum* of the two chemical shifts during a delay, t:

$$DQ_x \xrightarrow{\Omega_I I_z t + \Omega_S S_z t} DQ_x \cos[(\Omega_I + \Omega_S)t] + DQ_y \sin[(\Omega_I + \Omega_S)t]; \qquad [2.244]$$

$$DQ_y \xrightarrow{\Omega_I I_z t + \Omega_S S_z t} DQ_y \cos[(\Omega_I + \Omega_S)t] - DQ_x \sin[(\Omega_I + \Omega_S)t]. \qquad [2.245]$$

Similarly, pure zero quantum coherence is represented by

$$\frac{1}{2}(I^+S^- + I^-S^+) = \frac{1}{2}(2I_xS_x + 2I_yS_y) = ZQ_x; \qquad [2.246]$$

$$\frac{1}{2i}(I^+S^- - I^-S^+) = \frac{1}{2}(2I_yS_x - 2I_xS_y) = ZQ_y. \qquad [2.247]$$

and evolution occurs at the *difference* of the chemical shifts of the spins involved:

$$ZQ_x \xrightarrow{\Omega_I I_z t + \Omega_S S_z t} ZQ_x \cos[(\Omega_I - \Omega_S)t] + ZQ_y \sin[(\Omega_I - \Omega_S)t]; \quad [2.248]$$

$$ZQ_y \xrightarrow{\Omega_I I_z t + \Omega_S S_z t} ZQ_y \cos[(\Omega_I - \Omega_S)t] - ZQ_x \sin[(\Omega_I - \Omega_S)t]. \quad [2.249]$$

Two-spin multiple-quantum coherence such as that noted above *does not evolve under the influence of the scalar coupling of the two spins involved in the coherence (the active coupling)*. However, multiple-quantum coherence can evolve under the influence of a scalar coupling to a third *passive* spin. For example, consider the three-spin system I, S, R where the couplings present are J_{IS}, J_{IR}, and J_{SR}. A multiple-quantum coherence term can be identified by the appearance of more than one transverse Cartesian operator in the product; therefore, the operator, $4I_yS_xR_z$, is a multiple-quantum coherence with respect to I and S. This operator evolves under the J_{IR} and J_{SR} scalar coupling interactions but not under the J_{IS} scalar coupling interaction. Evolution under the J_{IR} scalar coupling interaction is given by

$$4I_yS_xR_z \xrightarrow{2\pi J_{IR} t I_z R_z} 4I_yS_xR_z \cos(\pi J_{IR}t) - 2I_xS_x \sin(\pi J_{IR}t). \quad [2.250]$$

Evolution of multiple-quantum coherences under the scalar coupling interaction proceeds at the sum and difference frequencies of the passive scalar coupling constants in a manner analogous to that of the chemical shift evolution. For example, consider the zero-quantum term $ZQ_y^{IS} = \frac{1}{2}(2I_yS_x - 2I_xS_y)$ evolving under the passive coupling effects J_{IR} and J_{SR} for a time t

$$ZQ_y^{IS} \xrightarrow{2\pi J_{IR} t I_z R_z + 2\pi J_{SR} t S_z R_z} ZQ_y^{IS} \cos(\pi K_{IS}t) - 2ZQ_x^{IS}R_z \sin(\pi K_{IS}t), \quad [2.251]$$

in which $K_{IS} = |J_{SR} - J_{IR}|$ is known as the *zero-quantum splitting* and

$$2ZQ_x^{IS}R_z = \frac{1}{2}(2I_xS_x + 2I_yS_y)R_z. \qquad [2.252]$$

2.7.6 COHERENCE TRANSFER AND GENERATION OF MULTIPLE-QUANTUM COHERENCE

Coherence transfer is a vital effect in multidimensional NMR spectroscopy, and, most notably, an effect that cannot be described in the Bloch model. Suppose that an antiphase component, $2I_xS_z$, of the density operator has been generated in some manner. As will be discussed below, antiphase operators can be produced by the use of a spin echo pulse sequence. The effect of applying a 90_y° pulse to both spins is

$$2I_xS_z \xrightarrow{\frac{\pi}{2}I_y} -2I_zS_z \xrightarrow{\frac{\pi}{2}S_y} -2I_zS_x. \qquad [2.253]$$

The original antiphase coherence on the I spin (containing a single transverse operator) is transferred to antiphase coherence on the S spin. *Coherence has been transferred from one spin to another under the influence of the rf pulse.*

In contrast, application of a 90_x° to both spins gives

$$2I_xS_z \xrightarrow{\frac{\pi}{2}(I_x+S_x)} -2I_xS_y. \qquad [2.254]$$

This operator represents multiple-quantum coherence (containing more than one transverse operator). The same result would be obtained if a pulse is applied to the S spin alone:

$$2I_xS_z \xrightarrow{\frac{\pi}{2}S_x} -2I_xS_y. \qquad [2.255]$$

The two examples represented by [2.254] and [2.255] represent the generation of multiple-quantum coherences in homonuclear and heteronuclear spin systems, respectively.

2.7.7 EXAMPLES OF PRODUCT OPERATOR CALCULATIONS

Some simple examples using product operators to follow evolution during a spin-echo and polarization transfer pulse sequences will be presented. Although these examples may appear trivial, each one can play an important part as an individual component of a more complicated pulse sequence. These pulse sequence elements will be encountered in many of the multidimensional NMR experiments discussed later (Chapters 6 and 7).

2.7.7.1 The Spin Echo   The spin-echo pulse sequence must be examined in three cases: (1) one spin, (2) two coupled spins of the same nuclear type (homonuclear case), and (3) two coupled spins of different nuclear types (heteronuclear case).

Starting from equilibrium magnetization proportional to I_z, an initial 90°_x pulse yields

$$I_z \xrightarrow{\frac{\pi}{2}I_x} -I_y. \qquad [2.256]$$

The spin-echo pulse sequence for an isolated spin is written as

$$-t-180^\circ_x-t-. \qquad [2.257]$$

Evolution during the period of free precession, t, yields

$$-I_y \xrightarrow{\Omega_I I_z t} -I_y \cos(\Omega_I t) + I_x \sin(\Omega_I t). \qquad [2.258]$$

The 180°_x pulse converts this density operator to

$$-I_y \cos(\Omega_I t) + I_x \sin(\Omega_I t) \xrightarrow{\pi I_x} I_y \cos(\Omega_I t) + I_x \sin(\Omega_I t). \qquad [2.259]$$

The 180°_x pulse inverts the I_y term but does not affect the I_x term. The final part of the spin-echo sequence is another delay of duration t:

$$I_y \cos(\Omega_I t) + I_x \sin(\Omega_I t) \xrightarrow{\Omega_I I_z t} I_y \cos^2(\Omega_I t) - I_x \cos(\Omega_I t) \sin(\Omega_I t)$$
$$+ I_x \sin(\Omega_I t) \cos(\Omega_I t) + I_y \sin^2(\Omega_I t). \qquad [2.260]$$

Using the identity $\cos^2\theta + \sin^2\theta = 1$, we can write [2.260] as

$$I_y \cos(\Omega_I t) + I_x \sin(\Omega_I t) \xrightarrow{\Omega_I I_z t} I_y. \qquad [2.261]$$

The overall effect of the spin-echo segment, $-t-180^\circ_x-t-$, is seen to take an initial state $-I_y$ and generate a final state I_y. Apart from a sign change, no net evolution of the chemical shift occurs during the spin-echo sequence: *evolution under the chemical shift Hamiltonian is refocused.*

The same result can be demonstrated more elegantly as follows. The density operator at the end of the pulse sequence is given by $\sigma(t) = \mathbf{U}\,\sigma(0)\,\mathbf{U}^{-1}$ with

$$\mathbf{U} = \exp[-i\Omega_I t I_z]\, \exp[-i\pi I_x]\, \exp[-i\Omega_I t I_z], \qquad [2.262]$$

in which each factor in $\mathbf{U}$ represents the propagator for one segment of the spin-echo sequence. Applying the identity of [2.117] yields

$$\mathbf{U} = \exp[-i\Omega_I t I_z]\, \exp[-i\pi I_x]\, \exp[-i\Omega_I t I_z]$$
$$= \exp[-i\Omega_I t I_z]\, \exp[-i\pi I_x]\, \exp[-i\Omega_I t I_z]\, \exp[i\pi I_x]\, \exp[-i\pi I_x]$$

$$= \exp[-i\Omega_I t I_z] \exp[-i\Omega_I t e^{-i\pi I_x} I_z e^{i\pi I_x}] \exp[-i\pi I_x]$$

$$= \exp[-i\Omega_I t I_z] \exp[i\Omega_I t I_z] \exp[-i\pi I_x]$$

$$= \exp[-i\pi I_x]. \tag{2.263}$$

Therefore, the *net* evolution during the spin-echo sequence is given by

$$-I_y \xrightarrow{\pi I_x} I_y, \tag{2.264}$$

in agreement with [2.261]. Considerable simplification of propagators for pulse sequences containing 180° pulses is often possible by use of [2.117].

The same spin-echo pulse sequence can be applied to a homonuclear two-spin system. The pulses are assumed to be nonselective and affect both the I and S spins equally. As for the isolated spin, the chemical-shift evolution of the I and S spins is refocused over the spin-echo sequence and can be neglected. Therefore, evolution during the pulse sequence is due to the scalar coupling interaction only. The initial $90°_x$ pulse generates the $-I_y$ operator from the equilibrium operator I_z (the similar S spin term is omitted for clarity). The coupling develops during t:

$$-I_y \xrightarrow{2\pi J_{IS} I_z S_z t} -I_y \cos(\pi J_{IS} t) + 2I_x S_z \sin(\pi J_{IS} t). \tag{2.265}$$

The 180° pulse, regarded as a 180° pulse on one spin followed by a 180° pulse on the other spin, yields

$$-I_y \cos(\pi J_{IS} t) + 2I_x S_z \sin(\pi J_{IS} t) \xrightarrow{\pi I_x} I_y \cos(\pi J_{IS} t) + 2I_x S_z \sin(\pi J_{IS} t)$$

$$\xrightarrow{\pi S_x} I_y \cos(\pi J_{IS} t) - 2I_x S_z \sin(\pi J_{IS} t). \tag{2.266}$$

The $180°_x$ pulse applied to the I spin does not affect the S spin and vice versa. Evolution during the second delay, t, yields

$$I_y \cos(\pi J_{IS} t) - 2I_x S_z \sin(\pi J_{IS} t) \xrightarrow{2\pi J_{IS} I_z S_z t}$$

$$I_y \cos^2(\pi J_{IS} t) - 2I_x S_z \sin(\pi J_{IS} t) \cos(\pi J_{IS} t)$$

$$- 2I_x S_z \cos(\pi J_{IS} t) \sin(\pi J_{IS} t) - I_y \sin^2(\pi J_{IS} t), \tag{2.267}$$

which, using the identities $\cos(2\theta) = \cos^2\theta - \sin^2\theta$ and $\sin(2\theta) = 2\sin\theta \cos\theta$, reduces to

$$I_y \cos(\pi J_{IS} t) - 2I_x S_z \sin(\pi J_{IS} t) \xrightarrow{2\pi J_{IS} I_z S_z t}$$

$$I_y \cos(2\pi J_{IS} t) - 2I_x S_z \sin(2\pi J_{IS} t). \tag{2.268}$$

The overall effect of the $-t-180_x^\circ-t$-pulse sequence on the initial $-I_y$ magnetization is given by

$$-I_y \xrightarrow{\;t-\pi(I_x+S_x)-t\;} I_y \cos(2\pi J_{IS}t) - 2I_x S_z \sin(2\pi J_{IS}t). \qquad [2.269]$$

The result obtained for initial S_z magnetization is obtained by exchanging I and S operators in [2.269]. Setting the delay $t = 1/(4J_{IS})$ generates the purely antiphase term $2I_x S_z$, whereas having $t = 1/(2J_{IS})$ produces the in-phase term $-I_y$ (i.e., the magnetization is inverted). The generation of an antiphase state by this method is a common feature in many pulse sequences.

If the two scalar coupled spins belong to different nuclear species, then the rf pulses in the spin-echo sequence can be applied to *only one* of the scalar-coupled spins (the I spins in the following example). For example, the spin-echo sequence can be applied selectively to the proton spins in a ^{1}H–^{15}N scalar coupled spin system. Again I spin chemical shift is refocused and can be ignored. As before, following the 90_x° pulse on the I spin, evolution occurs as follows:

$$-I_y \xrightarrow{\;2\pi J_{IS}I_z S_z t\;} -I_y \cos(\pi J_{IS}t) + 2I_x S_z \sin(\pi J_{IS}t)$$

$$\xrightarrow{\;\pi I_x\;} I_y \cos(\pi J_{IS}t) + 2I_x S_z \sin(\pi J_{IS}t). \qquad [2.270]$$

Only the I_y term is inverted by the 180_x° pulse; the S spin is unaffected. The second delay generates

$$I_y \cos(\pi J_{IS}t) + 2I_x S_z \sin(\pi J_{IS}t) \xrightarrow{\;2\pi J_{IS}I_z S_z t\;}$$

$$I_y \cos^2(\pi J_{IS}t) - 2I_x S_z \sin(\pi J_{IS}t) \cos(\pi J_{IS}t)$$

$$+ 2I_x S_z \cos(\pi J_{IS}t) \sin(\pi J_{IS}t) + I_y \sin^2(\pi J_{IS}t), \qquad [2.271]$$

which reduces to

$$I_y \cos(\pi J_{IS}t) + 2I_x S_z \sin(\pi J_{IS}t) \xrightarrow{\;2\pi J_{IS}I_z S_z t\;} I_y. \qquad [2.272]$$

So for the heteronuclear spin echo,

$$-I_y \xrightarrow{\;-t-\pi I_x-t\;} I_y, \qquad [2.273]$$

and *both* the chemical shift and the heteronuclear coupling are refocused. In essence, the S spins have been decoupled from the I spins by use of the echo sequence.

2.7.7.2 Insensitive Nuclei Enhanced by Polarization Transfer Pulse sequence elements can be combined to produce more complex sequences designed to perform specific tasks. An important experiment that takes advantage of the basic schemes is the insensitive nuclei enhanced by polarization transfer (INEPT) sequence (*14*). The INEPT sequence is a crucial component of many multidimensional NMR experiments. The aim of the INEPT sequence is to transfer magnetization from a sensitive nucleus with a high gyromagnetic ratio (usually protons) to a less sensitive nucleus with a lower gyromagnetic ratio (e.g., nitrogen or carbon) by means of the scalar coupling interaction. By doing this, the detected signal from the heteronucleus will be increased. As discussed in Section 2.7.7.1, simply applying a spin echo sequence, t–180°–t to the I spin causes the decoupling of the S and I spins. The scalar coupling interaction evolves over the entire duration of the spin echo sequence in a *homonuclear spin echo* because the 180° echo pulse affects both the I and S spins equally. By analogy, heteronuclear scalar coupling interaction evolves over the duration of a spin echo sequence if *180° pulses are applied to both the I and S spins simultaneously.* With this insight, the INEPT sequence can be written as

$$I \text{ spin:} \qquad 90^\circ_x \quad -t- \quad 180^\circ_x \quad -t- \quad 90^\circ_y$$

$$S \text{ spin:} \qquad\qquad\qquad 180^\circ_x \qquad\qquad 90^\circ_x \quad -detect. \qquad [2.274]$$

Up to the final pair of 90° pulses, the sequence is a spin echo in which both spins have been affected by 180°_x pulses, so that chemical shift is refocused during the echo, but scalar coupling evolves fully. Beginning with equilibrium magnetization $K_I I_z$, in which $K_I = \omega_I/(2k_B T)$ [2.122]:

$$K_I I_z \xrightarrow{\;\frac{\pi}{2} I_x - t - \pi (I_x + S_x) - t -\;} K_I \{ I_y \cos(2\pi J_{IS} t) - 2 I_x S_z \sin(2\pi J_{IS} t) \}.$$
$$[2.275]$$

A 90°_y pulse is applied to the I spin, and a 90°_x pulse is applied to the S spin:

$$K_I \{ I_y \cos(2\pi J_{IS} t) - 2 I_x S_z \sin(2\pi J_{IS} t) \} \xrightarrow{\;\frac{\pi}{2}(I_y + S_x)\;}$$
$$K_I \{ I_y \cos(2\pi J_{IS} t) - 2 I_z S_y \sin(2\pi J_{IS} t) \}. \qquad [2.276]$$

If the delay $t = 1/(4 J_{IS})$, then the final signal is given by

$$K_I \{ -2 I_z S_y \}. \qquad [2.277]$$

The antiphase magnetization has been transferred to the S spin. In addition, the antiphase term is scaled by a factor of K_I. Instead, if a single 90°_x

pulse is applied to equilibrium S spin magnetization, then the observable magnetization is given by

$$K_S\{-S_y\}, \qquad\qquad [2.278]$$

which is an in-phase doublet. The constant K_S is proportional to the gyromagnetic ratio of the S spin. The intensity ratio between the INEPT and conventional experiment is given by

$$\frac{\text{INEPT}}{\text{Conventional}} = \frac{K_I}{K_S} = \frac{\gamma_I}{\gamma_S}. \qquad\qquad [2.279]$$

The advantage of performing the INEPT experiment becomes enormous as the gyromagnetic ratio of the S spin decreases. INEPT procedures are used with great effect in multidimensional heteronuclear NMR experiments. An additional advantage of the INEPT experiment, sometimes overlooked, is that the repetition rate of the experiment is set by the relaxation time constants of the I spin rather than the S spin. Typically, the I spin is proton, and the time constants can be notably shorter than the relaxation time constants for the S spins (see Chapter 5).

2.7.7.3 Refocused INEPT NMR spectroscopy is a relatively insensitive technique because, as has been noted in Section 1.1, the differences in populations between stationary states of a nuclear spin are very small numbers. Maximizing the sensitivity of NMR experiments consequently is a major concern. The amplitudes of the resonance signals in a scalar-coupled heteronuclear spin system can be increased dramatically by *decoupling* the spins involved in the scalar coupling interaction. Decoupling reduces the size of the scalar coupling constant effectively to zero with the result that the signal normally observed as a multiplet is collapsed into a singlet resonance at the Larmor frequency of the unperturbed spin. Sensitivity is increased because the amplitude of the singlet is given by the sum of the amplitudes of the multiplet components. As will be discussed in Section 3.4.3, decoupling can be achieved by the application of a suitable rf field on one of the spins in a heteronuclear scalar-coupled spin system.

As the following example indicates, increased sensitivity is not necessarily obtained by application of a decoupling field to an arbitrary coherence. If a decoupling field is applied to the I spins during detection of the S spins following the INEPT sequence introduced in the previous section, the resonance signal disappears completely. Following the INEPT sequence, the density operator is proportional to $2I_z S_y$. Decoupling prevents evolution of this operator into observable in-phase single quantum S_x coherence under the influence of the scalar coupling Hamiltonian (Section 2.7.4). Viewed another way, the $2I_zS_y$ operator represents an *antiphase* doublet, in which one multiplet component of the doublet is *positive* and the other component is *negative*. Collapsing the doublet by decoupling the I spins results in the mutual cancellation of the doublet components of opposite sign. Constructive interference between multiplet components is obtained only if the decoupling field is applied to an in-phase operator (with respect to the decoupled spin).

Reexamination of the INEPT experiment indicates that the antiphase coherence $2I_zS_y$ can be converted into in-phase coherence by an appropriate extension to the INEPT pulse sequence. The ensuing refocused INEPT experiment (*15*) can now be written as

I spin: 90°_x $-t-$ 180°_x $-t-$ 90°_y $-\tau-$ 180°_x $-\tau-$ *decouple*;

S spin: 180°_x 90°_x 180°_x *detect*.

$$[2.280]$$

By setting $\tau = 1/(4J_{IS})$, the final echo component of the sequence yields

$$-K_I 2I_z S_y \xrightarrow{\ \tau - \pi(I_x + S_x) - \tau\ } K_I S_x. \qquad [2.281]$$

This is an in-phase doublet and can now be decoupled to give enhanced sensitivity in the spectrum. As will be discussed in Section 7.1.1.3, the value of t required for a refocused INEPT sequence depends on the nature of the spin system and must be adjusted appropriately for spin systems other than the two-spin system considered in this example.

The S_x operator obtained following the refocussed INEPT sequence can be converted to an S_z operator by application of a 90°_{-y} pulse to the S spin:

$$K_I S_x \xrightarrow{\ \frac{\pi}{2}I_{-y}\ } K_I S_z. \qquad [2.282]$$

Since the equilibrium magnetization for the S spin is proportional to $K_S S_z$, the remarkable result is obtained that the Boltzmann population difference for the I spin has been transferred to the S spin by the refocused INEPT pulse sequence.

References

1. K. Blum, "Density Matrix Theory and Applications," pp. 1–217. Plenum Press, New York, 1981.

2. R. R. Ernst, G. Bodenhausen, and A. Wokaun, "Principles of Nuclear Magnetic Resonance in One and Two Dimensions," pp. 1–610. Clarendon Press, Oxford, 1987.
3. E. Merzbacher, "Quantum Mechanics," pp. 1–621. Wiley, New York, 1970.
4. I. N. Levine, "Quantum Chemistry," pp. 1–566. Allyn & Bacon, Boston, 1983.
5. D. A. McQuarrie, "Quantum Chemistry," pp. 1–517. University Science Books, Mill Valley, Calif., 1983.
6. P. A. M. Dirac, "The Principles of Quantum Mechanics," pp. 1–314. Oxford University Press, New York, 1967.
7. A. Abragam, "Principles of Nuclear Magnetism," pp. 1–599. Clarendon Press, Oxford, 1961.
8. P. L. Corio, "Structure of High-Resolution NMR Spectra," pp. 1–548. Academic Press, New York, 1967.
9. K. J. Packer and K. M. Wright, *Mol. Phys.* **50,** 797–813 (1983).
10. O. W. Sørensen, G. W. Eich, M. H. Levitt, G. Bodenhausen, and R. R. Ernst, *Prog. NMR Spectrosc.* **16,** 163–192 (1983).
11. F. J. M. van de Ven and C. W. Hilbers, *J. Magn. Reson.* **54,** 512–520 (1983).
12. J. Jeener, *Adv. Magn. Reson.* **10,** 1–51 (1982).
13. M. H. Levitt, *in* "Pulse Methods in 1D and 2D Liquid-Phase NMR," (W. S. Brey, ed.), pp. 111–147. Academic Press, San Diego, 1988.
14. G. A. Morris and R. Freeman, *J. Am. Chem. Soc.* **101,** 760–762 (1979).
15. D. P. Burum and R. R. Ernst, *J. Magn. Reson.* **39,** 163–168 (1980).

3

EXPERIMENTAL ASPECTS OF NMR SPECTROSCOPY

3.1 NMR Instrumentation

Figure 3.1 illustrates a block diagram of a pulsed Fourier transform NMR spectrometer. The main subsystems of a NMR spectrometer are

1. Superconducting magnet
2. Probe
3. Pulse programmer and rf transmitter
4. Receiver
5. Data acquisition and processing computer

Each of these components is described briefly below. Necessary adjustments of the spectrometer for routine use are described in subsequent sections of this chapter.

A schematic of a superconducting magnet system is illustrated in Fig. 3.2. The magnet consists of a superconducting solenoid and cryoshim coils immersed in liquid helium. The inner dewar is surrounded by a radiation shield and an outer dewar filled with liquid nitrogen. The room-

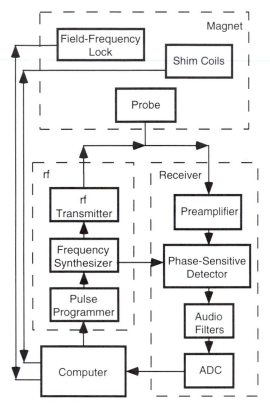

FIGURE 3.1 Block diagram of an NMR spectrometer. The major components, including the magnet, rf electronics, receiver and computer, and important subsystems are illustrated.

temperature bore of the magnet is centered on the z axis of the solenoid and houses the room-temperature shim coils and the probe. NMR spectroscopy requires enormous magnetic fields with extremely high homogeneity over macroscopic volumes. Present-generation magnets have field homogeneities on the order of 1 part in 10^9. As of 1995, the largest commercially available magnets have magnetic field strengths of 17.5 T with proton Larmor frequencies of 750 MHz. In the absence of other effects (such as increased contributions to the linewidth from chemical-shift anisotropy discussed in Section 5.4.4), the resolution in a NMR spectrum increases linearly with B_0 and the sensitivity increases as $B_0^{3/2}$ (*1*). Thus, a 750-MHz spectrometer should have 25% greater resolution and 40% greater sensitivity than a 600-MHz spectrometer. The impetus for continued development of higher-field magnets is therefore obvious.

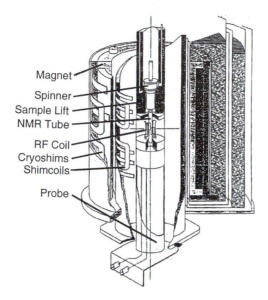

Magnet
Spinner
Sample Lift
NMR Tube
RF Coil
Cryoshims
Shimcoils
Probe

FIGURE 3.2 Cutaway diagram of a superconducting magnet. The probe, sample spinner, and room-temperature shim coils are positioned coaxially in the room-temperature bore of the magnet. The solenoid and cryoshim coils are immersed in liquid helium. The helium dewar is surrounded by a radiation shield and a liquid nitrogen dewar. Diagram courtesy of Bruker Instruments, Inc.

For high-resolution NMR spectroscopy, the temporal stability and spatial homogeneity of the magnetic field is critical. The stability of the static magnetic field is maintained using the field-frequency lock system. The lock circuitry is essentially a specifically tuned (usually to deuterium) NMR spectrometer that operates in parallel to the main spectrometer. The lock system continually measures the resonance frequency of deuterium, or other lock nuclei, in the sample. If the frequency begins to drift, then the electric current in a room-temperature electromagnet housed in the bore of the superconducting magnet is adjusted to return the frequency of the lock nucleus to its nominal value. In most cases, deuterated solvents provide a convenient method for introducing the necessary deuterium nuclei into the sample. The spatial homogeneity of the magnetic field is optimized by adjusting the currents in a set of room-temperature electromagnets called shims. Procedures for shimming are discussed in Section 3.6.2.2.

The probe, illustrated in Fig. 3.3, is positioned coaxially in the room-temperature bore of the magnet. Probe design strongly affects the sensitivity of the spectrometer, the homogeneity of the B_1 rf fields, and the quality

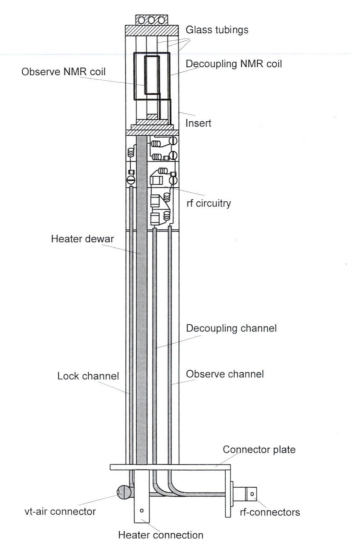

FIGURE 3.3 Probe assembly. Major components of a high-resolution NMR spectroscopy rf probe are illustrated. Diagram courtesy of Bruker Instruments, Inc.

of the solvent suppression. In its simplest manifestation, the probe consists simply of a rf circuit containing one or more wire coils in proximity to the NMR sample. In principle, quadrature detection of the precessing magnetization can be obtained by using two orthogonal coils (one for detecting the x magnetization and the other for detecting the y magnetization). Orthogonal crossed coils tuned to the same frequency are difficult to construct and optimize; modern high-resolution probes utilize a single Helmholtz coil design and obtain quadrature detection as described in Section 3.2.2. In most designs, the same coil is used for applying rf pulses and for detecting subsequent evolution of the magnetization. Depending on the probe, rf circuits may be tuned to a single frequency, may be double-tuned to be simultaneously sensitive to two different nuclei, or may be tunable over a wide frequency range (as in a so-called broadband probe). For example, in a triple-resonance ^{1}H–^{13}C–^{15}N probe, one coil is double tuned to ^{1}H and ^{2}H (for the lock system), and the other coil is double-tuned for ^{13}C and ^{15}N. The characteristics of the probe rf circuit are given by the quality factor, Q, and complex impedance, Z:

$$Q = \omega L/R; \tag{3.1}$$

$$Z = R + i[\omega L - 1/(\omega C)], \tag{3.2}$$

in which ω is the resonance frequency, L is the inductance, R is the resistance, and C is the capacitance of the coil circuit (for simplicity, the effects of the sample magnetism on the impedance have been omitted). The main task for the user is to tune the resonant frequency and match the impedance of the probe prior to use; other operating characteristics of the probe are difficult to alter without major reconstruction or retrofitting (Section 3.6.2.1).

The rf transmitter consists of frequency synthesizers, amplifiers, and associated electronics for producing pulses of highly monochromatic rf electromagnetic radiation with defined phases and amplitudes. Typically, one transmitter subsystem is dedicated to proton frequencies; one or more additional transmitters are used to generate rf frequencies for heteronuclear spectroscopy. The amplitude of the rf field measured in frequency units is given by $\omega_1 = \gamma B_1$; therefore, proportionally higher-power amplifiers are required for low-γ nuclei. Typical proton amplifiers have peak output powers of a few tens of watts; broadband amplifiers for heteronuclear spectroscopy have peak output powers on the order of a few hundred watts. The pulse programmer implements the pulse program necessary to perform a NMR experiment by controlling the timing, durations, amplitudes, and phases of the rf pulses. Radiofrequency pulses with arbitrary phase angles in the rotating frame are generated by applying a phase-shifted rf field that is linearly polarized along a fixed axis in the laboratory reference frame (see [1.18]).

The receiver includes the preamplifier, phase-sensitive detector, and analog-to-digital converter (ADC). The preamplifier provides an initial stage of amplification of the NMR signal prior to further detection and processing. The noise figure of the preamplifier is a critical parameter fixing the signal-to-noise level of the spectrometer because subsequent amplification and detection stages in the receiver unavoidably amplify the preamplifier noise along with the signal. To minimize losses, the preamplifier is located as close to the probe as practical. The phase-sensitive detector achieves quadrature detection of the signal as described in Section 3.2.2. The detector also includes audiofilters designed to restrict the frequency bandwidth of the spectrometer. As discussed in Section 3.2.1, the filters reduce the amount of noise power aliased into the spectrum. Unavoidably, the intensity of signals with frequencies near the cutoff of the filters will be attenuated. In addition, the time constants of the audiofilters are one of the significant sources of phase errors in NMR spectroscopy (2). The ADCs convert the amplified analog signal to digital form for subsequent digital processing. Current-generation NMR spectrometers use 16-bit digitizers as a compromise between conversion speed and dynamic range. A 16-bit digitizer can represent numbers between -2^8 to 2^8-1 ($-32,768$ to $32,767$). Clearly, the magnitude of the analog signal must not exceed the dynamic range of the ADC (or of earlier amplification stages). Similarly, if the magnitude of the analog signal is too small (less than approximately 0.5 bit), then the analog signal rarely registers on the ADC. In this case, extremely long acquisitions will be required to detect the signal and the results will contain distortions from digital quantization noise (i.e., the signal will be observed to take on only a limited number of digital values).

The data acquisition and processing computer, which may consist of a system of multiple computers, controls the operation of the various spectrometer systems. In particular, the data acquisition computer must implement a pulse programming language to permit the user to control the pulse programmer. The processing computer must permit digital signal processing of the recorded time-domain signal to produce the frequency-domain spectrum.

3.2 Data Acquisition

In modern pulsed Fourier transform NMR spectrometers, transverse magnetization is generated by a series of one or more rf pulses. The evolution in time of the magnetization generates a time-varying current in the probe coil. The current is amplified and digitized by the receiver and recorded by the NMR spectrometer. The resulting current vs time signal is called an interferogram or free-induction decay. The term free-induction decay (FID) refers specifically to the signal recorded during the acquisition period; the term interferogram may refer either to the FID or to the signals detected indirectly during evolution periods of multidimensional NMR experiments. The digitized time-domain signal is (generally) Fourier-transformed to generate the frequency-domain NMR spectrum. As discussed in the following sections, representation of a continuous time-varying signal by a discretely sampled, digitized sequence has profound consequences for NMR spectroscopy. Most of the considerations discussed below for acquisition and data processing of the observable magnetization signal apply equally well to acquisition and processing of the signals recorded indirectly during the evolution periods of multidimensional NMR experiments. Issues particularly important for multidimensional NMR spectroscopy are discussed in Chapter 4.

3.2.1 SAMPLING

The continuous NMR signal, $s(t)$, is sampled at evenly spaced time intervals and is represented as $s(k\Delta t)$ for $k = 0, 1, ...$, in which the sampling interval is Δt. The *Nyquist frequency*

$$f_n = 1/(2\Delta t) \qquad\qquad [3.3]$$

defines the highest-frequency sinusoidal signal that is sampled at least twice per period if the sampling rate is Δt^{-1}. The Nyquist frequency plays a central role in digital signal processing applications, including NMR spectroscopy, because of the sampling theorem (*3*):

> If a continuous function in time, $s(t)$, is bandwidth-limited to frequencies smaller in magnitude than some value f_c, then the continuous function is completely determined by the discretely sampled sequence, $s(k\Delta t)$, provided that the sampling interval Δt is such that $f_n \geq f_c$.

The sampling theorem requires that the sampling interval be $\Delta t \leq 1/(2f_c)$ or that the sampling rate be greater than or equal to $2f_c$. If the conditions of the sampling theorem are met, then the continuous function is given identically by

$$s(t) = \sum_{k=-\infty}^{\infty} s(k\Delta t)\, \text{sinc}[2\pi f_n(t - k\Delta t)], \qquad\qquad [3.4]$$

with $\text{sinc}(x) = \sin(x)/x$.

If a signal is recorded with a sampling interval Δt, then the frequency range accurately represented is given by

$$-f_n \leq \nu \leq f_n. \qquad\qquad [3.5]$$

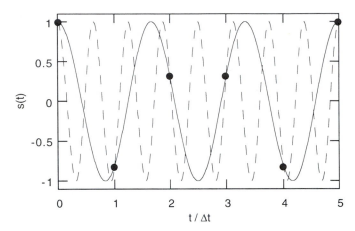

FIGURE 3.4 The Nyquist theorem. Sine waves with frequencies ν (solid line) and $\nu + f_n$ (dashed line) are illustrated. The two sine waves are sampled digitally at the Nyquist frequency Δt. The digital representations of the two sine waves are identical (solid dots). As a consequence, the two sine waves appear to have the same frequency in the digitally sampled data, and the high-frequency component is aliased to the lower frequency.

The total frequency interval is termed the spectral width, SW, and is given by

$$SW = 2f_n = 1/\Delta t. \qquad [3.6]$$

If the signal is not bandwidth-limited, then the signal components with frequencies $|\nu| > f_n$ appear artifactually within the frequency range $-f_n \le \nu \le f_n$. The spurious representation of frequencies greater than the Nyquist frequency is called folding or aliasing. As discussed in Section 4.3.4.3, conventional NMR usage ascribes slightly different meanings to the two terms. As a result of the sampling theorem, the frequency spectrum represented by the signal sequence must be periodic with a period equal to $2f_n = 1/\Delta t$. Thus, a frequency, $\nu_0 > f_n$ appears in the spectrum of a complex data sequence at an aliased frequency, ν_a, given by

$$\nu_0 = 2m f_n + \nu_a, \qquad [3.7]$$

in which m is an integer. Equation [3.7] indicates that frequencies greater than f_n (or less than $-f_n$) are "wrapped around" and appear at the other edge of the spectrum. By way of illustration, Fig. 3.4 shows two cosine waves with frequency ν_0 and $\nu_1 = \nu_0 + f_n$. The discretely sampled points are identical for each sine wave; thus, both signals will be represented identically in the sampled data, and the frequency ν_1 will be aliased to the frequency ν_0. In general, aliased or folded peaks have systematically altered phases because the frequency-dependent phase error is a function of ν_0 but the phase correction applied is a function of ν_a (Section 3.3.2.3). This property can be used to identify aliased peaks. Since ν_a depends on the spectral width, folded or aliased peaks will change their apparent positions in the spectrum if the spectral width is changed. Folding and aliasing is used to advantage in multidimensional NMR spectroscopy to minimize the spectral width in the indirectly detected dimensions (Section 4.3.4.3).

At first glance, the sampling theorem would appear to present a fatal flaw for Fourier transform NMR. Since the noise in the continuous signal would be expected to be nearly white (i.e., to have an infinite bandwidth), an infinite amount of noise power would be aliased into the frequency-domain NMR spectrum. To avoid this catastrophe, the receivers in NMR spectrometers incorporate analog or digital filters to limit the bandwidth of the signal to frequencies $-f_w \le \nu \le f_w$, in which $f_w > f_n$ is the cutoff frequency of the filters. Filters have two deleterious effects on the NMR spectrum: (1) the transient response of the filter to the incoming signal distorts the initial points of the FID and (2) the phase spectrum of the filter retards the phase evolution of the resonance signals and results in frequency-dependent phase errors in the NMR spectrum (see Section 3.3.2.3) (2). The Hahn-echo technique alleviates both of these effects (Section 3.6.4.2)

3.2.2 QUADRATURE DETECTION

The frequencies of magnetic resonance signals in NMR spectroscopy are measured as the offset from a rf reference frequency. Offset frequencies can be positive (resonance frequency greater than the reference) or negative (resonance frequency less than the reference). Characterization of a sinusoidal signal requires that both the sign and absolute magnitude of the offset frequency be determined. A single detector measures the trigonometric projection of the harmonic signal onto a reference axis. Thus, a single detector might measure the cosinusoidally varying component of the signal. The sign of the frequency cannot be determined from such a data sequence. As is well known, both the cosine and sine components of a harmonic signal must be recorded in order to determine the sign of the frequency. Sampling a signal in a manner such that both the sine and cosine components are recorded is known as quadrature detection.

In the earliest days of Fourier transform NMR spectroscopy, single-channel detection was the norm and the problem of determination of the

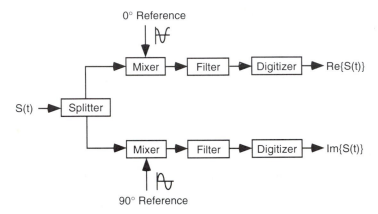

FIGURE 3.5 Experimental scheme for quadrature detection. The incoming signal recorded by the probe and preamplifier is split into two parallel channels. The signal in each channel is mixed with a reference signal, passed through a low-pass audiofilter, and digitized. Real (cosine-modulated) and imaginary (sine-modulated) components of the signal are obtained by shifting the relative phase of the reference signals by 90°.

sign of the offset was solved by placing the rf reference frequency at one edge of the frequency spectrum. In this case, all the resonance offset frequencies have the same sign so that quadrature detection is unnecessary. Almost without exception, modern NMR spectrometers record the signal in quadrature during acquisition of the FID; consequently, the rf reference can be set in the middle of the frequency spectrum. The latter approach offers some significant experimental advantages: (1) the frequency range that must be excited by the rf pulses is reduced by half, which reduces rf transmitter power requirements; (2) the required sampling rate is halved, which simplifies data acquisition hardware; and (3) aliasing of noise into the spectrum is minimized.

As illustrated in Fig. 3.5, quadrature detection during acquisition is accomplished by dividing the signal produced in a single coil into two channels. The high-frequency (MHz) signals in the two channels are mixed with rf reference frequencies to generate audiofrequency (kHz) signals. The two rf reference frequencies are 90° out of phase; therefore, the output of one channel consists of a cosine modulated signal at the frequency $\omega_0 - \omega_{ref}$ and the other channel consists of a sine-modulated signal at the same frequency. The two channels constitute the quadrature pair for frequency discrimination. If the signal produced at the output of the probe and preamplifier is sinusoidally modulated as $\cos \omega_0 t$, the detection process can be represented by

$$\cos(\omega_0 t) \xrightarrow{\text{splitter}} \cos(\omega_0 t) - i \cos(\omega_0 t)$$

$$\xrightarrow{\text{mixers}} \cos(\omega_0 t) \cos(\omega_{\text{ref}} t) - i \cos(\omega_0 t) \sin(\omega_{\text{ref}} t)$$

$$= \frac{1}{2} \cos\left[\left(\omega_0 + \omega_{\text{ref}}\right) t\right] + \frac{1}{2} \cos\left[\left(\omega_0 - \omega_{\text{ref}}\right) t\right]$$

$$- \frac{1}{2} i \sin\left[\left(\omega_0 + \omega_{\text{ref}}\right) t\right] + \frac{1}{2} i \sin\left[\left(\omega_0 - \omega_{\text{ref}}\right) t\right]$$

$$\xrightarrow{\text{audiofilters}} \frac{1}{2} \cos\left[\left(\omega_0 - \omega_{\text{ref}}\right) t\right] + \frac{1}{2} i \sin\left[\left(\omega_0 - \omega_{\text{ref}}\right) t\right]$$

$$= \frac{1}{2} \exp\left[i\left(\omega_0 - \omega_{\text{ref}}\right) t\right], \tag{3.8}$$

in which $i = \sqrt{-1}$ is used as a mathematical mechanism to distinguish between the signals in the two detection channels. The signal in the first (real) channel is modulated as $\cos\Omega t = [\exp(i\Omega t) + \exp(-i\Omega t)]/2$; therefore, the frequency-domain spectrum will contain two signals with positive amplitudes at frequencies $+\Omega$ and $-\Omega$. The signal in the second (imaginary) channel is modulated as $\sin\Omega t = [\exp(i\Omega t) - \exp(-i\Omega t)]/(2i)$; therefore, the frequency-domain spectrum will consist of a positive amplitude signal at a frequency of $+\Omega$, and a negative-amplitude signal at $-\Omega$. Combining the frequency-domain signals from the two channels as shown by [3.8] cancels the signals at a frequency of $-\Omega$ and yields a final frequency-domain spectrum containing a single signal with a frequency of $+\Omega$. The signals present at each step of the detection process are illustrated in Fig. 3.6. If the sensitivity of the two quadrature detection channels are not identical, the signal at $-\Omega$ will not be identically nulled. The final frequency-domain spectrum will contain a small signal at a frequency of $-\Omega$ that is called a quadrature image. CYCLOPS phase cycling frequently is used to reduce quadrature images (Section 4.3.2.3). Techniques for quadrature detection during the evolution periods of multidimensional NMR experiments are discussed in Section 4.3.4.

3.3 Data Processing

The representation of the NMR signal as a discrete sampling sequence in digital form means that powerful numerical digital signal processing

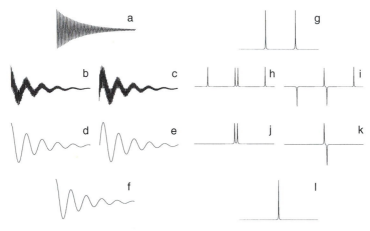

FIGURE 3.6 Quadrature detection. (a) The FID sampled by a single coil in the probe generates a signal modulated as $\cos(\omega_0 t)$ that yields (g) a frequency-domain spectrum consisting of signals at $\pm\omega_0$. To obtain quadrature detection, the signal is split into two channels. (b) The first channel is mixed with a reference signal modulated as $\cos(\omega_{ref} t)$ to generate a new FID. (h) The frequency-domain spectrum consists of signals at frequencies $\pm(\omega_0 - \omega_{ref})$ and $\pm(\omega_0 + \omega_{ref})$. (c) The second channel is mixed with a reference signal modulated as $\sin(\omega_{ref} t)$ to generate a new FID. (i) The frequency-domain spectrum consists of signals at frequencies $\pm(\omega_0 - \omega_{ref})$ and $\pm(\omega_0 + \omega_{ref})$; however, because the sine function is odd, the symmetrical signals are inverted relative to each other. (d,e) The filters remove the high-frequency components of the two signal channels. (j,k) The resulting frequency-domain spectrum contains only frequencies $\pm(\omega_0 - \omega_{ref})$. (f) The two channels are combined to yield a single complex data set. (l) The frequency-domain spectrum, obtained by summing (j) and (k) or by transforming the complex data set (f), contains a single-resonance signal at a frequency $\Omega = \omega_0 - \omega_{ref}$. For clarity, all sine-modulated signals have been phase-shifted by 90° in the frequency domain, which is equivalent to multiplication by i when forming the complex signal $s(t) = s_x(t) + is_y(t)$, in which $s_x(t)$ and $is_y(t)$ are the outputs of the two quadrature channels.

techniques can be used to extract the information content of the signal. The most common processing approach is to convert the time-domain signal into a frequency-domain spectrum by applying a Fourier transform. Various processing algorithms can be applied prior to or after the Fourier transformation to optimize the resulting spectrum. In addition, alternative techniques for spectral analysis, generally first applied in electronic or optical signal processing fields, are being applied increasingly to NMR spectroscopy to obviate the drawbacks to Fourier transformation (Section 3.3.4).

3.3.1 FOURIER TRANSFORMATION

The Fourier transformation defines a relationship between one function in the time domain and another function in the frequency domain (*4*):

$$S(\omega) = \mathcal{F}\{s(t)\} = \int_{-\infty}^{\infty} s(t)e^{-i\omega t}\, dt\,;$$

$$S(\nu) = \mathcal{F}\{s(t)\} = \int_{-\infty}^{\infty} s(t)e^{-i2\pi\nu t}\, dt\,,$$

[3.9]

in which $\omega = 2\pi\nu$. The two functions $s(t)$ and $S(\omega)$ [or $s(t)$ and $S(\nu)$] are said to form a Fourier transform pair. The inverse Fourier transformations are defined by

$$s(t) = \mathcal{F}^{-1}\{S(\omega)\} = \frac{1}{2\pi}\int_{-\infty}^{\infty} S(\omega)e^{i\omega t}\, d\omega\,;$$

$$s(t) = \mathcal{F}^{-1}\{S(\nu)\} = \int_{-\infty}^{\infty} S(\nu)e^{i2\pi\nu t}\, d\nu\,.$$

[3.10]

Fourier transformation and inverse Fourier transformation are linear operations and satisfy the relationships

$$\mathcal{F}\{c\, s(t)\} = c\, \mathcal{F}\{s(t)\};$$

[3.11]

$$\mathcal{F}\{s(t) + r(t)\} = \mathcal{F}\{s(t)\} + \mathcal{F}\{r(t)\},$$

[3.12]

in which c is a complex constant.

For completeness, some important theorems concerning Fourier transformations are listed below; proofs of these theorems can be found in standard texts (*4*):

1. Similarity:

$$\mathcal{F}\{s(at)\} = \frac{1}{|a|}\, S(\omega/a) = \frac{1}{|a|}\, S(\nu/a).$$

[3.13]

2. Time shifting:

$$\mathcal{F}\{s(t - \tau)\} = e^{-i\omega\tau}\, S(\omega) = e^{-i2\pi\nu\tau}\, S(\nu).$$

[3.14]

3. Frequency shifting:

$$\mathcal{F}\{s(t)e^{-i\omega_0\tau}\} = S(\omega - \omega_0);$$

$$\mathcal{F}\{s(t)e^{-i2\pi\nu_0\tau}\} = S(\nu - \nu_0).$$

[3.15]

4. Derivative theorem:

$$\mathscr{F}\left\{\frac{d^k}{dt^k} s(t)\right\} = (i\omega)^k S(\omega) = (i2\pi\nu)^k S(\nu). \qquad [3.16]$$

5. Convolution: If the *convolution integral* of two functions $r(t)$ and $s(t)$ is defined as

$$r(t)*s(t) = \int_{-\infty}^{\infty} r(\tau)s(t-\tau)\, d\tau, \qquad [3.17]$$

then

$$\mathscr{F}\{r(t)*s(t)\} = R(\omega)\, S(\omega) = R(\nu)\, S(\nu) \qquad [3.18]$$

6. Correlation: If the *correlation integral* of two functions $r(t)$ and $s(t)$ is defined as

$$Corr[r(t), s(t)] = \int_{-\infty}^{\infty} r(t+\tau)s(\tau)\, d\tau, \qquad [3.19]$$

then

$$\mathscr{F}\{Corr[r(t),s(t)]\} = R(\omega)\, S^*(\omega) = R(\nu)\, S^*(\nu), \qquad [3.20]$$

in which $S^*(\omega)$ and $S^*(\nu)$ are the complex conjugates of $S(\omega)$ and $S(\nu)$, respectively.

7. Parseval's theorem:

$$\int_{-\infty}^{\infty} |s(t)|^2\, dt = \int_{-\infty}^{\infty} |S(\omega)|^2\, d\omega = \int_{-\infty}^{\infty} |S(\nu)|^2\, d\nu. \qquad [3.21]$$

These theorems have important practical consequences for NMR spectroscopy. The similarity theorem demonstrates that broadening of a function in one dimension results in narrowing of the function in the other dimension. The time-shifting theorem demonstrates that delaying acquisition (intentionally or due to instrumental delays) in the time domain results in a frequency-dependent phase shift in the frequency domain. The frequency-shifting theorem permits the apparent frequencies in the frequency domain to be shifted after acquisition. The convolution and correlation theorems provide efficient means of calculating the convolution and correlation of two functions. In most cases, it is more efficient to Fourier-transform both functions, multiply their transforms, and inverse-Fourier-transform the result to obtain the convolution or correlation than by direct integration. As discussed below, apodization of the FID in the time domain is performed to convolute the signal in the frequency domain with a more desirable lineshape function. Parseval's theorem demonstrates that the signal energy is identical in the two domains and implies that the information content of the signal is identical in both time and frequency domains.

The most important operation for pulsed Fourier transform NMR spectroscopy in liquids is the Fourier transform of the damped oscillator with frequency ω_0 and decay constant λ_0:

$$s(t) = \exp[i\omega_0 t - \lambda_0 t] \qquad [3.22]$$

for $t \geq 0$; $s(t) = 0$ for $t < 0$. The Fourier transform of $s(t)$ is

$$S(\omega) = \int_0^\infty \exp\{[i(\omega_0 - \omega) - \lambda_0]t\}\, dt$$

$$= \frac{\exp\{[i(\omega_0 - \omega) - \lambda_0]t\}}{i(\omega_0 - \omega) - \lambda_0}\bigg|_0^\infty$$

$$= \frac{-1}{i(\omega_0 - \omega) - \lambda_0}$$

$$= \frac{-1}{i(\omega_0 - \omega) - \lambda_0} \times \frac{-i(\omega_0 - \omega) - \lambda_0}{-i(\omega_0 - \omega) - \lambda_0}$$

$$= \frac{i(\omega_0 - \omega) + \lambda_0}{(\omega_0 - \omega)^2 + \lambda_0^2}$$

$$= A(\omega) + iD(\omega), \qquad [3.23]$$

in which the absorption, $A(\omega)$, and dispersion, $D(\omega)$, Lorentzian lineshapes can be expressed as

$$A(\omega) = \frac{\lambda_0}{\lambda_0^2 + (\omega_0 - \omega)^2}; \qquad [3.24]$$

$$D(\omega) = \frac{(\omega_0 - \omega)}{\lambda_0^2 + (\omega_0 - \omega)^2}. \qquad [3.25]$$

The Lorentzian lineshapes are illustrated in Fig. 3.7. The linewidth of the absorptive Lorentzian is defined as the full-width at half-height (FWHH) and is given by $\Delta\omega_{FWHH} = 2\lambda_0$ or $\Delta\nu_{FWHH} = \lambda_0/\pi$. The maximum and minimum cusps of the dispersive lineshape are separated by exactly the absorptive linewidth. Note that for large frequency offsets, the decay of the absorptive Lorentzian lineshape is proportional to $1/(\omega_0 - \omega)^2$, but the decay of the dispersive Lorentzian lineshape is proportional to $1/(\omega_0 - \omega)$. Accordingly, absorptive phase lineshapes yield much more

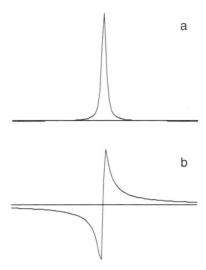

FIGURE 3.7 (a) Absorptive and (b) dispersive Lorentzian lineshapes. The Fourier transform of a damped sinusoid generates a frequency-domain signal with real and imaginary components described by the absorptive and dispersive Lorentzian functions, respectively.

highly resolved NMR spectra and are greatly preferred to dispersive line-shapes.

Since the FID is sampled digitally, the experimental frequency-domain spectrum is calculated using the discrete Fourier transform

$$S(\nu) = S[k/(N\Delta t)] = \mathcal{F}\{s(j\Delta t)\} = \sum_{j=0}^{N-1} s(j\Delta t)e^{-i2\pi jk/N}, \qquad [3.26]$$

in which N is the number of (complex) data points, Δt is the sampling interval, and $k = -N/2, ..., 0, ..., N/2$. The inverse transform is given by

$$s(j\Delta t) = \mathcal{F}^{-1}\{S[k/(N\Delta t)]\} = \frac{1}{N}\sum_{k=0}^{N-1} S[k/(N\Delta t)]e^{i2\pi jk/N}. \qquad [3.27]$$

The frequency range represented by the Fourier transformed signal is $-1/(2\Delta t) \le \nu \le 1/(2\Delta t)$ in discrete steps of $\Delta\nu = 1/(N\Delta t)$. In terms of the Nyquist frequency, $-f_n \le \nu \le f_n$. Equation [3.26] shows that the discrete Fourier transform of the N input signal points yield $N + 1$ frequency-domain data points. In fact, $S(f_n) = S(-f_n)$ so that only N unique points are obtained in the frequency-domain function. Most Fourier transformation algorithms provide as output the N points for $k = -N/2, ...,$

$N/2 - 1$; i.e., the point $S(f_n)$ is not returned. Consequently, the zero frequency point in the frequency-domain spectrum is not $k = N/2$ but rather $k = N/2 + 1$.

If $s(j\Delta t)$ is a real function, then

$$S[(N - k)/(N\Delta t)] = \sum_{j=0}^{N-1} s(j\Delta t)e^{-i2\pi j(N-k)/N}$$

$$= \sum_{j=0}^{N-1} s(j\Delta t)e^{-i2\pi j + i2\pi jk/N}$$

$$= \sum_{j=0}^{N-1} s(j\Delta t)e^{i2\pi jk/N}$$

$$= S^*[k/(N\Delta t)]. \qquad [3.28]$$

Equation [3.28] demonstrates that unique values of $S[k/(N\Delta t)]$ are obtained only for $k = 0, ..., N/2 - 1$. Thus, for a N point real time-domain signal, a unique $N/2$ point complex frequency-domain spectrum is obtained. Incidentally, $S(0) = S^*(1/\Delta t)$ and is consequently a real number.

The discrete Fourier transform is never calculated numerically by using [3.26]. Direct calculation of the Fourier transformation is an order N^2 process, which means that the computational burden increases as the square of the number of data points. Instead, the discrete Fourier transformation is calculated using the fast Fourier transformation (FFT) algorithm, which is an order $N \log_2 N$ process. The time savings afforded by the FFT algorithm are enormous. For a data sequence of 256 complex points, the FFT algorithm is on the order of 32 times more rapid; for a data sequence of 4096 complex points, the FFT is on the order of 300 times more efficient. From the standpoint of the spectroscopist, the use of the FFT algorithm has one important consequence—the number of data points N must be an integral power of 2, i.e., $N = 2^m$ with m an integer. In the acquisition dimension, the number of data points acquired is invariably a power of 2. In indirectly detected evolution periods, constraints on the total acquisition time may make acquiring the appropriate number of data points impractical. If the acquired number of data points is not a power of 2, then the data sequence must be extended to a power of 2 by zero-filling (Section 3.3.2.1) or by linear prediction (Section 3.3.4).

3.3.2 DATA MANIPULATIONS

Direct Fourier transformation of a recorded NMR signal rarely yields an optimal frequency-domain spectrum. Instead, a number of digital signal processing techniques are applied prior to (and after) Fourier transformation in order to maximize the information available from the spectrum.

3.3.2.1 Zero-Filling   Zero-filling or zero-padding is the process of appending a sequence of zeros to a data sequence prior to Fourier transformation. For example, as described in Section 3.3.1, FFT algorithms require that the number of data points, N, be equal to an integral power of 2. If $2^{m-1} < N \le 2^m$ for an integer m, then, prior to Fourier transformation, zero-filling is used to generate a new data sequence of 2^m points in which all points greater than N have the value zero.

NMR data obeys the *causality* principle because $s(t) = 0$ for $t < 0$; that is, the signal does not precede its cause (viz., the pulse sequence). Somewhat surprisingly, as a consequence of causality, the real and imaginary components of the complex frequency spectrum have a deterministic relationship embodied in the Kramers–Kronig relations (5):

$$\text{Re}\{S(\omega)\} = \frac{1}{\pi} \int_{-\infty}^{\infty} \frac{\text{Im}\{S(\omega')\}}{\omega - \omega'} \, d\omega'$$

$$\text{Im}\{S(\omega)\} = -\frac{1}{\pi} \int_{-\infty}^{\infty} \frac{\text{Re}\{S(\omega')\}}{\omega - \omega'} \, d\omega'$$

[3.29]

where Re and Im denote real and imaginary.

The mathematical operation indicated is called the Hilbert transform and permits the complex spectrum to be reconstructed given only the real component. The Hilbert transform finds frequent application in NMR spectroscopy. In many cases, particularly in multidimensional spectroscopy, the imaginary portion of the spectrum is discarded to reduce data storage requirements. Subsequently, the imaginary component of the spectrum can be regenerated by using [3.29]. The resulting complex spectrum can be phased normally (Section 3.3.2.3).

However, as noted by Bartholdi and Ernst, the Kramers–Kronig relations do not hold for discretely sampled NMR data unless the data sequence is extended by a factor of 2 by zero-filling, because the periodicity in the signal implicit in the discrete Fourier transform renders the real and imaginary components of the spectrum independent (5). Thus, if $2^{m-1} < N \le 2^m$, then real improvement in the information content of an NMR spectrum is obtained by zero-filling to obtain a sequence of 2^{m+1} data points. Additional zero-filling results only in cosmetic interpolation between data points in the frequency domain; no additional information is obtained.

3.3.2.2 Apodization   Direct Fourier transformation of an interferogram rarely yields a spectrum that is satisfactory in all respects. Most commonly, the spectrum will exhibit a number of shortcomings: truncation artifacts, low signal-to-noise ratios, limited resolution, or undesirable peak shapes. The properties of the spectrum can be improved by convoluting the spectrum with a more satisfactory lineshape function, $H(\omega)$:

$$S'(\omega) = H(\omega) * S(\omega). \qquad [3.30]$$

Because convolution in the frequency domain is equivalent to multiplication in the time domain, common practice is to multiply the interferogram prior to Fourier transformation by the time-domain filter function, $h(t)$, that represents the Fourier transform of the desired frequency-domain lineshape function,

$$S'(\omega) = \mathscr{F}\{h(t)s(t)\}. \qquad [3.31]$$

This process is variously termed *windowing*, *apodization*, or *filtering* in the time domain (*3,6*).

The digital signal processing literature contains a wealth of theoretical and empirical studies of apodization; nonetheless, relatively simple approaches have proved of greatest value in NMR spectroscopy. Of the theoretical results, only two will be mentioned:

1. Reduction of truncation artifacts requires that the time-domain signal be smoothly reduced to zero. The resulting frequency-domain lineshape is thereby broadened. The minimum truncation ripple for a given degree of broadening is given by the Dolph–Chebyshev window:

$$h(t) = \mathscr{F}^{-1} \left\{ \frac{\cos[2(N-1)\cos^{-1}\{z_0\cos(\omega\,\Delta t/2)\}]}{\cosh[2(N-1)\cosh^{-1}(z_0)]} \right\}, \qquad [3.32]$$

   in which N is the number of sample points, Δt is the sampling period,

$$z_0 = [\cos(\delta\Delta t/4)]^{-1}, \qquad [3.33]$$

   and δ is the broadening parameter measured in radians per second. The Dolph–Chebyshev window is not normally used because of the complexity of [3.32]; however, it serves as a benchmark for evaluating the efficacy of other filter functions.

2. Maximum signal-to-noise ratio is obtained in a spectrum if a matched-filter function is applied prior to Fourier transformation. The matched filter $h(t)$ is equal to the envelope function of the signal, $s_e(t)$. The envelope function is the function describing the decay of the signal (stripped of its harmonic content).

The acquisition time for the interferogram in NMR spectroscopy is limited to times $t \leq t_{max}$. Since the Fourier transformation algorithm assumes that data extends to $t = \infty$, the input signal for Fourier transformation can be represented as the product of the signal (extending to $t = \infty$) and the rectangle function:

$$s'(t) = s(t)r(t);$$

$$r(t) = \begin{cases} 1 & \text{for } 0 \leq t \leq t_{max} \\ 0 & \text{for } t > t_{max} \end{cases}. \qquad [3.34]$$

The resulting frequency spectrum is given by

$$S'(\omega) = \mathcal{F}\{s(\tau) * r(\tau)\} = S(\omega) * \text{sinc}(t_{max}\omega), \qquad [3.35]$$

in which $\text{sinc}(x) = \sin(x)/x$. As shown in Fig. 3.8, convolution of $S(\omega)$ with the sinc function produces severe oscillating *truncation artifacts*. The truncation artifacts can be reduced by apodization with a filter function that reduces the amplitude of the signal smoothly to zero at t_{max}. Figure 3.8 shows the lineshapes obtained for the cosine, Hamming and Kaiser filter functions, respectively:

$$h(t) = \cos(\pi t/2t_{max}); \qquad [3.36]$$

$$h(t) = 0.54 + 0.46 \cos(\pi t/t_{max}); \qquad [3.37]$$

$$h(t) = I_0\{\theta \sqrt{(1 - t^2/t_{max}^2)}\}/I_0\{\theta\}. \qquad [3.38]$$

In [3.38], $I_0\{\theta\}$ is the zero-order modified Bessel function, and θ is a parameter that determines the degree of apodization of the signal. Typical values of θ are π, 1.5π and 2π; increasing values of θ reduce the truncation ripples while increasing the degree of line broadening of the resonance signal. The cosine window perhaps is the window function most frequently applied to truncated NMR signals. Although the Hamming and Kaiser windows are used relatively infrequently, both are expected to more closely approach the performance of the Dolph–Chebyshev window (6). The Kaiser window has the added advantage that θ can be adjusted to optimize the trade off between apodization and line broadening in particular circumstances.

The signals recorded during an ideal solution-state NMR experiment are the sums of exponentially decaying sinusoidal functions. If sufficient data has been recorded to minimize truncation artifacts ($t_{max} > 3T_2$), then optimal sensitivity is obtained using the matched exponential filter function

$$h(t) = \exp(-\lambda t), \qquad [3.39]$$

in which λ is the line-broadening parameter. For matched filtering,

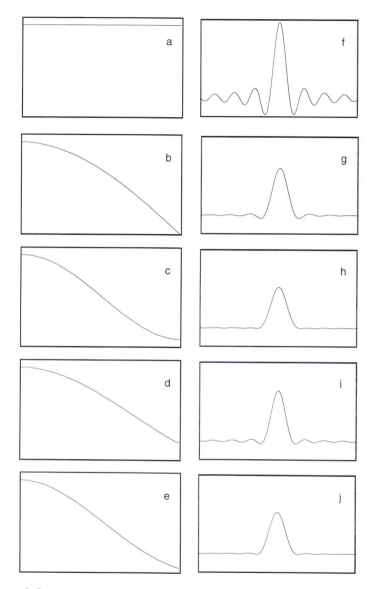

FIGURE 3.8 Window functions for apodization. In each case, the time-domain representation of the window function was zero-filled by a factor of 2 and Fourier-transformed to yield the frequency-domain representation. (a) A uniform square wave input yields a (f) sinc function on output. Other apodization functions illustrated include (b,g) cosine bell, (c,h) Hamming function, (d,i) Kaiser window with $\theta = \pi$, and (e,j) Kaiser window with $\theta = 2\pi$.

$\lambda \approx \lambda_0 = R_2$ (i.e., 2λ is the full-width at half-height of the Lorentzian lineshape measured in radians per second). Matched exponential filtering has the effect of doubling the linewidth in the frequency domain. Matched exponential filtering has two drawbacks. First, since different resonance signals in the spectrum frequently have different linewidths, λ cannot be optimized for all lines simultaneously; thus, $h(t)$ invariably is an approximation to the desired matched filter. Second, the lineshape in the frequency domain is Lorentzian; consequently, the absorption lineshape decays as $1/\omega^2$. The resulting tails degrade resolution in the spectrum and hinder accurate integration of peak intensities. Despite these drawbacks, exponential filtering generally can be recommended for application to the FID recorded during the acquisition dimension of NMR experiments because the signal is rarely truncated severely during acquisition. Exponential filtering is applied to indirectly detected evolution periods very infrequently because the interferograms are almost always severely truncated.

Certain experiments, such as COSY and multiple-quantum spectroscopy, yield antiphase peak shapes in the acquisition dimension. In these cases, the exponential filter is not an appropriate matched filter. Phase-shifted sine-bell functions frequently are applied in these cases:

$$h(t) = \sin\left[\pi\left(\frac{t + t_0}{t_{max} + t_0}\right)\right],$$ [3.40]

in which $\pi t_0 / (t_{max} + t_0)$ is the initial phase of the sine bell.

As noted above, the natural lineshape in solution NMR spectroscopy is Lorentzian. The spectrum can be given a new lineshape by use of the filter function

$$h(t) = s_e'(t)/s_e(t) = s_e'(t)\exp(\lambda t),$$ [3.41]

in which $s_e'(t) = \mathcal{F}^{-1}\{S'(\omega)\}$ and $S'(\omega)$ is the desired lineshape. Lineshape transformations frequently are used to enhance the resolution in a spectrum; however, resolution enhancement emphasizes later portions of the FID. As a consequence, truncation artifacts may become more prominent (unless t_{max} is very large or the filter function is suitably apodized) and the signal-to-noise ratio in the spectrum may be reduced. As a corollary, resolution can be enhanced only if signal has been recorded for long times. If the data is truncated, then little resolution enhancement is possible using digital filtering (but see Section 3.3.4). The Lorentzian-to-Gaussian transformation is obtained using

$$s_e'(t) = \exp\left[-\frac{\lambda_g^2 t^2}{4\ln 2}\right] = \mathcal{F}^{-1}\left\{\frac{\sqrt{4\pi\ln 2}}{\lambda_g}\exp\left[-\frac{\omega^2 \ln 2}{\lambda_g^2}\right]\right\}.$$ [3.42]

The resulting lineshape is Gaussian with a full-width at half-height equal to $\Delta\omega_{FWHH} = 2\lambda_g$ or $\Delta\nu_{FWHH} = \lambda_g/\pi$. The Gaussian lineshape decays exponentially; consequently, the tails do not degrade the resolution as much as the Lorentzian lineshape, and accurate integration of the signal intensity is facilitated. In principle, the lineshape can be arbitrarily narrowed by decreasing λ_g; in practice, $0.5\lambda < \lambda_g < 2.0\lambda$ provide adequate resolution enhancement without degrading signal-to-noise drastically. Maximum resolution enhancement with minimization of truncation artifacts can be obtained by using one of the filter functions recommended for removal of truncation artifacts (i.e., the Kaiser or Hamming functions) for $s'_e(t)$. The main disadvantage to this approach is that signal-to-noise ratios may be severely reduced. Examples of the results of matched filtering and resolution enhancement are given in Fig. 3.9 for a single-resonance signal and in Fig. 3.10 for the one-dimensional ^1H NMR spectrum of ubiquitin. In both examples, the highest signal-to-noise ratios are obtained for exponential matched filtering. The highest degree of resolution enhancement results in the smallest signal-to-noise ratios.

3.3.2.3 Phasing   If the digitized signal is represented as

$$s(k\Delta t) = \exp[(i\omega_0 - \lambda_0)(k\Delta t + \tau) + i\phi], \qquad [3.43]$$

in which τ is a sampling delay (which may arise from instrumental delays or may be intentionally set) and ϕ is the initial phase angle, then the spectrum displays a frequency-dependent phase error. Assuming that the number of data points $N \to \infty$ and $\Delta t \to 0$ (i.e., in the limit that the discrete Fourier transformation maps into the continuous transformation), the transformed spectrum is

$$S(\omega) = \exp[i(\omega_0\tau + \phi) - \lambda_0\tau][A(\omega) + iD(\omega)]$$

$$= \exp[-\lambda_0\tau]\{\cos(\omega_0\tau + \phi)A(\omega) - \sin(\omega_0\tau + \phi)D(\omega)$$

$$+ i[\sin(\omega_0\tau + \phi)A(\omega) + \cos(\omega_0\tau + \phi)D(\omega)]\}. \qquad [3.44]$$

This result is obtained by straightforward application of the time-shifting theorem, [3.14]. The real part of the spectrum is seen to be a mixture of absorptive and dispersive lineshapes, as shown in Fig. 3.11. The factor $\exp[-\lambda_0\tau]$ in [3.44] affects only the intensity of the resonance signal and is not written explicitly in the following. Mathematically, the absorptive and dispersive components of the spectrum can be separated by constructing a new data set by using the following prescription:

$$S'(\omega) = \exp[-i(\omega_0\tau + \phi)]S(\omega) = \cos(\omega_0\tau + \phi)\mathrm{Re}\{S(\omega)\}$$

$$+ \sin(\omega_0\tau + \phi)\mathrm{Im}\{S(\omega)\} - i\sin(\omega_0\tau + \phi)\mathrm{Re}\{S(\omega)\}$$

$$+ i\cos(\omega_0\tau + \phi)\mathrm{Im}\{S(\omega)\} = A(\omega) + iD(\omega). \qquad [3.45]$$

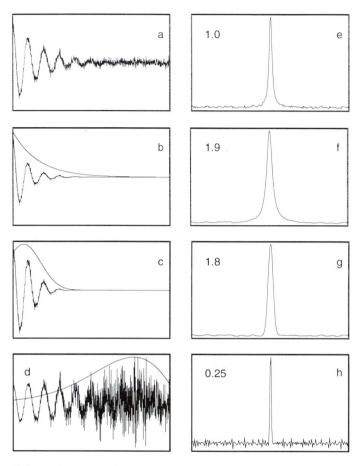

FIGURE 3.9 Digital resolution enhancement. (a) the unapodized FID and its (e) Fourier transform are illustrated. (b) A matched exponential window function and the resulting FID are shown together with (f) the resulting frequency-domain spectrum. (c) A Lorentzian-to-Gaussian transformation and the resulting FID are shown together with (g) the resulting frequency domain spectrum. (d,h) Maximum resolution enhancement is obtained by multiplying the FID with an increasing exponential and apodizing with a Kaiser window function. The signal-to-noise ratio in (e) is arbitrarily assigned a value of unity; relative signal-to-noise ratios for (f,g,h) are shown in the figure.

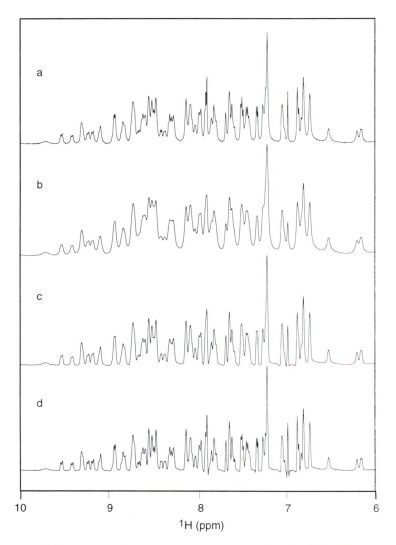

FIGURE 3.10 Digital resolution enhancement of ubiquitin ¹H NMR spectrum. The amide region from 6 to 10 ppm is illustrated for spectra obtained by Fourier transformation of (a) the unapodized FID, (b) an exponential window function, (c) a Lorentzian-to-Gaussian transformation, and (d) maximum resolution enhancement obtained by multiplying the FID with an increasing exponential and apodizing with a Kaiser window function. The window functions are similar to those used in Fig. 3.9. The signal-to-noise ratios for the resonance at 6.52 ppm are (a) 61, (b) 150, (c) 110, and (d) 73.

In practice, [3.45] cannot be used to phase a spectrum containing multiple resonances because ω_0 is different for each resonance. Instead, the phased spectrum is calculated as

$$S'(\omega) = u(\omega) + iv(\omega) \approx A(\omega) + iD(\omega), \qquad [3.46]$$

in which

$$u(\omega) = \cos[\theta(\omega)]\text{Re}\{S(\omega)\} + \sin[\theta(\omega)]\text{Im}\{S(\omega)\};$$
$$v(\omega) = -\sin[\theta(\omega)]\text{Re}\{S(\omega)\} + \cos[\theta(\omega)]\text{Im}\{S(\omega)\}, \qquad [3.47]$$

and $\theta(\omega) = \theta_0 + \theta_1(\omega)$ is determined empirically to minimize the phase error in the spectrum; θ_0 is called the zero-order phase correction and $\theta_1(\omega)$ is called the first-order phase correction. The zero-order phase correction is frequency-independent, while the first-order phase correction is frequency-dependent. In most processing software, $\theta_1(\omega) = \theta_1(\omega - \omega_{\text{pivot}})/SW$, in which the pivot frequency, ω_{pivot}, is selectable. Therefore, the frequency-dependent phase correction is zero at ω_{pivot} and has values of $-\theta_1(\frac{1}{2} + \omega_{\text{pivot}}/SW)$ and $\theta_1(\frac{1}{2} - \omega_{\text{pivot}}/SW)$ at the two edges of the spectrum. On modern NMR spectrometers, phasing can be performed interactively by adjusting θ_0 and $\theta_1(\omega)$ until the lineshapes in the real part of the spectrum are absorptive. An example of the use of zero- and first-order phase corrections is given in Fig. 3.12.

More detailed analyses of the discrete Fourier transform indicate that if $N < \infty$ and $\Delta t > 0$, the baseline of the frequency-domain spectrum displays a nonzero offset and curvature unless the initial signal phase is adjusted to be a multiple of $\pi/2$ and the sampling delay is adjusted such that $\tau = 0$ or $\tau = 1/(2SW)$ (7,8). The distortions in the baseline are particularly serious problems in multidimensional NMR spectroscopy. Fortunately, the receiver reference phase can be easily adjusted on modern NMR spectrometers in order to set the initial signal phase (usually equal to zero). A number of experimental techniques have been developed to ensure that $\tau = 0$ or $\tau = 1/(2SW)$. For example, the Hahn echo sequence can be used to adjust the sampling delay during the acquisition dimension (Section 3.6.4.2). Adjustment of the initial sampling delay for indirectly detected evolution periods in multidimensional NMR experiments must account for phase evolution during the preparation and mixing periods (see Chapter 4). The accrued phase depends on chemical-shift evolution during the evolution period and phase evolution during off-resonance rf pulses within or flanking the evolution period. The pulse sequence element $90°-t_1-90°$ is encountered frequently in homonuclear multidimensional pulse sequences. Utilizing the expression for phase evolution during rf

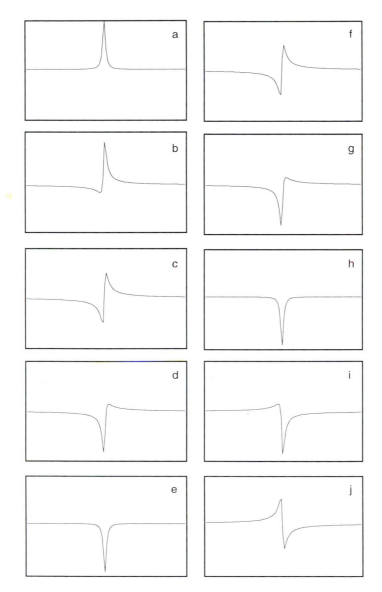

FIGURE 3.11 Phase dependence of lineshapes. (a–e) Real and (f–j) imaginary Lorentzian lineshapes are shown for phases of (a,f) 0°, (b,g) 45°, (c,h) 90°, (d,i) 135°, and (e,j) 180°.

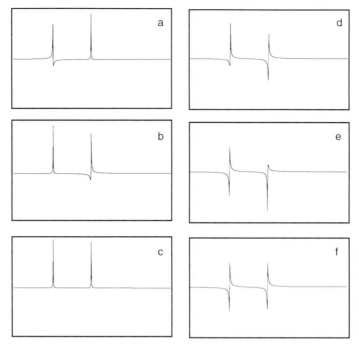

FIGURE 3.12 Phase corrections. (a–c) Real and (d–f) imaginary components of two signals of differing phase are shown. (a,d) The upfield resonance has been phased using a zero-order phase correction; however, the downfield resonance has a phase error. (b,e) The downfield resonance has been phased using a zero-order phase correction; however, the upfield resonance has a phase error. (c,f) Both signals have been phased simultaneously by applying zero- and first-order phase corrections.

pulses given by [3.65], to ensure an initial sampling delay of $\tau = 1/(2\,SW)$, the initial value of t_1 must be set to

$$t_1(0) = 1/(2\,SW) - 4\tau_{90}/\pi, \qquad [3.48]$$

in which τ_{90} is the length of a 90° pulse. If an initial sampling delay of zero is desired, the pulse sequence element $90°\text{-}t_1\text{-}\Delta\text{-}180°\text{-}\Delta\text{-}90°$, in which Δ is a fixed delay, can be used. The 180° pulse refocusses the effects of phase evolution during the flanking 90° pulses; thus

$$t_1(0) = \left\{ \begin{array}{c} 0 \\ 1/(2\,SW) \end{array} \right\} \qquad [3.49]$$

can be chosen as desired. In heteronuclear correlation NMR experiments, the pulse sequence element $90°(S)-t_1/2-180°(I)-t_1/2-90°(S)$ commonly is used to decouple I and S spins during the evolution period. In this case, the initial value of t_1 must be set to

$$t_1(0) = 1/(2\,SW) - 4\tau_{90(S)}/\pi - \tau_{180(I)} \qquad [3.50]$$

to ensure $\tau = 1/(2\,SW)$. Assuming that the initial signal phase and sampling delay have been properly adjusted, the required postacquisition phase corrections are given by

$$\theta_1 = -2\theta_0 = -360° \,\tau SW, \qquad [3.51]$$

in which the pivot is assumed to be set at the downfield edge of the spectrum (7). Equation [3.51] applies both to data sampled as complex sequences and as real sequences (Section 4.3.4). If $\tau = 0$, the initial point in the interferogram must be scaled by a factor of 0.5 prior to Fourier transformation because the discrete Fourier transform is periodic (9). An example of the baseline distortions observed if the signal phase is not correctly adjusted is given in Fig. 3.13.

In some cases, baseline distortions may be present even if sampling delays are properly taken into account. These distortions result from corruption of the first few points in the FID. If the receiver gain is set too high, the magnitude of the analog signal being detected may exceed the dynamic range of the ADC or earlier stages of the signal amplifiers. Invariably, points at the beginning of the FID are affected, and all appear with the same maximum value in the ADC. The FID is then said to be "clipped." Fourier transformation of the FID is essentially the Fourier transformation of the superposition of the uncorrupted FID and a square function. The resulting frequency-domain spectrum exhibits sinc-wiggles or truncation artifacts. The second problem, which is referred to as baseline roll, arises from the transient response of the audiofilters to the incoming signal (2). The digitized signal is the superposition of the uncorrupted FID and a set of points corresponding to the transient response of the filters. In practice only the first few points of the FID are affected by clipping or transient filter response. An error or distortion in the first sampled point of the FID gives rise to a constant baseline offset in the frequency-domain spectrum. Distortions in the second and subsequent points give rise to increasingly severe baseline effects. For example, distortion of the second point causes curvature in the baseline. Distortion of the third point results in a baseline with one node and two antinodes that resemble the superposition of the spectrum and a sine wave. Examples of these baseline distortions are shown in Fig. 3.14. Sizable reduction in baseline roll can be achieved by adjusting the time between the observe

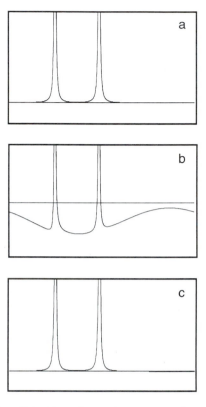

FIGURE 3.13 Baseline distortions from phase corrections. (a) A FID recorded with an initial sampling delay of zero generates a properly phased spectrum without baseline distortions. (b) A FID recorded with an arbitrary nonzero initial sampling delay generates a spectrum with baseline distortions after phase correction. (c) A FID recorded with an initial sampling delay adjusted to one-half the sampling time generates a spectrum without baseline distortions after phase correction.

pulse and the start of sampling so that sampling occurs close to the crossing point of the filter ringing pattern (2). The use of a so-called Hahn echo pulse sequence (Section 3.6.4.2) can alleviate many baseline distortions in ^{1}H-detected NMR spectroscopy. Linear prediction algorithms also can be used to correct the first few points of the FID to eliminate baseline distortions (Section 3.3.4).

3.3.3 SIGNAL-TO-NOISE RATIO

The frequency difference between adjacent points in the frequency-domain spectrum following discrete Fourier transformation is $\Delta\nu =$

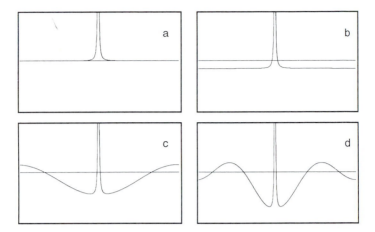

FIGURE 3.14 Baseline distortions from a corrupt FID. (a) An uncorrupted FID generates a spectrum without baseline distortion. Baseline distortions are observed if (b) the first point of the FID, (c) first two points of the FID, and (d) first three points of the FID are corrupted. For this figure, corrupted time-domain data points were set identically to zero.

$1/(N\,\Delta t) = 1/t_{max}$. The digital resolution in the final spectrum depends on the total acquisition time, and not on the sampling interval. Increasing the resolution in a spectrum requires that t_{max} be increased either by recording additional data points or by increasing Δt. Increasing Δt has the effect of reducing the spectral width, which may not be feasible. Increasing t_{max} is productive only if the signals of interest have sufficiently long T_2 values. Once the signals have decayed to zero, increasing t_{max} increases the noise in the spectrum without increasing the resolution between resonance signals.

NMR spectroscopy is an insensitive technique, and optimization of the signal-to-noise ratio has long been of concern. The initial impetus for the development of pulsed Fourier transform NMR spectroscopy was its increased sensitivity. For a simple one-pulse experiment, the sensitivity, defined as the signal-to-noise ratio per unit acquisition time, is given by (6)

$$\mathscr{S} = \frac{\langle sh \rangle}{\langle h^2 \rangle^{1/2}} \left(\frac{t_{max}}{T_c} \right)^{1/2} \frac{1}{\rho}, \qquad [3.52]$$

in which

$$\langle sh \rangle = \frac{s_0}{t_{max}} \int_0^{t_{max}} s_e(t) h(t)\, dt;$$

$$\langle h^2 \rangle = \frac{1}{t_{\max}} \int_0^{t_{\max}} h^2(t)\, dt. \qquad [3.53]$$

In [3.52] T_c is the total time between acquisitions (acquisition time plus the recycle delay), s_0 is the initial value of the signal, $h(t)$ is the apodization function, and ρ is the square root of the noise-power spectral density. As indicated by [3.52], the sensitivity depends on the ratio of the average-weighted signal amplitude and the root-mean-square amplitude of the apodization function. If $h(t)$ is chosen to be equal to $s_e(t)$ (see the discussion of matched filtering in Section 3.3.2.2), then

$$\mathscr{S} = s_0 \langle s_e^2 \rangle^{1/2} \left(\frac{t_{\max}}{T_c} \right)^{1/2} \frac{1}{\rho}. \qquad [3.54]$$

Optimal sensitivity depends, therefore, on the root-mean-square amplitude of the resonance signal. Since $s_0 \propto N_s$, and $\rho \propto \sqrt{N_s}$, in which N_s is the number of transients that are signal-averaged, $\mathscr{S} \propto \sqrt{N_s}$. More detailed analyses of the determinants of s_0 and ρ yield the result that (10)

$$\mathscr{S} \propto NQ\gamma^{5/2}B_0^{3/2}T_2^{1/2}T^{-3/2} \left(\frac{t_{\max}}{T_c} \right)^{1/2}, \qquad [3.55]$$

in which N is the number of nuclear spins, T is the temperature, and Q is the quality factor of the probe coil. Not surprisingly, the greatest sensitivity is obtained for nuclei with large values for γ and long T_2 relaxation times.

3.3.4 ALTERNATIVES TO FOURIER TRANSFORMATION

The Fourier transformation is fast, is numerically stable, and produces phase-sensitive frequency-domain spectra in a convenient representation. Nonetheless, the Fourier transformation is not without disadvantages; principally, for short data records, the resolution in the frequency-domain spectrum is reduced and truncation artifacts can become large (unless strong window functions are applied, which correspondingly reduces the resolution in the spectrum). As discussed in Chapter 4, the time required to acquire a multidimensional NMR data set is proportional to the number of points acquired in the indirectly detected dimensions. Therefore, data records in the indirectly detected dimensions are almost always truncated, and in the case of three- and four-dimensional data sets, severely so. Accordingly, extensive efforts have been made to develop alternative methods of producing frequency-domain spectra from truncated time-domain interferograms that are more satisfactory than Fourier transformation. The various methods proposed include linear prediction (*11*), maximum entropy reconstruction (*12*), maximum likelihood (*13*), and Baysian analysis (*14*); of these, linear prediction and maximum entropy reconstruction are the most frequently utilized. The review by Stephenson provides a detailed introduction to both linear prediction and maximum entropy methods in NMR spectroscopy (*15*).

3.3.4.1 Linear Prediction   Linear prediction algorithms model the time-domain signal (the interferogram) as (*3*)

$$s(k\,\Delta t) = -\sum_{m=1}^{M} a_m s([k-m]\,\Delta t) + \varepsilon_m, \qquad [3.56]$$

in which M is the *prediction order* or the number of signal *poles*, ε_m is the prediction error (distinct from the random noise), a_m is the mth linear prediction coefficient, and $k = 0, \ldots, N - 1$. In essence, the kth data point is modeled as a linear function of the previous M points. Although [3.56] could be postulated as a description of an arbitrary signal, a close connection exists between [3.56] and the damped sinusoidal signal [3.22]. Linear prediction algorithms attempt to find the set of coefficients a_m that minimizes, in the least-squares sense, the prediction error. Once the linear prediction coefficients have been determined, the frequency-domain spectrum can be calculated directly from the prediction coefficients or, more commonly, the linear prediction results can be used to calculate an extension to the interferogram prior to conventional processing by Fourier transformation. Linear prediction algorithms can predict the future behavior of a sinusoid over many periods; in contrast, polynomial expansions frequently fail to extend a sinusoid for more than a period accurately.

Implicit in the formulation of the linear prediction method are a number of important issues:

1. The maximum number of resonance signals that can be modeled is given by the prediction order.
2. Linear prediction algorithms generally require $M \le N/2$.
3. The optimal prediction order is difficult to determine rigorously.
4. Random noise is not incorporated into the linear prediction model. As a consequence, linear prediction methods generally work best for data with relatively high signal-to-noise ratios.
5. The interferogram usually cannot be extended by more than a factor of 2 without severely distorting the signal lineshapes.

6. Two- (and higher)-dimensional data sets are processed by linear-predicting each (t_1, ω_2) interferogram independently. Differences in numerical results from interferogram to interferogram can distort two-dimensional lineshapes.

The most common use for linear prediction methods is to extend the time-domain data for ^{13}C and ^{15}N resonance signals detected during evolution periods of multidimensional NMR experiments. If all ^{1}H dimensions are Fourier-transformed first, then the heteronuclear dimensions generally contain relatively few signals in each interferogram, which simplifies the linear prediction problem.

In some experimental situations, notably constant-time experiments or extremely truncated data, the signal interferogram is undamped by relaxation (limited damping by inhomogeneity broadening can be corrected by multiplication with an increasing exponential). If the phase has been properly adjusted by adjusting the initial value of the evolution period to 0 or $1/(2SW)$ (Section 3.3.2.3), then the complex signal satisfies the relationship $s(-t) = s^*(t)$, in which the asterisk indicates complex conjugation. This relationship can be used to generate a data sequence of length $2N$ from a sequence of length N in which the data points extend from $-N, -N + 1, \ldots, N - 1, N$ (if the initial value of t is 0, the extended sequence contains $2N - 1$ points). The longer data sequence can be used as the input for linear prediction after which the points from $-N$ to -1 are discarded. This technique has been called *mirror-image* linear prediction and frequently allows higher resolution estimates of the frequency-domain spectrum to be obtained (*16*).

At the present time, the algorithms based on single-value decomposition, such as the LPSVD and HSVD algorithms, appear to be the most robust (*17*). An example in which the HSVD algorithm was used to linear-predict the t_1 interferograms in a two-dimensional constant-time ^{1}H–^{13}C HSQC spectrum of ubiquitin (Section 7.1.3.1) is shown in Figs. 3.15 and 3.16.

Linear prediction also can be used to correct the magnitude of incorrectly sampled points (e.g., the initial points of the FID may be corrupted by pulse breakthrough and filter response; see Section 3.3.2.3). For these applications, computationally less demanding algorithms, such as the Burg or Levinson–Durbin methods, are satisfactory (*15,18*).

3.3.4.2 Maximum Entropy Reconstruction   Maximum entropy methods reconstruct the frequency-domain NMR spectrum directly (no subsequent Fourier transformation is necessary) by determining the spectrum $S[k/(N\,\Delta t)]$ for $k = 0, 1, \ldots, N - 1$ that maximizes the entropy function

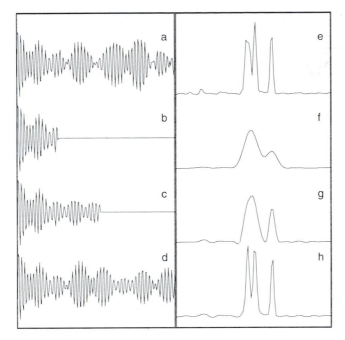

FIGURE 3.15 Linear prediction using the HSVD algorithm. A constant-time ^{1}H–^{13}C HSQC spectrum of ubiquitin was linear predicted in the $t_1(^{13}$C) dimension following conventional processing in the acquisition dimension. The t_1 interferogram at a $F_2(^1$H) shift of 4.52 ppm was used for illustration. Shown are the (a) complete 240-point interferogram, (b) truncated 64-point interferogram, (c) truncated interferogram extended to 128 points by HSVD linear prediction, and (d) truncated interferogram extended to 240 points by mirror-image HSVD linear prediction. The spectra obtained by Fourier transformation of the interferograms (a)–(d) are shown in (e)–(h).

$$\mathscr{S} = \sum_{k=0}^{N-1} S[k/(N\,\Delta t)] \ln S[k/(N\,\Delta t)], \qquad [3.57]$$

subject to the constraint

$$\mathscr{C} = \sum_{j=0}^{M-1} [s(j\,\Delta t) - \hat{s}(j\,\Delta t)]^2/\sigma_j^2 = M, \qquad [3.58]$$

in which M is the number of experimental time-domain points, $s(j\,\Delta t)$ is the jth experimental time-domain point, $\hat{s}(j\,\Delta t) = \mathscr{F}^{-1}\{S[k/(N\,\Delta t)]\}$ is the inverse Fourier transform of the reconstructed spectrum, and σ_j is the

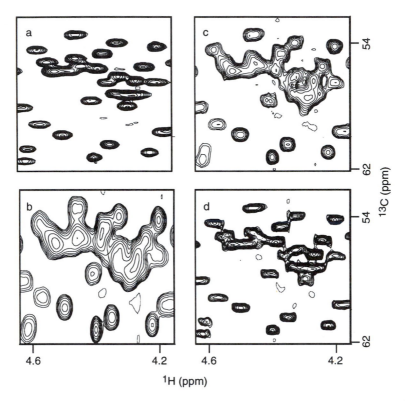

FIGURE 3.16 Linear prediction of a HSQC spectrum. A constant-time ^{1}H–^{13}C HSQC spectrum of ubiquitin was linear predicted in the $t_1(^{13}$C) dimension following conventional processing in the acquisition dimension. Shown are regions of the two-dimensional spectra obtained from (a) 240 t_1 points, (b) 64 t_1 points, (c) 64 t_1 points extended to 128 points by HSVD linear prediction, and (d) 64 t_1 points extended to 240 points by mirror-image HSVD linear prediction.

experimental noise level in the time-domain data. The entropy and constraint equations can be combined into a single objective or cost function, $\mathcal{V}$:

$$\mathcal{V} = \mathcal{S} + \lambda\mathcal{C}, \qquad\qquad [3.59]$$

in which λ is a Lagrange multiplier. The objective function is maximized by iteratively refining an initial guess for $S[k/(N\,\Delta t)]$ using standard numerical algorithms.

In essence, maximum entropy reconstruction amounts to the following prescription: From the (possibly infinite) set of candidate frequency-domain spectra whose time-domain interferograms reproduce the experimental interferogram within experimental uncertainty as described by the constraint function in [3.58], select the single spectrum for which the entropy defined by [3.57] is a maximum. The preceding formulations have a number of important corollaries:

1. Unlike linear prediction algorithms, the functional form of the NMR signals (damped sinusoids) are not assumed a priori by the maximum entropy methods; however, experimental uncertainty is considered explicitly by maximum entropy algorithms.
2. The spectral estimates are constrained to be positive by [3.57]. Negative intensity in NMR spectra is treated by representing the spectrum as

$$S[k/(N\,\Delta t)] = S^+[k/(N\,\Delta t)] - S^-[k/(N\,\Delta t)], \qquad [3.60]$$

   in which $S^+[k/(N\,\Delta t)]$ and $S^-[k/(N\,\Delta t)]$ respectively represent subspectra with positive and negative intensities (*19*).
3. Because M can be less than N, a high-resolution frequency-domain spectrum can be calculated from a limited number of data points.
4. Maximum entropy reconstruction can be performed for multidimensional data in straightforward fashion by generalizing [3.57] and [3.58] to include multiple summations (one summation for each dimension).
5. Justification of the use of [3.57] for NMR spectroscopy is not straightforward. Laue *et al.* discuss the rationale for using [3.59] in spectral reconstruction (*20*).

Maximum entropy reconstruction of the truncated t_1 interferograms in a two-dimensional constant-time ^{1}H–^{13}C HSQC spectrum of ubiquitin (Section 7.1.3.1) is shown in Fig. 3.17.

Examination of Figs. 3.15–3.17 demonstrate that for severely truncated data, both linear prediction and maximum entropy reconstruction can improve the resolution of the frequency-domain spectrum compared with Fourier transformation. In neither case are the resulting spectra as highly resolved and distortion-free as the spectrum obtained by Fourier transformation of the untruncated interferogram.

3.4 Pulse Techniques

3.4.1 OFF-RESONANCE EFFECTS

In practical situations, the nuclei in a molecule will possess a range of chemical shifts. Since a rf pulse can be applied at only one frequency

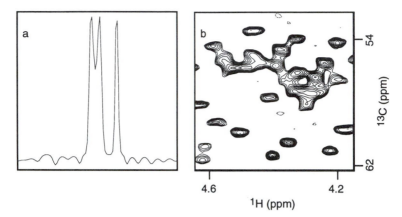

FIGURE 3.17 Maximum entropy reconstruction of a HSQC spectrum. A constant-time ^{1}H–^{13}C HSQC spectrum of ubiquitin was processed by maximum entropy reconstruction in the $t_1(^{13}$C) dimension following conventional processing in the acquisition dimension. Reconstruction was based on 64 t_1 points. Shown are (a) a one-dimensional slice at a F_2 shift of 4.52 ppm and (b) the same region of the two-dimensional spectrum shown in Fig. 3.16.

(the transmitter frequency, ω_{rf}), some nuclei will have resonant frequencies that are close to ω_{rf} whereas other nuclei will have resonant frequencies that are very different from ω_{rf}. Consequently, all nuclei cannot be expected to respond to the effect of a rf pulse in an ideal fashion. Those nuclei on-resonance with the pulse will respond ideally; other nuclei, near resonance with $\omega_0 \approx \omega_{rf}$, will precess around an effective field, B^r, that will be similar to that of B_1. Further and further *off-resonance*, where $\omega_0 \neq \omega_{rf}$ and the magnitude of the offset Ω increases, the effective field B^r will be very different from B_1. In many cases, the offset of some nuclei may be comparable to the strength of the pulse measured in frequency units (γB_1). In these circumstances the effective field is tilted away from the x–y plane toward the z axis (Fig. 1.2).

For the off-resonance case, a pulse of y phase applied to equilibrium z magnetization yields ([1.33])

$$M_x = M_0 \sin \alpha \sin \theta;$$

$$M_y = M_0(1 - \cos \alpha) \sin \theta \cos \theta; \qquad [3.61]$$

$$M_z = M_0(\cos^2\theta + \cos \alpha \sin^2\theta),$$

in which α and θ are defined by [1.23] and [1.21]. In practical terms, two effects must be noted that yield both phase and intensity anomalies for resonance lines that have large offsets from the transmitter frequency. The phase anomalies can be described in terms of a phase shift, β, where

$$\tan \beta = \frac{M_y}{M_x} = \frac{(1 - \cos \alpha) \cos \theta}{\sin \alpha}. \qquad [3.62]$$

Using [1.21], [3.62] can be expressed as

$$\tan \beta = \frac{(1 - \cos \alpha) \sin \theta}{\sin \alpha} \cdot \frac{\Omega}{(-\gamma B_1)}. \qquad [3.63]$$

This equation is a convenient form in which to view phase problems because the dependence on resonance offset is indicated directly. The amplitude of the resonance signal also changes with offset as given by

$$M_{x,y} = \sqrt{(M_x)^2 + (M_y)^2}. \qquad [3.64]$$

The magnetization, phase angle, and effective rotation angle for an off-resonance 90° pulse are illustrated in Fig. 3.18; the resulting lineshapes are shown in Fig. 3.19. As the offset, Ω, increases, the amplitude of the transverse magnetization remains approximately constant, and the phase of the transverse magnetization increases linearly until the offset equals the rf field strength, $\Omega = \gamma B_1$. At larger offsets, the signal amplitude decreases to zero. At the null point, the bulk magnetization vector rotates about the effective field such that it is aligned along the z axis once more at the end of the pulse. At even larger offsets, the signal oscillates from positive to negative and decreases in amplitude.

For offsets $\Omega \leq \gamma B_1$, nearly ideal results are obtained for a 90° pulse if the linear phase dependence is compensated. Expanding [3.63] in Taylor series about $\Omega = 0$ yields

$$\beta = -\Omega/(\gamma B_1) = -\Omega \tau_{90}/(\gamma B_1 \tau_{90}) = -2\tau_{90}\Omega/\pi, \qquad [3.65]$$

in which τ_{90} is the length of the on-resonance 90° pulse. Thus, an off-resonance 90° pulse can be treated as an ideal pulse followed by an evolution period given by

$$\tau = 2\tau_{90}/\pi. \qquad [3.66]$$

In contrast, Fig. 3.18 demonstrates that a 180° inversion pulse has a highly nonideal off-resonance excitation profile. Magnetization will not be fully inverted, and a notable amount of transverse magnetization will be generated. Because nonideal effects of 180° pulses cause significant

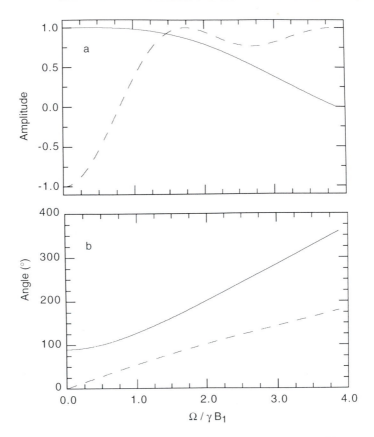

FIGURE 3.18 Off-resonance effects for 90° and 180° pulses. (a) Shown are (—) the magnitude of the transverse magnetization, $M_{x,y} = (M_x^2 + M_y^2)^{1/2}$, following a nominal 90° pulse and (- - -) the M_z magnetization following a nominal 180° pulse. (b) Shown are the (—) effective rotation angle and (- - -) phase of the transverse magnetization following a nominal 90° pulse.

effects in NMR spectra, considerable effort has been devoted to improving the performance of 180° pulses by use of EXORCYCLE phase cycles (Section 4.3.2.3), composite pulses (Section 3.4.2), and rf field gradients (Section 4.3.3.2).

The creation of a null in the frequency excitation profile of a pulse can be used to advantage in many experiments, including water-suppression methods and in some multidimensional NMR experiments. In some instances one set of spins must be excited while leaving others unperturbed.

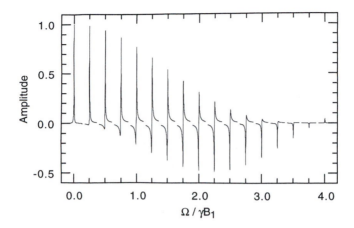

FIGURE 3.19 Resonance lineshapes for off-resonance 90° pulses. The lineshapes are calculated using the magnitudes and initial phases of the resonance signals given in Fig. 3.18.

Using [3.61], a null in the excitation profile for a pulse with on-resonance rotation angle $\alpha = 90°$ is achieved at an offset from resonance equal to

$$\Omega = \pm\sqrt{15}\,\gamma B_1. \qquad [3.67]$$

For a 180° pulse, a null is achieved when

$$\Omega = \pm\sqrt{3}\,\gamma B_1. \qquad [3.68]$$

Significant off-resonance effects can be observed even for nonresonant spins with $\Omega \gg \gamma B_1$ if rf pulses are applied to coherent states of the density operator, as opposed to thermal equilibrium magnetization. Off-resonance effects can lead to phase errors and frequency shifts in multidimensional NMR experiments, particularly when long, low-amplitude continuous or shaped rf pulses are employed. Off-resonance, or nonresonant, effects are attributed commonly, but incorrectly, to Bloch–Siegert shifts arising from the counterrotating component of the rf field [1.15] (21). The phase shift of transverse magnetization during an off-resonance pulse train arises from a z rotation and is given by (22,23)

$$\phi_{NR} = \langle \omega_1(t)^2 \rangle \tau_p/(2\Omega), \qquad [3.69]$$

in which the angle brackets represent an average over the pulse train. If τ_p is not a constant but varies with the evolution period of a multidimen-

NMR experiment, the off-resonance effect is manifested as a frequency shift, ω_{NR}, given by

$$\omega_{NR} = \langle \omega_1(t)^2 \rangle / (2\Omega). \qquad [3.70]$$

In addition to the nonresonant phase shift, a rotation occurs about the axis along which the rf field is applied and is given by

$$\zeta_{NR} = \phi_{NR} \langle \omega_1(t) \rangle / \Omega. \qquad [3.71]$$

The nonresonant rotation is smaller than the phase shift by a factor of at least $\gamma B_1 / \Omega$ and is eliminated for amplitude- or phase-modulated pulse trains with a net rotation angle of zero $[\langle \omega_1(t) \rangle = 0]$, such as decoupling sequences.

Nonresonant effects are common when performing frequency-selective homonuclear decoupling. For example, ^{13}CO and $^{13}C^\alpha$ spins can be decoupled by applying a selective 180° pulse to the ^{13}CO or C^α spins at the midpoint of the desired evolution period or by applying a frequency-selective decoupling sequence during the evolution period. Four techniques have been proposed for compensating for the consequent nonresonant phase shifts. These techniques will be illustrated for a $t_1/2 - 180°(CO) - t_1/2$ evolution period for aliphatic ^{13}C magnetization. The 180°(CO) pulse is applied as a weak rectangular pulse with $\gamma B_1 = \Omega_\alpha / \sqrt{3}$, in which Ω_α is the frequency difference between ^{13}CO and $^{13}C^\alpha$ spins [3.68]. For a rectangular 180° pulse, $\langle \omega_1(t)^2 \rangle = (\gamma B_1)^2$, and the nonresonant phase shift of the aliphatic spins is given by

$$\phi_{NR} = (\gamma B_1)^2 \tau_p / (2\Omega) = \pi \gamma B_1 / (2\Omega) = \pi / (2\sqrt{3}), \qquad [3.72]$$

in which the last equality obtains for $\Omega = \Omega_\alpha$. In the simplest approach, the phase errors are corrected during processing of the acquired spectrum. In the second approach, the phases of the aliphatic ^{13}C pulses subsequent to the evolution period are shifted by ϕ_{NR} to account for the non-resonant phase shift of the aliphatic spins. In the third approach, compensatory rf fields are produced by modulating the amplitude of the selective pulse with a cosine function, $\cos(\Omega_\alpha t)$ for $t = 0$ to τ_p (22). For a rectangular pulse, the resulting rf field is (assuming x phase) given by [1.15]:

$$\mathbf{B}_{rf}(t) = 2B_1 \cos(\Omega_\alpha t) \cos(\omega_{rf} t) \mathbf{i}$$
$$= B_1 \cos[(\omega_{rf} + \Omega_\alpha)t]\mathbf{i} + B_1 \cos[(\omega_{rf} - \Omega_\alpha)t]\mathbf{i}, \qquad [3.73]$$

in which ω_{rf} is the transmitter frequency of the $^{13}C^\alpha$ spins. Because the cosinusoidal modulation generates two effective rf fields, the amplitude of the B_1 field must be doubled (i.e., $\gamma B_1 = 2\Omega_\alpha / \sqrt{3}$). The component of the field with frequency $\omega_{rf} + \Omega_\alpha$ is resonant with the ^{13}CO spins and generates the 180°(CO) pulse. The component of the field with frequency $\omega_{rf} - \Omega_\alpha$ is not resonant with any spins of interest. The non-resonant phase shifts of the $^{13}C^\alpha$ spins caused by the two field components are equal and opposite; therefore, the net phase shift is zero. Because [3.69] depends nonlinearly on the resonance offset, nonresonant phase shifts can be corrected by these three approaches only if the spins of interest resonate in a narrow frequency range near Ω_α. In the fourth approach, compensatory rf pulses are incorporated into the pulse sequence. For example, if the simple $t_1/2-180°(CO)-t_1/2$ evolution period is replaced with the sequence

$$t_1/2-180°(CO)-t_1/2-180°(^{13}C)-\Delta-180°(CO)-\Delta, \qquad [3.74]$$

in which Δ is a short fixed delay, then the $180°(^{13}C)$ pulse applied to the aliphatic spins serves to refocus the phase evolution occuring during the two flanking 180°(CO) pulses. This method does not require that the spectral region of interest be narrow and consequently is generally applicable. In addition, the initial sampling delay can be adjusted to zero by setting $t_1(0) = \Delta$ (Section 3.3.2.3).

3.4.2 COMPOSITE PULSES

The usual goal of applying an rf pulse to the sample is to achieve a rotation of coherences around a defined axis by a specified angle. The actual performance of a rf pulse can be degraded by any number of factors, including finite rise times of the pulse, amplitude droop during the pulse, phase instability during the pulse, spatial inhomogeneity of the rf field across the sample, resonance offset effects, or relaxation during the pulse. Certain of these effects are minimized by improvements in NMR spectrometer hardware (such as digital frequency synthesis), phase cycling, and signal averaging. Other effects can be compensated by replacement of a single rf pulse by an extended pulse train that is designed to achieve the same ideal rotation as the single pulse but which is more resistant to nonideal influences. These pulse trains, called composite pulses, have been extensively developed for reduction of the effects of rf field inhomogeneity and resonance offset effects. Although composite pulses have been designed to meet a large number of objectives, for multidimensional NMR spectroscopy of biomolecules, three objectives are of the most relevance: 90° rotation of z magnetization into the transverse plane, inversion of z magnetization, and refocusing of transverse magnetization. In general, the design of composite pulses that compensate for both rf inhomogeneity and resonance offset effects is more difficult than the design of composite pulses that compensate for one of the two effects. Shorter composite pulse trains generally give less satisfactory compensation; however, long composite pulse trains may be impractical (due to amplifier droop and evolution of the spin system during the extended pulse sequence). The initial state of the magnetization is important to the design of the composite pulse; a composite 180° pulse that produces accurate inversion of z magnetization may not necessarily produce accurate refocusing of transverse magnetization. The detailed mathematical derivations of composite pulses will not be presented in this section; the interested reader is referred to the comprehensive review by Levitt (24). One should also note that the design of composite pulses is closely related to the design of spin decoupling techniques discussed in the next section.

Numerous composite pulses have been developed for effecting 90° and 180° rotations. Composite 180° pulses are utilized in multidimensional NMR spectroscopy much more frequently than composite 90° pulses because of the nearly ideal performance of a 90° pulse (Fig. 3.18). The most successful composite pulses are designed to invert z magnetization and refocus scalar coupling evolution. Composite pulses often do not perform appreciably better than a single 180° pulse for refocusing chemical-shift evolution because many composite pulses introduce offset-dependent phase shifts of the refocused coherences that are difficult to compensate in multidimensional NMR experiments (24). Early efforts at designing composite pulses relied on some combination of guesswork, insight, average Hamiltonian theory, Fourier analysis, and numerical integration of the Bloch equations. Increasingly, new composite pulse sequences are obtained by computer optimization of initial trial pulse sequences generated by the aforementioned techniques. Calculating the performance of a particular composite pulse is considerably simpler than designing the pulse sequence. A windowless composite pulse (i.e., a series of consecutive pulses without intervening delays) of rotation angle, α, and rf phase, ϕ, that consists of N rf pulses can be described by

$$\tilde{P}_\phi(\alpha) = \prod_{i=1}^{N} P_{\phi_i}(\alpha_i). \qquad [3.75]$$

The net rotation induced by the composite pulse is given by

$$\tilde{R}_\phi(\alpha) = \prod_{i=1}^{N} R_z(\phi_i) R_y(\theta_i) R_z(\alpha_i) R_y(-\theta_i) R_z(-\phi_i), \qquad [3.76]$$

in which ϕ_i, θ_i, and α_i respectively represent the values of the phase, tilt angle, and effective rotation angle for the ith pulse element. The rotation matrices are given by [1.34]. The effective rotation angle and the tilt angle are described by [1.23] and [1.21]. The magnetization following the composite pulse is

$$\mathbf{M}(0_+) = \tilde{\mathbf{R}}_\phi(\alpha)\mathbf{M}(0_-),$$ [3.77]

in which $\mathbf{M}(0_-)$ and $\mathbf{M}(0_+)$ are the magnetizations before and after the pulse, respectively.

As an example, product operator calculations indicate that the z magnetization obtained from a simple, on-resonance pulse of length $2\beta \approx 180°$ is

$$I_z \xrightarrow{2\beta_x} -I_y \sin 2\beta + I_z \cos 2\beta$$
$$= -I_y \sin 2\beta - I_z(1 - 2\cos^2\beta).$$ [3.78]

The pulse sequence $\beta_x(2\beta)_y\beta_x$ with $2\beta \approx 180°$ is a commonly used composite $180°$ pulse. The evolution of I_z is given by

$$I_z \xrightarrow{\beta_x} -I_y \sin\beta + I_z \cos\beta$$

$$\xrightarrow{2\beta_y} I_x \cos\beta \sin 2\beta - I_y \sin\beta + I_z \cos\beta \cos 2\beta$$

$$\xrightarrow{\beta_x} I_x \cos\beta \sin 2\beta - I_y \sin 2\beta(1 + \cos 2\beta)/2$$
$$- I_z(\sin^2\beta - \cos^2\beta \cos 2\beta)$$
$$= 2I_x \cos^2\beta \sin\beta - 2I_y \cos^3\beta \sin\beta - I_z(1 - 2\cos^4\beta).$$ [3.79]

Because $|\cos\beta| < 1$, $\cos^4\beta < \cos^2\beta$ and the composite pulse sequence yields more accurate inversion of I_z than does a simple pulse with rotation angle $2\beta \approx 180°$, the composite pulse is compensated for rf inhomogeneity or missetting of the pulse lengths.

For more detailed analysis, numerical calculations of the performance of a composite pulse are more convenient. Plots showing the theoretical performance of some selected composite pulses are shown in Fig. 3.20. As indicated by the plots, design of a composite $180°$ pulse that improves inversion of z magnetization is considerably easier than design of a composite $180°$ pulse for refocusing transverse magnetization. Inclusion of the composite pulse (most commonly, $90_x180_y90_x$ or $90_x240_y90_x$) into a pulse sequence when inversion of a z operator is required frequently will improve the performance of the pulse sequence. In addition, composite pulses are more useful for ^{13}C and ^{15}N than for ^1H because the larger chemical-shift dispersion of the heteronuclei results in larger off-resonance effects. Therefore, the most common application of composite pulses in multidi-

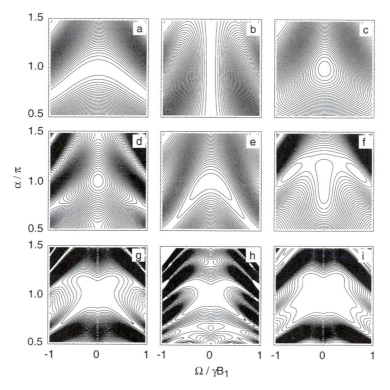

FIGURE 3.20 Excitation profiles for 180° composite pulses. Resulting (a,d,g) M_x, (b,e,h) M_y, and (c,f,i) M_z magnetization components obtained following application of (a–c) rectangular, (d–f) $90_x 180_y 90_x$ composite, and (g–i) $360_0 180_{120} 180_0 180_{120}$ composite 180° pulses to (a,d,g) M_x, (b,e,h) M_y, and (c,f,i) M_z magnetization. Results are shown as contour plots with resonance offset effects displayed on the x axis and effects of pulse mismeasure (or rf inhomogeneity) displayed on the y axis. (a–f) Results are calculated relative to an ideal y-phase pulse that inverts M_x and M_z magnetization and does not affect M_y magnetization. (g–i) Results are calculated relative to an ideal x-phase pulse that inverts M_y and M_z magnetization and does not affect M_x magnetization.

mensional NMR spectroscopy is for achieving heteronuclear spin decoupling during a delay Δ by applying a 180° pulse at $\Delta/2$. Amplifier droop, sample heating, and other instrumental imperfections may limit the total number or overall duration of composite pulses that can be included in a particular pulse sequence. Determination of which 180° pulses can be replaced profitably by composite pulses in a particular sequence is an empirical process.

3.4.3 Spin Decoupling

As has been discussed extensively in Chapter 2, if two spins I and S are scalar-coupled with a coupling constant J_{IS}, evolution of transverse magnetization of the S spin leads to splitting of the resonance signal into two multiplet components at frequencies $\omega_S + \pi J_{IS}$ and $\omega_S - \pi J_{IS}$ (or $\nu_S + J_{IS}/2$ and $\nu_S - J_{IS}/2$ in hertz). Product operator calculations demonstrate that if a series of 180° pulses are applied selectively at the frequency of the I spin, the effect of the scalar coupling interaction is refocused and the net evolution of the S spin yields a single resonance signal at ω_S. This process is called spin decoupling. In the limit that the pulse spacing approaches zero, decoupling is obtained by applying a continuous rf field at the resonance frequency of the I spin while observing the S spin.

This simple picture of spin decoupling suffers from the problem that when applying the product operator formalism, pulses are assumed to be infinitely short so that evolution under the scalar coupling and Zeeman Hamiltonians during the pulse may be ignored. In the event that an rf field is applied for an extended period of time, this assumption is not valid; thus, a more complex theoretical analysis is required. Also, as discussed in the previous section, the performance of a nominal 180° pulse in inverting magnetization is a strong function of resonance offset and rf inhomogeneity. Since resonance offset effects are minimized by increasing the rf power, obtaining effective decoupling using a continuous rf field necessitates the use of very high rf powers. The theory of spin decoupling has been reviewed elegantly by Shaka and Keeler (*25*); in the present section, an outline of the principles used to develop improved spin decoupling sequences is presented.

One measure of the efficiency of a decoupling sequence is the relative scalar coupling observed in the decoupled NMR spectrum as a function of the resonance offset and rf field strength. Shaka and Keeler show that for continuous rf irradiation, the scaling factor is

$$\lambda_c = \frac{\Omega}{\sqrt{(\gamma B_1)^2 + \Omega^2}}. \qquad [3.80]$$

Perfect decoupling ($\lambda_c = 0$) is obtained only for on-resonance irradiation. For off-resonance irradiation, effective decoupling requires that $\gamma B_1 \gg \Omega$, which is a condition that is extremely difficult to achieve experimentally. For example, if $|\Omega/(\gamma B_1)| = 1$, then $\lambda_c = 0.707$, which provides marginal reduction in the coupling constant.

The discussion in the previous section suggests that the spin decoupling may be improved by replacing the series of 180° pulses with a series of composite 180° pulses. In the limit that the pulse spacing approaches zero, a phase-modulated rf field is obtained that can be described as the repetition of some phase-modulated sequence R representing the composite 180° pulse. Coherent averaging theory (26) demonstrates that this approach in fact improves decoupling and also leads to the conclusion that, given a sequence element R, improved sequence elements can be developed by recursive expansion of R into supercycles. For example, if R is a composite 180° pulse and $\overline{R}$ is obtained by inverting the phases of the pulses comprising R, then the supercycles

$$R$$

$$RR\overline{RR}$$

$$RR\overline{RR}\quad \overline{RR}RR$$

$$RR\overline{RR}\quad \overline{RR}RR\quad \overline{RR}RR\quad RR\overline{RR} \tag{3.81}$$

yield progressively better decoupling performance. In addition, if R is a composite pulse sequence, then $R_p\overline{R}_p$ performs better than R. The element R_p is obtained by cyclically permuting a 90° pulse with the element R. For example, if R is

$$R = 90_x 180_{-x} 270_x = 1\overline{2}3, \tag{3.82}$$

then

$$RR\overline{RR} = 1\overline{2}3\quad 1\overline{2}3\quad \overline{1}2\overline{3}\quad \overline{1}2\overline{3} \tag{3.83}$$

improves on R. If the preceding element is cyclically permuted from left to right, then

$$R_p = \overline{2}3\quad 1\overline{2}3\quad \overline{1}2\overline{3}\quad \overline{1}2\overline{3}\quad 1$$

$$= \overline{2}4\overline{2}3\overline{1}\quad 2\overline{4}2\overline{3}1 \tag{3.84}$$

and

$$R_p\overline{R}_p = \overline{2}4\overline{2}3\overline{1}\quad 2\overline{4}2\overline{3}1\quad \overline{2}4\overline{2}3\overline{1}\quad 2\overline{4}2\overline{3}\overline{1}. \tag{3.85}$$

The composite pulse given in [3.82] is the primitive element for the WALTZ family of decoupling sequences (27). Equation [3.83] is the WALTZ-4 sequence, and [3.85] is the WALTZ-8 sequence. The WALTZ-16 sequence is obtained by a cyclic permutation of [3.85] from right to left:

$$\overline{RR}RR = \overline{3}42\overline{3}1\overline{2}42\overline{3}\quad \overline{3}42\overline{3}1\overline{2}42\overline{3}\quad \overline{3}42\overline{3}1\overline{2}42\overline{3}\quad \overline{3}42\overline{3}1\overline{2}42\overline{3}. \tag{3.86}$$

Each expansion of the basic element results in an improvement in the decoupling efficiency by approximately an order of magnitude. The decoupling scaling factors for the WALTZ family of decoupling sequences are

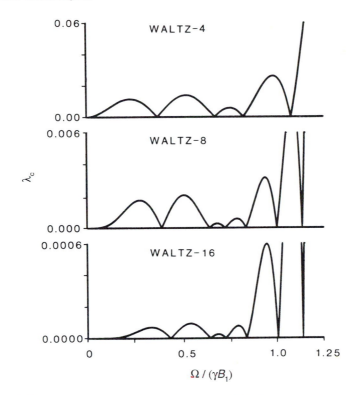

FIGURE 3.21 Scalar coupling scaling factor for WALTZ decoupling sequences.

shown in Fig. 3.21. WALTZ-16 yields $\lambda_c < 0.0006$ for resonance offsets $|\Omega/(\gamma B_1)| \leq 1$, an improvement of three orders of magnitude compared with continuous-wave decoupling.

The search for new decoupling sequences at present is heavily dependent on computer optimization. An early example of this approach lead to the GARP decoupling sequence (28), $RR\overline{RR}$, in which the element R is given by

$$R = 30.5 \quad \overline{55.2} \quad 257.8 \quad \overline{268.3} \quad 69.3 \quad \overline{62.2} \quad 85.0 \quad \overline{91.8} \quad 134.5$$

$$\overline{256.1} \quad 66.4 \quad \overline{45.9} \quad 25.5 \quad \overline{72.7} \quad 119.5 \quad \overline{138.2} \quad 258.4$$

$$\overline{64.9} \quad 70.9 \quad \overline{77.2} \quad 98.2 \quad \overline{133.6} \quad 255.9 \quad \overline{65.5} \quad 53.4, \qquad [3.87]$$

in which pulse lengths are given in degrees and overbars indicate 180° phase shifts. GARP yields $\lambda_c < 0.002$ for resonance offsets $|\Omega/(\gamma B_1)| \leq 2.5$. The decoupling scaling factors for the GARP decoupling sequence is shown in Fig. 3.22.

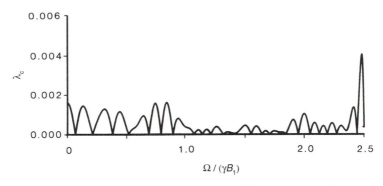

FIGURE 3.22 Scalar coupling scaling factor for GARP-1 decoupling sequence.

The WALTZ-16 and GARP sequences are the two most commonly used techniques for spin decoupling in macromolecules. Both of these sequences utilize phase modulation of the rf field only. Other families of decoupling sequences have been developed in which the rf field is windowed or in which the rf field amplitude and phase are modulated. Amplitude- and phase-modulated sequences show promise for application to decoupling of spins with moderately close resonance frequencies, i.e., for decoupling carbonyl and C^α spins from each other (*29*). For example, the SEDUCE-1 selective decoupling sequence is obtained by replacing each rectangular pulse in the WALTZ-16 pulse sequence with an amplitude-modulated selective pulse.

Incidentally, spin decoupling is somewhat related to the problem of obtaining an isotropic effective Hamiltonian for generating total correlation spectroscopy (TOCSY) coherence transfer via strong coupling. These applications are discussed in Section 4.2.1.2.

3.4.4 SELECTIVE PULSES

One of the original objectives in the development of pulsed NMR spectroscopy was to replace the time-consuming swept selective excitation of continuous-wave spectroscopy by broadband excitation using strong rf pulses. As spectrometer hardware has become more sophisticated and as the experimental systems studied have become more complex, the incorporation of selective pulses that excite resonances within a narrow frequency range into otherwise broadband pulse sequences has been shown to have some advantages; selective excitation of the solvent resonance can improve solvent suppression, spins with moderately different Larmor frequencies can be manipulated independently (i.e. carbonyl and C^α carbon spins), and the digitization requirements for multidimensional spectroscopy can be reduced (*30*).

The simplest selective pulse is obtained by reducing the amplitude and lengthening the duration of a rectangular pulse, i.e., a rf field of constant amplitude, B_1, that is turned on at a time t and turned off at time $t + \tau_p$. As shown by [3.61], a rectangular pulse produces excitation over the range $-2\omega_1 \leq \Omega \leq 2\omega_1$. Such a simple approach is unsatisfactory because the excitation profile has lobes that extend over an extensive frequency range, the phase of the transverse magnetization varies strongly with offset, and the excitation profile is not uniform in the neighborhood of the rf carrier frequency. The performance of selective (or soft) rectangular pulses can be improved by utilizing more complex amplitude or phase modulation of the rf field. Just as for composite pulses, integration of the Bloch equations to ascertain the excitation profile of a given selective pulse shape is straightforward. The inverse problem, determining the necessary pulse shape from the desired excitation profile, is much more difficult because of the nonlinear character of the Bloch equations. Accordingly, many selective pulses have been developed initially using approximate techniques (such as Fourier analysis of the desired excitation profile) and subsequently refined. The development of selective excitation pulses and their incorporation into multidimensional pulse sequences remains an active area of research, and significant future progress can be expected.

The desirable characteristics of a selective pulse are

1. Uniform excitation profile over the desired frequency range.
2. Minimal excitation of resonances outside the desired frequency range.
3. Phase of the transverse magnetization should be a smooth (preferably linear) function of offset.
4. The pulse should have a short duration to minimize relaxation effects and phase evolution during the pulse.
5. The pulse modulation scheme should be simple to implement experimentally.

These objectives are not satisfied simultaneously for known pulse shapes. For example, some selective pulses that yield uniform excitation profiles include short high-power pulse segments; these pulse waveforms can be difficult to implement without special rf transmitters. Consequently, an optimal selective pulse cannot be determined without reference to the particular application. The selective pulses commonly used in multidimensional NMR spectroscopy of biological macromolecules are the soft rectangular pulse, the Gaussian pulse, the half-Gaussian pulse, and the sinc(sin x/x) pulse. The durations of the Gaussian pulses are truncated at the point that the amplitude of the rf field is equal to a few percent of its maximum value. The sinc pulse normally is truncated to include only the central lobe of the pulse shape.

For the purpose of calculations (and for implementation in spectrometer hardware), the continuous selective pulse shape is approximated by a large number, N, of rectangular pulses of length Δt. The amplitudes and phases of the rectangular pulses are adjusted to mimic the desired shape. Thus, the total duration of the selective pulse is $\tau_p = N\Delta t$.

$$\tilde{P}(\tau_p) = \prod_{i=1}^{N} P_{\phi_i}(\alpha_i). \qquad [3.88]$$

The net rotation induced by the selective pulse is given by

$$\tilde{R}(\tau_p) = \prod_{i=1}^{N} R_z(\phi_i)R_y(\theta_i)R_z(\alpha_i)R_y(-\theta_i)R_z(-\phi_i), \qquad [3.89]$$

in which ϕ_i, θ_i, and α_i respectively represent the values of the phase, tilt angle, and effective rotation angle for the ith period Δt. The rotation matrices are given by [1.34]. The effective rotation angle and tilt angle are described by [1.23] and [1.21]. The magnetization following the selective pulse is

$$\mathbf{M}(0_+) = \tilde{R}(\tau_p)\mathbf{M}(0_-), \qquad [3.90]$$

in which $\mathbf{M}(0_-)$ and $\mathbf{M}(0_+)$ are the magnetization vectors before and after the pulse, respectively. In contrast to [3.77], the duration τ_p for a selective pulse may not be short relative to either relaxation or evolution of scalar coupling interactions; accordingly, [3.90] describes the effect of a selective pulse on an uncoupled, isolated spin in the absence of relaxation (31).

The pulse shapes, excitation profiles, and phase profiles for common selective pulses are shown in Figs. 3.23–3.25. The most common, and simplest, applications of selective pulses in protein NMR spectroscopy are (1) decoupling of carbonyl and $^{13}C^{\alpha}$ spins by using selective 180° pulses (Section 7.1.2.2) and (2) selective excitation of the solvent resonance to obtain solvent suppression without deleterious effects of presaturation (Section 3.5.1). For these undemanding applications, any of the sinc, Gaussian, and Hermitian pulse shapes are satisfactory. More sophisticated uses of selective pulses in multidimensional NMR spectroscopy are described by Emsley (30).

3.5 Water-Suppression Techniques

The overwhelming majority of the studies of biological macromolecules using high-resolution NMR spectroscopy are performed in aqueous

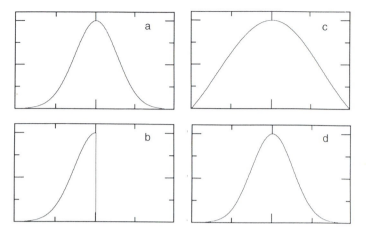

FIGURE 3.23 (a) Gaussian, (b) half-Gaussian, (c) sinc, and (d) Hermitian amplitude-modulated selective pulse shapes. Amplitudes are proportional to $\exp(-at^2)$,
$\sin(\pi at)/(\pi at)$, and $(1 - 2at^{2/3}) \exp(-at^2)$ for the Gaussian, sinc, and Hermitian
pulses, respectively. Gaussian, half-Gaussian, and Hermitian pulse shapes are
truncated at the 1% amplitude level. The sinc pulse is truncated at the first zero
of the sinc function.

solutions. The concentration of protons in water is approximately 110 M,
compared with typical concentrations of macromolecules of 1–2 mM.
Thus, the equilibrium magnetization of the water protons is approximately
10^4–10^5 greater than the equilibrium magnetization of a single proton in
a macromolecule. Detection of the solute signal in the presence of the
solvent signal presents a difficult problem because, inevitably, the dynamic
range of the electronic components of the spectrometer is limited. If, for
example, the signal from a 1-mM protein solution is to be digitized with
4-bit precision utilizing a 16-bit analog-to-digital converter, then the water
signal must be reduced by a factor of at least 50 prior to acquisition to
avoid overflowing the receiver ADC. In addition, even if the water and
solute signal are digitized adequately, the solute resonances may be obscured by the tails of the large, broad solvent peak. This problem is
particularly severe for multidimensional spectra in which changes in the
phase of the solvent signal from experiment to experiment lead to severe
distortions of the final spectrum.

NMR spectroscopy in aqueous solution also suffers from *radiation
damping* of the solvent signal (*32,33*). Following a rf pulse, the water
magnetization precesses in the transverse plane and induces a time-varying
oscillating current in the coil. The current, in turn, induces an electromagnetic field of the same frequency that acts to rotate the water magnetization

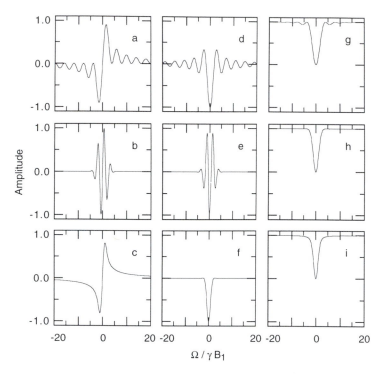

FIGURE 3.24 Selective 90° pulses. Resulting (a–c) M_x, (d–f) M_y, and (g–i) M_z magnetization components obtained following application of (a,d,g) rectangular, (b,e,h) Gaussian, and (c,f,i) half-Gaussian 90° pulses to equilibrium M_z magnetization. All pulses have x phase.

back toward the z axis. Following a 90° pulse, the time constant for radiation damping is given approximately by (*32*)

$$\tau_{RD} = (2\pi M_0 Q \gamma)^{-1}. \qquad [3.91]$$

Thus, radiation damping is important for the water magnetization and not the solute magnetization because the (unsuppressed) solvent magnetization is approximately two orders of magnitude larger. Although water protons have T_1 relaxation times of seconds, for a typical high-Q NMR probe, radiation damping following a 90° pulse will return the transverse magnetization to the z axis in tens of milliseconds. Radiation damping severely interferes with the expected evolution of the water magnetization through a pulse sequence.

A straightforward method of reducing the resonance signal from H_2O is to use D_2O as the solvent. Deuterium oxide with a deuteron content of

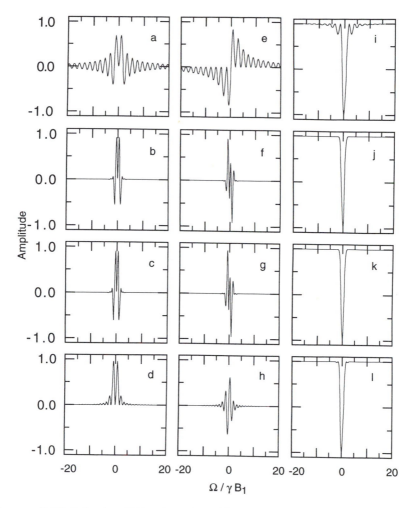

FIGURE 3.25 Selective 180° pulses. Resulting (a–d) M_x, (e–h) M_y, and (i–l) M_z magnetization components obtained following application of (a,e,i) rectangular, (b,f,j) Gaussian, (c,g,k) Hermitian, and (d,h,l) sinc 180° pulses to equilibrium M_z magnetization. All pulses have x phase.

as much as to 99.999% is commercially available and offers theoretical reduction in the water proton signal by a factor of 10^5. The main limitation to this approach is that signals from exchangeable protons in the macromolecule are reduced as well. For example, signals from the amide groups in proteins usually are absent in spectra acquired using D_2O solutions.

Because scalar coupling and dipolar interactions of the backbone amide protons with amide nitrogens and with α-protons are critical for backbone resonance assignments and secondary structure analyses (Chapter 8), at least some protein spectra must be acquired in H_2O rather than D_2O solution. A secondary problem is that inevitable differences in conditions between samples in H_2O and D_2O (principally, pH differences or isotope effects) can complicate comparisons of data acquired in the two solvents. The usual course of action will be to acquire a series of experiments from H_2O solution initially (with 5–10% D_2O present for the field-frequency lock system), and then to transfer the sample to D_2O for a second series of experiments. Even in D_2O solution, the HDO resonance resulting from residual protons frequently is further suppressed by one of the techniques described below.

The general subject of solvent suppression techniques can be divided into three stages: (1) techniques for reducing the solvent signal detected during a single transient acquisition to within the dynamic range of the spectrometer; (2) techniques for further reduction of the solvent signal that results from a single, signal-averaged, experiment; and (3) postacquisition digital signal processing to improve the solvent suppression. The commonly used methods for solvent suppression (dynamic range reduction) in biological samples are (1) presaturation of the solvent resonance during the recycle delay between transient acquisitions, (2) (semi)selective excitation of macromolecule resonances, and (3) dephasing of the solvent magnetization using rf or field-spoiling pulses. Experimental techniques for solvent suppression have been reviewed (*34,35*) and are discussed briefly below.

3.5.1 PRESATURATION

The most commonly used solvent-suppression technique is presaturation of the solvent signal during the recycle delay using a weak rf field. Presaturation is very simple to implement and is very effective. The main disadvantages are that signals that resonate very close to the solvent signal (principally the α-protons in proteins) may be partially saturated by the rf field and that saturation transfer may partially saturate exchangeable protons. As shown in Fig. 3.26, the rates of exchange between amide protons in proteins and water protons is pH-dependent; accordingly, saturation transfer to the amide protons in proteins is particularly deleterious above pH 7. The quality of the water suppression obtained by presaturation depends very critically on the homogeneity of the magnetic field and hence on the quality of the shimming. Solvent signals originating outside the main sample volume also degrade solvent suppression; consequently,

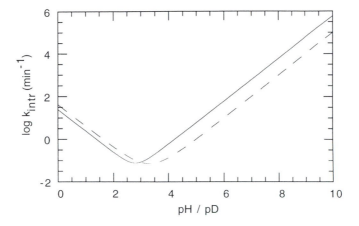

FIGURE 3.26 Intrinsic backbone amide proton exchange rates calculated according to Connelly *et al.* (*63*). The intrinsic exchange rate, k_{intr}, is shown for exchange of a backbone amide proton with (—) H_2O or (- - -) D_2O as a function of pH or pD. The pD values are corrected for isotope effects; uncorrected pH meter readings would be 0.4 units smaller.

modern probes designed for use with water solutions have rf coils with shielded leads.

Irradiation times on the order of 1–2 s using rf fields with amplitudes of approximately 50 Hz usually provide adequate water suppression. Short irradiation times minimize the effects of saturation transfer, but require higher power rf fields and worsen saturation of signals resonating near the irradiation frequency. For each new sample, the shimming, irradiation frequency, and irradiation power must be adjusted to optimize water suppression.

3.5.2 JUMP–RETURN AND BINOMIAL SEQUENCES

A variety of techniques have been developed that selectively excite the resonances from the solute while leaving equilibrium magnetization of the water protons relatively undisturbed. Of these techniques, only the $1-\bar{1}$ "jump–return" and $1-\bar{3}-3-\bar{1}$ binomial sequence have achieved much popularity for protein NMR spectroscopy (*36,37*). These techniques can be incorporated into nuclear Overhauser-effect spectroscopic (NOESY) pulse sequences in order to observe cross-relaxation between protons that exchange rapidly with solvent.

In the jump–return technique, the final read pulse in a pulse sequence is replaced by the pulse element $90_x-\tau-90_{-x}$. The carrier is placed on the solvent resonance and $\tau = 1/(4\Delta\nu_{max})$, in which $\Delta\nu_{max}$ is the offset from the carrier at which excitation is maximized. Jump–return and binomial sequences provide a nice illustration of the technique of [2.117] in the analysis of propagators. For the jump–return sequence, the propagator is given by

$$
\begin{aligned}
\mathbf{U} &= \exp[i(\pi/2)I_x]\exp[-i\Omega\tau I_z]\exp[-i(\pi/2)I_x] \\
&= \exp[-i\Omega\tau e^{i(\pi/2)I_x}I_z e^{-i(\pi/2)I_x}] \\
&= \exp[-i\Omega\tau I_y],
\end{aligned}
\qquad [3.92]
$$

in which the final two lines are obtained by applying [2.117]. Evolution during the jump–return element is given by

$$
I_z \xrightarrow{\ \Omega\tau I_y\ } I_z\cos(\Omega\tau) + I_x\sin(\Omega\tau). \qquad [3.93]
$$

The resonance offset of the solvent protons is zero because the rf carrier is placed on the solvent resonance. No transverse solvent magnetization is generated, and, in theory, complete suppression of the solvent signal is obtained. The amplitude of the detected signal for coherences that are not resonant with the rf carrier is proportional to I_x and depends on the delay τ through the factor $\sin(\Omega\tau)$. The spectrum that results has opposite phase both above and below the carrier position because the sine is an odd function; however, no linear phase correction is required.

In the $1-\bar{3}-3-\bar{1}$ technique, the read pulse is replaced by the pulse element $\alpha_x-\tau-3\alpha_{-x}-\tau-3\alpha_x-\tau-\alpha_{-x}$, in which $8\alpha = 90°$ and $\tau = 1/(2\Delta\nu_{max})$. The propagator is

$$
\begin{aligned}
\mathbf{U} &= \exp[i\alpha I_x]\exp[-i\Omega\tau I_z]\exp[-i3\alpha I_x]\exp[-i\Omega\tau I_z]\exp[i3\alpha I_x] \\
&\quad \times \exp[-i\Omega\tau I_z]\exp[-i\alpha I_x].
\end{aligned}
\qquad [3.94]
$$

Clearly, for $\Omega = 0$, $\mathbf{U} = \mathbf{E}$ and no transverse magnetization is generated. The $1-\bar{3}-3-\bar{1}$ sequence offers better theoretical suppression of the solvent resonance than the jump–return sequence. The evolution of the density operator for coherences that are not resonant with the rf carrier is quite complicated; for simplicity, consider only the offset at which maximum excitation occurs. In this case, $\Omega\tau = \pi$ and the propagator is given by

$$
\begin{aligned}
\mathbf{U} &= \exp[i\alpha I_x]\exp[-i\pi I_z]\exp[-i3\alpha I_x]\exp[-i\pi I_z]\exp[i3\alpha I_x] \\
&\quad \times \exp[-i\pi I_z]\exp[-i\alpha I_x] \\
&= \exp[i\alpha I_x]\exp[-i3\alpha e^{-i\pi I_z}I_x e^{i\pi I_z}]\exp[i3\alpha e^{-i2\pi I_z}I_x e^{i2\pi I_z}]
\end{aligned}
$$

$$\times \exp[-i\alpha e^{-i3\pi I_z}I_x e^{i3\pi I_z}]\exp[-i3\pi I_z]$$
$$= \exp[i\alpha I_x]\exp[i3\alpha I_x]\exp[i3\alpha I_x]\exp[i\alpha I_x]\exp[-i3\pi I_z]$$
$$= \exp[i8\alpha I_x]\exp[-i3\pi I_z]$$
$$= \exp[i(\pi/2)I_x]\exp[-i\pi I_z], \qquad\qquad [3.95]$$

in which the last four lines are obtained by applying [2.117]. The evolution of the equilibrium density operator is given by

$$I_z \xrightarrow{\pi I_z} I_z \xrightarrow{-(\pi/2)I_x} I_y. \qquad\qquad [3.96]$$

Unfortunately, for other offsets, the amplitude of the detected signal depends on the delay τ, and a strong linear phase gradient exists across the spectrum. The linear phase gradient leads to baseline distortions that are a particular problem in multidimensional NMR spectroscopy. Consequently, the jump–return sequence has been much more widely applied in multidimensional NMR spectroscopy. Excitation profiles for the jump–return and binomial sequences are shown in Fig. 3.27.

Jump–return and binomial pulse sequences are sensitive to pulse and phase imperfections. Optimal water suppression requires careful adjustment of the spectrometer. Pulse lengths and phases usually are adjusted slightly around the theoretical values to maximize water suppression (typically by 0.1–0.3 μs and 1–3°). In practice, suppression factors of 50–100 are generally obtained and are adequate when combined with postacquisition signal processing techniques. These sequences are most successful when used as the read pulses for relatively simple experiments with few pulses and no extended rf mixing sequences, such as NOESY and HMQC experiments. These sequences have an advantage over presaturation and the techniques described in Section 3.5.3 in that the net excitation of the water signal is zero, or nearly so. Since the water signal remains nearly at equilibrium, saturation transfer from the water to the solute is minimized.

3.5.3 SPIN LOCK AND GRADIENT PULSES

Recent technological advances in NMR spectrometer hardware have resulted in the development of new methods of solvent suppression using combinations of selective pulses, spin-lock purge pulses, and field-gradient pulses. These techniques can generate nearly ideal excitation profiles with high degrees of water suppression. Large linear phase gradients are avoided and the techniques can be implemented into nearly all NMR experiments. To date, the majority of the new methods saturate the water

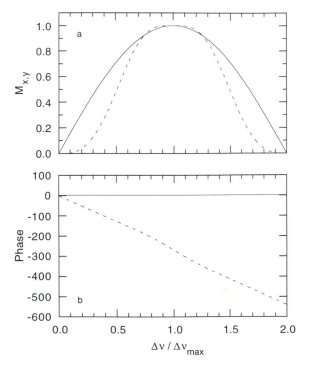

FIGURE 3.27 Binomial excitation profiles. (a) Magnitude and (b) phase of the transverse magnetization, $M_{x,y} = (M_x^2 + M_y^2)^{1/2}$ excited by the (—) jump–return and (---) $1-\bar{3}-3-\bar{1}$ pulse sequences are plotted as function of resonance offset. Resonances on opposite sides of the carrier have inverted phases (not shown).

signal within a few milliseconds. Saturation transfer effects are much smaller than for presaturation; however, unless the recycle delay is very long, some attenuation of the water resonance and consequent saturation transfer to the solute molecule is unavoidable. More recently, techniques that attempt to maintain the water magnetization close to its equilibrium value have been developed (*38,39*). For illustrative purposes, a variety of these water-suppression techniques have been incorporated into homonuclear Hahn echo pulse sequences (Section 3.6.4.2) in Fig. 3.28.

In the simplest approach, shown in Fig. 3.28a, a selective 90° pulse is used to rotate the water magnetization into the transverse plane. A hard (nonselective) 90° pulse rotates solute magnetization into the transverse plane and rotates the water magnetization back to the longitudinal axis. Finally, a long spin lock purge pulse is applied with phase shifted by 90° relative to the other pulses. Assuming that the selective and nonselective

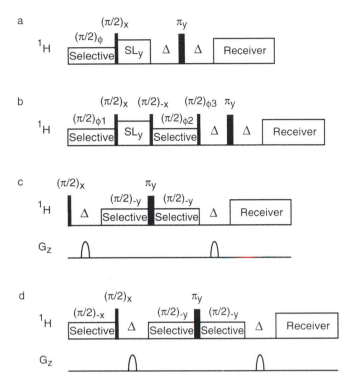

FIGURE 3.28 Water suppression using spin-lock purge pulses and field-gradient pulses incorporated into Hahn echo pulse sequences. Rectangular or shaped selective pulses are applied to the solvent magnetization and minimally perturb the solute magnetization. Thin and thick solid bars represent 90° and 180° nonselective pulses, respectively. SL represents a spin-lock purge pulse of 1–2 ms duration. Delays Δ are adjusted as described in Section 3.6.4.2 for the basic Hahn echo experiment and must be long enough to permit recovery following gradient pulses in (c) and (d). The phase cycles are (a) $\phi = (x, -x)$, receiver $= (x, x)$; (b) $\phi_1 = (x, -x, x, -x)$, $\phi_2 = (x, x, -x, -x)$, $\phi_3 = (-x, -x, x, x)$, receiver $= (x, x, -x, -x)$. Basic phase cycles can be elaborated using CYCLOPS and EXCORCYLE as described for the basic Hahn echo pulse sequence in Section 3.6.4.2.

90° pulses are applied with x phase, the solute magnetization is described by an operator $-I_y$, and the solvent magnetization is described by an operator $-S_z$ prior to the purge pulse. If the purge pulse is applied with y phase, then the solute operator commutes with the rotation operator for the pulse and no evolution occurs (other than relaxation). Solute magnetization is said to be *spin-locked* by the pulse. However, solvent magnetization is orthogonal to the spin-lock pulse and is rapidly dephased by the inhomogeneity of the rf field as described below (hence the name spin-lock purge pulse). The original implementation of this technique used a rectangular soft pulse for the selective 90° pulse; subsequent elaborations of the principle have used shaped selective pulses and have extended the sequence as shown in Fig. 3.28b to improve the water-suppression property (40).

The *inhomogeneity* of the rf field describes the variation in the amplitude of the B_1 field as a function of position in the sample. Thus

$$B_1(\mathbf{r}) = B_1(0) + \Delta B_1(\mathbf{r}), \qquad [3.97]$$

in which $B_1(0)$ is the amplitude of the $\mathbf{B}_1$ field at the center of the sample and $\Delta B_1(\mathbf{r})$ characterizes the inhomogeneity of the field. The net evolution of the S_z operator during a pulse of length t averaged over the sample volume, V, is given by

$$S_z \xrightarrow{[-\gamma B_1(\mathbf{r})t]S_y} S_z \int_V \cos(\gamma B_1(\mathbf{r})t) \, d\mathbf{r}/V + S_x \int_V \sin(\gamma B_1(\mathbf{r})t) \, d\mathbf{r}/V$$

$$= S_z \left\{ \cos(\gamma B_1(0)t) \int_V \cos(\gamma \Delta B_1(\mathbf{r})t) \, d\mathbf{r} \right.$$

$$\left. - \sin(\gamma B_1(0)t) \int_V \sin(\gamma \Delta B_1(\mathbf{r})t) \, d\mathbf{r} \right\} \Big/ V$$

$$+ S_x \left\{ \sin(\gamma B_1(0)t) \int_V \cos(\gamma \Delta B_1(\mathbf{r})t) \, d\mathbf{r} \right.$$

$$\left. + \cos(\gamma B_1(0)t) \int_V \sin(\gamma \Delta B_1(\mathbf{r})t) \, d\mathbf{r} \right\} \Big/ V$$

$$= \{S_z \cos(\gamma B_1(0)t) + S_x \sin(\gamma B_1(0)t)\} \int_V \cos(\gamma \Delta B_1(\mathbf{r})t) \, d\mathbf{r}/V$$

$$- \{S_z \sin(\gamma B_1(0)t) - S_x \cos(\gamma B_1(0)t)\} \int_V \sin(\gamma \Delta B_1(\mathbf{r})t) \, d\mathbf{r}/V. \qquad [3.98]$$

Because the cosine and sine functions are oscillatory, the integrals in [3.98] tend to zero provided that $\Delta B_1(\mathbf{r})$ is nonuniform throughout the sample and t is sufficiently long to ensure that $\gamma \Delta B_1(\mathbf{r})t$ is spatially randomized between 0 and 2π. For modern NMR probes, $|\Delta B_1(\mathbf{r})/B_1(0)| \approx 10$–$20\%$, and significant dephasing of the solvent signal can be achieved in a few milliseconds.

Spin-lock purge pulses also dephase protein coherences orthogonal to the rf field (I_x or I_z coherences) while preserving the coherence locked along the rf field (I_y coherence). Purge pulses can be used to eliminate artifacts in NMR spectra arising from undesired operators orthogonal to the operator of interest (see Section 7.2.4.3 for an example). However, product operators containing two spin operators orthogonal to the rf field (e.g., homonuclear $2I_{1z}I_{2z}$ or $2I_{1z}I_{2x}$ operators) contain components that behave as zero-quantum coherences in the tilted rotating reference frame of the rf field (see Sections 5.2.3 and 5.4.3 for additional discussion of tilted rotating reference frames) and consequently are dephased inefficiently by the inhomogenouse rf field (41,42).

The purpose of the selective pulses in Fig. 3.28a,b is to produce orthogonal magnetization components from solute protons and solvent protons. If the solute is isotopically enriched with ^{13}C or ^{15}N, evolution under the heteronuclear scalar coupling Hamiltonian can be used to produce solute magnetization orthogonal to the solvent (and uncoupled solute protons). Spin-lock purge pulses can be applied to preferentially dephase the solvent magnetization without recourse to selective pulses (43). Such techniques are commonly utilized in multidimensional heteronuclear NMR spectroscopy (Chapter 7).

The development of actively shielded probes that can produce high-power field-gradient pulses has resulted in new classes of experiments that utilize field gradients rather than phase cycling for coherence selection (see Section 4.3.3). Field gradients have also been used to aid in water suppression and mitigation of radiation damping effects. Commonly utilized experimental techniques developed to date fall into two categories. The first approach is similar in spirit to the spin-lock purge pulse techniques in that a strong field-gradient pulse is used to preferentially dephase the solvent signal (44). The second approach is similar in spirit to jump–return and binomial sequences in that the water magnetization is returned to the z axis prior to acquisition (38).

An example of the former technique, which has been designated WATERGATE (44), is illustrated in Fig. 3.28c. Following the initial nonselective pulse, a strong gradient pulse dephases both solvent and solute magnetization. Solute magnetization is unaffected by the selective pulses (assuming that the solute resonances of interest are outside the bandwidth of the selective pulses). The nonselective 180° pulse inverts the coherence order of the solute magnetization; therefore, the second gradient pulse rephases the solute magnetization to form a gradient echo. In contrast, the combination of the selective 90° pulses and the nonselective 180° pulse leaves the coherence order of the solvent magnetization unchanged; therefore, the second gradient pulse continues to dephase the solvent magnetization, and no gradient echo is formed.

An example of the second technique, which has been called "water flip-back'' (38), is shown in Fig. 3.28d. The sequence is similar to the WATERGATE sequence except that a selective 90° pulse is inserted prior to the initial nonselective 90° pulse. Following the nonselective 90° pulse, the solute magnetization is described by an operator $-I_y$, and the solvent magnetization is described by an operator S_z. As in the WATERGATE sequence, the field gradient flanking the nonselective 180° pulse results in formation of a gradient echo for solute magnetization (45). Any transverse solvent magnetization created from nonideal performance of the selective 90° pulses and the nonselective 180° pulse is dephased by the field-gradient pulses. The bulk of the water magnetization is aligned along the z axis and is never saturated in this experiment. Radiation damping effects are minimized because any transverse solvent magnetization is dephased by the gradient pulses. In practice, 60–80% of the equilibrium water magnetization can be maintained along the z axis, and >30% increases in sensitivity are obtained for exchangeable solute protons due to reduced saturation transfer effects (38,39).

A subtle difference between the WATERGATE and flip-back experiments concerns the strength of the gradient pulses. In the WATERGATE experiment, the bulk solvent signal must be dephased and strong (20–30 G/cm) gradients must be employed. In the flip-back experiments, only the fraction of the solvent signal that is not returned to the z axis must be dephased and relatively weaker (5 G/cm) gradients are satisfactory.

Examples of various water-suppression techniques are given in Fig. 3.29. As described in Section 3.6.4.2, improved water suppression frequently is obtained if Hahn echo sequences are incorporated into the experimental pulse sequence. An example of the improved water suppression obtained is given in Fig. 3.37 (later).

3.5.4 POSTACQUISITION SIGNAL PROCESSING

Although numerous techniques have been proposed for postacquisition water suppression, the only technique in common usage at present is the convolution difference low-pass filter technique (46). In this approach, the low-frequency components are filtered from the signal by constructing the data sequence

$$\bar{s}(k\,\Delta t) = \sum_{j=-m}^{m} b_j s([k-j]\,\Delta t) \bigg/ \sum_{j=-m}^{m} b_j, \qquad [3.99]$$

in which b_j are filter coefficients. The cosine-bell filter function, $b_j = \cos[\pi j/(2m+2)]$, works well in practice. Typical values of m range between 8 and 64. The filtered data set is then obtained as $s(k\,\Delta t) - \bar{s}(k\,\Delta t)$. If the experimental data consists of points $s(k\,\Delta t)$ for $k = 0, \ldots, N-1$, the

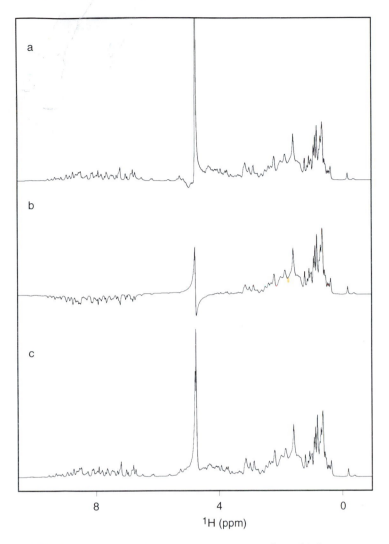

8   4   0

^{1}H (ppm)

FIGURE 3.29 Water suppression using (a) presaturation, (b) jump-return, and (c) spin-lock purge pulses. Additional postacquisition water suppression can be obtained using a digital low-pass filter as illustrated in Fig. 3.30.

filtered data set consists of points $m, \ldots, N - m - 1$. The missing m points at the end of the data sequence are generally unimportant because the original data set can be truncated; however, the first m points must be reconstructed to avoid large distortions in the resulting filtered spectrum. In the original approach, the first points were obtained by linear extrapolation (46); other approaches have included linear prediction of the first m points (47) or acquisition of data points for $k < 0$ using a Hahn echo pulse sequence (48). In most applications, the carrier is positioned at the frequency of the water signal and the protocol described above filters the water signal from the spectrum. If the carrier is positioned elsewhere in the spectrum, the water frequency is shifted digitally to zero by multiplication by a complex exponential function as shown by [3.15]. The convolution filter is applied to suppress the water signal, and the original frequency reference is restored by multiplication by the complex conjugate of the original exponential function. The use of the digital low-pass filter for water suppression is illustrated in Fig. 3.30. As shown, the water signal is almost completely removed, although some distortion of the baseline is obtained near the location of the water signal. In all likelihood, improved digital filtration techniques will be developed (or adapted) for use in NMR spectroscopy.

3.6 One-Dimensional Proton NMR Spectroscopy

3.6.1 SAMPLE PREPARATION

Sophisticated NMR experiments only rarely can compensate for an ill-behaved or ill-prepared NMR sample. Accordingly, in nearly all investigations, preliminary experiments must be performed to determine sample conditions that satisfy the following criteria:

1. The protein must be in a native, functional conformation (unless unfolded or intermediate states of the protein are the focus of investigation). Ideally, the pH and solvent composition should be close to physiological conditions so that the structural and dynamic observations reflect a functionally relevant state of the protein.
2. Solubility must be sufficient to permit spectra with satisfactory signal-to-noise ratios to be acquired in reasonable time periods. For the majority of spectrometer configurations in use today, 0.4–0.6 ml of a protein solution of approximately 1 mM concentration (in a 5 mm NMR tube) will be required for studies at 500 MHz.
3. Protein samples used for NMR spectroscopy must be free of contaminants arising from the NMR tube or the protein preparation. The protein should be monodisperse (unaggregated) at the concentrations required for NMR spectroscopy and stable for time periods longer than those required for the desired NMR experiments.

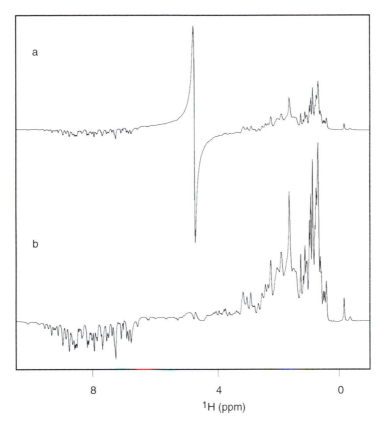

FIGURE 3.30 Low-pass filter. An example of postacquisition water suppression using a low-pass digital filter is illustrated for ubiquitin. (a) Jump–return spectrum of ubiquitin displays a large amplitude residual solvent peak with dispersive tails that obscure nearby resonances. (b) Following filtering in the time domain, the water signal has been suppressed dramatically. Only a small distorted region exists at the position of the water signal.

4. Ideally, NMR spectroscopy should be performed under sample conditions that yield optimal spectra (maximal resonance dispersion and minimal linewidths).

Temperature, pH, concentration, and buffer composition affect the solubility, aggregation state, and stability of proteins dramatically, and must be optimized empirically. Sample preparation for NMR spectroscopy has been the subject of extensive review articles (49,50).

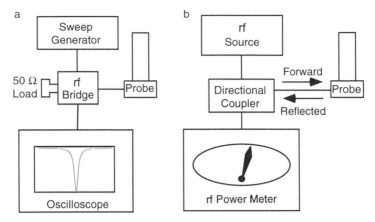

FIGURE 3.31 Tuning circuits employing (a) sweep generator and (b) monochromatic rf source. Details on the use of these circuits are given in the text.

3.6.2 INSTRUMENT SETUP

3.6.2.1 Tuning   In order to efficiently deliver rf energy into the sample volume and to sensitively detect precessing transverse magnetization, the probe circuitry must be *tuned* so that the *resonant frequency* of the circuit is equal to the rf frequency. In addition, the *impedance* of the coil must be *matched* to the impedance of the spectrometer electronics [3.2]. Tuning and matching are performed by adjusting capacitors in the probe rf circuitry. The principle is simple: The coil is driven by an rf input and the response is observed as the fraction of reflected power (i.e., the rf power not transmitted into the sample volume). The capacitors are adjusted interactively to optimize the response. Two methods are commonly used to tune and match the probe; the two methods differ in the way in which rf is applied and the response detected. Increasingly, the capability to tune and match the probe is built into NMR spectrometers.

In the first method, illustrated in Fig. 3.31a, a sweep generator, a 50-Ω load, an oscilloscope, and the probe are connected to the terminals of an rf bridge. The sweep generator generates an rf field with a frequency that is cyclically varied in time. As the rf frequency "sweeps" through the resonance frequency of the coil circuit, the sides of the rf bridge become more balanced, and a dip (or peak, depending on the input polarity) is observed on the scope. The horizontal position of the dip indicates the resonance frequency of the coil (typically, an rf reference source is also displayed on the oscilloscope in order to calibrate the display); the depth of the dip is a measure of the match between the impedance of the circuit and the 50-Ω load. The tuning and matching capacitors are adjusted until the resonance frequency of the coil equals the desired value and the impedance is optimally matched (as indicated by a maximum in the depth of the dip).

In the second method, illustrated in Fig. 3.31b, a fixed-frequency rf source and a voltage standing wave ratio (VSWR) rf power meter are connected to the probe through a directional coupler. The signal measured on the power meter is proportional to the power reflected from the probe; as the tuning and matching of the probe improves, the amount of reflected power decreases, because more power is being transmitted into the sample volume. Accordingly, the tuning and matching capacitors are adjusted to minimize the reflected power.

If the probe is very poorly tuned, or if one is searching for a new nuclear resonance signal on a broadband probe, the sweep generator configuration is superior because tuning and matching responses are monitored independently. Once the probe is nearly tuned and matched, the second approach typically is more sensitive.

Minimum pulse lengths for a given amplifier power level are obtained only if the probe is properly tuned and matched. A historical record of pulse lengths previously measured for the same sample and probe should be maintained (pulse lengths for different samples are a function of ionic strength). A measured pulse length that deviates by more than a few percent from previous values indicates that either the probe has been improperly tuned or that a problem has developed with the transmitter circuitry.

3.6.2.2 Shimming The homogeneity of the static magnetic field is of paramount importance for high-resolution NMR spectroscopy. The natural magnetic field of a superconducting magnet is not sufficiently homogeneous for high-resolution spectroscopy. Accordingly, a necessary task prior to performance of an NMR experiment is the careful adjustment of the magnetic fields produced by a set of auxiliary room-temperature electromagnets to compensate the inhomogeneity of the main static field. This process is known as shimming. Two articles devoted to the process of shimming form the basis for the present discussion (*51*).

The spatial variation of the static magnetic field within the bore of the magnet satisfies the Laplace equation and can be described by an expansion in orthogonal spherical harmonic functions, $r^n P_{nm}(\cos \theta) \cos[m(\varphi - \varphi_{nm})]$:

$$B_0(r, \theta, \varphi) = \sum_{n=0}^{\infty} \sum_{m=0}^{n} c_{nm} \left(\frac{r}{a}\right)^n P_{nm}(\cos \theta) \cos[m(\varphi - \varphi_{nm})], \quad [3.100]$$

TABLE 3.1
Shim Coil Spherical Harmonic Functions

Order (n)	Degree (m)	Shim name	Functions
1	0	z	z
2	0	z^2	$2z^2 - (x^2 + y^2)$
3	0	z^3	$z[2z^2 - 3(x^2 + y^2)]$
4	0	z^4	$8z[z^2 - 3(x^2 + y^2)] + 3(x^2 + y^2)^2$
5	0	z^5	$48z^3[z^2 - 5(x^2 + y^2)] + 90z(x^2 + y^2)^2$
1	1	x	x
1	1'	y	y
2	1	zx	zx
2	1'	zy	zy
3	1	z^2x	$x[4z^2 - (x^2 + y^2)]$
3	1'	z^2y	$y[4z^2 - (x^2 + y^2)]$
2	2	$x^2 - y^2$	$x^2 - y^2$
2	2'	xy	xy
3	2	$z(x^2 - y^2)$	$z(x^2 - y^2)$
3	2'	zxy	zxy
3	3	x^3	$x(x^2 - 3y^2)$
3	3'	y^3	$y(3x^2 - y^2)$

in which a is the radius of the magnet bore, c_{nm} and φ_{nm} are constants, and $P_{nm}(\cos \theta)$ are the associated Legendre polynomials. The values of n and m for which c_{nm} are nonzero, and the particular values of c_{nm} and φ_{nm} are determined (depending on one's point of view) by the empirical variation in the magnetic field or by the solution to the Laplace equation subject to the (complicated) appropriate boundary conditions inside the magnet bore. If the field were perfectly homogenous, the only nonzero c_{nm} in [3.100] would be c_{00}, the amplitude of the static field.

In principle, the inhomogeneity in the static magnetic field can be compensated identically by generation of additional magnetic fields with spatial variation described by the spherical harmonic functions and amplitude given by $-c_{nm}$. Each additional magnetic field would negate one of the terms in [3.100], and the number of auxiliary fields needed would be determined by the desired degree of compensation. In practice, the auxiliary fields are produced by specially designed electromagnets and the magnitude of the fields are proportional to the applied electric currents. The shim coils are conventionally described using Cartesian rather than spherical coordinates; the spherical harmonic functions corresponding to shim coils frequently encountered on NMR spectrometers are given in

Table 3.1. The functions of degree $m > 0$ appear in pairs with x and y interchanged in order to reproduce the phase dependence, φ_{nm}, in [3.100].

The difficulties in shimming a magnet arise because the shim magnets cannot be designed or fabricated to produce a pure spherical harmonic magnetic field; consequently, the field produced by the shim coils is described by

$$B'(r, \theta, \varphi) = \sum_{k=1}^{N} b_k B_k(r, \theta, \varphi), \qquad [3.101]$$

in which N is the number of shim coils, $B_k(r, \theta, \varphi)$ is the spatial field produced by the kth shim coil, and b_k is the amplitude of the field determined by the electric current applied to the electromagnet. Each $B_k(r, \theta, \varphi)$ is given by an expansion similar to [3.100]; as a result, the terms in [3.101] are not orthogonal and no one-to-one correspondence exists between the terms in [3.101] and the terms in [3.100]. Fortunately, the problem is simplified somewhat because a given shim field is contaminated primarily by shim fields of lower orders with the same parity (i.e., z^3 contains contributions primarily from z, and z^4 contains contributions primarily from z^2). In addition, the resonance lineshape obtained is frequently indicative of the order of the shim coil that must be adjusted. Misadjustment of the even-order shims results in asymmetrical lineshapes; misadjustment of the odd-order shims result in symmetrical, but non-Lorentzian, lineshapes. Furthermore, the effect on the lineshape is observed closer to the baseline for higher-order shims. The problem of shimming reduces to implementation of a reasonable strategy for optimizing the coefficients b_k in [3.101] to approximate to [3.100].

Empirical adjustment of the shim coils is performed using the magnitude of the field-frequency lock signal, the decay envelope of the FID, and the lineshape in the frequency-domain spectrum as measures of the homogeneity of the magnetic field. The lock signal is the simplest parameter to observe; since the integral of the resonance signal is constant, the lock signal increases in magnitude as the field homogeneity improves and the resonance lineshape becomes narrower. However, the magnitude of the lock signal does not indicate the quality of the lineshape obtained. In addition, the lock system must be properly calibrated if proper shimming is to be obtained. The phase of the lock receiver must be adjusted to yield a purely dispersive signal. Drifts in the magnetic field are detected most sensitively if the lock signal is dispersive, because the dispersive lineshape has a null at the exact resonance frequency [3.25]. The power level must be adjusted to yield a signal with adequate signal-to-noise ratio without saturating the lock resonance. If the FID can be displayed in real time, the magnetic field can be shimmed by observing the shape of the decay envelope. Optimal homogeneity implies that the decay envelope is exponential with a maximal decay-time constant. Unlike the lock signal, the FID is a direct indication of the quality of the lineshape. Shimming using the FID is easiest if the decay is dominated by a single-resonance line; thus, for samples in water, the magnet can be shimmed using the water resonance. In this case, the probe should be detuned while shimming to avoid radiation damping that distorts the shape of the FID (Section 3.5). After the magnet is shimmed, the probe is retuned (after tuning, the shims may need additional minor adjustments).

The ultimate measures of the homogeneity of the magnetic field are the lineshape and resolution in the spectrum and quality of the solvent suppression. When beginning a research project on a new biomolecule, a pair of closely spaced resonances, preferably representing a scalar-coupled multiplet, should be identified that can be used subsequently to monitor shimming. That is, if a particular multiplet is resolvable when the magnet is properly shimmed, the quality of the shimming can quickly be checked by examining the degree of resolution of the multiplet lineshape.

External samples also can be used to check the homogeneity of the magnetic field. Samples of basic pancreatic trypsin inhibitor (BPTI), ubiquitin, or tryptophan are useful in this regard. Unlike resolution test samples used by instrument manufacturers, these molecules can be dissolved in aqueous solutions with buffers and ionic strength matched to the conditions of the sample of interest. For BPTI, resolution can be evaluated by examining the tyrosine 23 $^1H^\varepsilon$ multiplet at 6.3 ppm. For ubiquitin, resolution can be evaluated by examining the leucine 50 $^1H^\delta$ multiplet at -0.17 ppm or the phenylalanine 45 $^1H^\delta$ multiplet at 7.33 ppm. For tryptophan, resolution can be evaluated by examining the $^1H^{\eta 2}$ triplet at 7.24 ppm. In a high-resolution spectrum, the ^4J coupling to the $^1H^{\varepsilon 3}$ proton should be resolvable almost to baseline and the small (~0.5 Hz) ^6J coupling to $^1H^{\delta 1}$ should be discernible.

The exact protocol that is optimal for shimming a particular magnet depends on the complement of shim coils provided by the spectrometer manufacturer. In addition, automated shimming software is continually being improved by the spectrometer manufacturers and can remove some of the initial drudgery of shimming, particularly when shimming a new probe. Accordingly, the protocol given in Fig. 3.32 should be taken as guidelines appropriate for the shim set given in Table 3.1. The increased availability of three-axis gradient probes has lead to new methods of shimming superconducting magnets (52). Field-gradient pulses can be used to map the spatial distribution of the inhomogeneity of the field and the actual fields produced by each shim. Numerical optimization of the shim currents is then considerably simplified.

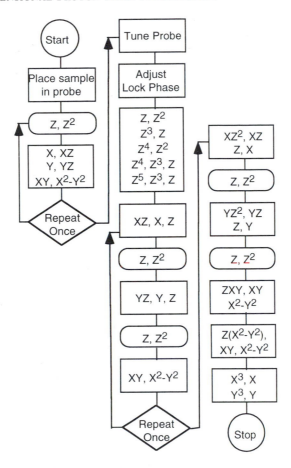

FIGURE 3.32 Shimming protocol.

3.6.2.3 Temperature Calibration   Most commercial spectrometers have some means of controlling the probe temperature, and this hardware provides a coarse estimate of the actual sample temperature. More accurate schemes have been developed that make use of the temperature-dependent chemical shifts of methanol and ethylene glycol to calibrate the actual temperature of a sample in the probe (*53*). Over the range 250–320 K, the difference in chemical shift between the methyl and hydroxyl resonances of 100% methanol, $\Delta\delta$ in ppm (parts per million), is given by

$$T(\text{K}) = 403.0 - 29.53\,\Delta\delta - 23.87(\Delta\delta)^2. \qquad [3.102]$$

Over the range 300–370 K, the difference in chemical shift between the methylene and hydroxyl resonances of 100% ethylene glycol, $\Delta\delta$ in ppm, is given by

$$T(K) = 466.0 - 101.6\ \Delta\delta. \qquad [3.103]$$

3.6.2.4 Pulse-Width Calibration   For a spatially homogenous rf field applied on-resonance, the nominal rotation angle is $\alpha = -\gamma B_1 \tau_p$. The pulse duration yielding a particular rotation angle must be determined empirically each time the spectrometer is to be used for an experiment. Even for the same sample in the same spectrometer, small variations in τ_p will be observed on a day-to-day basis.

Modern NMR pulse sequences frequently use pulses with reduced amplitudes for solvent saturation, extended rf mixing or spin-locking periods, and selective pulses. Accordingly, the mathematical relationship between attenuation, transmitter power, pulse lengths, and B_1 field strength is useful for approximating the pulse lengths at some attenuation given the measured pulse lengths at another attenuation. Attenuation is commonly measured on the decibel scale:

$$P = P_{ref} \cdot 10^{dB/10}. \qquad [3.104]$$

For the dBm (decibels >1 mW) scale, $P_{ref} = 1$ mW and for the dBW (decibels >1 W) scale, $P_{ref} = 1$ W. Thus, an rf amplifier with a maximum power output rated at 20 dBW produces an output of $P = 10^{20/10}$ W = 100 W; a preamplifier with a noise figure of -20 dBm produces a noise output of $P = 10^{-20/10}$ mW = 0.01 mW. The attenuation difference between two different power levels P_1 and P_2 is given by

$$dB = 10\ \log_{10}(P_1/P_2). \qquad [3.105]$$

The B_1 field produced depends on the voltage in the coil, V, not on the power. Since

$$P = V^2/R, \qquad [3.106]$$

in which R is the resistance (typically 50 Ω in a tuned probe),

$$dB = 20\ \log_{10}(V_1/V_2). \qquad [3.107]$$

The difference between voltage and power is critical; changing the attenuation by 3 dB changes the power by a factor of 2, but the attenuation must be changed by 6 dB to change the voltage by a factor of 2. Since the B_1 field strength depends on the voltage, doubling the strength of the B_1 field (which is equivalent to halving the 90° pulse length) requires that the power output of the transmitter be quadrupled.

The voltage V in [3.106] and [3.107] is the root-mean-square voltage produced by the rf field. The peak-to-peak voltage, V_{pp}, is more easily measured using an oscilloscope. For a time-dependent voltage, $V(t) = (V_{pp}/2) \cos \omega_{rf} t$, the two quantities are related by

$$V^2 = \frac{V_{pp}^2}{4} \int_0^{2\pi} \cos^2 \theta \, d\theta \bigg/ \int_0^{2\pi} d\theta = \frac{V_{pp}^2}{8}. \qquad [3.108]$$

Combining this result with [3.106] yields the following useful relationship for power measured in watts, voltage measured in volts, and an assumed 50-Ω resistance:

$$P = V_{pp}^2/400. \qquad [3.109]$$

If direct observation of the signal from a particular nucleus is feasible (e.g., protons), the length of a 360° pulse is determined by searching for the null in the signal observed after application of the pulse. First, a FID is acquired with a pulse–acquire sequence using a short pulse length. The spectrum is Fourier-transformed and phased. The result of this experiment is

$$I_z \xrightarrow{\alpha I_x} I_z \cos \alpha - I_y \sin \alpha, \qquad [3.110]$$

where the pulse applied for a time t produces a net rotation of I_z by α radians. The intensity of the signal will depend on the rotation angle α (and the time t) in a sinusoidal manner. Determining the length of time required to produce a specific rotation is most accurately accomplished when the observed signal is at a null, i.e., where the pulse produces a rotation by a multiple of π radians. After every pulse–acquire experiment, the system should be allowed to reach equilibrium, so that I_z is at a maximum at the start of the next experiment. For this reason, measurement of a 2π pulse is most accurate: as the null is approached, the magnetization will be rotated almost back to the $+z$ axis, and will therefore require less time to return to equilibrium. If the approximate pulse length is not known, care must be exercised to ensure that the null corresponds to a 2π rotation, rather than a different multiple of π rotations. Once the length of time required for a 2π pulse has been determined, the length of time required for other pulses can be calculated from the proportionality between α and t.

As described in Section 3.4.1, when the offset is large compared to the B_1 field strength, the magnetization does not behave as predicted by [3.110]. Thus, when calibrating the weak pulses required for spin-lock or composite pulse decoupling schemes, only the magnitude of signals on resonance should be considered. However, the best solvent suppression is achieved when the H_2O signal is on-resonance, in which case protein peaks that are close to resonance will be obscured by the incompletely suppressed solvent signal. Thus, weak B_1 fields are best calibrated by shifting the transmitter frequency, after the presaturation pulse and prior to the excitation pulse, to be on-resonance with a well-resolved peak, such as that of an upfield-shifted methyl group. In this way, the pulse length required to achieve a null can be accurately determined without influence of offset effects.

In principle, pulse lengths for the heteronuclear channels can be measured in an analogous way; however, this is rarely practical because of the insensitivity of these nuclei. Further, the heteronuclear experiments described in this book make use of indirect detection of ^{13}C and ^{15}N spins by transfer of coherence to the directly bonded ^1H nuclei. Acquisition of such spectra may require probe and preamplifier configurations different from those required for a direct-detect experiment. Since the pulse lengths depend on the exact rf circuitry utilized, pulse lengths for an indirect–detect experiment should be determined by an indirect method. If the protein sample contains a well-resolved proton signal scalar coupled to the desired heteronucleus, the protein sample may be used directly for calibration, although the amount of signal averaging to achieve acceptable signal-to-noise ratios may make such calibrations time-consuming. In many cases, heteronuclear pulse lengths can be obtained more readily on a test sample of higher concentration that contains a single labeled moiety for each heteronucleus to be calibrated. Small peptides or amino acids are useful for this purpose. The solution conditions of this sample (particularly solvent and ionic strength) should be as close to the protein sample as possible, in order to minimize differences in tuning and pulse length between the two.

Pulse sequences for indirect calibration rely on the coherence transfer properties of an IS ($I = {}^1H$ and $S = {}^{13}C$ or ^{15}N) spin system. Three pulse schemes for indirect measurements of heteronuclear pulse lengths are given in Fig. 3.33. If the resonance for the I spin attached to the S spin is well resolved, the pulse sequence of Fig. 3.33a is satisfactory. The product operator analysis of the pulse sequence yields

$$I_z \rightarrow 2I_x S_z \cos \alpha - 2I_x S_y \sin \alpha, \qquad [3.111]$$

in which $\Delta = 1/(2J_{IS})$ and J_{IS} is the scalar coupling constant between the I and S spins. For $\alpha \rightarrow 0$, an antiphase lineshape is obtained; the signal is nulled when $\alpha = \pi/2$, because the $2I_x S_y$ operator does not evolve into observable coherence during acquisition. The experiment is first run with $\alpha = 0$ so that the phase parameters can be determined to give an antiphase

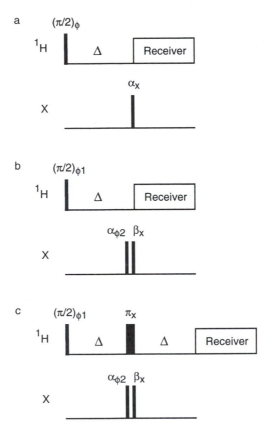

FIGURE 3.33 Pulse sequences for indirect calibration of heteronuclear pulse lengths. In all sequences, $\Delta = 1/(2J_{XH})$. Phase cycles are (a) $\phi_1 = \{x, y, -x, -y\}$, receiver $= \{x, y, -x, -y\}$; (b,c) $\phi_1 = 2\{x, y, -x, -y\}$, $\phi_2 = \{x, x, x, x, -x, -x, -x, -x\}$, receiver $= \{x, y, -x, -y, -x, -y, x, y\}$. Additional details are given in the text.

doublet; α is then systematically increased until a null is found for the intensity of the doublet. If the spectrum is very crowded, overlapping antiphase components may lead to large degree of cancellation of the signals and difficulty in accurately determining the pulse length.

If the signal from the S-bound proton is not well resolved from signals of protons not bound to heteronuclei (e.g., in a peptide containing a single site of ^{15}N incorporation), the signal can be more clearly discerned by using the sequence of Fig. 3.33b. Product operator analysis yields

$$I_z \rightarrow -2I_x S_z \sin \alpha \sin \beta - 2I_x S_y \sin \alpha \cos \beta. \qquad [3.112]$$

This experiment yields antiphase observable signals proportional to $\sin \alpha$ $\sin \beta$. The maximum signal is obtained for $\alpha = \beta \approx \pi/2$, and a null is observed for $\beta = \pi$. This sequence incorporates an isotope filter, by alternating the phase of the first pulse on the heteronuclear channel, so that signals from protons not coupled to an S spin are suppressed. Initially, both α and β are set equal to an estimate of the $\pi/2$ pulse length for the heteronuclear channel, resulting in an antiphase doublet for the S-bound proton. The value of β is then systematically increased, until a null is obtained. Alternatively, in-phase signals can be obtained by using the more elaborate pulse sequence of Fig. 3.33c, which yields

$$I_z \rightarrow -I_y \sin \alpha \sin \beta - 2I_x S_y \sin \alpha \cos \beta. \qquad [3.113]$$

Because of the wide spectral widths encountered in ^{13}C and ^{15}N spectra, offset effects can be very severe. Thus, having the S nucleus on-resonance is very important if accurate pulse lengths are to be determined. In cases where the chemical shifts of the S nucleus are not known, the pulses may be roughly measured by estimating the S nucleus frequency. A short HSQC or HMQC experiment (Section 7.1) is acquired to ascertain the exact chemical shift (these experiments are reasonably tolerant to imperfect pulse lengths). With this information available, the pulse length can be remeasured accurately.

3.6.2.5 Recycle Delay   The optimal combination of the recycle delay between transients and the pulse rotation angle depends on the rate at which equilibrium magnetization recovers after a perturbation. For signal averaging in a one-pulse experiment with a delay between acquisitions of T (equal to the sum of the recycle delay and acquisition time), the initial amplitude of the FID is proportional to

$$\varepsilon = \frac{1 - e^{-T_c/T_1}}{1 - e^{-T_c/T_1} \cos \alpha} \sin \alpha. \qquad [3.114]$$

Equation [3.114] is plotted in Fig. 3.34. For a time T_c, the maximum signal is obtained for a rotation angle, α_e, known as the Ernst angle:

$$\cos \alpha_e = \exp(-T_c/T_1). \qquad [3.115]$$

Thus, if $T_c < \infty$, then $\alpha_e < 90°$ and $\varepsilon < 1$. Essentially complete recovery of the equilibrium longitudinal magnetization occurs for $T_c > 3T_1$. Recycle delays of $T_1 < T_c < 1.5\ T_1$ yield superior sensitivity per unit time, and are frequently used in multidimensional experiments, because the reduced recycle delay permits increased signal averaging, which offsets the loss of sensitivity due to incomplete relaxation recovery.

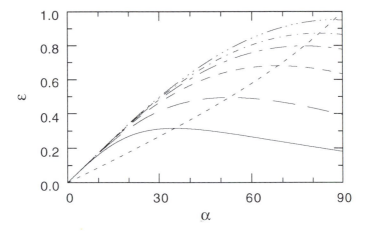

FIGURE 3.34 Ernst angle. The fractional signal intensity, ε, is shown as a function of pulse length, α, in a pulse–acquire NMR experiment. Results are shown for normalized recycle delays, T_c/T_1, equal to (——) 0.2, (— —) 0.5, (– – –) 1.0, (– – –) 1.5, (– · – ·) 2.0, and (– · · · –) 3.0. The optimal curve yielding the highest value of ε is shown also (- - - -).

An approximate value of the proton T_1 can be determined using the inversion recovery pulse sequence

$$\text{Recycle}–180°–\tau–90°–\text{acquire,} \qquad [3.116]$$

in which the 90° pulse and receiver are phase cycled in 90° increments. The recycle delay should be greater than $3T_1$ to ensure complete relaxation between transients (which may require repeating the experiment following an initial approximate determination of T_1). The magnitude of the transverse magnetization following the 90° pulse varies with the delay, τ, as

$$M(\tau) = M(0)[1 - 2 \exp(-\tau/T_1)]. \qquad [3.117]$$

The delay is systematically varied until the signal is nulled. The value of T_1 is approximately given by

$$T_1 = \tau_{\text{null}}/\ln 2. \qquad [3.118]$$

An example of a nonselective inversion recovery experiment for a ubiquitin sample in D_2O solution is shown in Fig. 3.35. As is evident, the upfield methyl protons in ubiquitin have shorter T_1 values than do the H^α and aromatic protons and recover to equilibrium more rapidly. Approximate values of T_1 determined using [3.118] are 0.7, 1.4, and 1.8 s for methyl,

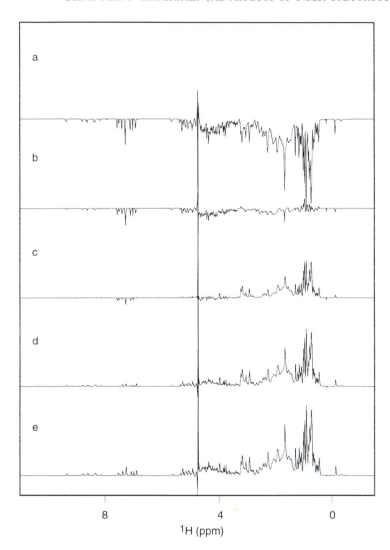

FIGURE 3.35 ¹H inversion recovery spectra of ubiquitin. Partially relaxed nonselective inversion recovery spectra of ubiquitin are shown for recovery delays of 3 μs (a), 0.5 s (b), 1.0 s (c), 1.5 s (d), and 2.0 s (e). Spectra were recorded with a 12-s recycle delay using a ubiquitin sample in 100% D_2O solution.

H^α, and aromatic protons, respectively. A small number of persistent amide protons with $T_1 \approx 1.0$ s are evident in the spectra.

3.6.3 Referencing

The concept of the chemical shift was introduced in Section 1.5. To facilitate comparisons of resonance positions between different samples and spectrometers, chemical shifts are measured by reference to a standard compound using units of ppm, as described by [1.47]. Tetramethylsilane (TMS) is the universal reference for ^{1}H NMR. In studies of organic molecules, TMS may be added directly to the solvent (e.g., deuterated chloroform spiked with TMS is commercially available), thereby providing an internal reference from which the chemical shifts of sample resonances can be determined. The situation is less straightforward in studies of proteins as TMS is not soluble in aqueous solutions. Instead, either a different internal reference species or an external reference must be used.

The ideal internal chemical-shift marker should not interact with the protein under investigation, and should have a single resonance whose chemical shift varies with temperature and pH in a known manner. The resonance should be well resolved from the resonances of the protein, because the reference signal will have a long T_1, and t_1 noise emanating from it may obscure cross-peaks in multidimensional spectra. Suitable internal markers include water-soluble compounds containing a trimethylsilyl group [e.g., 3-(trimethylsilyl) propionate sodium salt (TSP), perdeuterated, except for the trimethylsilyl group, so that the only resonance is at 0.00 ppm], dioxane (chemical shift = 3.75 ppm), 2,2-dimethyl-2-silapentane-5-sulfonic acid (DSS; chemical shift 0.00), residual protons in deuterated buffer components (e.g., acetate methyl resonance at 2.05 ppm.), or the water resonance itself.

Most of the alternative standards have significant drawbacks (54). The resonance frequency of TSP is pH-dependent. Nonspecific binding of TSP to hydrophobic patches on the surface of proteins can lead to a chemical shift that is not equal to zero and that is dependent on the concentration of protein present because of the fast exchange between free and protein-bound species. Similarly, dioxane may cause problems due to overlap with protein resonances. The position of the H_2O peak provides one of the least perturbing, but least accurate, methods for referencing the chemical shift. The shift has a slight dependence on pH (0.02 ppm per pH unit), and a more dramatic dependence with temperature, so the chemical shift of the water must be calculated in each case from (55,56)

$$\delta(H_2O) = 7.83 - T/96.9 \text{ ppm}, \qquad [3.119]$$

TABLE 3.2
Indirect Chemical-Shift References

Compound	γ_C/γ_H	γ_N/γ_H
DSS	0.25144952	0.10132905
TMS	0.25145002	0.10132914
TSP	0.25144954	0.10132900

Reported values are from Refs. *57, 58,* and *62.*

in which the temperature is measured in kelvins and pH = 5.5. As a consequence of these considerations, Wishart and Sykes have recommended the use of 10–20-μM DSS as an internal chemical-shift reference standard for protein ^{1}H NMR spectroscopy (*54*).

The insensitivity of heteronuclei (^{13}C or ^{15}N) makes calibration via direct observation troublesome. In addition, particularly for ^{15}N, satisfactory internal reference standards do not exist (*54*). As described by Live *et al.* (*57*) and Bax and Subramanian (*58*), once the proton shifts are calibrated, the heteronuclear chemical shifts can be indirectly calibrated by using the following relationship:

$$\nu_0^X = \nu_0^H \gamma_X / \gamma_H,$$ [3.120]

where ν_0^X is the absolute frequency of 0 ppm for the X nucleus, ν_0^H is the absolute frequency of 0 ppm for the H nucleus, and γ_X/γ_H is the ratio of gyromagnetic ratios for the X and H nuclei. Values of γ_X/γ_H for DSS, TMS, and TSP are given in Table 3.2. Given that the proton shifts have been referenced, the absolute ^{1}H frequency of 0 ppm can be determined, and the offset required for a particular ^{13}C or ^{15}N shift can be calculated from [3.120].

As an example, if the absolute frequency of DSS is measured to be 500.136662 MHz at 0.0 ppm, the absolute zero frequency of ^{15}N is determined from [3.120] to be

$$500.136662 \, \gamma_X/\gamma_H = 500.136662 \times 0.10132905$$

$$= 50.678373 \text{ MHz.}$$ [3.121]

Therefore, to perform an HSQC experiment with the ^{15}N region centered in the middle of the amide nitrogen resonances (say, at 115.0 ppm), the required experimental offset would be

$$50.678373 \, (1 + 115.0 \times 10^{-6}) = 50.684201 \text{ MHz.}$$ [3.122]

3.6.4 Acquisition and Data Processing

3.6.4.1 One-Pulse Experiment   The basic NMR experiment is the so-called one-pulse experiment in which a rf pulse of rotation angle α is applied to the system and the resulting transverse signal detected:

$$\text{Recycle–pulse–acquire.} \qquad\qquad [3.123]$$

As discussed in Section 4.3.2.3, CYCLOPS phase cycling commonly is used during the one-pulse experiment to suppress quadrature images. In CYCLOPS phase cycling, the phase of the pulse and the phase of the receiver are shifted in 90° steps between transients. At minimum, the following parameters must be adjusted in setting up the one-pulse experiment: recycle delay, rf carrier position, pulse duration, sampling interval or spectral width, number of digitized data points in the time domain, number of transients to be acquired, and receiver amplifier gain. The rf carrier frequency is set to the center of the spectrum or to the frequency of the solvent resonance. Normally the sampling interval, Δt, is adjusted such that the Nyquist frequency is larger than the maximum resonance frequency arising from the sample; however, in some instances, resonances may be deliberately aliased in order to minimize the sampling rate (Section 7.1.2.5). The number of data points acquired, N, is a power of two such that $N \Delta t > 3T_2$ in order to minimize truncation artifacts. The receiver gain is adjusted so that the signal arising from the FID does not underflow or overflow the dynamic range of the receiver. The number of transients acquired depends on the signal averaging required to achieve the desired signal-to-noise ratio in the spectrum; the number of transients must be a multiple of 4 because of the CYCLOPS phase cycling.

3.6.4.2 Hahn Echo Experiment   The simple one-pulse experiment may not be satisfactory for detailed examination of the one-dimensional spectrum of a biological macromolecule. As was noted in Section 3.3.2.3, baseline distortions can hamper the interpretation of NMR spectra. Removal of baseline distortions can be achieved by the use of a so-called Hahn echo pulse sequence (*59,60*). This technique, originally introduced by Rance and Byrd for wideline NMR spectroscopy in anisotropic media (*59*), is well suited to application in high-resolution ^{1}H-detected NMR spectroscopy of biomolecules.

In NMR spectroscopy, the entire FID must be recorded to obtain spectra free of distortions. Accurate detection of the initial part of the FID is crucial in wideline NMR spectroscopy since the signal decays rapidly (the resonance linewidth can be of the order of several kilohertz). In the simple pulse–acquire detection scheme, a dead-time follows the high-power rf pulse as the receiver is saturated and ringing effects are introduced in the tuned circuits. The FID decays markedly during this period; consequently, when the receiver is eventually actuated, the first part of the FID is absent. Rance and Byrd used the Hahn echo pulse sequence

$$90°-\tau_1-180°-\tau_2-\text{acquire} \qquad [3.124]$$

to create a spin echo at τ_2 (59). By having the echo form beyond the receiver dead-time period, the Hahn echo sequence avoids distortions due to finite receiver recovery time.

A similar situation is encountered in high-resolution ^{1}H NMR spectroscopy. Baseline roll in high-resolution spectra, as was discussed in Section 3.3.2.3, commonly is caused by the transient response of the spectrometer to the incoming signal that distorts the first few points of the FID. The Hahn echo sequence allows these distortions to be avoided by the same mechanism as for wideline spectra.

The delays in the Hahn echo sequence can be calculated as

$$\tau_1 = \tau_2 - 2\tau_{90}/\pi - \tau_{\text{gate}}, \qquad [3.125]$$

in which τ_2 is greater than the filter response time, τ_{90} is the length of the 90° pulse, and τ_{gate} is the receiver gating delay. In each case, τ_2 is adjusted empirically by small amounts to ensure that acquisition is initiated at the exact top of the echo. As shown in Fig. 3.36, phase errors in the spectrum are eliminated when τ_2 is adjusted accurately. Once τ_2 is optimized with respect to the phase of the spectrum, the value of τ_2 is reduced by enough sample times to allow the filter transient response to decay prior to the top of the echo. Data points acquired before the top of the echo are not included in the Fourier transformation (48). A wide spectral width typically is used for oversampling of the FID (61). The basic phase cycle incorporates CYCLOPS phase cycling for the 90° pulse (ϕ) and EXORCYCLE phase cycling for the 180° pulse (ψ) to yield a 16-step phase cycle: $\phi = \{xxxxyyyy \ -x-x-x-x-y-y-y-y\}$, $\psi = 4 \{xy-x-y\}$, and receiver $= \{x-xx-xy-yy-y-xx-xx-yy-yy\}$. CYCLOPS and EXORCYCLE are discussed in Section 4.3.2.3.

The Hahn echo sequence is slightly longer than a simple pulse–acquire sequence. Relaxation during the τ_1 and τ_2 delays can reduce the signal intensity; however, the delays are sufficiently short that relaxation effects generally can be safely ignored. Similarly, evolution of the homonuclear scalar coupling occurs during the τ_1 and τ_2 delays. For typical values of $J_{HH} < 15$ Hz, $\cos[\pi J_{HH}(\tau_1 + \tau_2)] > 0.9999$, and $\sin[\pi J_{HH}(\tau_1 + \tau_2)] < 0.016$; consequently, scalar coupling effects also can be ignored. In addition to its other benefits, the Hahn echo sequence also has demonstrated

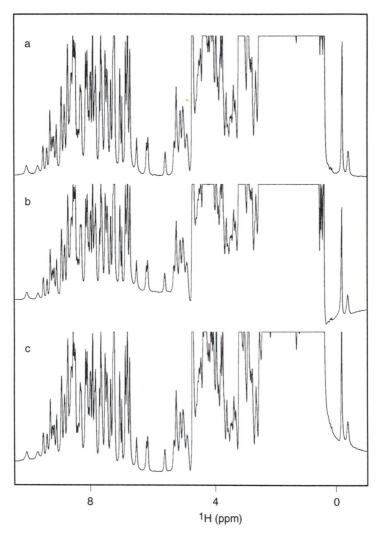

FIGURE 3.36 Hahn echo ^{1}H NMR spectra acquired with a spectral width of 12,500 Hz and a filter width of 30,000 Hz. (a) Echo delays $\tau_1 = 140$ μs and $\tau_2 = 174$ μs are adjusted to eliminate phase errors in the spectrum, (b) τ_2 is misadjusted to be 10 μs shorter than optimal, and (c) τ_2 is misadjusted to be 10 μs longer than optimal.

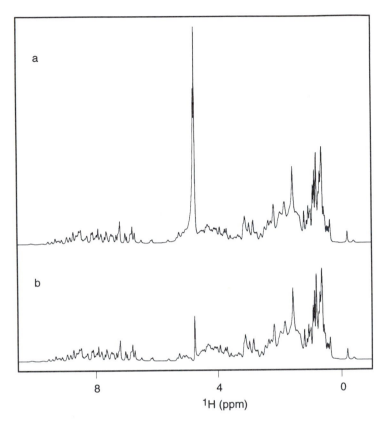

FIGURE 3.37 (a) Pulse-acquire and (b) Hahn echo ^{1}H NMR spectra of ubiquitin. The Hahn echo spectrum has better water suppression and flatter baseline than the pulse–acquire spectrum. Both spectra were acquired with identical spectral and filter widths.

significantly improved water suppression due principally to the refocusing properties of the 180° pulse, which reduces the effects of inhomogeneous broadening at the base of the residual water peak. Example one-dimensional NMR spectra acquired with pulse–acquire and Hahn echo pulse sequences are shown in Fig. 3.37.

References

1. A. Abragam, "Principles of Nuclear Magnetism," pp. 1–599. Clarendon Press, Oxford, 1961.

2. D. I. Hoult, C.-N. Chen, H. Eden, and M. Eden, *J. Magn. Reson.* **51,** 110–117 (1983).

3. S. L. Marple, "Digital Signal Processing," pp. 1–492. Prentice-Hall, Englewood Cliffs, N.J. 1987.

4. R. N. Bracewell, "The Fourier Transform and Its Applications," pp. 1–474. McGraw-Hill, New York, 1986.

5. E. Bartholdi and R. R. Ernst, *J. Magn. Reson.* **11,** 9–19 (1973).

6. R. R. Ernst, G. Bodenhausen, and A. Wokaun, "Principles of Nuclear Magnetic Resonance in One and Two Dimensions," pp. 1–610. Clarendon Press, Oxford, 1987.

7. A. Bax, M. Ikura, L. E. Kay, and G. Zhu, *J. Magn. Reson.* **91,** 174–178 (1991).

8. G. Zhu, D. A. Torchia, and A. Bax, *J. Magn. Reson., Ser. A* **105,** 219–222 (1993).

9. G. Otting, H. Widmer, G. Wagner, and K. Wüthrich, *J. Magn. Reson.* **66,** 187–193 (1986).

10. R. K. Harris, "Nuclear Magnetic Resonance Spectroscopy," pp. 1–260, Longman Scientific and Technical, Harlow (UK), 1986.

11. H. Barkhuijsen, R. de Beer, W. M. M. J. Bovée, and D. van Ormondt, *J. Magn. Reson.* **61,** 465–481 (1985).

12. S. Sibisi, J. Skilling, R. G. Brereton, E. D. Laue, and J. Staunton, *Nature* (London) **311,** 446–447 (1984).

13. R. E. Hoffman, A. Kumar, K. D. Bishop, P. N. Borer, and G. C. Levy, *J. Magn. Reson.* **83,** 586–594 (1989).

14. J. J. Kotyk, N. G. Hoffman, W. C. Hutton, G. L. Bretthorst, and J. J. H. Ackerman, *J. Magn. Reson.* **98,** 483–500 (1992).

15. D. S. Stephenson, *Prog. Nucl. Magn. Reson. Spectrosc.* **20,** 515–626 (1988).

16. G. Zhu and A. Bax, *J. Magn. Reson.* **90,** 405–410 (1990).

17. H. Barkhuijsen, R. de Beer, and D. van Ormondt, *J. Magn. Reson.* **73,** 553–557 (1987).

18. D. Marion and A. Bax, *J. Magn. Reson.* **83,** 205–211 (1989).

19. E. D. Laue, M. R. Mayger, J. Skilling, and J. Staunton, *J. Magn. Reson.* **68,** 14–29 (1986).

20. E. D. Laue, J. Skilling, and J. Staunton, *J. Magn. Reson.* **63,** 418–424 (1985).

21. F. Bloch and A. Siegert, *Phys. Rev.* **57,** 522–527 (1948).

22. M. A. McCoy and L. Mueller, *J. Magn. Reson.* **99,** 18–36 (1992).

23. M. A. McCoy and L. Mueller, *J. Magn. Reson.* **98,** 674–679 (1992).

24. M. H. Levitt, *Prog. NMR Spectrosc.* **18,** 61–122 (1986).

25. A. J. Shaka and J. Keeler, *Prog. NMR Spectrosc.* **19,** 47–129 (1987).

26. J. S. Waugh, *J. Magn. Reson.* **50,** 30–49 (1982).

27. A. J. Shaka, J. Keeler, T. Frenkiel, and R. Freeman, *J. Magn. Reson.* **52,** 335–338 (1983).

28. A. J. Shaka, P. B. Barker, and R. Freeman, *J. Magn. Reson.* **64,** 547–552 (1985).

29. M. McCoy and L. Mueller, *J. Magn. Reson., Ser. A* **101,** 122–130 (1993).

30. L. Emsley, *Meth. Enzymol.* **239,** 207–246 (1994).

31. J. Shen and L. E. Lerner, *J. Magn. Reson., Ser. A* **112,** 265–269 (1995).

32. N. Bloembergen and R. V. Pound, *Phys. Rev.* **95,** 8–12 (1954).

33. S. Bloom, *J. Appl. Phys.* **28,** 800–805 (1957).

34. P. J. Hore, *Meth. Enzymol.* **176,** 64–77 (1989).

35. M. Guéron, P. Plateau, and M. Decorps, *Prog. NMR Spectrosc.* **23,** 135–209 (1991).

36. P. Plateau and M. Guéron, *J. Am. Chem. Soc.* **104,** 7310–7311 (1982).

37. P. J. Hore, *J. Magn. Reson.* **55,** 283–300 (1983).

38. S. Grzesiek and A. Bax, *J. Am. Chem. Soc.* **115,** 12593–12594 (1993).

39. Y.-C. Li and G. T. Montelione, *J. Magn. Reson., Ser. B* **101,** 315–319 (1993).

40. V. Sklenár and A. Bax, *J. Magn. Reson.* **75,** 378–383 (1987).

41. M. H. Levitt, G. Bodenhausen, and R. R. Ernst, *J. Magn. Reson.* **53,** 443–461 (1983).

42. A. L. Davis, G. Estcourt, J. Keeler, E. D. Laue, and J. J. Titman, *J. Magn. Reson., Ser. A* **105,** 167–183 (1993).

43. B. A. Messerle, G. Wider, G. Otting, C. Weber, and K. Wüthrich, *J. Magn. Reson.* **85**, 608–613 (1989).
44. M. Piotto, V. Saudek, and V. Sklenár, *J. Biomol. NMR* **2**, 661–665 (1992).
45. A. Bax and S. S. Pochapsky, *J. Magn. Reson.* **99**, 638–643 (1992).
46. D. Marion, M. Ikura, and A. Bax, *J. Magn. Reson.* **84**, 425–430 (1989).
47. F. Ni, *J. Magn. Reson.* **99**, 391–397 (1992).
48. J. P. Waltho and J. Cavanagh, *J. Magn. Reson., Ser. A* **103**, 338–348 (1993).
49. N. J. Oppenheimer, *Meth. Enzymol.* **176**, 78–89 (1989).
50. W. U. Primrose, *in* "NMR of Macromolecules, a Practical Approach" (G. C.K. Roberts, ed.), pp. 7–34, IRL Press, Oxford, 1993.
51a. W. W. Conover, *in* "Topics in Carbon-13 NMR Spectroscopy" (G. C. Levy, ed.), Vol. 4, pp. 37–51. Wiley, New York, 1984.
51b. G. N. Chmurny and D. I. Hoult, *Concepts Magn. Reson.* **2**, 131–149 (1990).
52. P. C. M. van Zijl, S. Sukumar, M. O'Neil Johnson, P. Webb, and R. E. Hurd, *J. Magn. Reson., Ser. A* **111**, 203–207 (1994).
53. D. S. Raiford, C. L. Fisk, and E. D. Becker, *Anal. Chem.* **51**, 2050–2051 (1979).
54. D. S. Wishart and B. D. Sykes, *Meth. Enzymol.* **239**, 363–392 (1994).
55. A. J. Hartel, P. P. Lankhorst, and C. Altona, *Eur. J. Biochem.* **129**, 343–357 (1982).
56. L. P. M. Orbons, G. A. van der Marel, J. H. van Boom, and C. Altona, *Eur. J. Biochem.* **170**, 225–239 (1987).
57. D. H. Live, D. G. Davis, W. C. Agosta, and D. Cowburn, *J. Am. Chem. Soc.* **106**, 1939–1941 (1984).
58. A. Bax and S. Subramanian, *J. Magn. Reson.* **67**, 565–569 (1986).
59. M. Rance and R. A. Byrd, *J. Magn. Reson.* **52**, 221–240 (1983).
60. D. G. Davis, *J. Magn. Reson.* **81**, 603–607 (1989).
61. M. Delsuc and J. Lallemand, *J. Magn. Reson.* **69**, 504–507 (1986).
62. A. S. Edison, F. Abildgaard, W. M. Westler, E. S. Mooberry, and J. L. Markley, *Meth. Enzymol.* **239**, 3–79 (1994).
63. G. P. Connelly, Y. Bai, M. F. Jeng, and S. W. Englander, *Proteins: Struct. Funct. Genet.* **17**, 87–92 (1993).

CHAPTER

4

MULTIDIMENSIONAL NMR SPECTROSCOPY

In the simplest pulsed NMR experiment, transverse magnetization excited by a rf pulse is sampled and stored at regular intervals during the *acquisition period* to generate a digital representation of the FID. Fourier transformation of the digitized signal yields the conventional one-dimensional spectrum (Chapter 3). In more complex one-dimensional NMR experiments, perturbations, which usually take the form of applied rf fields, are imposed on the spin system during the acquisition period or during a *preparation period* that precedes the acquisition period. Comparison of the spectra obtained in the presence and absence of the perturbations then yields information on the properties of the spin system affected. For example, weak irradiation of a particular spin during the acquisition period of a spin-tickling experiment (*1,2*) alters the natural multiplet patterns of spins that are scalar coupled to the irradiated spin. As another example, selective saturation of the resonance of a particular spin during the preparation period of a NOE difference experiment (*3,4*) alters the normal intensities of nearby, dipolar-coupled spins.

Unfortunately, one-dimensional NMR techniques, such as spin tickling, selective decoupling (*2,5–7*) and NOE difference experiments, which yield extremely useful information in small molecules, are of limited applicability to the complex, highly overlapped spectra of biological macromol-

183

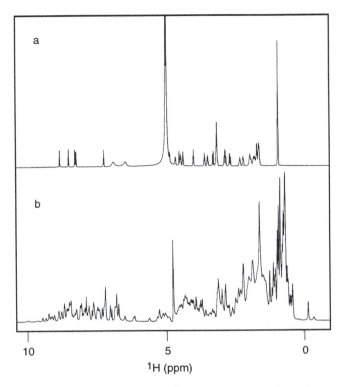

FIGURE 4.1 One-dimensional 500-MHz ^{1}H NMR spectra of (a) a hexapeptide at 280 K and (b) ubiquitin at 300 K. Samples were prepared in 90%/10% H_2O/D_2O. The two data sets were recorded at different temperatures; therefore, the water resonance signal appears at different chemical shifts in the two spectra.

ecules. By way of illustration, Fig. 4.1a shows the one-dimensional ^{1}H spectrum of a hexapeptide. Virtually all of the proton (multiplet) resonances are resolved; consequently, the assignment of each resonance to a particular proton in the molecule is straightforward, and perturbations to the spectrum that result from selective irradiation of particular spins are easily detected. On the other hand, Fig. 4.1b shows a one-dimensional ^{1}H spectrum of the protein ubiquitin [relative molecular mass $M_r = 8565$ daltons (Da)]. Several hundred proton resonances are crowded into approximately the same spectral region as the few peptide resonances. Because many resonances are degenerate, the signals are impossible to assign by using one-dimensional techniques, and selective perturbations to particular spins are difficult to achieve or detect experimentally. To effectively

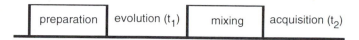

FIGURE 4.2 General scheme for two-dimensional NMR spectroscopy. The two-dimensional NMR experiment can be divided into four defined components: preparation, evolution, mixing, and acquisition.

utilize the information available from NMR spectroscopy of biological macromolecules, a general method is required for improving resolution, facilitating resonance assignments, and detecting the effects of interactions between nuclei in complex NMR spectra. The explosive growth in the application of NMR spectroscopy to biological macromolecules in the past two decades attests to the success of multidimensional experiments in achieving these objectives.

4.1 Two-Dimensional NMR Spectroscopy

Initially, multidimensional NMR spectroscopy will be introduced by concentrating on two-dimensional (2D) spectroscopy. The overall structure of 2D NMR experiments will be presented, the separation of interactions into more than one frequency dimension will be discussed, and techniques for selection of coherence transfer pathways will be introduced. All the methods and principles presented can be extended into higher dimensions in a straightforward manner. Specific multidimensional NMR experiments are discussed in Chapters 6 and 7.

In two-dimensional spectroscopy, two new elements, known as the *evolution* and *mixing periods*, are introduced into the NMR experiment between the preparation and acquisition periods. Thus, a general scheme for recording two-dimensional spectra can be segmented into the four distinct parts illustrated in Fig. 4.2. The evolution period contains a variable time delay that is increased during the course of a two-dimensional NMR experiment from an initial value to a final value in m (usually equal) increments. For each m value of the incrementable delay, the same pulse sequence is executed and a FID consisting of n digitized data points is recorded. The FID consists of p coadded transients for signal averaging or phase cycling. Thus, a total of $m \times p$ separate transients are recorded during the two-dimensional experiment, and a data array, which takes the form of a $m \times n$ matrix, is generated as a function of the two separate, independent time periods. Herein, the acquisition time (when the receiver is turned on and actually detects signal) is designated t_2, and the *indirect evolution* time is designated t_1. The rows of the data matrix represent data collected for a fixed t_1 value and different t_2 values; the columns represent data collected for a fixed t_2 value and varying t_1 values. Fourier transformation with respect to these two time domains generates a two-dimensional spectrum with two independent frequency dimensions, F_1 (from t_1) and F_2 (from t_2).

Most importantly, the signal eventually recorded during t_2 is modulated by events occurring during the evolution time t_1. As an example, consider a single isolated spin with a Larmor frequency Ω and a simple pulse sequence consisting of two 90°_x pulses separated by the variable period, t_1. Using the product operator formalism introduced in Chapter 2, the evolution of the density operator through the pulse sequence is

$$I_z \xrightarrow{\left(\frac{\pi}{2}\right)_x} -I_y$$

$$\xrightarrow{t_1} -I_y \cos(\Omega t_1) + I_x \sin(\Omega t_1)$$

$$\xrightarrow{\left(\frac{\pi}{2}\right)_x} -I_z \cos(\Omega t_1) + I_x \sin(\Omega t_1)$$

$$\xrightarrow{t_2} -I_z \cos(\Omega t_1) + I_x \sin(\Omega t_1) \cos(\Omega t_2) + I_y \sin(\Omega t_1) \sin(\Omega t_2). \quad [4.1]$$

Thus, the complex signal detected during the acquisition period is proportional to $\sin(\Omega t_1)\exp(-i\Omega t_2)$ and, as a result, depends parametrically on the value of t_1. As shown in Fig. 4.3, following Fourier transformation of the data recorded during t_2, a null signal is obtained if $\Omega t_1 = k\pi$, a signal with maximal amplitude is obtained if $\Omega t_1 = k\pi/2$, and an inverted signal is obtained if $\Omega t_1 = 3k\pi/2$, with k an integer. Formally, a *correlation* is established between the two time domains. The amplitude of the resonance signal obtained from the Fourier transformation of the data recorded during t_2, when displayed as a function of t_1, forms an *interferogram* similar in appearance to the FID. The t_1 interferogram is indirectly sampled and differs in this respect from the FID, which is directly detected by the spectrometer during t_2. Fourier transformation of the interferogram with respect to t_1 yields the F_1 dimension of the two-dimensional spectrum.

The correlation between the two time domains in the two-dimensional spectrum represented by [4.1] is trivial because the I spin operators precess at the same frequencies during t_1 and t_2 under the free-precession Hamiltonian. Consequently, signals are observed only at positions satisfying the relationship $F_1 = F_2$, and the conventional one-dimensional spectrum is reproduced along the diagonal of the two-dimensional spectrum. To provide additional information, a two-dimensional spectrum must contain resonance signals for which $F_1 \neq F_2$; this condition requires that the

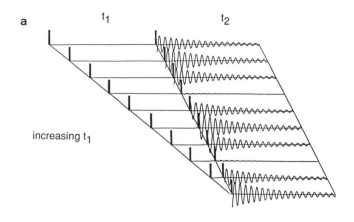

a

t_1 t_2

increasing t_1

Fourier transformation

with respect to t_2

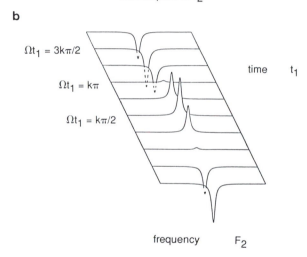

b

$\Omega t_1 = 3k\pi/2$

time t_1

$\Omega t_1 = k\pi$

$\Omega t_1 = k\pi/2$

frequency F_2

FIGURE 4.3 Dependence of the NMR signal on the evolution period, t_1. (a) The complex signal detected during the acquisition period, t_2, is modulated as a function of the evolution period, t_1. The two vertical bars represent 90° pulses. The separation between the pulses is equal to t_1. (b) Following Fourier transformation of the recorded data with respect to t_2, the amplitude of the resonance signal varies periodically as a function of t_1.

components of the density operator that eventually give rise to the observed resonance must evolve under different Hamiltonians during t_1 and t_2. Fortunately, because the components of the density operator that evolve during t_1 (or in fact any time before t_2) are never actually recorded, the mixing period can serve to transfer magnetization, or more generally, coherence, between spins prior to acquisition. The presence of a signal in the two-dimensional spectrum at the frequency of one spin in F_1 and at the frequency of a second spin in F_2 is direct evidence for transfer of coherence between the two spins during the mixing period. With carefully constructed sequences of rf pulses and delays during the mixing period, correlations can be established between the coherences present during t_1 and t_2 that result in chemically useful information.

To illustrate the importance of the mixing period, the example discussed above is extended to include a second spin, S. The Larmor frequencies of the I and S spins are now designated Ω_I and Ω_S; the two spins are assumed to have a scalar coupling interaction with a coupling constant, J. Focusing on magnetization that originates on the I spin, we obtain

$$I_z \xrightarrow{\left(\frac{\pi}{2}\right)_x} -I_y$$

$$\xrightarrow{t_1} -I_y \cos(\Omega_I t_1) \cos(\pi J t_1) + 2I_x S_z \cos(\Omega_I t_1) \sin(\pi J t_1)$$

$$+ I_x \sin(\Omega_I t_1) \cos(\pi J t_1) + 2I_y S_z \sin(\Omega_I t_1) \sin(\pi J t_1)$$

$$\xrightarrow{\left(\frac{\pi}{2}\right)_x} -I_z \cos(\Omega_I t_1)_1 \cos(\pi J t_1) - 2I_x S_y \cos(\Omega_I t_1) \sin(\pi J t_1)$$

$$+ I_x \sin(\Omega_I t_1) \cos(\pi J t_1) + 2I_z S_y \sin(\Omega_I t_1) \sin(\pi J t_1). \qquad [4.2]$$

Only the two terms proportional to I_x and $2I_z S_y$ on the last line of [4.2] result in detectable signals during the acquisition period. The I_x term leads to a detected signal proportional to $\sin(\Omega_I t_1) \cos(\pi J t_1) \cos(\pi J t_2)$ $\exp(-i\Omega_I t_2)$; the $2I_z S_y$ term leads to a signal proportional to $\sin(\Omega_I t_1)$ $\sin(\pi J t_1) \sin(\pi J t_2) \exp(-i\Omega_S t_2)$. In the present example, a complementary coherence transfer pathway also exists, whereby magnetization originating on spin S can be transferred to the I spin during the mixing period. Ignoring multiplet structures and peak shapes for the present discussion (see Section 6.2.1), Fourier transformation with respect to t_1 and t_2 generates a (schematic) spectrum of the form shown in Fig. 4.4. The two peaks, **D**, known as *diagonal peaks*, result from magnetization that remains on the same spin throughout the experiment and essentially form a one-dimensional spectrum. On the other hand, the two peaks, **X**, known as *cross-peaks*, result from magnetization that has been transferred from one spin to the other during the mixing period.

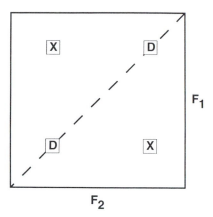

FIGURE 4.4 Schematic two-dimensional NMR spectrum for two spins, I and S. **D** symbols represent diagonal peaks that occur at the same frequency in both dimensions, and **X** symbols represent cross-peaks that appear at different frequencies in each dimension. Cross-peaks result from the transfer of coherence from one spin to the other during the mixing period of the experiment. Diagonal peaks result from coherence that is not transferred between spins during the mixing period.

Inspection of [4.2] shows that the cross-peak results from the $2I_zS_y$ operator present at the beginning of t_2. This operator is generated by the second 90°_x pulse in the sequence acting on the $2I_yS_z$ operator. The action of this pulse results in the conversion of an antiphase I operator, which evolves with Larmor frequency Ω_I, into an antiphase S operator, which evolves with Larmor frequency Ω_S; thereby coherence is transferred from the I spin to the S spin. The second pulse in this experiment comprises the entire mixing period. In the present case, the mixing sequence was designed to effect transfer of magnetization via the scalar coupling between two spins; consequently, the appearance of cross-peaks between spins I and S in the spectrum unambiguously indicates that the two spins are scalar coupled and establishes a through-bond correlation between the spins. In actual fact, this imaginary experiment is the basis of the original two-dimensional NMR technique, the *correlated spectroscopy* (COSY) experiment (8,9). Detection of a scalar coupling interaction between the two spins does not depend on observation of changes in the multiplet structure or intensity of the diagonal resonances as in one-dimensional NMR experiments, but rather on the appearance of cross-peaks. This property gives multidimensional spectroscopy its immense power; not only are important correlations established, but the two independent frequency coordinates effectively increase the resolution in the spectrum.

Overlapping signals in the conventional one-dimensional spectrum, which arise, for example, from multiple scalar coupling interactions, are dispersed into the additional frequency dimension in a process called *separation of interactions*.

The frequency in the F_2 dimension (recorded during t_2) of each peak in the two-dimensional spectrum must correspond to the frequency of a peak in the conventional one-dimensional spectrum. The converse is not true, because certain resonances in the one-dimensional spectrum can be suppressed in the two-dimensional spectrum. Removal of undesirable resonances by using multiple-quantum and isotope filters is discussed in Chapters 6 and 7. The F_1 frequency of a peak in the two-dimensional spectrum need not correspond to a frequency in the conventional spectrum; the exact form of the relationship between the F_1 and F_2 frequencies of a peak in the two-dimensional spectrum will depend on the particular manipulations of the spins before acquisition. Two very useful features arise because the evolution of a coherence during t_1 is never actually physically detected. First, the coherence that is present during t_1 can be of a type that cannot normally be recorded, such as multiple-quantum coherence. A typical experiment of this kind would proceed by preparing multiple-quantum coherence, frequency labeling the coherence during t_1, and returning the multiple-quantum coherence to single-quantum coherence for detection. Second, the apparent frequencies of peaks in F_1 can be manipulated to be different from the actual frequencies with which coherence evolves during t_1. Examples of this type of experiment include the removal of chemical-shift evolution and heteronuclear scalar couplings by the application of a 180° pulse at the midpoint of t_1.

The functions of the different periods of a two-dimensional NMR experiment can be summarized as follows:

1. *Preparation*: The desired nonequilibrium state of the spin system is prepared from the initial (equilibrium) state of the spin system. The preparation period in its simplest form consists of a single pulse that generates transverse magnetization, but more complex sequences of pulses can be used to prepare other coherences, such as multiple-quantum coherences, and to perform solvent suppression.

2. *Evolution*: The off-diagonal components of the density operator prepared in (1) evolve under the Hamiltonian, $\mathcal{H}_e$. During the course of the experiment, the incrementable time t_1 begins at an initial value and increases in discrete steps to a maximum value, $t_{1,max}$. The Hamiltonian, $\mathcal{H}_e$, may be the free-precession Hamiltonian or may include applied rf fields. The frequencies with which the detected coherence evolves during t_1 results in signals appearing at those frequencies in the F_1

dimension of the final two-dimensional spectrum. This process is known as F_1 frequency labeling of the coherence.

3. *Mixing*: During the mixing period, coherence is transferred from one spin to another. The mixing period is the key to establishing the type of correlation between the two dimensions and consequently dictates the information content of the spectrum. Depending on the type of experiment, the mixing period consists of one or more pulses and delays.

4. *Acquisition*: The FID is recorded in the conventional fashion. As discussed in Section 4.3, if more than one coherence transfer pathway is feasible, phase cycling or field-gradient pulses are used to determine which coherence transfer processes contribute to the final spectrum.

The evolution of the density operator of the spin system during the pulse sequence can be described, using the product operator formalism, as a transformation of the density operator from its initial to final value during each of the four parts of the experiment. Thus

$$\sigma^{eq} \xrightarrow{\text{preparation}} \sigma(0)$$

$$\sigma(0) \xrightarrow{\text{evolution}} \sigma(t_1)$$

$$\sigma(t_1) \xrightarrow{\text{mixing}} \sigma(t_1,0) \tag{4.3}$$

$$\sigma(t_1,0) \xrightarrow{\text{detection}} \sigma(t_1,t_2)$$

The modular nature of two-dimensional NMR experiments evident from the preceding discussion facilitates the construction of new experiments from combinations of prefabricated "building block" pulse sequences that effect particular transformations of the density operator. This approach is even more powerful for the design of three- and four-dimensional experiments, as will be discussed in subsequent chapters.

4.2 Coherence Transfer and Mixing

The key to obtaining useful chemical or structural information from two-dimensional NMR spectroscopy is the transfer of coherence from one spin to another during the mixing period. As the preceding discussion illustrates, bilinear components of the Hamiltonian, specifically, interactions that depend on the state of two different spins, are responsible for coherence transfer between spins. In the following, two mechanisms for transferring coherence between spins in multidimensional NMR spectroscopy are discussed. Coherence transfer in homonuclear spin systems is discussed first; the generalization to heteronuclear spin systems follows directly. This presentation is not meant to represent a detailed comprehensive account of magnetization transfer processes in multidimensional spectroscopy. Rather, the idea that *through-bond* scalar coupling and *through-space* dipolar interactions are responsible for transfer of coherence between spins should be appreciated.

4.2.1 THROUGH-BOND COHERENCE TRANSFER

A crucial mixing process in many multidimensional experiments is migration of coherence between scalar coupled nuclei. This process, known as coherence transfer, was introduced in Section 2.7.6. Coherence transfer can be produced by evolution of the spin system under a series of rf pulses and free-precession delays ("pulse-interrupted free precession" or COSY-type coherence transfer), or by cross-polarization of the spin system by using continuous, phase-modulated rf fields [TOCSY- or homonuclear Hartmann–Hahn (HOHAHA)-type coherence transfer]. Scalar coupling interactions are mediated by covalent bonding interactions; therefore, COSY and TOCSY mixing generate through-bond coherence transfer.

4.2.1.1 COSY-Type Coherence Transfer   The density operator must contain antiphase terms in order to transfer coherence from one spin to another by COSY techniques (8,9). Thus, before coherence transfer can be effected, antiphase coherence must develop from in-phase coherence by evolution under the scalar coupling interaction between spins. The antiphase coherence with respect to one spin is transferred by the mixing pulse into antiphase coherence with respect to the scalar-coupled partner. This coherence transfer process is illuminated by considering the effect of the pulse sequence:

$$90_x^\circ - \tau/2 - 180_y^\circ - \tau/2 - 90_y^\circ \qquad [4.4]$$

on a homonuclear *IS* spin system, with a mutual scalar coupling constant J_{IS}. Using [2.117], the propagator for the pulse sequence is

$$\mathbf{U} = \exp\left[i\frac{\pi}{2}(I_y + S_y)\right] \exp[-i2\pi J_{IS}\tau I_z S_z] \exp\left[-i\frac{\pi}{2}(I_x + S_x)\right]. \qquad [4.5]$$

As discussed in Section 2.7.7.1, the 180° pulse refocuses chemical-shift evolution; in addition, this pulse effectively inverts the phase of the second 90° pulse. The principles of the experiment can be deduced by concentrating on the fate of just one of the spins. Starting with equilibrium I spin magnetization, I_z, the analysis proceeds as follows:

$$I_z \xrightarrow{\left(\frac{\pi}{2}\right)_x} -I_y$$

$$\xrightarrow{\ \tau\ } -I_y \cos(\pi J_{IS}\tau) + 2I_x S_z \sin(\pi J_{IS}\tau)$$

$$\xrightarrow{\left(\frac{\pi}{2}\right)_{-y}} -I_y \cos(\pi J_{IS}\tau) - 2I_z S_x \sin(\pi J_{IS}\tau). \qquad [4.6]$$

The term proportional to I_y represents coherence of spin I that is not transferred to spin S during the sequence. The term $2I_zS_x$ corresponds to antiphase single-quantum coherence of the S spin and represents coherence transferred to the S spin from the I spin by the mixing sequence. Notice that the $2I_zS_x$ operator was generated by the application of the 90°_y pulse to the antiphase operator $2I_xS_z$. Of course, the analogous treatment for the S spin results in transfer of coherence from the S spin to the I spin during the mixing sequence. The coefficients of the operators depend on the rate at which the antiphase coherence evolves. The coherence transfer amplitude, given by $\sin(\pi J_{IS}\tau)$, is maximal when $\tau = 1/(2J_{IS})$. As stated above, for coherence transfer to proceed at all, the system must have evolved to an antiphase state with respect to the scalar coupling during the period τ.

Now consider the extension to the situation in which spin I is coupled to two other spins, R and S, with scalar coupling constants J_{IR} and J_{IS}. The R and S spins are assumed to lack a scalar coupling interaction (*i.e.* $J_{RS} = 0$). The product operator analysis of the mixing sequence yields

$$I_z \xrightarrow{\left(\frac{\pi}{2}\right)_x} -I_y$$

$$\xrightarrow{\ \tau\ } -I_y \cos(\pi J_{IR}\tau)\cos(\pi J_{IS}\tau) + 2I_x R_z \sin(\pi J_{IR}\tau)\cos(\pi J_{IS}\tau)$$

$$+ 2I_x S_z \cos(\pi J_{IR}\tau)\sin(\pi J_{IS}\tau) + 4I_y R_z S_z \sin(\pi J_{IR}\tau)\sin(\pi J_{IS}\tau)$$

$$\xrightarrow{\left(\frac{\pi}{2}\right)_{-y}} -I_y \cos(\pi J_{IR}\tau)\cos(\pi J_{IS}\tau) - 2I_z R_x \sin(\pi J_{IR}\tau)\cos(\pi J_{IS}\tau)$$

$$- 2I_z S_x \cos(\pi J_{IR}\tau)\sin(\pi J_{IS}\tau) + 4I_y R_x S_x \sin\pi(\pi J_{IR}\tau)\sin(\pi J_{IS}\tau).$$

$$[4.7]$$

Again, the term proportional to I_y represents coherence of spin I that is not transferred to either coupled spin during the sequence. The term $2I_z R_x$ corresponds to antiphase single-quantum coherence of the R spin and represents coherence transferred to the R spin from the I spin by the mixing sequence. Similarly, the term $2I_zS_x$ corresponds to antiphase single-quantum coherence of the S spin and represents coherence transferred to the S spin from the I spin by the mixing sequence. The term $4I_yR_xS_x$ represents a linear combination of zero- and multiple-quantum coherences and is not of further interest here. Analogous treatments for the R and S spins result in transfers of coherence from the R and S spins to the I spin. In this example, no coherence is transferred between spins R and S because they are not scalar coupled to each other, even though they are mutually coupled to the I spin.

In the present example, the spin I evolves under the influence of two scalar coupling interactions during τ. In the evolution leading to the term $2I_xR_z$, the scalar coupling to spin R is called the active coupling and the scalar coupling to spin S is called the passive coupling; in the evolution leading to the term $2I_xS_z$, the scalar coupling to spin R is called the passive coupling and the scalar coupling to spin S is called the active coupling. As shown, each active coupling contributes a factor of $\sin(\pi J_a\tau)$, in which J_a is the active scalar coupling constant, and each passive coupling contributes a factor of $\cos(\pi J_p\tau)$, in which J_p is the passive scalar coupling constant, to the magnitude of the product operators. If a given spin is scalar coupled to N other spins, the operators that lead to coherence transfer in the COSY-type mixing sequences have a single active coupling and $N - 1$ passive couplings; operators with no active couplings represent operators for which no coherence transfer occurs; and operators with greater than one active coupling represent the creation of multiple-quantum coherences. As before, the coefficients of the operators depend on the rate at which antiphase coherence is generated. However, maximization of coherence transfer from I to R or S requires knowledge of the values of J_{IR} and J_{IS}. In general, coherence transfers from I to R and from I to S are not maximized for the same value of τ unless $J_{IR} = J_{IS}$. In addition, coherence transfer efficiencies cannot be simultaneously maximized for two and three (or more) spin systems because of the different trigonometric expressions encountered in [4.6] and [4.7].

The final operator of interest in [4.6] is an antiphase S operator. In some circumstances, coherence transfer to an in-phase operator is desirable. In the COSY-style mixing sequences, a second delay period must be used to refocus the antiphase operator. The entire mixing sequence is

$$90^{\circ}_x-\tau/2-180^{\circ}_y-\tau/2-90^{\circ}_y-\tau_2/2-180^{\circ}_y-\tau_2/2. \qquad [4.8]$$

By the same reasoning as above, the effects of the pulse sequence can be obtained by analysis of the propagator:

$$\mathbf{U} = \exp[-i2\pi J_{IS}\tau_2 I_z S_z]\exp\left[i\frac{\pi}{2}(I_y + S_y)\right]$$
$$\times \exp[-i2\pi J_{IS}\tau I_z S_z]\exp\left[-i\frac{\pi}{2}(I_x + S_x)\right], \qquad [4.9]$$

in which only the scalar coupling Hamiltonian is effective during τ and τ_2. The evolution up to the 90_y° pulse was presented in [4.6] and [4.7]; only the analysis of the additional effects of the τ_2 period must be considered here. For simplicity, only the antiphase term $2I_z S_x$, that results from the coherence transfer step analyzed in [4.6], is treated:

$$2I_z S_x \sin(\pi J_{IS}\tau) \xrightarrow{\tau_2} 2I_z S_x \sin(\pi J_{IS}\tau)\cos(\pi J_{IS}\tau_2)$$
$$+ S_y \sin(\pi J_{IS}\tau)\sin(\pi J_{IS}\tau_2). \qquad [4.10]$$

The second term on the right-hand side of [4.10] represents in-phase transverse magnetization of the S spin. Complete refocusing of the antiphase operator is obtained for $\tau_2 = 1/(2J_{IS})$. Thus, coherence transfer from an in-phase state on one spin to an in-phase state on a coupled spin requires a total time of $1/J_{IS}$ when employing pulse interrupted free-precession methods.

The principal limitation of COSY-type coherence transfer arises from the antiphase multiplet structure of the resulting cross-peaks in the spectrum. If the size of the active coupling is comparable to the linewidth, partial cancellation of the multiplet occurs as a result of destructive interference between the antiphase components of the peak. To avoid self-cancellation, the antiphase components can be refocused, so that the resulting cross-peak multiplet is composed of peaks entirely of the same algebraic sign. The destructive interference effects are then eliminated, but only at the expense of an additional refocusing period of duration $1/(2J_{IS})$. A more detailed discussion of COSY experiments is given in Section 6.2.

4.2.1.2 TOCSY Transfer through Bonds   Thus far, coherence transfer has been limited to pulse interrupted free-precession techniques, or COSY-type transfer via evolution under the scalar coupling Hamiltonian in the weak coupling limit. To begin the present discussion, consider the evolution of the density operator under the strong scalar coupling Hamiltonian between two spins I and S (*10*):

$$\mathcal{H} = 2\pi J_{IS}\mathbf{I}\cdot\mathbf{S} = 2\pi J_{IS}(I_x S_x + I_y S_y + I_z S_z). \qquad [4.11]$$

The evolution of the I_z operator is given by

$$\exp(-i\mathcal{H}\tau_m)I_z\exp(-i\mathcal{H}\tau_m)$$

$$= \exp[-i2\pi J_{IS}\tau_m(I_xS_x + I_yS_y + I_zS_z)]I_z$$

$$\times \exp[i2\pi J_{IS}\tau_m(I_xS_x + I_yS_y + I_zS_z)]$$

$$= \exp(-i\zeta 2I_xS_x)\exp(-i\zeta 2I_yS_y)\exp(-i\zeta 2I_zS_z)I_z$$

$$\times \exp(i\zeta 2I_zS_z)\exp(i\zeta 2I_yS_y)\exp(i\zeta 2I_xS_x)$$

$$= \exp(-i\zeta 2I_xS_x)\exp(-i\zeta 2I_yS_y)I_z\exp(i\zeta 2I_yS_y)\exp(i\zeta 2I_xS_x), \quad [4.12]$$

in which $\zeta = \pi J_{IS}\tau_m$, τ_m is the mixing time, and the third line is obtained because the operators I_xS_x, I_yS_y, and I_zS_z commute with each other. Evolution under the propagator $\mathbf{U} = \exp(-i2\zeta I_yS_y)$ can be evaluated by applying a similarity transformation to $\exp(-i2\zeta I_zS_z)$:

$$\exp(-i\zeta 2I_yS_y)I_z\exp(i\zeta 2I_yS_y)$$

$$= \exp\left(-i\frac{\pi}{2}I_x\right)\exp\left(-i\frac{\pi}{2}S_x\right)\exp(-i\zeta 2I_zS_z)$$

$$\times \exp\left(i\frac{\pi}{2}S_x\right)\exp\left(i\frac{\pi}{2}I_x\right)I_z\exp\left(-i\frac{\pi}{2}I_x\right)\exp\left(-i\frac{\pi}{2}S_x\right)$$

$$\times \exp(i\zeta 2I_zS_z)\exp\left(i\frac{\pi}{2}S_x\right)\exp\left(i\frac{\pi}{2}I_x\right)$$

$$= \exp\left(-i\frac{\pi}{2}I_x\right)\exp\left(-i\frac{\pi}{2}S_x\right)\exp(-i\zeta 2I_zS_z)I_y\exp(i\zeta 2I_zS_z)$$

$$\times \exp\left(i\frac{\pi}{2}S_x\right)\exp\left(i\frac{\pi}{2}I_x\right)$$

$$= \exp\left(-i\frac{\pi}{2}I_x\right)\exp\left(-i\frac{\pi}{2}S_x\right)(I_y\cos\zeta - 2I_xS_z\sin\zeta)$$

$$\times \exp\left(i\frac{\pi}{2}S_x\right)\exp\left(i\frac{\pi}{2}I_x\right)$$

$$= \exp\left(-i\frac{\pi}{2}I_x\right)(I_y\cos\zeta + 2I_xS_y\sin\zeta)\exp\left(i\frac{\pi}{2}I_x\right)$$

$$= I_z\cos\zeta + 2I_xS_y\sin\zeta. \quad [4.13]$$

The same approach for $\mathbf{U} = \exp(-i2\zeta I_x S_x)$ gives:

$\exp(-i\zeta 2 I_x S_x)(I_z \cos\zeta + 2 I_x S_y \sin\zeta)\exp(i\zeta 2 I_x S_x)$

$$= \exp\left(-i\frac{\pi}{2}I_y\right)\exp\left(-i\frac{\pi}{2}S_y\right)\exp(-i\zeta 2 I_z S_z)\exp\left(i\frac{\pi}{2}S_y\right)$$

$$\times \exp\left(i\frac{\pi}{2}I_y\right)(I_z \cos\zeta + 2 I_x S_y \sin\zeta)\exp\left(-i\frac{\pi}{2}I_y\right)$$

$$\times \exp\left(-i\frac{\pi}{2}S_y\right)\exp(i\zeta 2 I_z S_z)\exp\left(i\frac{\pi}{2}S_y\right)\exp\left(i\frac{\pi}{2}I_y\right)$$

$$= I_z \cos^2\zeta + (-2I_y S_x + 2 I_x S_y)\cos\zeta\sin\zeta + S_z \sin^2\zeta, \qquad [4.14]$$

to yield the final result:

$$I_z \xrightarrow{\mathcal{H}\tau_{\mathrm{m}}} I_z \cos^2(\pi J_{IS}\tau_{\mathrm{m}}) + S_z \sin^2(\pi J_{IS}\tau_{\mathrm{m}})$$
$$+ 2(I_x S_y - I_y S_x)\cos(\pi J_{IS}\tau_{\mathrm{m}})\sin(\pi J_{IS}\tau_{\mathrm{m}}) \qquad [4.15]$$

$$S_z \xrightarrow{\mathcal{H}\tau_{\mathrm{m}}} I_z \sin^2(\pi J_{IS}\tau_{\mathrm{m}}) + S_z \cos^2(\pi J_{IS}\tau_{\mathrm{m}})$$
$$- 2(I_x S_y - I_y S_x)\cos(\pi J_{IS}\tau_{\mathrm{m}})\sin(\pi J_{IS}\tau_{\mathrm{m}}),$$

in which the evolution of the S_z operator is obtained by exchanging the I and S labels. Equation [4.15] predicts that the sum, $I_z + S_z$, is a constant and that the difference, $I_z - S_z$, is given by

$$(I_z - S_z)\xrightarrow{\mathcal{H}\tau_{\mathrm{m}}} (I_z - S_z)\cos(2\pi J_{IS}\tau_{\mathrm{m}}) + 2(I_x S_y - I_y S_x)\sin(2\pi J_{IS}\tau_{\mathrm{m}}).$$
$$[4.16]$$

If $\tau_{\mathrm{m}} = 1/(2J_{IS})$, I_z magnetization is transferred completely to S_z magnetization and vice versa. Evolution under the strong coupling Hamiltonian transfers in-phase magnetization between spins in a time of $1/(2J_{IS})$, compared with the time of $1/J_{IS}$ required for in-phase coherence transfer in weakly coupled systems via free-precession techniques. In addition, in a three-spin IRS system, magnetization can be transferred from R to S even if $J_{RS} = 0$ by the two-step transfer $R_z \rightarrow I_z \rightarrow S_z$.

In real situations, the Hamiltonian for a spin system contains chemical shift and rf terms in addition to the scalar coupling interaction. Magnetization transfer via the strong scalar coupling interaction is efficient only if all chemical-shift and rf terms of the Hamiltonian governing the spin system have identical values for each of the two spins I and S. This is the *Hartmann–Hahn* condition (*11*). Coherence transfer via Hartmann–Hahn cross-polarization has been used extensively in heteronuclear NMR experiments in the solid state. Braunschweiler and Ernst first demonstrated the feasibility of Hartmann–Hahn cross-polarization in homonuclear solution-phase NMR spectroscopy (*10*).

Hartmann–Hahn matching in the rotating reference frame can be achieved by application of a rf field sufficiently strong that any offset and chemical-shift effects are negligible in comparison. Müller and Ernst (*12*) demonstrated that transfer of I spin magnetization to the S spin proceeds as

$$I_z \xrightarrow{\mathcal{H}\tau_m} I_z a_{II}(\tau_m) + S_z a_{IS}(\tau_m);$$

$$S_z \xrightarrow{\mathcal{H}\tau_m} I_z a_{SI}(\tau_m) + S_z a_{SS}(\tau_m);$$

[4.17]

in which

$$a_{II}(\tau_m) = 1 - \sin^2 \phi \sin^2(q\tau_m);$$

$$a_{SS}(\tau_m) = 1 - \sin^2 \phi \sin^2(q\tau_m);$$

[4.18]

$$a_{IS}(\tau_m) = a_{SI}(\tau_m) = \sin^2 \phi \sin^2(q\tau_m),$$

and

$$q = [(\Omega_{I(\text{eff})} - \Omega_{S(\text{eff})})^2 + (\pi J_{IS} \sin \theta_I \sin \theta_S)^2]^{1/2};$$

$$\tan \phi = \frac{1}{2}\left(\frac{2\pi J_{IS} \sin \theta_I \sin \theta_S}{\Omega_{I(\text{eff})} - \Omega_{S(\text{eff})}}\right)$$

[4.19]

$$\Omega_{I(\text{eff})} = [\Omega_I^2 + \omega_{1I}^2]^{1/2} \quad \text{and} \quad \Omega_{S(\text{eff})} = [\Omega_S^2 + \omega_{1S}^2]^{1/2};$$

$$\theta_I = \tan^{-1}\left(\frac{\omega_{1I}}{\Omega_I}\right) \quad \text{and} \quad \theta_S = \tan^{-1}\left(\frac{\omega_{1S}}{\Omega_S}\right),$$

where θ_I and θ_S are the tilt angles of the effective field at the I and S spins, respectively; ω_{1I} and ω_{1S} are the rf field strengths experienced by the I and S spins, respectively; Ω_I is the I spin offset; Ω_S is the S spin offset; and J_{IS} is the scalar coupling constant between the I and S spins. In [4.17], $a_{II}(\tau_m)$ is the amount of magnetization remaining on spin I, $a_{SS}(\tau_m)$ is the amount of magnetization remaining on spin S, and $a_{IS}(\tau_m)$ is the amount of magnetization transferred to spin S at time τ_m. The functions $a_{II}(\tau_m)$, $a_{SS}(\tau_m)$, and $a_{IS}(\tau_m)$ frequently are called the mixing coefficients. If $\Omega_I = \pm\Omega_S$, the Hartmann–Hahn matching condition is satisfied because $\Omega_{I(\text{eff})} = \Omega_{S(\text{eff})}$. For the special case $\Omega_I = \Omega_S = 0$, [4.17] reduces to [4.15]. If the two scalar-coupled spins have different offsets from the rf carrier frequency, $\Omega_I \neq |\Omega_S|$, Hartmann–Hahn matching becomes more difficult

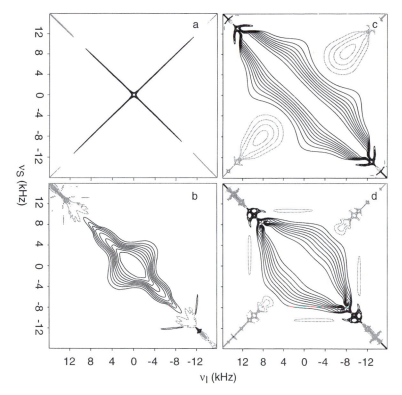

FIGURE 4.5 Efficiency of coherence transfer between two scalar coupled spins under the influence of different Hartmann–Hahn mixing schemes. The two-dimensional contour plots show the coherence transfer efficiency as a function of the relative offsets of the two scalar coupled spins for (a) a continuous rf field, (b) MLEV-17, (c) WALTZ-16, and (d) DIPSI-2. DIPSI-2, showing the more uniform profile, is the preferred sequence of these three for mediating efficient coherence transfer over a range of differing offsets. The MLEV-17 sequence in particular is much less satisfactory than DIPSI-2. In all cases, the simulations were performed for rf field strengths of 10 kHz, and contours are plotted at identical intervals.

and magnetization transfer is reduced drastically. The efficiency of magnetization transfer between two spins is graphed in Fig. 4.5a. In practice, the power required to accomplish efficient Hartmann–Hahn matching over a significant frequency range $\Omega_I \neq |\Omega_S|$ by using a continuous rf field would produce disastrous sample and probe heating effects.

As shown by [4.17], differences in chemical shift between two coupled spins prevents efficient Hartmann–Hahn matching by application of a

continuous rf field. Ideally, an effective Hamiltonian is required that eliminates the chemical-shift terms over a significant frequency range while the
rf field is applied. In effect, the spin Hamiltonian, $\mathcal{H}$, must be reduced from

$$\mathcal{H} = \sum_i - \omega_i I_{iz} + 2\pi \sum_{i \neq j} J_{ij} \mathbf{I}_i \cdot \mathbf{I}_j \qquad [4.20]$$

to an average Hamiltonian, $\mathcal{H}_{av}$,

$$\mathcal{H}_{av} = 2\pi \sum_{i \neq j} J_{ij} \mathbf{I}_i \cdot \mathbf{I}_j \qquad [4.21]$$

for a period τ_m, which is of the order of $1/J_{ij}$ (10). The last equation
consists of only the pure scalar coupling term, in which all shift terms or
linear operators are removed, leaving only bilinear operators. The absence
of chemical-shift terms means that the Hartmann–Hahn condition is always satisfied. A pulse sequence that generates an average Hamiltonian
given by [4.21] is said to be *isotropic*. Magnetization transfer under the
influence of such a sequence is a continuous mixing process, with the
magnetization moving in a periodic fashion among all the spins in the
scalar coupled network. Such pulse sequences shall be referred to as
isotropic mixing sequences. An important practical consequence of an
isotropic mixing sequence is that the transfer of magnetization occurs
equally well for all angular-momentum components. That is, coherent
exchange of difference magnetization will occur under the isotropic scalar
coupling Hamiltonian according to [4.16] with similar relations for the x
and y components, obtained by cyclic permutation of the indices:

$$(I_x - S_x) \xrightarrow{\mathcal{H}\tau_m} (I_x - S_x)\cos(2\pi J_{IS}\tau_m) + (I_y S_z - I_z S_y)\sin(2\pi J_{IS}\tau_m);$$
$$[4.22]$$

$$(I_y - S_y) \xrightarrow{\mathcal{H}\tau_m} (I_y - S_y)\cos(2\pi J_{IS}\tau_m) + (I_x S_z - I_z S_x)\sin(2\pi J_{IS}\tau_m);$$
$$[4.23]$$

The first successful mixing sequences were derived from phase-modulated irradiation schemes used for spin decoupling, such as MLEV-17 (13)
and WALTZ-16 (14,15) (Section 3.4.3). Recently, many pulse sequences
have been developed that are isotropic over wide frequency ranges with
minimal power requirements. The most popular of these, at present, are
the DIPSI family of pulse sequences developed by Shaka and coworkers
(16,17). The DIPSI-2 sequence is given by the pulse sequence element

$$R = 320 \quad \overline{410} \quad 290 \quad \overline{285} \quad 30 \quad \overline{245} \quad 375 \quad \overline{265} \quad 370, \qquad [4.24]$$

in which pulse lengths are given in degrees and overbars indicate 180° phase shifts. Just as in the case for spin decoupling (Section 3.4.3), the performance of the DIPSI-2 sequence is improved by combining the basic pulse sequence element into a supercycle $R\overline{R}\,\overline{R}R$. The performance of mixing sequences is difficult to establish by product operator analysis. Instead, the efficiency of coherence transfer is calculated numerically as a function of resonance offsets using the full-density matrix formalism. Figure 4.5b–d illustrates the efficiency of coherence transfer for the MLEV-17, WALTZ-16, and the DIPSI-2 pulse sequences. DIPSI-2 provides significantly larger coherence transfer efficiencies than MLEV-17 and WALTZ-16.

Until now, the second terms on the right-hand sides of [4.16], [4.22], and [4.23] have not been considered. These terms contain bilinear product operators that are orthogonal to the in-phase terms of interest. The bilinear term in [4.16] is a multiple-quantum coherence term; bilinear terms in [4.22] and [4.23] represent antiphase single-quantum coherence operators. The significance of the bilinear terms will be discussed for [4.22], although the arguments are applicable equally to the other cases. Simple inspection shows that the antiphase term has y phase whereas the in-phase term has x phase; thus a 90° phase difference exists between the terms. If the signal resulting from the in-phase magnetization is phased to be absorptive, the signal resulting from the antiphase term automatically becomes dispersive. The dispersive antiphase multiplets occur in the same spectral position as the in-phase absorptive peaks and can disrupt the lineshapes in the two-dimensional NMR spectrum. Indeed, for small molecules with narrow linewidths, interference from the dispersive antiphase components can be observed. In contrast, the resonance peaks for large biological macromolecules usually appear to be completely absorptive, because the linewidths are notably larger, and the dispersive antiphase components self-cancel very efficiently. Dispersive antiphase resonances can be suppressed further by z filtration (Section 6.5) (*18,19*).

To summarize, coherence transfer can be obtained when two scalar-coupled spins are subjected simultaneously to a rf field that effectively removes the chemical shifts of the spins. When used for homonuclear coherence transfer, this technique often is referred to as a homonuclear Hartmann–Hahn (HOHAHA) (*13*) experiment to indicate the required Hartmann–Hahn matching condition. Throughout the remainder of this text, isotropic mixing sequences that automatically satisfy the Hartmann–Hahn condition will be used to mediate coherence transfer. Pulse sequences utilizing isotropic mixing will be referred to as total correlation spectroscopy (TOCSY) experiments, as originally suggested by Braunschweiler and Ernst (*10*). The use of the word "total" in deriving the acronym implies that all spins belonging to a scalar-coupled network can be connected by such an experiment.

4.2.2 THROUGH-SPACE COHERENCE TRANSFER

As will be discussed in Sections 5.1.2 and 5.5, perturbing the populations of stationary states within a dipolar coupled spin system causes time-dependent changes in the intensities of dipolar-coupled resonance signals via the nuclear Overhauser effect (NOE) (3,4). Dipolar cross-relaxation is an extremely useful mixing process in multidimensional NMR spectroscopy, because the efficiency of mixing depends on the distance between interacting spins. Thus, *through-space,* rather than *through-bond,* magnetization transfer generates cross-peaks in the NOE mixing process.

Consider the effect of the following pulse sequence on a pair of dipolar coupled spins, I and S, which have no scalar coupling between them:

$$90_x^\circ - t_1 - 90_x^\circ - \tau_m - 90_x^\circ - t_2. \qquad [4.25]$$

This pulse sequence is known as the nuclear Overhauser enhancement spectroscopy (NOESY) (20) experiment and is the most powerful and important technique available for structural investigations of biomolecules by NMR spectroscopy. A more detailed account of this experiment will be presented in Section 6.6.1. Concentrating solely on the I spin, the following product operators are present after the second 90_x° pulse:

$$I_z \xrightarrow{(\pi/2)_x - t_1 - (\pi/2)_x} - I_z \cos(\Omega_I t_1) + I_x \sin(\Omega_I t_1). \qquad [4.26]$$

Experimentally the I_x term can be suppressed by phase cycling or by application of a field-gradient pulse (Section 4.3), which leaves only the $-I_z \cos(\Omega_I t_1)$ term. A close analogy to the one-dimensional transient NOE experiment (Section 5.1.2) now is apparent. The term $-I_z \cos(\Omega_I t_1)$ represents a perturbation of the I spin from the equilibrium $+I_z$ state. The perturbation depends on the value of t_1; for example, whenever $\Omega_I t_1 = 2k\pi$, with k an integer, the populations are inverted across the I spin transitions. Consequently an NOE will be induced on the S spin during the fixed delay, τ_m, because the I and S spins have a dipolar coupling. The delay, τ_m, is known as the mixing time and is set to a suitable value (of the order of $1/R_1$) to allow a significant NOE to develop. The effect of cross-relaxation during the mixing time can be conveniently represented as

$$-I_z \cos(\Omega_I t_1) \xrightarrow{\tau_m} -I_z \cos(\Omega_I t_1) a_{II}(\tau_m) - S_z \cos(\Omega_I t_1) a_{IS}(\tau_m), \qquad [4.27]$$

in which $a_{II}(\tau_m)$ represents the fraction of the original magnetization remaining on the I spin, and $a_{IS}(\tau_m)$ represents the fraction of the original magnetization transferred from the I spin to the S spin during the mixing time by dipolar cross-relaxation. The functional forms of $a_{II}(\tau_m)$ and $a_{IS}(\tau_m)$ are discussed more fully in Section 5.1.2. The final pulse generates an observable term on the S spin of the form $S_y \cos(\Omega_I t_1)$; on two-dimensional Fourier transformation, a cross-peak will be generated at frequency $(F_1, F_2) = (\Omega_I, \Omega_S)$. For any magnetization that remains on the I spin, the final pulse will result in an observable term of the form $I_y \cos(\Omega_I t_1)$, which will yield a diagonal peak with frequency coordinates of $(F_1, F_2) = (\Omega_I, \Omega_I)$. Identical pathways also exist for magnetization transfer from S to I and corresponding diagonal and cross-peaks result.

4.2.3 Heteronuclear Coherence Transfer

In multidimensional NMR spectroscopy, different spins in the molecule are correlated by separating their interactions into more than one frequency dimension; however, the interacting spins do not necessarily have to be of the same nuclear species. Coherence can be transferred between different nuclear species using techniques analogous to those presented above for homonuclear spin systems. The corollaries to the COSY-style homonuclear coherence transfer sequences are the INEPT (21,22) and DEPT (23,24) family of pulse sequences. Heteronuclear cross-polarization corresponds to the TOCSY-style homonuclear sequences (12). Heteronuclear NOESY magnetization transfer via the heteronuclear dipolar coupling is analogous to the homonuclear experiment (25). In heteronuclear experiments, rf pulses are applied at more than one frequency (typically differing by hundreds of megahertz) in order to manipulate both the heteronuclei and the protons. Thus, in contrast to homonuclear experiments in which nonselective pulses affect all nuclei, different nuclear species can be manipulated independently by rf pulses. As has been shown in Section 2.7.7.2, the product operator approach can be used to describe manipulations of spin systems that contain operators corresponding to different nuclear species. As will be seen in Chapter 7, some of the most powerful multidimensional NMR methods rely on heteronuclear coherence transfer between $^{13}C/^{15}N$ and ^1H.

4.3 Coherence Selection, Phase Cycling, and Field Gradients

Modern NMR experiments consist of the application of multiple rf pulses to the system under investigation and detection of the resulting resonance signals. These multipulse NMR techniques are described by the pulse sequences used to generate the observed signal and by the evolution of the density operator through the pulse sequence. If an experiment consists of multiple pulses and delays, then more than one coherence transfer pathway that leads to observable signals may exist for the spin system of interest. A spectrum derived from many different but simultaneously occurring coherence transfer pathways would be extremely complex and difficult to interpret. Phase cycling or field gradients are used to select a specific pathway and provide an interpretable spectrum. The term phase cycling refers to the process of repeating a pulse sequence several times with a systematic variation of the relative phases of the pulses within the sequence. Coherence selection by means of phase cycling normally is implemented during the process of signal averaging. Field gradients are spatially inhomogenous magnetic fields that can be activated for specific periods within a pulse sequence. Coherence selection using pulsed field gradients can be achieved during a single repetition of the pulse sequence.

In the following sections, the principles of coherence selection with phase cycles and pulsed field gradients are illustrated. The text follows closely the excellent approach presented by Keeler (*26*) and employs the coherence transfer pathway methods of Bodenhausen *et al.* (*27*).

4.3.1 COHERENCE-LEVEL DIAGRAMS

The ideas of coherence developed in Section 2.6 are crucial to an understanding of phase cycling and field gradient techniques. Coherence obeying the selection rule $\Delta m = \pm 1$, commonly referred to as transverse magnetization, is termed single-quantum coherence. The word *single* is used to indicate that only one spin in the system changes its spin state (e.g., $\alpha \rightarrow \beta$). As previously noted (Section 2.4.1), only single-quantum coherence gives rise to directly observable signal. In coupled spin systems, processes occur that involve more than one spin, and these coherences are defined in an analogous fashion. For example, in a coupled two-spin system, double-quantum coherence corresponds to a transition $\Delta m = \pm 2$, with two spins flipping in the same sense (e.g., $\alpha\alpha \rightarrow \beta\beta$), whereas zero-quantum coherence corresponds to a transition $\Delta m = 0$ in which the two spins flip in opposite senses (e.g., $\alpha\beta \rightarrow \beta\alpha$). Longitudinal magnetization, although not strictly a coherence, has properties in common with zero-quantum coherence and is treated as such for phase cycling procedures. The value of $p = \Delta m$ is called the order of the coherence; the absolute value of Δm must be less than or equal to the number of spins involved in the coupling network.

At any particular point during a pulse sequence, various coherences can be present simultaneously. Normally, only one, or a small number, of the possible coherences are retained to generate a useful signal. At any time the coherences present can be classified according to their various orders (double, single, zero, etc.), and each coherence order is said to correspond to a different coherence level. For example, double-quantum coherence has a coherence level of ± 2, and longitudinal magnetization has a coherence level of zero. Formally the density operator, σ, can be written as the expansion

$$\sigma = \sum_{p=-p_{max}}^{p_{max}} \sigma_p, \qquad [4.28]$$

in which σ_p is the component of the density operator associated with a particular coherence level, p. The term p_{max} is the maximum possible coherence level available to the spin system; for example, $p_{max} = 3$ for a three-spin system. The sum in [4.28] contains $2p + 1$ terms and consequently does not represent an expansion of the density operator into a complete set of basis functions because a complete set consists of 4^p elements. Each term in [4.28] would have to be further subdivided to generate a complete basis set. For example longitudinal magnetization and zero-quantum coherence are part of the same component σ_0 in [4.28] but are represented by different basis operators. Each term σ_p is given in terms of single transition operators by

$$\sigma_p = \sum_{a,b} c_{ab} |a\rangle\langle b|, \qquad [4.29]$$

in which the sum extends over all combinations of eigenstates $|a\rangle$ and $\langle b|$ for which the magnetic quantum numbers satisfy the relationship $m_a - m_b = p$. If an arbitrary operator is expressed in the spherical operator basis (Section 2.7.2), then the coherence order is given by the number of raising operators minus the number of lowering operators constituting the representation.

The effects of a pulse sequence (i.e., pulses and periods of free precession) on coherence order are encapsulated in two rules: (1) rf pulses can cause the transfer of coherence from one level to another, and (2) periods of free precession conserve the order of coherence. Indeed, an rf pulse can transfer coherence to *all* coherence levels available to the spin system. The generation of different coherence orders during a NMR experiment is subject to the following three corollaries: (1) the coherence transfer pathway must start at coherence level $p = 0$, as this corresponds to thermal

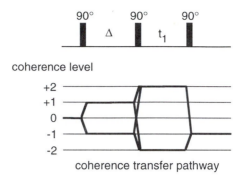

FIGURE 4.6 Coherence transfer pathway for a double quantum experiment. Double-quantum coherence is selected during the t_1 period. The pathway indicated is only one of many pathways that can be generated during the pulse sequence; unwanted coherence transfer pathways are rejected by phase cycling.

equilibrium; (2) only coherence orders $p = \pm 1$ can be created by an rf pulse acting on the thermal equilibrium density operator; and (3) if the complex signal is observed using quadrature detection, the pathway must end at $p = -1$.

As has been stated several times, most coherences generated during a pulse sequence are suppressed, and only the coherence that will generate the desired NMR spectrum is retained. Coherence is transferred in a specific manner between other coherences by rf pulses during a pulse sequence. The trace of coherence-level changes that results in the desired NMR spectrum is known as the coherence transfer pathway. The objective is to use an appropriately designed phase cycle or application of an appropriate set of field-gradient pulses to detect only those signals that follow the chosen coherence transfer pathway. For example, in a two-dimensional double-quantum experiment (Section 6.4.1), the intention is to have double-quantum coherence evolve during t_1. The coherences present at each point in the pulse sequence are expressed conveniently using a coherence-level diagram, as shown in Fig. 4.6 for a double-quantum experiment. The feasible coherence levels (-2, -1, 0, 1, 2 for a two-spin system) are shown as horizontal lines. The heavy solid lines trace the desired coherence transfer pathway by showing the desired coherence levels at every point in the pulse sequence. The indicated trace is only one of many possible coherence pathways that can be generated by this particular pulse sequence. Pathways (not shown) that have coherence levels of -1, 0, or $+1$ during t_1 must be suppressed by the phase cycle or field-gradient pulses.

4.3.2 Phase Cycles

Single-quantum coherence between two nuclear spin angular-momentum states, or transverse magnetization, is responsible for the induction of a voltage in the receiver coil. Coherence is an oscillating function of time and can be conveniently represented by a vector rotating in a circle (at least for an isolated spin treatable by the Bloch formalism). The angular position of this coherence "vector" at the beginning of the free-induction decay (FID) determines the phase of the corresponding line in the spectrum. Conventionally, one axis (the reference axis) is chosen such that an absorption mode line is produced when the coherence vector is aligned with this axis at the start of the acquisition period; other orientations of the vector give different phases of the resonance signal. Figure 4.7 illustrates the relationship between the phase of the rf pulse, the initial orientation of the coherence vector, and the phase of the resonance signal for a one-pulse experiment. An alternative way to change the phase of a resonance line is to shift the reference axis while keeping the pulse phase fixed. Figure 4.8 illustrates the relationship between the receiver phase and the phase of the resonance signal. Comparison of Figs. 4.7 and 4.8 indicates that a given phase difference in the signal can be achieved equally well by adjusting the phase of the rf pulses or of the receiver.

If the pulse is cycled through the four phases, x, y, $-x$, and $-y$ on four successive experiments and the transients added together (in either the time or frequency domain), a null signal is obtained because the two absorptive signals are exactly 180° out of phase with each other as are the two dispersive signals. A similar result is obtained for the transients shown in Fig. 4.8. However, if the phase of the pulse is cycled and the receiver phase is moved in concert to track the change in the phase of the coherence, each transient results in an absorption signal with the same phase, as shown in Fig. 4.9. If the four experiments are combined, the signals add constructively and a final spectrum is obtained that contains a single absorption line.

This simple example, which forms the basis of the CYCLOPS technique (Section 4.3.2.3), illustrates the basic principle of phase cycling. The signal of interest is forced to change phase, by shifting the phase of rf pulses, in conjunction with the receiver, so as to cause the signal recorded from different transients to accumulate. In the same manner, unwanted signals are suppressed by ensuring that signals recorded from a series of transients cancel.

4.3.2.1 Selection of a Coherence Transfer Pathway   The property used to select a specific coherence transfer pathway by phase cycling can be simply stated:

Pulse phase   Type of spectrum

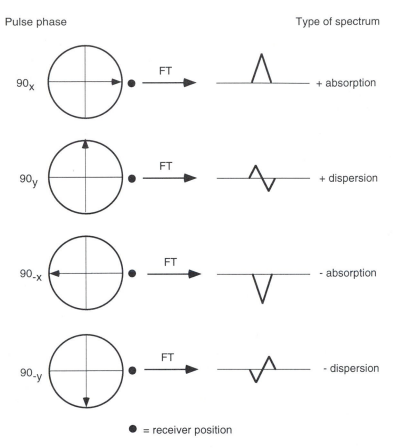

● = receiver position

FIGURE 4.7 Pulse phase. The phase of a 90° pulse is shifted in increments of 90° while keeping the phase of the receiver constant. The phase of the resulting signal in a one-pulse experiment changes in conjunction with the phase change of the pulse. Coaddition of the four resonance signals results in the cancellation of the signal.

If a pulse is changed in phase by an amount, ϕ, then a coherence undergoing a change in coherence level of Δp due to that pulse, acquires a phase shift of $-\Delta p\phi$.

For example, consider a coherence at level $+3$ being transferred to level $+1$ by the action of a pulse. If during the experiment the pulse changes phase by ϕ, the coherence will acquire phase $-\Delta p\phi$, where $\Delta p = (+1) - (+3) = -2$. Thus the coherence acquires phase $+2\phi$. The

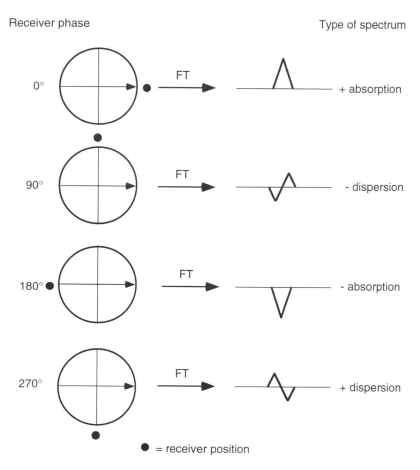

FIGURE 4.8 Receiver phase. The phase of the receiver is shifted in increments of 90° while keeping the phase of the 90° pulse constant. The phase of the resulting signal in a one-pulse experiment changes in conjunction with the phase change of the receiver. Coaddition of the four resonance signals results in the cancellation of the signal.

coherences are labeled with phase shifts during the pulse sequence and the accumulated phase angles of the desired coherence transfer pathway enable its selection. Coherence selection can be accomplished simply by changing the phases of the relevant pulses in the sequence by the appropriate amounts. The ultimate phase of the observed magnetization depends on the total phase angle that the coherence acquires during the coherence transfer steps in the sequence.

Pulse phase   Type of spectrum

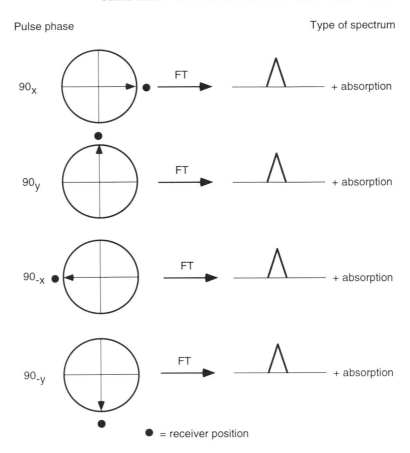

FIGURE 4.9 Pulse and receiver phase. The phase of the 90° pulse is shifted in increments of 90° while simultaneously shifting the phase of the receiver in increments of 90°. In this case, the resulting signal retains the same phase in each experiment (shown as absorptive here). Coaddition of the four resonance signals results in the coherent summation of the signal.

To prove the preceding statement, the effect on a particular coherence of a nonselective z rotation (affecting all spins identically) is proved first:

$$\exp(-i\phi\mathbf{F}_z)\sigma_p \exp(i\phi\mathbf{F}_z)$$

$$= \sum_{a,b} c_{ab} \exp(-i\phi\mathbf{F}_z)|a\rangle\langle b| \exp(i\phi\mathbf{F}_z)$$

$$= \sum_{a,b} c_{ab}\exp(-i\phi m_a)|a\rangle\langle b|\exp(i\phi m_b)$$

$$= \sum_{a,b} c_{ab} |a\rangle\langle b| \exp(-i\phi p)$$

$$= \sigma_p \exp(-i\phi p), \qquad [4.30]$$

in which

$$\mathbf{F}_z = \sum_{j=1}^{N} I_{jz}. \qquad [4.31]$$

Next, suppose that some propagator $\mathbf{U}$ transfers coherence from an element σ_p to an element $\sigma_q = \mathbf{U}\sigma_p\mathbf{U}^{-1}$. If all rf pulses are shifted in phase by an angle ϕ, then (note that the Zeeman and scalar coupling interactions are unaffected by z rotations):

$$\exp(-i\phi\mathbf{F}_z)\mathbf{U} \exp(i\phi\mathbf{F}_z)\sigma_p \exp(-i\phi\mathbf{F}_z)\mathbf{U}^{-1} \exp(i\phi\mathbf{F}_z)$$

$$= \exp(-i\phi\mathbf{F}_z)\mathbf{U}\sigma_p \exp(i\phi p)\mathbf{U}^{-1} \exp(i\phi\mathbf{F}_z)$$

$$= \exp(-i\phi\mathbf{F}_z)\sigma_q \exp(i\phi\mathbf{F}_z) \exp(i\phi p)$$

$$= \sigma_q \exp(-i\phi q) \exp(i\phi p)$$

$$= \sigma_q \exp(-i\phi\,\Delta p), \qquad [4.32]$$

with $\Delta p = q - p$, which provides the desired proof.

To illustrate the use of phase cycling to retain one coherence transfer pathway while rejecting another, suppose that in some pulse sequence a mixture of double-quantum coherence ($p = \pm 2$) and zero-quantum coherence ($p = 0$) has been created. Application of a 90° pulse to the initial zero and double quantum states causes the desired and undesired transfers between coherence levels shown in Fig. 4.10. The goal of the phase cycle is to convert double-quantum coherence to observable single-quantum coherence ($p = -1$) while suppressing any signal from the zero-quantum coherence.

The experiment is repeated four times; on each repetition, the phase of the pulse is incremented by 90° to yield a phase cycle of 0°, 90°, 180°, 270° (which conventionally is written as x, y, $-x$, $-y$). The desired coherence ($p = \pm 2$) goes through two pathways, with $\Delta p = -3$ ($p = +2$ to $p = -1$) and $\Delta p = +1$ ($p = -2$ to $p = -1$). The coherence that undergoes the level change $\Delta p = -3$ changes phase by an amount $-\Delta p\phi = -(-3)\phi = +3\phi$. The acquired phase of the observable coherence for the four transients will be 0°, 270°, 540°, 810°, or in terms of a 0° to 360° rotations, 0°, 270°, 180°, 90°.

In a similar way, the coherence undergoing a level change of $\Delta p = +1$ acquires a phase of $-\Delta p\phi = -(+1\phi) = -\phi$. The coherence acquires

a
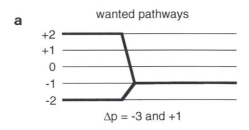

b
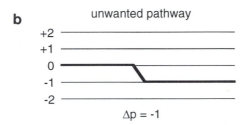

FIGURE 4.10 Selection of double-quantum coherence. Panel (a) illustrates the desired coherence-transfer pathways corresponding to coherence-level changes from double-quantum ($p = +2$ and -2) to observable single-quantum ($p = -1$) coherence. These pathways have coherence order changes of $\Delta p = -3$ and $+1$, respectively. Panel (b) illustrates the coherence transfer pathway to be rejected corresponding to a coherence level change from zero-quantum ($p = 0$) to observable single-quantum ($p = -1$) coherence. This pathway has a coherence order change of $\Delta p = -1$.

phase 0°, -90°, -180°, -270°, which in the 0° to 360° reference frame is equivalent to 0°, 270°, 180°, 90°. Thus, in the experiment described, the two coherence level changes of $\Delta p = -3$ and $\Delta p = +1$ result in the observable coherences acquiring phase in an identical fashion. To ensure that the required signal (from both pathways) accumulates constructively on successive transients, the receiver phase must follow exactly the phase shifts of the wanted coherence. Therefore, the receiver phase cycle is 0°, 270°, 180°, 90°. This information is contained in Table 4.1.

The coherence-level change from zero-quantum coherence to observable single-quantum coherence has $\Delta p = -1$. Employing the same analysis as above, the coherence acquires phase $-\Delta p \phi = -(-1\phi) = +\phi$; or 0°, 90°, 180°, 270°. This result, along with the receiver phase cycle determined above, is tabulated in Table 4.2. While steps 1 and 3 of the receiver phase cycle follow the coherence, steps 2 and 4 are exactly opposite to the coherence phase. Consequently, the same phase cycle, including pulse and receiver phase shifts, that retains the wanted pathways ($\Delta p = -3$ and $\Delta p = +1$), also serves to eliminate the unwanted pathway ($\Delta p = -1$), exactly as required.

Table 4.1

Selection of Double-Quantum Coherence

Pulse phase (ϕ)	$-\Delta p \phi$	Equivalent cycle	Receiver phase
Coherence change $\Delta p = -3$			
0	0	0	0
90	270	270	270
180	540	180	180
270	810	90	90
Coherence change $\Delta p = +1$			
0	0	0	0
90	-90	270	270
180	-180	180	180
270	-270	90	90

In the preceding example a four-step phase cycle with increments in phase of 90° was able to discriminate between coherence transfer pathways of $\Delta p = -3$, +1 and $\Delta p = -1$. The effect of this phase cycle can be conveniently represented using the nomenclature of Bodenhausen and coworkers (27), as **−3** (−2) (−1) (0) **+1** (+2) (+3), where the pathways passed by the cycle are set in bold and the pathways blocked by the cycle are set in brackets. Under the proposed scheme, *two* pathways are allowed to pass. In general, if a phase cycle uses steps of 360/N degrees, then along with the pathway Δp selected, pathways $\Delta m = \Delta p \pm nN$, where $n = 1, 2, 3, \ldots$, will also be selected. Bodenhausen showed that the length of the phase cycle is related to the degree of selectivity of that phase cycle. If a particular value of Δp is to be chosen from r consecutive values, N must be at least r. What this means in practical terms is that increased selectivity in choosing a specific coherence transfer pathway requires a larger number of smaller steps in the phase cycle.

Table 4.2

Rejection of Zero-Quantum Coherence

Pulse phase (ϕ)	$-\Delta p \phi$	Equivalent cycle	Receiver phase
Coherence change $\Delta p = -1$			
0	0	0	0
90	90	90	270
180	180	180	180
270	270	270	90

Table 4.3
Distinguishing $\Delta p = +1$ from $\Delta p = -3$

Pulse phase (ϕ)	$-\Delta p \phi$	Equivalent cycle (A)	Receiver phase (B)
	Coherence change $\Delta p = +1$		
0	0	0	0
60	−60	300	300
120	−120	240	240
180	−180	180	180
240	−240	120	120
300	−300	60	60
	Coherence change $\Delta p = -3$		
0	0	0	0
60	180	180	300
120	360	0	240
180	540	180	180
240	720	0	120
300	900	180	60

Continuing with the example above, now consider discriminating between the two pathways $\Delta p = -3$ and $\Delta p = +1$, both of which were retained by the original phase cycle. For instance, suppose that only the $\Delta p = +1$ pathway is to be retained and the $\Delta p = -3$ pathway is to be rejected. Table 4.3 shows the effects of extending the phase cycle to six steps rather than four and using a phase increment of 60° rather than 90°. The analysis proceeds exactly as before, making sure that the receiver shifts in phase so as to follow the phase acquired by the coherence going through the $\Delta p = +1$ pathway.

According to Table 4.3, the signal from the $\Delta p = +1$ pathway will add on all transients. However, the effect of this receiver phase cycle on the signal arising from the $\Delta p = -3$ pathway is not obvious. In the approach adopted by Keeler (26), the net effect of the phase cycle is represented by subtracting the phase acquired by the coherence from the receiver phase (or vice versa) and representing the differences as a vector diagram. Subtracting column A from column B in Table 4.3 gives the results shown in Table 4.4. Figure 4.11 shows that the net effect of the opposing vectors is zero. Thus, any signal resulting from the $\Delta p = -3$ pathway will be canceled.

As anticipated, a longer phase cycle with smaller phase increments

TABLE 4.4
Rejecting $\Delta p = -3$

Coherence phase (A)	Receiver phase (B)	(B) − (A)	Equivalent phase (B) − (A)
0	0	0	0
180	300	120	120
0	240	240	240
180	180	0	0
0	120	120	120
180	60	−120	240

on each step allows greater selectivity. Remembering the rules noted earlier, this phase cycle will also retain pathways where $\Delta p = +1 \pm 6n$. All other pathways are rejected. Using the nomenclature of Bodenhausen *et al.* (*27*), the effect of this phase cycle can be written as (-6) **−5** (-4) (-3) (-2) (-1) (0) **+1** $(+2)$ $(+3)$ $(+4)$ $(+5)$ $(+6)$ **+7**. In fact, the six-step phase cycle used in this example is overly selective; as the notation of Bodenhausen makes clear, a five-step phase cycle with 72° increments would also have been satisfactory (although more difficult to visualize).

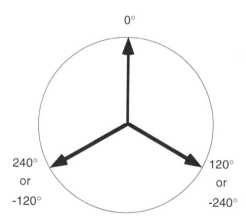

FIGURE 4.11 Vectorial picture of phase cycling. From the results in Table 4.4, vectors can be drawn corresponding to the difference in phase acquired by the coherence and the receiver phase, for the coherence transfer pathway $p = +2$ to $p = -1$ ($\Delta p = -3$). In this case the goal is to discriminate between the coherence order change $\Delta p = +1$ and $\Delta p = -3$. The sum of these vectors is zero, and any signal from the $\Delta p = -3$ pathway is eliminated.

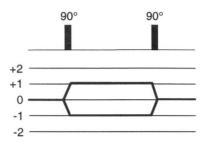

FIGURE 4.12 Two-pulse segment used to generate zero-quantum coherence.

These principles can be restated as follows:

1. A phase cycle (affecting a single rf pulse or a group of pulses) that consists of N steps with $360°/N$ increments selects coherence transfer pathways satisfying $\Delta m = \Delta p \pm nN$, in which n is an integer.
2. The value of Δp is selected, from the N consecutive possible values Δp, $\Delta p + 1$, $\Delta p + 2$, ..., $\Delta p + N - 1$, by shifting the receiver phase by $-360°\ \Delta p/N$ synchronously with the pulse phase cycle.

4.3.2.2 Saving Time   To unambiguously select a definite coherence transfer pathway in a pulse sequence, each pulse must have a specific phase cycle and each phase cycle must be executed independently. Unfortunately, strict application of this rule generates extremely long phase cycles, and some mechanisms must be found to reduce the phase cycle to an acceptable length.

   Rather than considering a coherence transfer step as being mediated by just one pulse, several pulses and intervening delays can be grouped together and regarded as a single unit that causes the desired coherence transfer. The phases of all the pulses in the unit can be shifted simultaneously in order to reduce the number of steps in the overall phase cycle. For example, consider the two pulse experiment shown in Fig. 4.12. The overall aim is to select the overall pathway $\Delta p = 0$ (coherence that starts at level 0 and ends at 0). Cycling each of the pulses independently in 90° increments to select the coherence transfer pathway shown results in a phase cycle of 16 steps. Phase cycling both the pulses simultaneously through the four phases 0°, 90°, 180°, 270° while holding the receiver phase constant selects the pathway $\Delta p = 0$ as required (this assertion can be checked by using the identity described above, $-\Delta p\phi$). In this way the phase cycle has been reduced from 16 steps to 4 steps. Any other pathways that have $\Delta p = 0$ will also be retained as this approach leaves the coherence level between the pulses unrestricted. In many circumstances, these other pathways may be disregarded when additional considerations discussed below are taken into account. However, if an undesired coherence with $\Delta p = 0$ results in observable signals when the 4-step phase cycle is used, then the full 16-step phase cycle must be used.

Normally the first pulse in a sequence is applied to equilibrium magnetization, and only coherences with $\Delta p = \pm 1$ are generated. Indeed, a phase cycling scheme will work only if the same initial state exists for successive transients and consequently suitable delays must be inserted between each transient to enable the system to return to equilibrium. Unless a specific reason exists for distinguishing the coherence levels $+1$ and -1 following the first pulse, phase cycling of that pulse is unnecessary. If an unambiguous coherence transfer pathway has been chosen by the phase cycles of earlier pulses in the sequence, the last pulse does not need to be phase-cycled. Although application of the last pulse to the system may generate many different coherence orders and therefore many coherence pathways, only those pathways that lead to a final coherence level of $p = -1$ (single-quantum coherence) are observable. The experiment, in essence, chooses the last coherence transfer step itself.

Certain coherence transfer pathways are improbable in a given spin system. The maximum coherence available in a system is restricted by the number of nuclear spins in that system. For spin-$\frac{1}{2}$ nuclei, at least N coupled spins are required to produce N-quantum coherence. In principle, phase cycles that discriminate against coherence orders higher than N are unnecessary; in practice, generating large amounts of high-order coherences is difficult, and thus coherence transfer pathways containing these coherence levels can be ignored. For proton spin systems, coherence orders greater than four or five may not require consideration and the resulting phase cycles can be correspondingly shorter.

4.3.2.3 Artifact Suppression   Before continuing, three simple phase cycling procedures that can be employed to reduce instrumental artifacts are discussed: CYCLOPS, EXORCYCLE, and axial peak suppression.

Quadrature detection is obtained during acquisition of the NMR signal by two phase-sensitive detection channels (Section 3.2.2). The two channels, in principle, should be identical except for a relative phase shift of 90°. Anomalies arise if differences exist between the two phase-sensitive detector channels. If the two quadrature channels have different sensitivities, the NMR spectrum contains spurious peaks called *quadrature images*. The images are located in symmetrical positions with respect to the center of the spectrum as genuine peaks (i.e., if a resonance has an offset of Ω, the quadrature image appears at $-\Omega$). Also, if the electrical direct-current (dc) (nonsinusoidal) baseline offset between the two channels differs, then a spike appears in the middle of the spectrum. This artifact most commonly is referred to as a *quadrature glitch*.

In order to remove these artifacts, a simple phase cycling routine known as CYCLOPS can be used (*28*). For the case of a simple pulse–acquire experiment, CYCLOPS consists of cycling the pulse and receiver through the phases 0°, 90°, 180°, 270° synchronously. Any gain difference between the two channels is compensated for by the 90° phase incrementation, while baseline offset errors are eliminated by the 180° phase inversions. For longer and more complicated pulse sequences, a phase cycle normally is employed that closely mimics the action of CYCLOPS. If this is not the case, CYCLOPS can be implemented by adding the phase incrementations 0°, 90°, 180°, 270° to *all* pulses in the sequence along with the receiver. The one drawback to this procedure is that the length of the phase cycle, and therefore the minimal experimental time, is increased by a factor of 4. If phase cycling limitations preclude full CYCLOPS phase cycling in a particular pulse sequence, a two-step phase cycle consisting of 0° and 90° phase shifts may be satisfactory in reducing quadrature image artifacts (but not quadrature glitches).

To illustrate the CYCLOPS technique more fully, an imbalanced detection operator is defined as

$$F^+(\varepsilon) = I_x + i(1 + \varepsilon)I_y = I^+ + \varepsilon/2\,(I^+ - I^-) = (1 + \varepsilon/2)I^+ - \varepsilon/2I^-,$$

$$[4.33]$$

in which $\varepsilon \neq 0$ is the imbalance term. The detected signal is given as usual by the trace of the density operator with the detection operator:

$$s^+(t) = \mathrm{Tr}\{\sigma(t)F^+(\varepsilon)\} = \sum_{-p_{max}}^{p_{max}} \mathrm{Tr}\{\sigma_p(t)F^+(\varepsilon)\}$$

$$[4.34]$$

$$= (1 + \varepsilon/2)\mathrm{Tr}\{\sigma_{-1}(t)I^+\} - (\varepsilon/2)\mathrm{Tr}\{\sigma_{+1}(t)I^-\}.$$

If $\varepsilon \neq 0$, any components of the density operator that are proportional to I^+, which has $p = +1$, will have a nonzero trace with the I^- operator and generate artifacts in the NMR spectrum. Referring back to the concepts of coherence transfer pathways, CYCLOPS phase cycling is equivalent to selecting $\Delta p = -1$ for the entire sequence. Thus, assuming that the initial density operator is given by the thermal equilibrium operator,

$$s^+(t) = \mathrm{Tr}\{\sigma_{-1}(t)F^+\} = (1 + \varepsilon/2)\mathrm{Tr}\{\sigma_{-1}(t)I^+\},$$

$$[4.35]$$

and the artifacts in the spectrum are suppressed.

The spin-echo sequence, τ–180°–τ, is a vital component of a large number of NMR experiments. The coherence-level diagram for this pulse

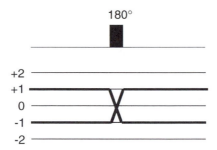

FIGURE 4.13 EXORCYCLE. Coherence transfer pathway for a spin-echo sequence. The 180° pulse serves to change the sign of the coherence order.

sequence is shown in Fig. 4.13. Spin-echo sequences are sensitive to common imperfections in 180°, such as miscalibrated pulse lengths and off-resonance effects, that can generate spurious responses. The EXORCYCLE phase cycling scheme is designed to compensate for imperfect 180° pulses (*29*). An ideal 180° refocusing pulse simply has the effect of changing the sign of the coherence level. For example, if the initial density operator has coherence level $p = +1$, then following the 180° pulse the coherence level would be $p = -1$. The desired coherence transfer pathway has $\Delta p = -2$, so the appropriate phase cycle for the 180° pulse would be 0°, 90°, 180°, 270°, while the receiver is cycled 0°, 180°, 0°, 180°. This is the EXORCYCLE phase cycle. This phase cycle also selects the mirror-image pathway $\Delta p = +2$ so that the EXORCYCLE procedure can be employed if the spin-echo segment is part of a more complicated sequence (Section 4.3.4). The undesired pathway of $\Delta p = 0$, which corresponds to unrefocussed magnetization and $\Delta p = \pm1$, which corresponds to a coherence transfer process, are both suppressed by the EXORCYCLE procedure. If suppression of $\Delta p = 0$ is not important in a particular application, a two-step phase cycle can be utilized, in which the 180° pulse is cycled by 0° and 180° and the receiver phase is not altered.

Axial peaks occur in multidimensional NMR experiments because magnetization relaxes toward equilibrium during free precession evolution periods, such as the t_1 interval. This magnetization is not frequency-labeled during the t_1 period, and is not sensitive to the phase cycling of pulses occurring earlier in the sequence. If the relaxed longitudinal magnetization is converted to observable magnetization prior to the acquisition period, spurious signals will be generated along the line $F_1 = 0$ in the NMR spectrum (see Section 4.3.4.3). Axial peaks can be eliminated by phase cycling a pulse or pulses prior to the t_1 period and the receiver by phase angles of 0° and 180°. The most common procedure for obtaining axial peak suppression is by phase cycling the initial pulse in the pulse sequence in conjunction with the receiver; additionally, many multidimensional NMR experiments have complex phase cycles that function incidentally to suppress axial peaks.

4.3.2.4 Limitations of Phase Cycling A phase cycle works by requiring signals arising from the desired coherence transfer pathway to add constructively on successive transients, whereas signals from unwanted pathways cancel. Evidently, phase cycling is simply a difference method and consequently will work only if experimental conditions remain constant from transient to transient. Unfortunately, as a practical matter, slight variations that occur between transients reduce the effectiveness of phase cycling and also generate t_1 noise. For example, amplitude or phase changes in the pulses or field-frequency variations in the lock circuitry can contribute to variability from transient to transient. One of the most common sources of instability is temperature fluctuations that cause resonances (including the lock resonance) to shift slightly.

The magnitude of the signals derived from the desired coherence transfer pathway compared to the magnitude of signals from unwanted pathways is an important determinant in the success of phase cycling. If signals from unwanted pathways are expected to be large, errors in the difference procedure may produce artifacts of intensity comparable to that of desired signals. Although the deleterious effects of random fluctuations on coherence selection would be expected to cancel after extended signal averaging, instrumental instabilities are frequently periodic, and even for random effects convergence is generally slow. From a practical point of view, the order in which the individual steps of the phase cycle are employed may result in better or worse suppression of undesirable signals. Unfortunately, the optimal order that the steps in a phase cycle should be applied can vary from one spectrometer to another and must be determined empirically.

Phase cycles also assume that at the beginning of each transient the system is in thermal equilibrium and only longitudinal magnetization exists. Leaving a long recycle time between successive transients is the optimum way to ensure this condition, but this approach can cause lengthy experiments, and recycle times on the order of $1/R_1$ to $1.5/R_1$ are commonly employed. In some instances, the phase cycle can be designed to suppress artifacts that arise from rapid repetition of a pulse sequence.

4.3.3 PULSED FIELD GRADIENTS

Recently, an alternative method for coherence transfer pathway selection has been developed that makes use of pulsed field gradients and avoids many of the problems associated with phase cycling procedures

(30–40). A field-gradient pulse is a pulse or a period during which the B_0 field is made deliberately inhomogeneous. In such an inhomogeneous field, transverse magnetization and other coherences dephase across the sample. However, the coherence can be refocused by another appropriately applied gradient to generate gradient echoes (41). The critical principle is the following:

> Coherence dephases in an inhomogenous magnetic field at a rate proportional to the coherence order and the gyromagnetic ratios of the affected nuclei.

For many years the disadvantages of this approach were that the strong field gradients required disturbed the field-frequency lock and generated eddy currents. However, with the advent of gradient coils that are actively shielded and advances in lock-blanking methods, these problems have been overcome. The use of field-gradient techniques for coherence transfer pathway selection can be expected to become increasingly popular, especially in multidimensional NMR spectroscopy.

In the simplest application, the B_0 field is varied linearly in the z direction during a gradient pulse. Conceptually, the NMR sample can be envisaged as a column of thin slices (called isochromats) along the z direction. The spins in each slice experience different magnetic fields and thus different Larmor resonance frequencies. Initially, the spins in each isochromat are phase-coherent. As the gradient is applied, the phase coherence between slices is lost as a result of Larmor precession. After a sufficiently long time, complete dephasing occurs and the net magnetization of the sample becomes zero. The dephasing process caused by the gradient pulse results because the coherences acquire *spatially dependent phase*.

More formally, the magnetic field produced by the gradient pulse, $B_g(z)$ varies linearly along the z axis and can be written

$$B_g(z) = zG_z, \qquad [4.36]$$

where G_z is the gradient strength given in teslas per meter (T/m) or, more often, gauss per centimeter (G/cm). With the origin of the z axis being taken as the center of the sample, the Larmor frequency at any point in the sample, $\omega(z)$ is given by

$$\omega(z) = -\gamma[B_0 + B_g(z)] = -\gamma[B_0 + zG_z] = \omega_0 - \gamma zG_z. \qquad [4.37]$$

In the rotating frame, the ω_0 term vanishes. If the gradient is applied for a time t, then the *spatially dependent phase* at any position in the sample, $\phi(z)$, is given by

$$\phi(z) = \gamma zG_zt. \qquad [4.38]$$

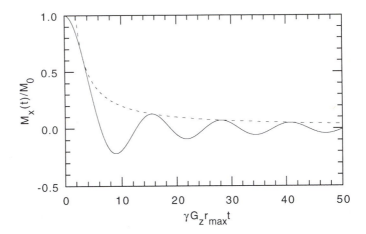

FIGURE 4.14 Dephasing of transverse magnetization by pulsed field gradients. The relative amplitude of transverse magnetization is plotted against $\gamma Gr_{max}t$ to show the effect of applying a pulsed field gradient in the z direction to transverse magnetization. The solid line represents decay of the oscillatory signal according to [4.40]. The dotted line shows the asymptotic decay $M_x(t)/M_0 = 2/(\gamma Gr_{max}t)$.

Consider the case of applying a gradient pulse of strength G_z for a time t to a system consisting of in-phase single quantum coherence, e.g., I_x. At any point in the sample the evolution of I_x is given by

$$I_x \xrightarrow{\;-\gamma z G_z t I_z\;} \cos(\gamma z G_z t)I_x + \sin(\gamma z G_z t)I_y. \qquad [4.39]$$

The net x magnetization across the whole sample is determined by summing (integrating) over all the slices through the sample:

$$M_x(t) = \frac{1}{r_{max}} \int_{-r_{max}/2}^{r_{max}/2} \cos(\gamma z G_z t)\, dz = \frac{2\sin(\gamma G_z t r_{max}/2)}{\gamma G_z t r_{max}}$$

$$= \mathrm{sinc}(\gamma G_z t r_{max}/2), \qquad [4.40]$$

where the sample extends over a region $\pm r_{max}/2$. This equation represents the decay of the x-magnetization oscillations during the gradient pulse. First and most obviously, [4.40] demonstrates that application of a stronger gradient causes the magnetization to decay at a faster rate. Second, [4.40] shows that magnetization from nuclei with higher gyromagnetic ratios decays faster. In addition, the strengths of gradients required to suppress coherences to a defined level can be estimated. Figure 4.14 shows a plot of $M_x(t)$ versus $\gamma Gr_{max}t$ for this equation. The overall decay for long times

is given by $2/(\gamma G r_{max} t)$. For example, in the case noted above the amount of x magnetization will have decayed to a fraction α of its initial value after a time of the order $2/(\gamma G r_{max} \alpha)$. Consequently, if the requirement were to suppress proton x magnetization to 0.1% of its original value, then a gradient pulse of 2.5 ms with strength 30 G/cm (0.3 T/m) over a sample region $r_{max} = 1$ cm would be sufficient. For a typical biological sample in a 5 mm NMR tube with a sample volume of approximately 450 μl, giving $r_{max} \sim 3.25$ cm, a 30-G/cm gradient pulse would need to be applied for only 0.77 ms for the same level of suppression. Of course, this is the idealized calculation and in practice, fine-tuning of the duration and the gradient strength is always required.

In practice, field gradients are not applied as rectangular pulses, because a rectangular pulse with an infinite slope at time zero induces currents that generate torque on the gradient coils. If the gradient pulse is strong enough, the gradient coils can be damaged physically. Consequently, field-gradient pulses normally are shaped such that the amplitude envelope of the gradient pulse is a smooth function of time. Generalizing the preceding discussion, the spatially dependent phase generated by a shaped gradient pulse of duration t applied to a (possibly multiple-spin heteronuclear) coherence is

$$\phi(\mathbf{r}, t) = sB_g(\mathbf{r})t \sum_i p_i \gamma_i, \qquad [4.41]$$

in which γ_i and p_i are the gyromagnetic ratios and coherence level of each nuclear species, i, contributing to the coherence. The shape factor, s, is defined as

$$s = \frac{1}{t} \int_0^t A(t)\, dt, \qquad [4.42]$$

in which the amplitude profile of the gradient pulse is given by $|A(t)| \leq 1$. Opposing gradients, i.e., those that either increase or decrease as the z coordinate increases, have values of s that are opposite in sign. The overall amplitude of the gradient is represented by $B_g(\mathbf{r})$. For convenience, the dependence of ϕ on $(\mathbf{r}, t)$ and B_g on $\mathbf{r}$ will be implied in the following.

Practically, field gradients in the z direction can be generated in two ways. First, the shim coils can be used to produce a spatially inhomogeneous field, but this method produces gradient pulses only of ~ 1 G/cm. Gradients produced with the shim coils are called homospoil pulses. Second, and much more preferable, specialized actively shielded gradient coils can be built directly into the high-resolution probe. Gradient strengths of 30 G/cm are easily obtainable by this setup. Probe manufacturers now also offer the option of three-axis gradient systems, containing coils

which can apply simultaneous gradients, if required, along the x, y and z directions. Dephasing of magnetization depends on the function $1/(\gamma^3 G_x G_y G_z v t^3)$, where G_i is the strength of the gradient in direction i and v is the sample volume, and consequently is much more rapid than for a single z gradient.

4.3.3.1 Selection of a Coherence Transfer Pathway A particular coherence transfer pathway is selected by using field-gradient pulses to generate gradient echoes for specific coherences, while leaving unwanted coherences randomized. Consider the case in which field-gradient pulses are applied on either side of a mixing period that mediates coherence transfer between coherence levels p_i and p_f. The first gradient pulse induces a spatially dependent phase of ϕ_i, and the second gradient pulse induces a phase of ϕ_f, where

$$\phi_i = s_1 p_i B_{g1} \tau_1 ; \qquad \phi_f = s_2 p_f B_{g2} \tau_2 . \qquad [4.43]$$

Following the second gradient pulse, the net phase accrued by the final coherence is $\phi_i + \phi_f$. Selection of a *particular* coherence transfer pathway $p_i \rightarrow p_f$ occurs by ensuring that the overall phase change is zero; therefore, the durations and amplitudes of the two gradient pulses must be adjusted such that $\phi_i = -\phi_f$. The second gradient can be thought to "unwind" the effects of the first gradient to form an echo. Coherence orders for which $\phi_i + \phi_f \neq 0$ remain dephased and do not contribute to the resulting signal.

An illustrative example of coherence selection by gradient pulses is shown in Fig. 4.15. In this example a gradient is applied for a time t prior to the pulse, which causes the coherence transfer. The coherence $p = +2$ dephases by an amount that is proportional to $+2t$. The pulse transfers the coherence to $p = -1$, a gradient, in the same sense, is reapplied, but this time for a time $2t$. The coherence at this level dephases by an amount proportional to $-2t$, which is in a sense opposite to that induced by the first gradient pulse. Consequently, after a time $2t$, the required coherence is fully rephased. By using strong field gradients, all other coherences involved in other pathways are dephased and coherence selection is achieved.

The real advantage of gradient pulses compared with phase cycling is that signals arising from unwanted pathways are removed by the gradients in each individual transient rather than relying on subtraction processes after digitization of the signal. Consequently, artifacts from instrumental instabilities can be significantly smaller than in experiments using phase cycling for coherence selection. In addition, for some experiments, the number of transients that must be accumulated to achieve a particular signal-to-noise ratio is smaller than the number of transients required to

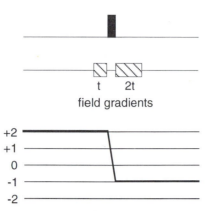

t   2t

field gradients

+2
+1
0
-1
-2

FIGURE 4.15 Selection of the coherence transfer pathway $p = +2$ to $p = -1$ by the use of pulsed field gradients. The double-quantum coherence $p = +2$ accrues a phase proportional to $+2t$, in which t is the duration of the first gradient pulse. The following rf pulse (solid bar) transfers the coherence to $p = -1$. To rephase *only* the final coherence that originated at the $p = +2$ coherence level, application of the second gradient for duration $2t$ serves to "unwind" exactly the required coherence.

select a particular coherence transfer pathway by phase cycling procedures; the experiment can be executed in less time if gradient pulses are used for coherence selection. This attribute becomes important when recording higher-dimensional experiments that require large amounts of spectrometer time.

4.3.3.2 Artifact Suppression   Artifact suppression using CYCLOPS and EXORCYCLE phase cycling was discussed in Section 4.3.2.3. Implementation of these schemes requires a minimum number of phase cycle steps, and employing more than one of these schemes can increase the length of a phase cycle enormously (i.e., independent EXORCYCLE phase cycling of n 180° pulses requires 4^n steps). To reduce such prolonged phase cycles, very simple combinations of gradients can be used for artifact suppression. In practice, three- and four-dimensional experiments, with correspondingly large numbers of pulses, coherence transfer steps, and refocusing periods, exhibit the most artifacts and are subject to the most severe restrictions on the overall length of the phase cycle. The performance of a number of common components of heteronuclear multidimensional NMR experiments can be augmented by the introduction of appropriate pulsed field gradients *(42)*.

The 180° refocusing pulse in a spin-echo sequence is notoriously prone to pulse artifacts that historically would be removed by the EXORCYCLE phase cycle. The gradient-enhanced homonuclear spin echo sequence is

$$\begin{array}{ccc} \tau & - \quad 180°(I) \quad - & \tau \\ +z \text{ gradient} & & +z \text{ gradient.} \end{array} \qquad [4.44]$$

A transverse operator, for example, either I_x or I_y, is dephased by the first gradient, the coherence order is inverted by the $180°(I)$ pulse, and then the operator is rephased by a gradient *of the same sign and the same strength*. The gradient pulses eliminate the effect of pulse imperfections that lead to transfer between transverse and longitudinal magnetization, and additionally, any transverse operator of a different spin not affected by the 180° pulse is effectively removed. This is all accomplished *in a single transient*.

In another common application, a $180°(I)$ heteronuclear decoupling pulse is used to invert the longitudinal I spin component of antiphase heteronuclear coherence, $2I_zS_x$ or $2I_zS_y$, from $+z$ to $-z$ (or vice versa). This serves to decouple the two spins during the time period 2τ:

$$\begin{array}{ccc} \tau & - \quad 180°(I) \quad - & \tau \\ +z \text{ gradient} & & -z \text{ gradient.} \end{array} \qquad [4.45]$$

The z component inverted by the 180° pulse is unaffected by the application of either gradient. The $+z$ and $-z$ gradients refocus the transverse S magnetization. Conversion of I_z magnetization into transverse magnetization by an imperfect $180°(I)$ pulse is eliminated because such transverse magnetization will be irreversibly dephased by the second gradient. Again, artifact suppression is accomplished in one transient.

Longitudinal two-spin order, $2I_zS_z$, is created from antiphase coherence, $2I_xS_z$ or $2I_yS_z$, by application of a $90°(I)$ pulse with y or x phase, respectively. Two-spin order has a coherence order of zero and is unperturbed by gradient pulses. Accordingly, other coherences (except for longitudinal magnetization proportional to I_z and S_z) can be selectively dephased by applying a field-gradient pulse after the creation of two-spin order.

4.3.3.3 Limitations of Pulsed Field Gradients   The main limitation of pulsed field gradients for coherence selection is evident from [4.43]; if a coherence pathway $p_i \to p_f$ is selected by gradient techniques, the corresponding pathway $-p_i \to p_f$ cannot be selected simultaneously. As will be discussed in Section 4.3.4, frequency discrimination in indirect evolution periods requires that the signals be recorded for both the $p_i \to p_f$ and $-p_i \to p_f$ pathways (in which p_i is the coherence order during t_1 and $p_f = -1$ for observable magnetization during t_2). In most pulsed field-gradient experiments, signals from the two pathways must be acquired sequentially (i.e., in two separate experiments) with the result that the sensitivity of the pulsed field-gradient experiment is reduced by a factor of $\sqrt{2}$ compared to the corresponding phase-cycled experiment (*37,43,44*).

A second limitation is that the use of pulsed field gradients usually requires lengthening of the pulse sequence. Even with actively shielded gradients, delays on the order of 0.2–1.0 ms may be required to permit the spectrometer system to stabilize following a gradient pulse. Additional spin-echo sequences also may be necessary to refocus the effects of chemical-shift evolution arising from the unperturbed Larmor frequency during a gradient pulse [4.37]. Inevitable relaxation during the inserted delays reduces sensitivity.

4.3.4 FREQUENCY DISCRIMINATION

As noted in Section 4.3.3, coherence order is a signed quantity. The sign indicates the sense of precession of the coherence relative to a reference frame rotating at the transmitter frequency; differentiating between evolution frequencies higher or lower than the transmitter frequency is called frequency discrimination or quadrature detection (Section 3.2.2). In high-resolution multidimensional NMR spectroscopy, spectra are desired in which frequency discrimination is obtained and absorption lineshapes are retained in all dimensions. Outlined below are the methods that have been designed to do this. The two basic techniques for frequency discrimination during evolution periods are termed the *hypercomplex* (or *States*) method and the *time-proportional phase-incrementation* (TPPI) method. The analysis here is based on the work of Bodenhausen *et al.* (*27,45*) and on the seminal paper by Keeler and Neuhaus (*46*). The following discussion is for two-dimensional spectroscopy; extension to higher dimensions is straightforward.

As was discussed in Section 3.2.2, frequency discrimination during the acquisition period is obtained by quadrature detection: The sine- and cosine-modulated components of the evolving magnetization are recorded independently by orthogonal detectors (the exact method depends on the construction of the spectrometer) and treated as a complex signal during subsequent processing. In a two-dimensional experiment, since the signal during t_1 is never actually recorded, conventional quadrature detection cannot be used to determine the relative sense of precession of magnetization in that dimension. Nonetheless, the fundamental result that both cosine- and sine-modulated components of the appropriate coherences must be recorded is equally valid for the evolution period as for the acquisition period.

To continue with the analysis, some important and useful relationships must be developed. Adopting the conventions and notations of Keeler and Neuhaus, the *cosine modulated* time domain (t_1) function $S_c(t_1, t_2)$ is

$$S_c(t_1, t_2) = \cos(\Omega_1 t_1) \exp(i\Omega_2 t_2)$$

$$= \tfrac{1}{2}[\exp(i\Omega_1 t_1) + \exp(-i\Omega_1 t_1)] \exp(i\Omega_2 t_2), \qquad [4.46]$$

where Ω_1 and Ω_2 are the chemical shifts in the first and second dimensions, respectively. The *sine-modulated* time domain (t_1) function, $S_s(t_1, t_2)$, is

$$S_s(t_1, t_2) = \sin(\Omega_1 t_1) \exp(i\Omega_2 t_2)$$

$$= -\tfrac{i}{2}[\exp(i\Omega_1 t_1) - \exp(-i\Omega_1 t_1)] \exp(i\Omega_2 t_2). \qquad [4.47]$$

In both $S_c(t_1, t_2)$ and $S_s(t_1, t_2)$, the evolution during the t_1 period modulates the *amplitude* of the signal recorded during t_2. Data in which evolution during the t_1 period modulates the phase of the signal recorded during t_2 are referred to as *P-type* and *N-type* signals, respectively:

$$S_P(t_1, t_2) = \exp[i(\Omega_2 t_2 + \Omega_1 t_1)]; \qquad [4.48]$$

$$S_N(t_1, t_2) = \exp[i(\Omega_2 t_2 - \Omega_1 t_1)]. \qquad [4.49]$$

In P-type modulation, the sense of the frequency modulation is the same in t_1 and t_2, whereas in N-type modulation, the sense of the frequency modulation is opposite in t_1 and t_2 (47,48). The following relationships are obtained trivially:

$$S_c(t_1, t_2) = [S_P(t_1, t_2) + S_N(t_1, t_2)]/2; \qquad [4.50]$$

$$S_s(t_1, t_2) = -i[S_P(t_1, t_2) - S_N(t_1, t_2)]/2. \qquad [4.51]$$

For amplitude-modulated signals, the precession of magnetization during t_1 is described by a superposition of two complex signals with opposite frequency, $-\Omega_1$ and $+\Omega_1$. These signals result from evolution of the shift operators I^- and I^+, respectively, which in turn correspond to the coherence levels $p = -1$ and $p = +1$ (more generally, for multiple-quantum coherences of order p_i during t_1, $p = p_i$ and $p = -p_i$). Amplitude-modulated data sets require that both positive and negative coherence levels are selected during t_1. In contrast, for phase-modulated signals, precession of magnetization during t_1 is described by a complex signal with frequency given by either $-\Omega_1$ (for P-type signals) or $+\Omega_1$ (for N-type signals). These signals result from selection of only one of the coherence levels $p = -1$ and $p = +1$ during t_1.

The Fourier transform of the function $\exp(i\Omega t - R_2 t)$ is given by

$$\mathcal{F}\{\exp(i\Omega t - R_2 t)\} = A + iD, \qquad [4.52]$$

in which

$$A = \frac{R_2}{(\omega - \Omega)^2 + R_2^2} \qquad [4.53]$$

is an absorptive Lorentzian line,

$$D = \frac{\omega - \Omega}{(\omega - \Omega)^2 + R_2^2} \qquad [4.54]$$

is a dispersive Lorentzian line, and relaxation of the form $\exp(-R_2 t)$ has been assumed. The shorthand notation A_2, D_2 and A_1, D_1 will be used to represent the absorption and dispersion parts of the signal in the F_2 and F_1 dimensions, respectively. Depending on whether the peaks are located at $+\Omega_1$ or $-\Omega_1$ in the F_1 dimension, the signals will be noted as A_1^+, D_1^+ or A_1^-, D_1^-, respectively.

Fourier transformation of [4.46] with respect to t_2 yields

$$S_c(t_1, F_2) = [\exp(i\Omega_1 t_1) + \exp(-i\Omega_1 t_1)][A_2 + iD_2]/2. \qquad [4.55]$$

Performing a real (cosine) Fourier transform of [4.55] with respect to t_1 yields the two-dimensional spectrum:

$$S_c(F_1, F_2) = [A_1^+ A_2 + A_1^- A_2]/2. \qquad [4.56]$$

Alternatively, if the imaginary part of [4.55] is discarded to give S_c',

$$S_c'(t_1, F_2) = [\exp(i\Omega_1 t_1) + \exp(-i\Omega_1 t_1)]A_2/2, \qquad [4.57]$$

and a complex Fourier transform of [4.57] is performed with respect to t_1, then the real part of the resulting spectrum is exactly as represented by [4.56]. Equation [4.56] represents two signals, one at $+\Omega_1$, the other at $-\Omega_1$, that are absorptive in both dimensions, as shown in Fig. 4.16a.

Fourier transformation of [4.47] with respect to t_2 yields

$$S_s(t_1, F_2) = -i[\exp(i\Omega_1 t_1) - \exp(-i\Omega_1 t_1)][A_2 + iD_2]/2. \qquad [4.58]$$

Performing a real Fourier transform of [4.58] with respect to t_1 yields the two-dimensional spectrum:

$$S_s(F_1, F_2) = -i[A_1^+ A_2 - A_1^- A_2]/2. \qquad [4.59]$$

Alternatively, if the imaginary part of [4.58] is discarded to give S_s', then

$$S_s'(t_1, F_2) = -i[\exp(i\Omega_1 t_1) - \exp(-i\Omega_1 t_1)]A_2/2, \qquad [4.60]$$

and a complex Fourier transform of [4.60] is performed with respect to t_1, then the real part of the resulting spectrum is exactly as represented

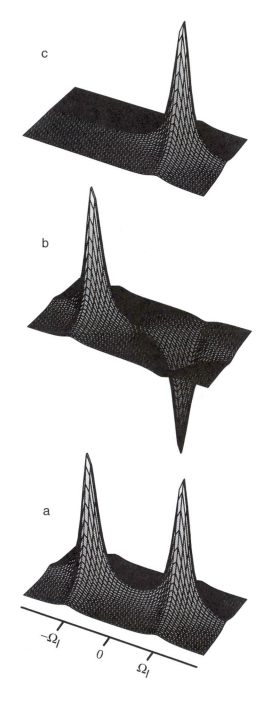

by [4.59]. Equation [4.59] represents two signals, one at $+\Omega_1$, the other at $-\Omega_1$, that are absorptive in both dimensions; however, one peak is *positive* and the other is *negative* (Fig. 4.16b).

Combining the results of [4.56] and [4.59] as a complex pair generates a frequency discriminated spectrum of the form

$$S(F_1, F_2) = S_c(F_1, F_2) + iS_s(F_1, F_2) = A_1^+ A_2, \qquad [4.61]$$

which provides frequency discrimination with retention of a pure double-absorption lineshape. The process is shown schematically in Fig. 4.16.

In contrast, two-dimensional Fourier transformations of [4.48] and [4.49] with respect to t_1 and t_2 yield

$$S_P(F_1, F_2) = [A_1^- + iD_1^-][A_2 + iD_2]$$
$$= [A_1^- A_2 - D_1^- D_2] + i[A_1^- D_2 + D_1^- A_2]; \qquad [4.62]$$
$$S_N(F_1, F_2) = [A_1^+ - iD_1^+][A_2 + iD_2]$$
$$= [A_1^+ A_2 + D_1^+ D_2] + i[A_1^+ D_2 - D_1^+ A_2]. \qquad [4.63]$$

The real parts of these spectra represent frequency-discriminated spectra, as desired; however, the lineshape is a superposition of doubly absorptive and doubly dispersive signals. This lineshape is called *phase-twisted* and is extremely undesirable in high-resolution NMR spectroscopy because the dispersive tails in the lineshape degrade the resolution in the spectrum. The P and N signals are phase twisted in the opposite sense. Absorption and phase-twisted lineshapes are compared in Fig. 4.17. As will be seen in Section 4.3.4.2, [4.50] and [4.51] can be used to generate amplitude-modulated data from separately recorded P-type and N-type signals, and the resulting amplitude modulated data can be used to generate a phase-sensitive spectrum as described above.

FIGURE 4.16 Frequency discrimination in the F_1 dimension. (a) A real Fourier transform of [4.55] yields a spectrum that is not frequency-discriminated in the F_1 dimension. Instead, the spectrum consists of positive pure absorption signals at $+\Omega_1$ and $-\Omega_1$. (b) A real Fourier transform of [4.58] is a spectrum that is not frequency-discriminated in the F_1 dimension. The spectrum consists of a negative pure absorption signal at $+\Omega_1$, and a positive pure absorption signal at $-\Omega_1$. (c) Subtraction of the two signals (a) and (b) results in a single pure absorption signal that is frequency-discriminated in the F_1 dimension. A single pure absorption signal is obtained at $+\Omega_1$.

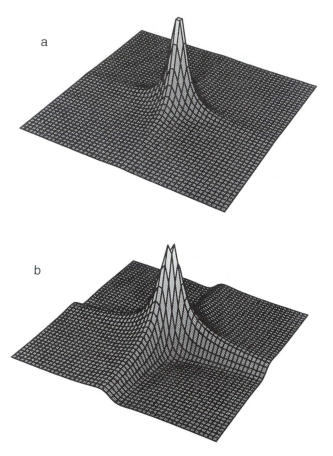

FIGURE 4.17 Comparison of (a) absorptive and (b) phase-twisted lineshapes.

4.3.4.1 Frequency Discrimination by Phase Cycling   Both positive and negative coherence levels can be selected during t_1 in a natural fashion in phase-cycled experiments by using the periodicity of phase-cycled coherence filters to select both desired pathways simultaneously. The practical problem is then to separately record the two data sets $S_c(t_1, t_2)$ and $S_s(t_1, t_2)$. The key is to recognize that, if a given pulse sequence selects a coherence of order p during t_1 and results in a cosine-modulated signal, the sine-modulated signal can be recorded in a second experiment by shifting the phase of an appropriate pulse acting on the coherence (usually the pulse preceding the evolution period) by $\pi/(2|p|)$ [4.32].

In the hypercomplex method (*49*), cosine- and sine-modulated data sets are recorded using a sampling interval in t_1 of $1/SW_1$, where SW_1 is the required spectral width in the F_1 dimension and processed exactly as described above. The combination of the two separate data sets can be performed before the second complex Fourier transform (with respect to t_1) since the Fourier transform is linear.

An alternative method was designed by Marion and Wüthrich (50). This procedure achieves identical results but employs real Fourier transformations. The idea finds its roots in Redfield's method for quadrature-detected spectra using a single analog-to-digital converter (ADC) (51). Some NMR instruments were originally designed with only one ADC, rather than two, as this was deemed an expensive component. One-dimensional quadrature detection on such spectrometers requires recording data points every $1/(2SW)$, where SW is the spectral width, twice the normal rate of data acquisition. In addition, the phase of the receiver is incremented by 90° after each data point is recorded. For this reason, the method is widely known as *time-proportional phase incrementation* (TPPI). The spectrum is subsequently obtained by application of a real Fourier transform. Overall, the effect of the TPPI procedure is to add a frequency of $SW/2$ to each data point in the transformed spectrum, thus achieving frequency discrimination. Those peaks in the spectrum below the transmitter frequency (at zero), between $-SW/2$ and zero, are shifted to between zero and $SW/2$, whereas those peaks between zero and $+SW/2$ are shifted to between $+SW/2$ and SW. Therefore, all resonances in the spectrum appear with the same sign of precession.

In two dimensions the same method can be employed. The incrementable period, t_1, is incremented in steps of $1/2SW_1$. Between each successive t_1 increment, the phase of the coherence during t_1 is shifted by 90° (by shifting the phases of the pulses prior to t_1 appropriately). Consider the effect of this phase incrementation for the odd and even t_1 increments for a sampling interval of $\delta = 1/2SW_1$. For example, data points 1, 3, and 5 are all recorded on x, but with alternating sign, $+x$, $-x$, $+x$,... and $t_1 = 0, 2\delta, 4\delta,...$. The cosine-modulated signal for the odd data points, $S_{\text{odd}}(t_1)$, is represented by

$$S_{\text{odd}}(t_1) = \cos(\Omega_1 t_1) \cos(\pi t_1 / 2\delta)$$

$$= \cos(\Omega_1 t_1) \cos(\pi t_1 SW_1)$$

$$= \tfrac{1}{2}[\cos(\Omega_1 - \Omega_{\text{TPPI}})t_1 + \cos(\Omega_1 + \Omega_{\text{TPPI}})t_1], \qquad [4.64]$$

where the phase incrementation has introduced an extra modulation $\cos(\pi t_1 SW_1)$ and $\Omega_{\text{TPPI}} = \pi SW_1$. For the even data points recorded on y

with alternating sign, y, $-y$, y, $-y$, ... and $t_1 = \delta$, 3δ, 5δ, ..., the signal, $S_{\text{even}}(t_1)$, is represented by

$$S_{\text{even}}(t_1) = \sin(\Omega_1 t_1) \sin(\pi t_1 / 2\delta)$$
$$= \sin(\Omega_1 t_1) \sin(\pi t_1 SW_1)$$
$$= \tfrac{1}{2}[\cos(\Omega_1 - \Omega_{\text{TPPI}})t_1 - \cos(\Omega_1 + \Omega_{\text{TPPI}})t_1]. \qquad [4.65]$$

Real Fourier transformation of $S_{\text{odd}}(t_1)$ produces a spectrum of two lines at frequencies $(\Omega_1 - \Omega_{\text{TPPI}})$ and $(\Omega_1 + \Omega_{\text{TPPI}})$, both with absorption phase and both positive. Real Fourier transformation of $S_{\text{even}}(t_1)$ produces a spectrum of two lines at frequencies $(\Omega_1 - \Omega_{\text{TPPI}})$ and $(\Omega_1 + \Omega_{\text{TPPI}})$, both in absorption, but one positive and one negative. Adding the two signals together therefore retains just one absorption-mode, frequency-discriminated line. This procedure is analogous to that proposed by States and coworkers; in particular, the TPPI method requires exactly the same total number of transients and total acquisition time as the States method.

Salient features of the States (hypercomplex), TPPI, and hybrid TPPI–States (52) protocols are summarized in Table 4.5. Although the preceding discussions and derivations have been applied solely to two-dimensional NMR examples, all the methodology can be extended into higher dimensions.

4.3.4.2 Frequency Discrimination by Pulsed Field Gradients   Coherence selection by pulsed field gradients naturally results in P-type or N-type modulation of the observed signal. Either P-type or N-type data is selected depending on the relative sense of the initial dephasing gradient pulse and the second, refocusing gradient. To obtain frequency-discriminated phase-sensitive spectra by using the gradient approach, both the P-type and the N-type data are recorded separately (37,43,44). Cosine and sine modulated data are obtained by combining the P- and N-type data using [4.50] and [4.51]. The resulting amplitude-modulated data is processed as complex data by using the States method of frequency discrimination.

Although this procedure produces frequency discrimination in the indirect dimension of the spectrum while keeping absorption lineshapes, the signal-to-noise ratio is reduced by a factor of $\sqrt{2}$ compared to the phase-cycled approaches to frequency discrimination. However, in many experiments this reduction is avoided by the use of so-called PEP (preservation of equivalent pathways) experiments described in Section 7.1.3.3.

Table 4.5

Quadrature Detection Methods

Experiment	Increment	Pulse phase	Receiver phase
	TPPI		
$(4k + 1)$	$t_1(0) + (4k)\Delta$	x	x
$(4k + 2)$	$t_1(0) + (4k + 1)\Delta$	y	x
$(4k + 3)$	$t_1(0) + (4k + 2)\Delta$	$-x$	x
$(4k + 4)$	$t_1(0) + (4k + 3)\Delta$	$-y$	x
	States		
$(4k + 1)$	$t_1(0) + (4k)2\Delta$	x	x
$(4k + 2)$	$t_1(0) + (4k)2\Delta$	y	x
$(4k + 3)$	$t_1(0) + (4k + 1)2\Delta$	x	x
$(4k + 4)$	$t_1(0) + (4k + 1)2\Delta$	y	x
	TPPI–States		
$(4k + 1)$	$t_1(0) + (4k)2\Delta$	x	x
$(4k + 2)$	$t_1(0) + (4k)2\Delta$	y	x
$(4k + 3)$	$t_1(0) + (4k + 1)2\Delta$	$-x$	$-x$
$(4k + 4)$	$t_1(0) + (4k + 1)2\Delta$	$-y$	$-x$

The index $k = 0, 1, \ldots, N/4 - 1$, in which N is the number of experiments acquired in the t_1 dimension, and $\Delta = 1/(2SW_1)$. The initial sampling delay is $t_1(0)$ and is usually set to 0 or Δ. The t_1 interferogram consists of N real points for TPPI method and $N/2$ complex points for States and TPPI–States methods. The t_1 increment is twice as large for the States and TPPI–States methods as for the TPPI method, but SW_1 and $t_{1,\max}$ are identical for all methods.

4.3.4.3 Aliasing and Folding in Multidimensional NMR Spectroscopy   The location of axial peaks in a multi-dimensional NMR spectrum depends on the manner in which frequency discrimination is performed in the F_1 dimension. If the hypercomplex (States) method is used, the axial peaks occur at $F_1 = 0$ and result in a ridge of axial peaks across the center of the spectrum, parallel to the F_2 axis. If the TPPI method is used, the axial peaks occur at the edge of the spectrum for the following reason. The TPPI phase cycle adds a frequency Ω_{TPPI} to the resonance frequencies in F_1; however, the axial peaks are unaffected by the phase cycle and consequently appear in the spectrum at the apparent frequency $-\Omega_{\text{TPPI}}$. Generally, even if axial peak suppression phase cycling is employed, placement of axial peaks at the edge of the spectrum minimizes artifacts. The hybrid TPPI–States protocol is identical to the conventional States method, except that the axial peaks are shifted to the edge of the spectrum.

As described in Section 3.2.1, resonance frequencies outside the spectral width appear artifactually within the spectral width. For complex data

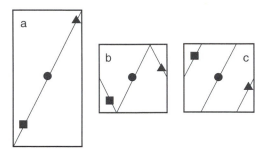

FIGURE 4.18 Folding and aliasing in the F_1 dimension. (a) Schematic spectrum with a full spectral width. (b) The F_1 spectral width is halved, and frequency discrimination is performed using real (TPPI) acquisition. Resonances outside the spectral width are folded into the spectrum. (c) The F_1 spectral width is halved and frequency discrimination is performed using hypercomplex (States or TPPI–States) acquisition. Resonances outside the spectral width are aliased into the spectrum.

(States or TPPI–States frequency discrimination), resonance signals are aliased: signals upfield (downfield) of the edge of the spectral region appear in the downfield (upfield) spectral region. The apparent resonance frequency, ν_a, and the true resonance frequency, ν_0, are related by

$$\nu_a = \nu_0 - 2mf_n, \qquad [4.66]$$

in which m is an integer equal to the number of times the signal has been aliased, and f_n is the Nyquist frequency. For real data (TPPI frequency discrimination), resonance signals are folded: signals upfield (downfield) of the edge of the spectral region reflect back across the upfield (downfield) edge of the spectral region. The apparent resonance frequency, ν_a, and the true resonance frequency, ν_0, are related by

$$\nu_a = (-1)^m (\nu_0 - 2mf_n). \qquad [4.67]$$

Aliasing and folding (53) are illustrated in Fig. 4.18. Aliasing is useful particularly when minimizing the spectral width in heteronuclear NMR experiments and is discussed further in Section 7.1.2.5.

4.4 Resolution and Sensitivity

The sensitivity of a multidimensional spectrum is close to that of an equivalent one-dimensional experiment recorded in the same total experiment time. Although a particular peak may appear only weakly in each

F_2 spectrum (recorded for each t_1 value), the Fourier transform with respect to t_1 concentrates all the signal into a few points in the final multidimensional spectrum. In effect, the signal-to-noise ratio of a peak in the multidimensional spectrum is a function of the time-average signal throughout the *entire* multidimensional NMR experiment.

Three factors lead to a reduction in sensitivity of multidimensional spectra when compared to their one-dimensional counterparts: (1) relaxation during the incrementable time periods result in a progressive loss of signal as the variable delay increases, (2) cancellation of antiphase signals due to overlap (Section 6.2.1.5), and (3) the integrated intensity of a single peak in the one-dimensional spectrum is associated with several peaks in the multidimensional spectrum (i.e., diagonal and cross-peaks). The tradeoff between sensitivity and resolution is the same as for one-dimensional spectroscopy: more t_1 increments must be recorded to increase resolution ($t_{1,max}$); however, relaxation during the added t_1 time periods causes signal decay and loss of sensitivity.

Because the signal nearly always is truncated in the incrementable time periods of multidimensional NMR experiments, suitable apodization functions must be applied to generate spectra containing the required information. Such functions and their effects on truncated data were discussed in Section 3.3.2.2.

4.5 Three- and Four-Dimensional NMR Spectroscopy

Two-dimensional NMR spectroscopy has proved to be one of the most important developments in high-resolution NMR. However, for proteins with masses greater than 10–12 kilodaltons (kDa), even the increased resolution of the two-dimensional spectra is insufficient, and so alternative approaches have been sought.

The approach that has now become widely adopted is to increase the number of frequency dimensions possessed by the spectrum. The principles and fundamental ideas discussed earlier for two-dimensional NMR extend into higher-dimensional experiments: the same type of magnetization transfer processes are active, the same principles concerning coherence selection, quadrature detection, resolution, and sensitivity are applicable, and product operator analysis can be employed in the same way. Higher-dimensional experiments are built from combinations of two-dimensional experiments and can combine the magnetization transfer capabilities of both dipolar and scalar coupling interactions in the same sequence.

Three-dimensional pulse sequences are derived from a combination

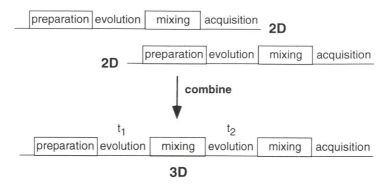

FIGURE 4.19 Schematic generation of a three-dimensional NMR experiment from the combination of two two-dimensional NMR experiments. The mixing period of one two-dimensional experiment and the preparation period of a second two-dimensional experiment are combined. The three-dimensional experiment contains three independent time periods. The FID is recorded during the acquisition time, t_3, as a function of two independently incremented evolution times, t_1 and t_2. A mixing period follows each evolution time, causing a potential two-step magnetization transfer process.

of two 2D pulse sequences *(54–56)* as shown schematically in Fig. 4.19. In three-dimensional experiments the signal is recorded conventionally during an acquisition time, denoted t_3, as a function of *two* evolution times, t_1 and t_2, which are incremented independently from one experiment to another. This procedure generates a three-dimensional time-domain data matrix to which a three-dimensional Fourier transformation is applied. The corresponding frequency dimensions are denoted F_1, F_2, and F_3. The spectrum can be represented as a three-dimensional cube, but analysis of the three-dimensional spectrum is more convenient if two-dimensional slices are taken from the cube as shown in Fig. 4.20. In this case the *tiers* or *planes* from the cube can be seen as sets of two-dimensional spectra (F_3, F_1) separated by another interaction along the F_2 dimension.

Again, as in two-dimensional spectroscopy, either similar or different nuclear types can appear in different dimensions, as required, and correlations between the separate dimensions can be established via NOE effects or through scalar couplings. Those experiments in which all three dimensions contain proton chemical shifts or couplings are referred to as homonuclear three-dimensional experiments. Those experiments in which one or more dimension is not proton (usually ^{13}C and/or ^{15}N) are referred to as heteronuclear three-dimensional experiments. Analyses of the most

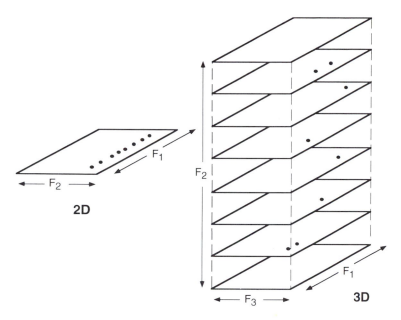

FIGURE 4.20 The development of a three-dimensional data set from a two-dimensional data set. The two-dimensional data set shown shows a set of resonances which, though resolved in the F_1 dimension, are not clearly determined in the F_2 dimension. The introduction of an additional evolution period generates a third frequency dimension perpendicular to the first two. The increased resolution afforded by virtue of a second magnetization transfer step can alleviate ambiguities in the two-dimensional spectrum.

important three-dimensional methods for biological studies are presented in Chapters 6 and 7.

In a similar fashion a four-dimensional experiment consists of a combination of three 2D sequences omitting the detection periods of the first and second experiments and the preparation stages of the second and third experiments (57). As shown in Fig. 4.21, the general experiment

| preparation | t_1 | mixing(1) | t_2 | mixing(2) | t_3 | mixing(3) | acquisition |

FIGURE 4.21 A schematic representation of a four-dimensional NMR experiment. The four-dimensional experiment contains four independent time periods. The FID is recorded during the acquisition time, t_4, as a function of three independently incremented evolution times, t_1, t_2, and t_3. A mixing period follows each evolution period, causing a potential three-step magnetization transfer process.

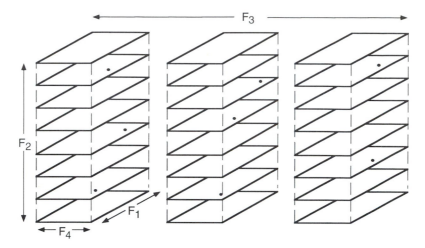

FIGURE 4.22 A four-dimensional experiment is visualized as a series of three-dimensional cubes. Each cube represents the three-dimensional F_1, F_2, F_4 subspectrum for a different value of F_3.

contains three independently incrementable time periods (t_1, t_2, t_3) and the acquisition time period, t_4, and consequently the resulting data is a function of these four time periods. Four-dimensional experiments are used in those cases when there is still ambiguity arising due to degeneracy and overlap even in three-dimensional spectra. Up to this point in time, four-dimensional experiments have been exclusively heteronuclear techniques. Detailed discussions of some useful four-dimensional experiments are presented in Chapter 7.

Visualizing a four-dimensional spectrum can be difficult, although one convenient method is to imagine that each two-dimensional slice of a three-dimensional spectrum expanded into an additional dimension by another type of interaction. The progression from two-dimensional spectra to four-dimensional spectra is represented schematically in Fig. 4.22.

References

1. R. Freeman and W. A. Anderson, *J. Chem. Phys.* **37**, 2053–2073 (1962).
2. R. A. Hoffman and S. Forsén, *Progr. NMR Spectrosc.* **1**, 15–204 (1966).
3. J. H. Noggle and R. E. Shirmer, "The Nuclear Overhauser Effect: Chemical Applications," pp. 1–259. Academic Press, New York, 1971.
4. D. Neuhaus and M. Williamson, "The Nuclear Overhauser Effect in Structural and Conformational Analysis," pp. 1–522. VCH Publishers, New York, 1989.

5. W. A. Anderson and R. Freeman, *J. Chem. Phys.* **37**, 85–103 (1962).
6. W. A. Anderson and F. A. Nelson, *J. Chem. Phys.* **39**, 183–189 (1963).
7. R. R. Ernst, *J. Chem. Phys.* **45**, 3845–3861 (1966).
8. J. Jeener (Lecture, Ampère Summer School, Basko Polje, Yugoslavia, 1971).
9. W. P. Aue, E. Bartholdi, and R. R. Ernst, *J. Chem. Phys.* **64**, 2229–2246 (1976).
10. L. Braunschweiler and R. R. Ernst, *J. Magn. Reson.* **53**, 521–528 (1983).
11. S. R. Hartmann and E. L. Hahn, *Phys. Rev.* **128**, 2042–2053 (1962).
12. L. Müller and R. R. Ernst, *Mol. Phys.* **38**, 963–992 (1979).
13. A. Bax and D. G. Davis, *J. Magn. Reson.* **65**, 355–360 (1985).
14. A. J. Shaka, J. Keeler, and R. Freeman, *J. Magn. Reson.* **53**, 313–340 (1983).
15. A. J. Shaka, J. Keeler, T. Frenkiel, and R. Freeman, *J. Magn. Reson.* **52**, 335–338 (1983).
16. A. J. Shaka, C. J. Lee, and A. Pines, *J. Magn. Reson.* **77**, 274–293 (1988).
17. S. P. Rucker and A. J. Shaka, *Mol. Phys.* **68**, 509–517 (1989).
18. M. Rance, *J. Magn. Reson.* **74**, 557–564 (1987).
19. R. Bazzo and I. D. Campbell, *J. Magn. Reson.* **76**, 358–361 (1988).
20. J. Jeener, B. H. Meier, P. Bachmann, and R. R. Ernst, *J. Chem. Phys.* **71**, 4546–4553 (1979).
21. G. A. Morris and R. Freeman, *J. Am. Chem. Soc.* **101**, 760–762 (1979).
22. D. P. Burum and R. R. Ernst, *J. Magn. Reson.* **39**, 163–168 (1980).
23. D. M. Doddrell, D. T. Pegg, and M. R. Bendall, *J. Magn. Reson.* **48**, 323–327 (1982).
24. M. R. Bendall and D. T. Pegg, *J. Magn. Reson.* **53**, 272–296 (1983).
25. K. E. Kövér and G. Batta, *Prog. NMR Spectrosc.* **19**, 223–266 (1987).
26. J. Keeler, in "Multinuclear Magnetic Resonance in Liquids and Solids—Chemical Applications" (P. Granger and R. K. Harris, eds.), pp. 103–129, Vol. 322, NATO ASI Series C. Klumer Academic Press, Netherlands, 1990.
27. G. Bodenhausen, H. Kogler, and R. R. Ernst, *J. Magn. Reson.* **58**, 370–388 (1984).
28. D. I. Hoult, *Proc. R. Soc. Lond., Ser. A* **344**, 311–340 (1975).
29. G. Bodenhausen, R. Freeman, and D. L. Turner, *J. Magn. Reson.* **26**, 373–378 (1977).
30. A. Bax, P. G. D. Jong, A. F. Mehlkopf, and J. Smidt, *Chem. Phys. Lett.* **69**, 567–570 (1980).
31. P. Barker and R. Freeman, *J. Magn. Reson.* **64**, 334–338 (1985).
32. R. E. Hurd, *J. Magn. Reson.* **87**, 422–428 (1990).
33. A. L. Davis, E. D. Laue, J. Keeler, D. Moskau, and J. Lohman, *J. Magn. Reson.* **94**, 637–644 (1991).
34. R. E. Hurd and B. K. John, *J. Magn. Reson.* **91**, 648–653 (1991).
35. R. E. Hurd and B. K. John, *J. Magn. Reson.* **92**, 658–668 (1991).
36. G. W. Vuister, R. Boelens, R. Kaptein, R. E. Hurd, B. John, and P. C. M. van Zijl, *J. Am. Chem. Soc.* **113**, 9688–9690 (1991).
37. A. L. Davis, J. Keeler, E. D. Laue, and D. Moskau, *J. Magn. Reson.* **98**, 207–216 (1992).
38. B. K. John, D. Plant, P. Webb, and R. E. Hurd, *J. Magn. Reson.* **98**, 200–206 (1992).
39. G. W. Vuister, R. Boelens, R. Kaptein, M. Burgering, and P. C. M. van Zijl, *J. Biomol. NMR.* **2**, 301–305 (1992).
40. J. Keeler, R. T. Clowes, A. L. Davis, and E. D. Laue, *Meth. Enzymol.* **239**, 145–207 (1994).
41. A. Wokaun and R. R. Ernst, *Mol. Phys.* **36**, 317–341 (1976).
42. A. Bax and S. S. Pochapsky, *J. Magn. Reson.* **99**, 638–643 (1992).
43. J. Boyd, N. Soffe, B. K. John, D. Plant, and R. E. Hurd, *J. Magn. Reson.* **98**, 660–664 (1992).
44. J. R. Tolman, J. Chung, and J. P. Prestegard, *J. Magn. Reson.* **98**, 462–467 (1992).

45. G. Bodenhausen, R. Freeman, R. Niedermeyer, and D. L. Turner, *J. Magn. Reson.* **26,** 133–164 (1977).

46. J. Keeler and D. Neuhaus, *J. Magn. Reson.* **63,** 454–472 (1985).

47. K. Nagayama, A. Kumar, K. Wüthrich, and R. R. Ernst, *J. Magn. Reson.* **40,** 321–334 (1980).

48. A. Bax and R. Freeman, *J. Magn. Reson.* **44,** 542–561 (1981).

49. D. J. States, R. A. Haberkorn, and D. J. Ruben, *J. Magn. Reson.* **48,** 286–292 (1982).

50. D. Marion and K. Wüthrich, *Biochem. Biophys. Res. Commun.* **113,** 967–974 (1983).

51. A. G. Redfield and S. D. Kunz, *J. Magn. Reson.* **19,** 250–254 (1975).

52. D. Marion, M. Ikura, R. Tschudin, and A. Bax, *J. Magn. Reson.* **85,** 393–399 (1989).

53. R. R. Ernst, G. Bodenhausen, and A. Wokaun, "Principles of Nuclear Magnetic Resonance in One and Two Dimensions," pp. 1–610. Clarendon Press, Oxford, 1987.

54. C. Griesinger, O. W. Sørensen, and R. R. Ernst, *J. Magn. Reson.* **73,** 574–579 (1987).

55. H. Oschkinat, C. Griesinger, P. J. Kraulis, O. W. Sørenson, R. R. Ernst, A. M. Gronenborn, and G. M. Clore, *Nature* **332,** 374–376 (1988).

56. C. Griesinger, O. W. Sørensen, and R. R. Ernst, *J. Magn. Reson.* **84,** 14–63 (1989).

57. L. E. Kay, G. M. Clore, A. Bax, and A. M. Gronenborn, *Science* **249,** 411–414 (1990).

RELAXATION AND DYNAMIC PROCESSES

Previous chapters have utilized the density matrix and product operator formalism to describe the evolution of the density operator under the chemical shift, scalar coupling, and rf Hamiltonians, which are responsible for the chemical shifts, multiplet structures, and coherence transfer phenomena observed by NMR spectroscopy. Although multiple-pulse and multidimensional NMR techniques permit generation of off-diagonal density matrix elements and observation of complex coherence transfer processes, eventually the density operator returns to an equilibrium state in which all coherences (off-diagonal elements of the density operator) have decayed to zero and the populations of the energy levels of the system (diagonal elements of the density operator) have been restored to the Boltzmann distribution. Analogously with similar phenomena in other areas of spectroscopy, the process by which an arbitrary density operator returns to the equilibrium operator is called nuclear magnetic, or spin, relaxation. The present chapter will describe the general features of spin relaxation; consequences of spin relaxation processes for particular multidimensional NMR experiments are described in Chapters 6 and 7. In addition, the NMR experiment may be affected by other dynamic processes, such as chemical reactions and conformational exchange that transfer nuclei between magnetic environments; these processes are also discussed in this chapter.

As relaxation is one of the fundamental aspects of magnetic resonance, an extensive literature on theoretical and experimental aspects of relaxation has developed since the earliest days of NMR spectroscopy (see Ref. *1* and references cited therein). At one level, relaxation has important consequences for the NMR experiment: the relaxation rates of single-quantum transverse operators determine the linewidths of the resonances detected during the acquisition period of an NMR experiment; the relaxation rates of the longitudinal magnetization and off-diagonal coherences generated by the pulse sequence determine the length of the recycle delay needed between acquisitions; and the relaxation rates of operators of interest during multidimensional experiments determine the linewidths of resonances in the indirectly detected dimensions and affect the overall sensitivity of the experiments. At a second level, relaxation provides experimental information on the physical processes governing relaxation, including molecular motions and intramolecular distances. In particular, cross-relaxation gives rise to the nuclear Overhauser effect (NOE) and makes possible the determination of three-dimensional molecular structures by NMR spectroscopy. Additionally, a variety of chemical kinetic processes can be studied through effects manifested in the NMR spectrum; in many cases, such phenomena can be studied while the molecular system remains in chemical equilibrium.

Because the theoretical formalism describing relaxation is more complicated mathematically than the product operator formalism emphasized in this text, the present treatment will emphasize application of semiclassical relaxation theory to cases of practical interest, rather than fundamental derivations. Semiquantitative or approximate results are utilized if substantial simplification of the mathematical formalism is obtained. More detailed descriptions of the derivation of the relaxation equations are presented elsewhere (*1,2,3*).

5.1 Introduction and Survey of Theoretical Approaches

Introductory theoretical treatments of optical spectroscopy emphasize the role of spontaneous and stimulated emission in relaxation from excited states back to the ground state of a molecule. The probability per unit time, W, for transition from the upper to lower energy state of an isolated magnetic dipole by spontaneous emission of a photon of energy $\Delta E = \hbar\omega$ is given by (*2*)

$$W = \frac{2\gamma^2\hbar}{3\lambda^3},$$

[5.1]

in which $\chi = c/\omega$ and c is the speed of light. For a proton with a Larmor frequency of 500 MHz, $W \approx 10^{-21}$ s^{-1}; thus, spontaneous emission is a completely ineffective relaxation mechanism for nuclear magnetic resonance. Calculation of stimulated emission transition probabilities is complicated by consideration of the coil in the probe; nonetheless, stimulated emission also can be shown to have a negligible influence on nuclear spin relaxation. Spontaneous and stimulated emission are important in optical spectroscopy because the relevant photon frequencies are several orders of magnitude larger.

Instead, nuclear spin relaxation is a consequence of coupling of the spin system to the *surroundings*. The surroundings have historically been termed the *lattice* following the early studies of NMR relaxation in solids where the surroundings were genuinely a solid lattice. The lattice includes other degrees of freedom of the molecule containing the spins (such as rotational degrees of freedom) as well as other molecules comprising the system (Section 2.2.2). The energy levels of the lattice are assumed to be quasicontinuous with populations that are described by a Boltzmann distribution. Furthermore, the lattice is assumed to be in thermal equilibrium at all times. The lattice modifies the local magnetic fields at the locations of nuclei and thereby (weakly) couples the lattice and the spin system. Stochastic Brownian rotational motions of molecules in liquid solutions render the local magnetic fields time-dependent. The time-dependent local magnetic fields can be decomposed by Fourier analysis into a superposition of harmonically varying magnetic fields with different frequencies. In addition, the fields can be resolved into components perpendicular and parallel to the main static field.

Transverse components of the stochastic local field are responsible for *nonadiabatic* contributions to relaxation. If the Fourier spectrum of the fluctuating transverse magnetic fields at the location of a nucleus contains components with frequencies corresponding to the energy differences between eigenstates of the spin system, then transitions between eigenstates can occur. In this case, transition of the spin system from a higher (lower) energy state to a lower (higher) energy state is accompanied by an energy-conserving transition of the lattice from a lower (higher) to higher (lower) energy state. A transition of the spin system from higher energy to lower energy is more probable because the lattice is always in thermal equilibrium and has a larger population in the lower energy state. Thus, exchange of energy between the spin system and the lattice brings the spin system into thermal equilibrium with the lattice and the populations of the stationary states return to the Boltzmann distribution. Furthermore, transitions between stationary states caused by nonadiabatic processes decrease the lifetimes of these states and introduces uncertainties in the energies of the nuclear spin states through a Heisenberg uncertainty relationship. As a result, the Larmor frequencies of the spins vary randomly and the phase coherence between spins is reduced over time. Consequently, nonadiabatic fluctuations that cause transitions between states result in both thermal equilibration of the spin state populations and decay of off-diagonal coherences.

Fluctuating fields parallel to the main static field are responsible for *adiabatic* contributions to relaxation. The fluctuating fields generate variations in the total magnetic field in the z direction, and consequently, in the energies in the nuclear spin energy levels. Thus, adiabatic processes cause the Larmor frequencies of the spins to vary randomly. Over time, the spins gradually lose phase coherence and off-diagonal elements of the density matrix decay to zero. The populations of the states are not altered, and no energy is exchanged between the spin system and the lattice because transitions between stationary states do not occur.

5.1.1 RELAXATION IN THE BLOCH EQUATIONS

In the simplest theoretical approach to spin relaxation, the relaxation of isolated spins is characterized in the Bloch equations [1.28] by two phenomenological first-order rate constants: the *spin–lattice* or *longitudinal* relaxation rate constant, R_1, and the *spin–spin* or *transverse* relaxation rate constant, R_2 (4). In the following, rate constants rather than time constants, are utilized; the two quantities are reciprocals of each other (e.g., $T_1 = 1/R_1$). The spin–lattice relaxation rate constant describes the recovery of the longitudinal magnetization to thermal equilibrium, or, equivalently, return of the populations of the energy levels of the spin system (diagonal elements of the density operator) to the equilibrium Boltzmann distribution. The spin–spin relaxation rate constant describes the decay of the transverse magnetization to zero, or equivalently, the decay of transverse single-quantum coherences (off-diagonal elements of the density matrix). Nonadiabatic processes contribute to both spin–lattice and spin–spin relaxation. Adiabatic processes only contribute to spin–spin relaxation; spin–lattice relaxation is not affected because adiabatic processes do not change the populations of stationary states.

The Bloch formulation provides qualitative insights into the effects of relaxation on the NMR experiment, and the phenomenological rate constants can be measured experimentally. For example, the Bloch equations predict that the FID is the sum of exponentially damped sinusoidal functions and that, following a pulse sequence that perturbs a spin system from equilibrium, R_2 governs the length of time that the FID can be observed and R_1 governs the minimum time required for equilibrium to be restored. The Bloch formulation does not provide a microscopic explanation of the origin or magnitude of the relaxation rate constants, nor is it extendible to more complex, coupled spin systems. For example, in dipolar-coupled two spin systems, multiple spin operators, such as zero-quantum coherence, have relaxation rate constants that differ from both R_1 and R_2.

In the spirit of the Bloch equations, the results for product operator analyses of the evolution of a spin system under a particular pulse sequence in many instances can be corrected approximately for relaxation effects by adding an exponential damping factor for each temporal period post hoc. Thus if product operator analysis of a two-dimensional pulse sequence yields a propagator $\mathbf{U} = \mathbf{U}_a(t_2)\mathbf{U}_m\mathbf{U}_e(t_1)\mathbf{U}_p$, in which $\mathbf{U}_p$ is the propagator for the preparation period, etc., relaxation effects approximately can be included by writing

$$\sigma(t_1, t_2) = \mathbf{U}\sigma(0)\mathbf{U}^{-1} \exp[-R_p t_p - R_e t_1 - R_m t_m - R_a t_2], \qquad [5.2]$$

in which R_p is the relaxation rate constant for the operators of interest during the preparation time, t_p, etc. Cross-correlation and cross-relaxation effects are assumed to be negligible.

For example, the signal recorded in a ^{15}N–^{1}H HSQC spectrum is found by product operator analysis to be proportional to $\cos(\omega_N t_1) \cos(\omega_H t_2) \cos(\pi J_{H^N H^\alpha} t_2)$, in which ω_N and ω_H are the Larmor frequencies of the ^{15}N and ^{1}H, respectively, and $J_{H^N H^\alpha}$ is the proton scalar coupling constant between the amide and α protons. The phenomenological approach modifies this expression to $\cos(\omega_N t_1) \cos(\omega_H t_2) \cos(\pi J_{H^N H^\alpha} t_2) \exp[-R_{2N} t_1 - R_{2H} t_2]$, in which R_{2N} and R_{2H} are the transverse relaxation rate constants for the ^{15}N and ^{1}H operators present during t_1 and t_2, respectively, and relaxation during the INEPT sequences has been ignored. Relaxation effects on HSQC spectra are discussed in additional detail in Section 7.1.2.4. As a second example, product operator analysis of the INEPT pulse sequence [2.274] in the absence of relaxation, yields a density operator term proportional to $-2I_z S_y \sin(2\pi J_{IS} t)$. Coherence transfer is maximized for $2t = 1/(2J_{IS})$ [2.276]. If relaxation is considered, the result is modified to $-2I_z S_y \sin(2\pi J_{IS} t) \exp(-2R_{2I} t)$, in which R_{2I} is the relaxation rate of the I spin operators present during the period $2t$. Maximum coherence transfer is obtained for

$$2t = (\pi J_{IS})^{-1} \tan^{-1}(\pi J_{IS}/R_{2I}) \le 1/(2J_{IS}). \qquad [5.3]$$

5.1.2 THE SOLOMON EQUATIONS

Spin–lattice relaxation for interacting spins can be treated theoretically by considering the rates of transitions of the spins between energy

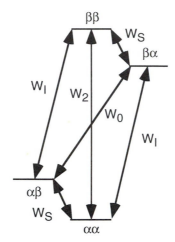

FIGURE 5.1 Transitions and associated rate constants for a two-spin system.

levels, as was demonstrated first by Bloembergen *et al.* (5). Figure 5.1 shows the energy levels for a two-spin system with transition frequencies labeled. The four energy levels are labeled in the normal way as $|m_I m_S\rangle$. The rate constants for transitions between the energy levels are denoted by W_0, W_I, W_S, and W_2 and are distinguished according to which spins change spin state during the transition. Thus, W_I denotes a relaxation process involving an I spin flip, W_S denotes a relaxation process involving an S spin flip, W_0 is a relaxation process in which both spins are flipped in opposite senses (flip-flop transition), and W_2 is a relaxation process in which both spins are flipped in the same sense (flip-flip transition). A differential equation governing the population of the state $|\alpha\alpha\rangle$ can be written by inspection:

$$\frac{dP_{\alpha\alpha}}{dt} = -(W_I + W_S + W_2)P_{\alpha\alpha} + W_I P_{\beta\alpha} + W_S P_{\alpha\beta} + W_2 P_{\beta\beta} + K, \quad [5.4]$$

in which $P_{\gamma\delta}$ is the population of the state $|\gamma\delta\rangle$ and K is a constant chosen to ensure that the population $P_{\gamma\delta}$ returns to the equilibrium value $P_{\gamma\delta}^0$. The value of K can be found by setting the left-hand side of [5.4] equal to zero:

$$K = (W_I + W_S + W_2)P_{\alpha\alpha}^0 - W_I P_{\beta\alpha}^0 - W_S P_{\alpha\beta}^0 - W_2 P_{\beta\beta}^0. \quad [5.5]$$

Thus, writing $\Delta P_{\gamma\delta} = P_{\gamma\delta} - P_{\gamma\delta}^0$ yields an equation for the deviation of the population of the $|\alpha\alpha\rangle$ state from the equilibrium population:

$$\frac{d\Delta P_{\alpha\alpha}}{dt} = -(W_I + W_S + W_2)\Delta P_{\alpha\alpha} + W_I\Delta P_{\beta\alpha} + W_S\Delta P_{\alpha\beta} + W_2\Delta P_{\beta\beta}.$$

[5.6]

Similar equations can be written for the other three states:

$$\frac{d\Delta P_{\alpha\beta}}{dt} = -(W_0 + W_I + W_S)\Delta P_{\alpha\beta} + W_0\Delta P_{\beta\alpha} + W_I\Delta P_{\beta\beta} + W_S\Delta P_{\alpha\alpha};$$

$$\frac{d\Delta P_{\beta\alpha}}{dt} = -(W_0 + W_I + W_S)\Delta P_{\beta\alpha} + W_0\Delta P_{\alpha\beta} + W_I\Delta P_{\alpha\alpha} + W_S\Delta P_{\beta\beta};$$

$$\frac{d\Delta P_{\beta\beta}}{dt} = -(W_I + W_S + W_2)\Delta P_{\beta\beta} + W_I\Delta P_{\alpha\beta} + W_S\Delta P_{\beta\alpha} + W_2\Delta P_{\alpha\alpha}.$$

[5.7]

Now recalling that $\langle I_z \rangle(t) = \mathrm{Tr}\{\sigma(t)I_z\} = \sigma_{11} + \sigma_{22} - \sigma_{33} - \sigma_{44} = P_{\alpha\alpha} + P_{\alpha\beta} - P_{\beta\alpha} - P_{\beta\beta}$ and $\langle S_z \rangle(t) = \mathrm{Tr}\{\sigma(t)S_z\} = \sigma_{11} - \sigma_{22} + \sigma_{33} - \sigma_{44} = P_{\alpha\alpha} - P_{\alpha\beta} + P_{\beta\alpha} - P_{\beta\beta}$ leads to

$$\frac{d\Delta I_z(t)}{dt} = -(W_0 + 2W_I + W_2)\Delta I_z(t) - (W_2 - W_0)\Delta S_z(t);$$

[5.8]

$$\frac{d\Delta S_z(t)}{dt} = -(W_0 + 2W_S + W_2)\Delta S_z(t) - (W_2 - W_0)\Delta I_z,$$

in which $\Delta I_z(t) = \langle I_z \rangle(t) - \langle I_z^0 \rangle$ and $\langle I_z^0 \rangle$ is the equilibrium magnitude of the I_z operator. Corresponding relationships hold for S_z. Making the identifications $\rho_I = W_0 + 2W_I + W_2$, $\rho_S = W_0 + 2W_S + W_2$, and $\sigma_{IS} = W_2 - W_0$ leads to the Solomon equations for a two-spin system (6):

$$\frac{d\Delta I_z(t)}{dt} = -\rho_I\Delta I_z(t) - \sigma_{IS}\Delta S_z(t);$$

[5.9]

$$\frac{d\Delta S_z(t)}{dt} = -\rho_S\Delta S_z(t) - \sigma_{IS}\Delta I_z(t).$$

The rate constants ρ_I and ρ_S are the *autorelaxation rate constants* (or the spin–lattice relaxation rate constants, R_{1I} and R_{1S}, in the Bloch terminology) for the I and S spins, respectively, and σ_{IS} is the *cross-relaxation rate constant* for exchange of magnetization between the two spins.

The Solomon equations can easily be extended to N interacting spins:

$$\frac{d\Delta I_{kz}(t)}{dt} = -\rho_k\Delta I_{kz}(t) - \sum_{j\neq k}\sigma_{kj}\Delta I_{jz}(t), \qquad [5.10]$$

in which

$$\rho_k = \sum_{k \neq j} \rho_{kj} \qquad [5.11]$$

reflects the relaxation of the kth spin by all other spins (in the absence of interference effects; see Section 5.2.1). Equation [5.10] written in matrix nomenclature as

$$\frac{d\Delta\mathbf{M}_z(t)}{dt} = -\mathbf{R}\Delta\mathbf{M}_z(t), \qquad [5.12]$$

in which $\mathbf{R}$ is a $N \times N$ matrix with elements $R_{kk} = \rho_k$ and $R_{kj} = \sigma_{kj}$, and $\Delta\mathbf{M}_z(t)$ is a $N \times 1$ column vector with entries $\Delta M_k(t) = \Delta I_{kz}(t)$. The Solomon equations in matrix form have the formal solution:

$$\Delta\mathbf{M}_z(t) = e^{-\mathbf{R}t}\,\Delta\mathbf{M}_z(0) = \mathbf{U}^{-1}e^{-\mathbf{D}t}\mathbf{U}\,\Delta\mathbf{M}_z(0), \qquad [5.13]$$

in which $\mathbf{D}$ is a diagonal matrix of the eigenvalues of $\mathbf{R}$, $\mathbf{U}$ is a unitary matrix, and

$$\mathbf{D} = \mathbf{U}\mathbf{R}\mathbf{U}^{-1} \qquad [5.14]$$

is the similarity transformation that diagonalizes $\mathbf{R}$. These differential equations show that if the populations of the energy levels of the spin system are perturbed from equilibrium, then relaxation of a particular spin is in general a multiexponential process.

For a two-spin system

$$\mathbf{R} = \begin{bmatrix} \rho_I & \sigma_{IS} \\ \sigma_{IS} & \rho_S \end{bmatrix};$$

$$\mathbf{D} = \begin{bmatrix} \lambda_+ & 0 \\ 0 & \lambda_- \end{bmatrix}; \qquad [5.15]$$

$$\lambda_{\pm} = \frac{1}{2}\left\{(\rho_I + \rho_S) \pm [(\rho_I - \rho_S)^2 + 4\sigma_{IS}^2]^{1/2}\right\};$$

$$\mathbf{U} = \begin{bmatrix} \dfrac{-\sigma_{IS}}{[(\rho_I - \lambda_+)^2 + \sigma_{IS}^2]^{1/2}} & \dfrac{-\sigma_{IS}}{[(\rho_I - \lambda_-)^2 + \sigma_{IS}^2]^{1/2}} \\[2em] \dfrac{\rho_I - \lambda_+}{[(\rho_I - \lambda_+)^2 + \sigma_{IS}^2]^{1/2}} & \dfrac{\rho_I - \lambda_-}{[(\rho_I - \lambda_-)^2 + \sigma_{IS}^2]^{1/2}} \end{bmatrix},$$

and on substituting into [5.13], the result obtained is

$$
\begin{bmatrix} \Delta M_I(t) \\ \Delta M_S(t) \end{bmatrix} = \begin{bmatrix} a_{II}(t) & a_{IS}(t) \\ a_{SI}(t) & a_{SS}(t) \end{bmatrix} \begin{bmatrix} \Delta M_I(0) \\ \Delta M_S(0) \end{bmatrix},
\tag{5.16}
$$

in which

$$
a_{II}(t) = \frac{1}{2}\left[\left(1 - \frac{\rho_I - \rho_S}{(\lambda_+ - \lambda_-)}\right)\exp(-\lambda_- t) + \left(1 + \frac{\rho_I - \rho_S}{(\lambda_+ - \lambda_-)}\right)\exp(-\lambda_+ t)\right];
$$

$$
a_{SS}(t) = \frac{1}{2}\left[\left(1 + \frac{\rho_I - \rho_S}{(\lambda_+ - \lambda_-)}\right)\exp(-\lambda_- t) + \left(1 - \frac{\rho_I - \rho_S}{(\lambda_+ - \lambda_-)}\right)\exp(-\lambda_+ t)\right];
$$

$$
a_{IS}(t) = a_{SI}(t) = \frac{-\sigma_{IS}}{(\lambda_+ - \lambda_-)}[\exp(-\lambda_- t) - \exp(-\lambda_+ t)].
\tag{5.17}
$$

These equations frequently are written in the form

$$
a_{II}(t) = \frac{1}{2}\left[\left(1 - \frac{\rho_I - \rho_S}{R_C}\right) + \left(1 + \frac{\rho_I - \rho_S}{R_C}\right)\exp(-R_C t)\right]\exp(-R_L t);
$$

$$
a_{SS}(t) = \frac{1}{2}\left[\left(1 + \frac{\rho_I - \rho_S}{R_C}\right) + \left(1 - \frac{\rho_I - \rho_S}{R_C}\right)\exp(-R_C t)\right]\exp(-R_L t);
$$

$$
a_{IS}(t) = a_{SI}(t) = \frac{-\sigma_{IS}}{R_C}[1 - \exp(-R_C t)]\exp(-R_L t)
\tag{5.18}
$$

by defining the cross-rate constant, R_C, and a leakage rate constant, R_L:

$$
R_C = \lambda_+ - \lambda_- = [(\rho_I - \rho_S)^2 + 4\sigma_{IS}^2]^{1/2};
$$

$$
R_L = \lambda_-.
\tag{5.19}
$$

If $\rho_I = \rho_S = \rho$, and $\sigma_{IS} = \sigma$, [5.17] simplifies to

$$
a_{II}(t) = a_{SS}(t) = \tfrac{1}{2}\exp\{-(\rho - \sigma)t\}[1 + \exp(-2\sigma t)];
$$

$$
a_{IS}(t) = a_{SI}(t) = -\tfrac{1}{2}\exp\{-(\rho - \sigma)t\}[1 - \exp(-2\sigma t)].
\tag{5.20}
$$

The time dependences of the matrix elements $a_{II}(t)$ and $a_{IS}(t)$ are illustrated in Fig. 5.2.

To illustrate aspects of longitudinal relaxation as exemplified by the Solomon equations, four different experiments are analyzed. For simplicity, a homonuclear spin system with $\gamma_I = \gamma_S$, $\rho_I = \rho_S = \rho$, and $\sigma_{IS} = \sigma$ is assumed. The experiments use the pulse sequence:

$$
180° - t - 90° - \text{acquire}.
\tag{5.21}
$$

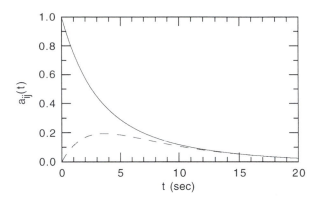

FIGURE 5.2 Time dependence of (—) $a_{II}(t)$ and (- - -) $a_{IS}(t)$ calculated using [5.20] with $\rho = 0.30$ s^{-1} and $\sigma = -0.15$ s^{-1}.

The initial state of the longitudinal magnetization is prepared by application of the 180° pulse to thermal equilibrium magnetization. The longitudinal magnetization relaxes according to the Solomon equations during the delay t. The final state of the longitudinal magnetization is converted into transverse magnetization by the 90° pulse and recorded during the acquisition period.

In the *selective inversion recovery* experiment, the 180° pulse is applied selectively to the I spin. The initial conditions are $\Delta I_z(0) = \langle I_z \rangle(0) - \langle I_z^0 \rangle = -2\langle I_z^0 \rangle$, and $\Delta S_z(0) = \langle S_z \rangle(0) - \langle S_z^0 \rangle = 0$. The time decay of the I spin magnetization is given by

$$\langle I_z \rangle(t)/\langle I_z^0 \rangle = 1 - \exp\{-(\rho - \sigma)t\}[1 + \exp(-2\sigma t)], \qquad [5.22]$$

and is generally biexponential. In the *initial-rate regime*, the slope of the recovery curve is given by

$$\left. \frac{d(\langle I_z \rangle(t)/\langle I_z^0 \rangle)}{dt} \right|_{t=0} = 2\rho. \qquad [5.23]$$

In the *nonselective inversion recovery* experiment, the 180° pulse is nonselective. The initial conditions are $\Delta I_z(0) = \langle I_z \rangle(0) - \langle I_z^0 \rangle = -2\langle I_z^0 \rangle$ and

$\Delta S_z(0) = \langle S_z \rangle(0) - \langle S_z^0 \rangle = -2\langle S_z^0 \rangle$. The time course of the I spin magnetization is given by

$$\langle I_z \rangle(t)/\langle I_z^0 \rangle = 1 - \exp\{-(\rho - \sigma)t\}[1 + \exp(-2\sigma t)]$$
$$+ (\langle S_z^0 \rangle/\langle I_z^0 \rangle) \exp\{-(\rho - \sigma)t\}[1 - \exp(-2\sigma t)]$$
$$= 1 - 2\exp\{-(\rho + \sigma)t\}, \qquad [5.24]$$

in which the final line is obtained by using $\langle S_z^0 \rangle/\langle I_z^0 \rangle = \gamma_S/\gamma_I = 1$. The recovery curve is monoexponential with rate constant $\rho + \sigma$. In the initial-rate regime,

$$\left. \frac{d(\langle I_z \rangle(t)/\langle I_z^0 \rangle)}{dt} \right|_{t=0} = 2(\rho + \sigma). \qquad [5.25]$$

In the *transient NOE experiment*, the S spin longitudinal magnetization is inverted with a selective 180° pulse to produce initial conditions $\Delta I_z(0) = \langle I_z \rangle(0) - \langle I_z^0 \rangle = 0$ and $\Delta S_z(0) = \langle S_z \rangle(0) - \langle S_z^0 \rangle = -2\langle S_z^0 \rangle$. The time course of the I spin magnetization is given by

$$\langle I_z \rangle(t)/\langle I_z^0 \rangle = 1 + (\langle S_z^0 \rangle/\langle I_z^0 \rangle) \exp\{-(\rho - \sigma)t\}[1 - \exp(-2\sigma t)]$$
$$= 1 + \exp\{-(\rho - \sigma)t\}[1 - \exp(-2\sigma t)], \qquad [5.26]$$

and is biexponential. In the initial-rate regime,

$$\left. \frac{d(\langle I_z \rangle(t)/\langle I_z^0 \rangle)}{dt} \right|_{t=0} = 2\sigma. \qquad [5.27]$$

Thus, the initial rate of change of the I spin intensity in the transient NOE experiment is proportional to the cross-relaxation rate, σ. In the *decoupled inversion recovery experiment*, the S spin is irradiated by a weak selective rf field (so as not to perturb the I spin) throughout the experiment in order to equalize the populations across the S spin transitions. In this situation, $\langle S_z \rangle(t) = 0$ for all t, and the S spins are said to be saturated. Equation [5.9] reduces to

$$\frac{d\langle I_z \rangle(t)}{dt} = -\rho[\langle I_z \rangle(t) - \langle I_z^0 \rangle] + \sigma\langle S_z^0 \rangle$$

$$= -\rho\left[\langle I_z \rangle(t) - \langle I_z^0 \rangle\left(1 + \frac{\sigma}{\rho}\right)\right]. \qquad [5.28]$$

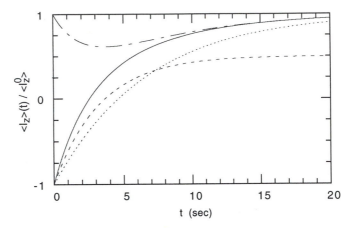

FIGURE 5.3 Magnetization decays for inversion recovery experiments. (—) Selective inversion recovery calculated using [5.22]; (· · ·) nonselective inversion recovery calculated using [5.24]; (− · −) transient NOE recovery calculated using [5.26]; and (− − −) decoupled inversion recovery calculated using [5.29]. Calculations were performed for a homonuclear IS spin system with $\gamma_I = \gamma_S$, $\rho = 0.30$ s^{-1}, and $\sigma = -0.15$ s^{-1}.

Following the 180° pulse, $\Delta I_z(0) = \langle I_z\rangle(0) - \langle I_z^0\rangle = -2\langle I_z^0\rangle$, and the time course of the I spin magnetization is given by

$$\langle I_z\rangle(t)/\langle I_z^0\rangle = 1 + \frac{\sigma}{\rho} - \left(2 + \frac{\sigma}{\rho}\right)\exp(-\rho t). \qquad [5.29]$$

In the initial-rate regime,

$$\left.\frac{d(\langle I_z\rangle(t)/\langle I_z^0\rangle)}{dt}\right|_{t=0} = 2\rho - \sigma. \qquad [5.30]$$

In this case, the recovery curve is monoexponential with rate constant ρ. The preceding analyses indicate that, even for an isolated two-spin system, the time dependence of the longitudinal magnetization usually is biexponential. The actual time course observed depends on the initial condition of the spin system prepared by the NMR pulse sequence. Examples of the time courses of the I spin magnetization for these experiments are given in Fig. 5.3.

The present derivation does not provide theoretical expressions for the transition rate constants, W_0, W_I, W_S, and W_2. Bloembergen *et al.*

(5) derived expressions for the transition rate constants; however, herein, the transition rate constants will be calculated using the semiclassical relaxation theory as described in Section 5.2. As will be shown, the transition rate constants depend on the different frequency components of the stochastic magnetic fields expressed in [5.91]. Thus, the transition characterized by W_I is induced by molecular motions that produce fields oscillating at the Larmor frequency of the I spin, and the transition characterized by W_S is induced by molecular motions that produce fields oscillating at the Larmor frequency of the S spin. The W_0 pathway is induced by fields oscillating at the *difference* of the Larmor frequencies of the I and S spins, and the W_2 pathway is induced by fields oscillating at the *sum* of the Larmor frequencies of the two spins. Most importantly, the cross-relaxation rate constant is nonzero only if $W_2 - W_0 \neq 0$; therefore, the relaxation mechanism must generate nonzero rate constants for the flip-flip (double-quantum) and flip-flop (zero-quantum) transitions. For biological macromolecules, dipolar coupling between nuclear spins is the main interaction for which W_2 and W_0 are nonzero. The Solomon equations are central to the study of the NOE and will be discussed in additional detail in Section 5.5.

5.1.3 BLOCH, WANGSNESS, AND REDFIELD THEORY

A microscopic semiclassical theory of spin relaxation was formulated by Bloch, Wangsness, and Redfield (BWR) and has proved to be the most useful approach for practical applications (7,8). In the semiclassical approach the spin system is treated quantum-mechanically, and the surroundings (the heat bath or lattice) are treated classically. This treatment suffers primarily from the defect that the spin system evolves toward a final state in which energy levels of the spin system are populated equally. Equivalently, the semiclassical theory is formally correct only for an infinite Boltzmann spin temperature; at finite temperatures, an ad hoc correction is required to the theory to ensure that the spin system relaxes toward an equilibrium state in which the populations are described by a Boltzmann distribution. A fully quantum-mechanical treatment of spin relaxation overcomes this defect and predicts the proper approach to equilibrium; however, the computational details of the quantum mechanical relaxation theory are outside the scope of this text (2,8).

5.2 The Master Equation

In the semiclassical theory of spin relaxation, the Hamiltonian for the system is written as the sum of a deterministic quantum-mechanical

Hamiltonian that acts only on the spin system, $\mathcal{H}_{det}(t)$, and a stochastic Hamiltonian, $\mathcal{H}_1(t)$, that couples the spin system to the lattice:

$$\mathcal{H}(t) = \mathcal{H}_{det}(t) + \mathcal{H}_1(t) = \mathcal{H}_0 + \mathcal{H}_{rf}(t) + \mathcal{H}_1(t). \qquad [5.31]$$

The Liouville equation of motion of the density operator is (Section 2.2.3)

$$d\sigma(t)/dt = -i[\mathcal{H}(t), \sigma(t)]. \qquad [5.32]$$

The Hamiltonians $\mathcal{H}_{rf}(t)$ and $\mathcal{H}_1(t)$ are regarded as time-dependent perturbations acting on the main time-independent Hamiltonian, $\mathcal{H}_0$. The explicit influence of $\mathcal{H}_0$ can be removed by transforming the Liouville equation into a new reference frame, which is called conventionally, the *interaction frame*. In the absence of an applied rf field (see Section 5.2.3 for the effects of rf fields), the density operator and stochastic Hamiltonian in the interaction frame are defined as

$$\sigma^T(t) = \exp\{i\mathcal{H}_0 t\}\sigma(t)\exp\{-i\mathcal{H}_0 t\};$$

$$\mathcal{H}_1^T(t) = \exp\{i\mathcal{H}_0 t\}\mathcal{H}_1(t)\exp\{-i\mathcal{H}_0 t\}. \qquad [5.33]$$

The form of the transformed Liouville equation is determined as follows:

$$\frac{d\sigma^T(t)}{dt} = i\exp(i\mathcal{H}_0 t)\mathcal{H}_0\sigma(t)\exp(-i\mathcal{H}_0 t) - i\exp(i\mathcal{H}_0 t)\sigma(t)\mathcal{H}_0\exp(-i\mathcal{H}_0 t)$$

$$+ \exp(i\mathcal{H}_0 t)\frac{d\sigma(t)}{dt}\exp(-i\mathcal{H}_0 t)$$

$$= i\exp(i\mathcal{H}_0 t)[\mathcal{H}_0, \sigma(t)]\exp(-i\mathcal{H}_0 t)$$

$$- i\exp(i\mathcal{H}_0 t)[\mathcal{H}_0 + \mathcal{H}_1(t), \sigma(t)]\exp(-i\mathcal{H}_0 t)$$

$$= -i\exp(i\mathcal{H}_0 t)[\mathcal{H}_1(t), \sigma(t)]\exp(-i\mathcal{H}_0 t)$$

$$= -i\exp(i\mathcal{H}_0 t)\mathcal{H}_1(t)\sigma(t)\exp(-i\mathcal{H}_0 t)$$

$$+ i\exp(i\mathcal{H}_0 t)\sigma(t)\mathcal{H}_1(t)\exp(-i\mathcal{H}_0 t)$$

$$= -i\mathcal{H}_1^T(t)\sigma^T(t) + i\sigma^T(t)\mathcal{H}_1^T(t), \qquad [5.35]$$

with the final result that

$$d\sigma^T(t)/dt = -i[\mathcal{H}_1^T(t), \sigma^T(t)]. \qquad [5.36]$$

The transformation into the interaction frame is isomorphous to the rotating-frame transformation; however, important differences exist between the two. The rotating frame transformation removes the explicit time dependence of the rf Hamiltonian and renders the Hamiltonian time-independent in the rotating frame. The Hamiltonian $\mathcal{H}_0$ is active in the rotating frame. The interaction frame transformation removes the explicit dependence on $\mathcal{H}_0$; however, $\mathcal{H}_1^T(t)$ remains time-dependent. As discussed in

Section 5.2.3, the rotating-frame and interaction-frame transformations are performed sequentially in some circumstances.

As described in more detail elsewhere (2,3,8), [5.36] can be integrated to second order to yield

$$\frac{d\overline{\sigma}^{T}(t)}{dt} = -\int_{0}^{\infty} d\tau \, \overline{[\mathcal{H}_{1}^{T}(t), [\mathcal{H}_{1}^{T}(t-\tau), \sigma^{T}(t) - \sigma_{0}]]}, \qquad [5.37]$$

in which the overbar indicates ensemble averaging and $\overline{\sigma}^{T}(t)$ now designates the ensemble average of the density matrix. Equation [5.37] is valid subject to the following assumptions:

1. The ensemble average of $\mathcal{H}_{1}^{T}(t)$ is zero. Any components of $\mathcal{H}_{1}^{T}(t)$ that do not vanish on ensemble averaging can be incorporated into $\mathcal{H}_{0}$.
2. $\mathcal{H}_{1}^{T}(t)$ and $\sigma^{T}(t)$ are uncorrelated so that the ensemble average can be taken independently for each quantity.
3. The characteristic correlation time for $\mathcal{H}_{1}^{T}(t)$, $\tau_{\mathcal{H}}$, is much shorter than t and $\{1/R_{ij}\}$, in which $\{R_{ij}\}$ are the relaxation rate constants for the density matrix elements, $\{\sigma_{ij}^{T}\}$. In liquids, $\tau_{\mathcal{H}}$, is on the order of a collision time, 10^{-14}–10^{-12} s, and t and $\{1/R_{ij}\}$ are on the order of 10^{-3} s or longer; therefore, this assumption is satisfied.
4. The term σ_{0} can be introduced into [5.37] to ensure that the spin system relaxes toward equilibrium.

Detailed discussion of the range of validity of these assumptions can be found elsewhere (2,3).

To proceed further, the stochastic Hamiltonian is decomposed as

$$\mathcal{H}_{1}(t) = \sum_{q=-k}^{k} F_{k}^{q}(t)\mathbf{A}_{k}^{q}, \qquad [5.38]$$

in which $F_{k}^{q}(t)$ is a random function of spatial variables and $\mathbf{A}_{k}^{q}$ is a tensor spin operator (2,9,10). Additionally, $\mathbf{A}_{k}^{-q} \equiv \mathbf{A}_{k}^{q\dagger}$ and $F_{k}^{-q}(t) \equiv F_{k}^{q*}(t)$. For the Hamiltonians of interest in NMR spectroscopy, the rank of the tensor operator, k, is 1 or 2, and the decomposition is always possible. To proceed, the operators $\mathbf{A}_{k}^{q}$ are expanded in terms of basis operators

$$\mathbf{A}_{k}^{q} = \sum_{p} \mathbf{A}_{kp}^{q} = \sum_{p} c_{p}^{q}\mathcal{H}_{p} \qquad [5.39]$$

that satisfy the relationship

$$[\mathcal{H}_{0}, \mathcal{H}_{p}] = \omega_{p}\mathcal{H}_{p}, \qquad [5.40]$$

where $\mathcal{H}_{p}$ and ω_{p} are the eigenfunctions and eigenfrequencies of the Hamiltonian commutation superoperator. Equation [5.40] implies the additional property

$$\exp(-i\mathcal{H}_0 t)\mathcal{H}_p \exp(i\mathcal{H}_0 t) = \exp(-i\omega_p t)\mathcal{H}_p, \qquad [5.41]$$

which can be proved as follows. First,

$$\frac{d}{dt}\{\exp(-i\mathcal{H}_0 t)\mathcal{H}_p \exp(i\mathcal{H}_0 t)\}$$

$$= -i\exp(-i\mathcal{H}_0 t)\mathcal{H}_0\mathcal{H}_p \exp(i\mathcal{H}_0 t) + i\exp(-i\mathcal{H}_0 t)\mathcal{H}_p\mathcal{H}_0 \exp(i\mathcal{H}_0 t)$$

$$= -i\exp(-i\mathcal{H}_0 t)[\mathcal{H}_0, \mathcal{H}_p]\exp(i\mathcal{H}_0 t)$$

$$= -i\omega_p \exp(-i\mathcal{H}_0 t)\mathcal{H}_p \exp(i\mathcal{H}_0 t), \qquad [5.42]$$

which implies

$$\frac{d^n}{dt^n}\{\exp(-i\mathcal{H}_0 t)\mathcal{H}_p \exp(i\mathcal{H}_0 t)\} = (-i\omega_p)^n \exp(-i\mathcal{H}_0 t)\mathcal{H}_p \exp(i\mathcal{H}_0 t).$$

$$[5.43]$$

Therefore, the Taylor series expansion of the left-hand side of [5.41] is

$$\exp(-i\mathcal{H}_0 t)\mathcal{H}_p \exp(i\mathcal{H}_0 t)$$

$$= \mathcal{H}_p - i\omega_p t\mathcal{H}_p + \tfrac{1}{2}\omega_p^2 t\mathcal{H}_p + \cdots$$

$$= \{1 - i\omega_p t + \tfrac{1}{2}\omega_p^2 t + \cdots\}\mathcal{H}_p$$

$$= \exp(-i\omega_p t)\mathcal{H}_p, \qquad [5.44]$$

which completes the proof.

For example, if $\mathcal{H}_0 = \omega_I I_z + \omega_S S_z$, then the single-element operator $2I_z S^+ = I^\alpha S^+ - I^\beta S^+ = |\alpha\alpha\rangle\langle\alpha\beta| - |\beta\alpha\rangle\langle\beta\beta|$ (see [2.204]) is an eigenoperator with eigenfrequency ω_S:

$$[\mathcal{H}_0, I^\alpha S^+ - I^\beta S^+]$$

$$= (\omega_I I_z + \omega_S S_z)(|\alpha\alpha\rangle\langle\alpha\beta| - |\beta\alpha\rangle\langle\beta\beta|)$$

$$\quad - (|\alpha\alpha\rangle\langle\alpha\beta| - |\beta\alpha\rangle\langle\beta\beta|)(\omega_I I_z + \omega_S S_z)$$

$$= \omega_I(I_z|\alpha\alpha\rangle\langle\alpha\beta| - I_z|\beta\alpha\rangle\langle\beta\beta| - |\alpha\alpha\rangle\langle\alpha\beta|I_z + |\beta\alpha\rangle\langle\beta\beta|I_z)$$

$$\quad + \omega_S(S_z|\alpha\alpha\rangle\langle\alpha\beta| - S_z|\beta\alpha\rangle\langle\beta\beta| - |\alpha\alpha\rangle\langle\alpha\beta|S_z + |\beta\alpha\rangle\langle\beta\beta|S_z)$$

$$= \tfrac{1}{2}\omega_I(|\alpha\alpha\rangle\langle\alpha\beta| + |\beta\alpha\rangle\langle\beta\beta| - |\alpha\alpha\rangle\langle\alpha\beta| - |\beta\alpha\rangle\langle\beta\beta|)$$

$$\quad + \tfrac{1}{2}\omega_S(|\alpha\alpha\rangle\langle\alpha\beta| - |\beta\alpha\rangle\langle\beta\beta| + |\alpha\alpha\rangle\langle\alpha\beta| - |\beta\alpha\rangle\langle\beta\beta|)$$

$$= \omega_S(|\alpha\alpha\rangle\langle\alpha\beta| - |\beta\alpha\rangle\langle\beta\beta|)$$

$$= \omega_S(I^\alpha S^+ - I^\beta S^+). \qquad [5.45]$$

Applying [5.41], in the interaction frame,

$$\mathbf{A}_k^{q\mathrm{T}} = \exp\{i\mathcal{H}_0 t\}\mathbf{A}_k^q \exp\{-i\mathcal{H}_0 t\} = \sum_p \exp\{i\mathcal{H}_0 t\}\mathbf{A}_{kp}^q \exp\{-i\mathcal{H}_0 t\}$$

$$= \sum_p \mathbf{A}_{kp}^q \exp\{i\omega_p t\}; \qquad\qquad [5.46]$$

$$\mathbf{A}_k^{-q\mathrm{T}} = \exp\{i\mathcal{H}_0 t\}\mathbf{A}_k^{-q} \exp\{-i\mathcal{H}_0 t\} = \sum_p \exp\{i\mathcal{H}_0 t\}\mathbf{A}_{kp}^{-q} \exp\{-i\mathcal{H}_0 t\}$$

$$= \sum_p \mathbf{A}_{kp}^{-q} \exp\{-i\omega_p t\}. \qquad\qquad [5.47]$$

Substituting [5.38], [5.46], and [5.47] into [5.37] yields

$$\frac{d\sigma^{\mathrm{T}}(t)}{dt} = -\sum_{q,q'}\sum_{p,p'} \exp\{i(-\omega_{p'} + \omega_p)t\}[\mathbf{A}_{kp'}^{q'}, [\mathbf{A}_{kp}^q, \sigma^{\mathrm{T}}(t) - \sigma_0]]$$

$$\times \int_0^\infty \overline{F_k^{q'}(t)F_k^q(t-\tau)} \exp\{-i\omega_p \tau\} \, d\tau. \qquad\qquad [5.48]$$

The imaginary part of the integral leads to second-order frequency shifts of the resonance lines, which are called dynamic frequency shifts; these shifts may be included in $\mathcal{H}_0$ and are not considered further. Considering only the real part of the integral, [5.48] can be written as

$$\frac{d\sigma^{\mathrm{T}}(t)}{dt} = -\frac{1}{2}\sum_q\sum_{p,p'} \exp\{i(-\omega_{p'} + \omega_p)t\}[\mathbf{A}_{kp'}^{-q}, [\mathbf{A}_{kp}^q, \sigma^{\mathrm{T}}(t) - \sigma_0]]j^q(\omega_p),$$

$$[5.49]$$

in which the power spectral density function, $j^q(\omega)$, is given by

$$j^q(\omega) = \mathrm{Re}\left\{\int_{-\infty}^\infty \overline{F_k^{-q}(t)F_k^q(t-\tau)} \exp\{-i\omega\tau\} \, d\tau\right\}$$

$$= \mathrm{Re}\left\{\int_{-\infty}^\infty \overline{F_k^q(t)F_k^{-q}(t+\tau)} \exp\{-i\omega\tau\} \, d\tau\right\}, \qquad [5.50]$$

and the random processes $F_k^q(t)$ and $F_k^{q'}(t)$ have been assumed to be statistically independent unless $q' = -q$; therefore, the ensemble average in [5.48] vanishes if $q' \neq -q$. Terms in [5.49] in which $|\omega_p - \omega_{p'}| \gg 0$ are *nonsecular* in the sense of perturbation theory, and do not affect the long-time behavior of $\sigma^{\mathrm{T}}(t)$ because the rapidly oscillating factors $\exp\{i(-\omega_{p'} + \omega_p)\}$ average to zero much more rapidly than relaxation occurs. Furthermore, if none of the eigenfrequencies are degenerate, terms in [5.49] are *secular* and nonzero only if $p = p'$; thus

$$\frac{d\sigma^{\mathrm{T}}(t)}{dt} = -\frac{1}{2}\sum_q \sum_p [\mathbf{A}_{kp}^{-q}, [\mathbf{A}_{kp}^q, \sigma^{\mathrm{T}}(t) - \sigma_0]]j^q(\omega_p). \qquad [5.51]$$

This equation can be transformed to the laboratory frame to yield the Liouville–von Neuman differential equation for the density operator:

$$\frac{d\sigma(t)}{dt} = -i[\mathscr{H}_0, \sigma(t)] - \hat{\Gamma}(\sigma(t) - \sigma_0), \qquad [5.52]$$

in which the relaxation superoperator (an operator that acts on operators) is

$$\hat{\Gamma} = \sum_q \sum_p j^q(\omega_p)[\mathbf{A}_{kp}^{-q}, [\mathbf{A}_{kp}^q, \ ]]. \qquad [5.53]$$

Two critical requirements for a stochastic Hamiltonian to be effective in causing relaxation are encapsulated in [5.52]: (1) the double commutator $[\mathbf{A}_{kp}^{-q}, [\mathbf{A}_{kp}^q, \sigma(t) - \sigma_0]]$ must not vanish, and (2) the spectral density function for the random process that modulates the spin interactions must have significant components at the characteristic frequencies of the spin system, ω_p. The former requirement can be regarded as a kind of selection rule for whether the term in the stochastic Hamiltonian that depends on the operator $\mathbf{A}$ is effective in causing relaxation of the density operator. In most cases, the stochastic random process is a consequence of molecular reorientational motions. This observation is central to the dramatic differences in spin relaxation and, thus, in NMR spectroscopy, of rapidly rotating small molecules and slowly rotating macromolecules.

Equation [5.52] can be converted into an equation for product operator, or other basis operators, by expanding the density operator in terms of the basis operators [2.190] to yield the matrix form of the master equation

$$db_r(t)/dt = \sum_s \{-i\Omega_{rs}b_s(t) - \Gamma_{rs}[b_s(t) - b_s^0]\}, \qquad [5.54]$$

in which

$$\Omega_{rs} = \langle \mathbf{B}_r | [\mathscr{H}_0, \mathbf{B}_s] \rangle / \langle \mathbf{B}_r | \mathbf{B}_r \rangle \qquad [5.55]$$

is a characteristic frequency,

$$\Gamma_{rs} = \langle \mathbf{B}_r | \hat{\Gamma} \mathbf{B}_s \rangle / \langle \mathbf{B}_r | \mathbf{B}_r \rangle$$

$$= \frac{1}{2}\sum_q \sum_p \{\langle \mathbf{B}_r | [\mathbf{A}_{kp}^{-q}, [\mathbf{A}_{kp}^q, \mathbf{B}_s]] \rangle / \langle \mathbf{B}_r | \mathbf{B}_r \rangle\}j^q(\omega_p) \qquad [5.56]$$

is the rate constant for relaxation between the operators $\mathbf{B}_r$ and $\mathbf{B}_s$, and

$$b_j(t) = \langle \mathbf{B}_j | \sigma(t) \rangle. \qquad [5.57]$$

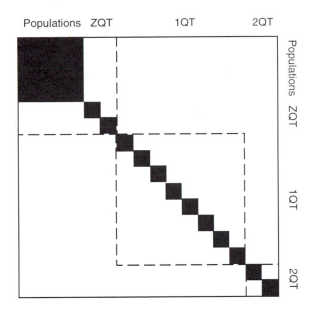

FIGURE 5.4 Redfield kite. Solid blocks indicate nonzero relaxation rate constants between operators in the absence of degenerate transitions. Populations have nonzero cross-relaxation rate constants, but all other coherences relax independently. If transitions are degenerate, the dashed blocks indicate the additional nonzero cross-relaxation rate constants observed between coherences with the same coherence level.

For normalized basis operators with $\mathrm{Tr}\{\mathbf{B}_r^2\} = \mathrm{Tr}\{\mathbf{B}_s^2\}$, $\Gamma_{rs} = \Gamma_{rs}$. Equations [5.54]–[5.57] are the main results of this section for relaxation in the laboratory reference frame. As shown by [5.54], the evolution of the base operators for a spin system is described by a set of coupled differential equations. Diagonal elements Γ_{rr} are the rate constants for auto- or self-relaxation of $\mathbf{B}_r$; off-diagonal elements Γ_{rs} are the rate constants for cross-relaxation between $\mathbf{B}_r$ and $\mathbf{B}_s$. Cross-relaxation between operators with different coherence orders is precluded as a consequence of restricting [5.54] to secular contributions; for example, cross-relaxation does not occur between zero- and single-quantum coherence. Furthermore, if none of the transitions in the spin system are degenerate (to within approximately a linewidth), then cross-relaxation rate constants between off-diagonal elements of the density operator in the laboratory reference frame are also zero. Consequently, the matrix of relaxation rate constants between operators has a characteristic block diagonal form, known as the Redfield kite, illustrated in Fig. 5.4.

Calculation of relaxation rate constants involves two steps: (1) calculation of the double commutator and trace formation over the spin variables and (2) calculation of the spectral density function. These two calculations are pursued in the following sections.

5.2.1 INTERFERENCE EFFECTS

In many instances, more than one stochastic Hamiltonian capable of causing relaxation of a given spin may be operative. In this circumstance, equation [5.38] is generalized to

$$\mathcal{H}_1(t) = \sum_m \sum_{q=-k}^{k} F_{mk}^q(t) \mathbf{A}_{mk}^q, \qquad [5.58]$$

in which the summation over the index m refers to the different relaxation interactions or stochastic Hamiltonians. Using [5.58] rather than [5.38] in the derivation given above leads once more to [5.54] with Γ_{rs} given by a generalization of [5.56]:

$$\Gamma_{rs} = \frac{1}{2} \sum_m \sum_q \sum_p \{\langle \mathbf{B}_r | [\mathbf{A}_{mkp}^{-q}, [\mathbf{A}_{mkp}^q, \mathbf{B}_s]] \rangle / \langle \mathbf{B}_r | \mathbf{B}_r \rangle \} j^q(\omega_p)$$

$$+ \frac{1}{2} \sum_{\substack{m,n \\ m \neq n}} \sum_q \sum_p \{\langle \mathbf{B}_r | [\mathbf{A}_{mkp}^{-q}, [\mathbf{A}_{nkp}^q, \mathbf{B}_s]] \rangle / \langle \mathbf{B}_r | \mathbf{B}_r \rangle \} j_{mn}^q(\omega_p)$$

$$= \sum_m \Gamma_{rs}^m + \sum_{\substack{m,n \\ m \neq n}} \Gamma_{rs}^{mn}, \qquad [5.59]$$

in which the cross-spectral density is

$$j_{mn}^q(\omega) = \mathrm{Re} \left\{ \int_{-\infty}^{\infty} \overline{F_{mk}^q(t) F_{nk}^{-q}(t+\tau)} \exp\{-i\omega\tau\} d\tau \right\} \qquad [5.60]$$

Γ_{rs}^m is the relaxation rate due to the mth relaxation mechanism and Γ_{rs}^{mn} is the relaxation rate constant arising from interference or cross-correlation between the mth and nth relaxation mechanisms.

Clearly, $j_{mn}^q(\omega) = 0$ unless the random processes $F_{mk}^q(t)$ and $F_{nk}^q(t)$ are correlated. In the absence of correlation between the different relaxation mechanisms, $\Gamma_{rs}^{mn} = 0$ for all m and n and each mechanism contributes additively to relaxation of the spin system.

The two most frequently encountered interference or cross-correlation effects in biological macromolecules are interference between dipolar and anisotropic chemical-shift (CSA) interactions, and interference between the dipolar interactions of different pairs of spins. The prototypical example of the former is the interference between the dipolar and CSA interactions for ^{15}N (*11*). The prototypical example of the latter is the interference between the dipolar interactions in a I_2S or I_3S spin system such as a methylene (I_2 represents the two methylene protons; S represents either a remote proton or the methylene ^{13}C) or methyl group (I_3 represents the three methyl protons; S represents either a remote proton or the methyl ^{13}C) (*12*). Most importantly, interference effects can result in cross-relaxation between pairs of operators for which cross-relaxation would not be observed otherwise. Thus, the observation of otherwise "forbidden" cross-relaxation pathways is one of the hallmarks of interference effects (*13*).

5.2.2 LIKE AND UNLIKE SPINS

A distinction frequently is made between like and unlike spins, and relaxation rate constants are derived independently for each case (*2*). Like spins are defined as spins with identical Larmor frequencies, and unlike spins are defined as spins with widely different Larmor frequencies. Such distinctions can obscure the generality of the theory embodied in [5.54]. In actuality, the presence of spins with degenerate Larmor frequencies has straightforward consequences for relaxation. First, particular operators $\mathbf{A}_{kp}^q$ in [5.38] may become degenerate (i.e., have the same eigenfrequency, ω_p) and are therefore secular with respect to each other. Thus, prior to applying the secular condition, the set of $\mathbf{A}_{kp}^q$ must be redefined as

$$\mathbf{A}_{kp}^q = \sum_m \mathbf{A}_{km}^q, \qquad [5.61]$$

in which the summation extends over the operators for $\mathbf{A}_{km}^q$ for which $\omega_p = \omega_m$. For example, operators $\mathbf{A}_{km}^q$ with eigenfrequencies of 0 and $\omega_I - \omega_S$ belong to different orders p for unlike spins; the eigenfrequencies are degenerate for like spins and the corresponding operators would be summed to yield a single operator with eigenfrequency of zero. Second, for spins that are magnetically equivalent, such as the three protons in a methyl group, basis operators that exhibit the maximum symmetry of the chemical moiety can be derived using group theoretical methods (*9,14*). Although such basis operators simplify the resulting calculations, the group theoretical treatment of relaxation of magnetically equivalent spins is beyond the scope of the present text; the interested reader is referred to the original literature. As the distinction between like and unlike spins is artificial within the framework of the semiclassical relaxation theory, the following discussions will focus on spin systems without degenerate transitions; results of practical interest that arise as a consequence of degeneracy will be presented as necessary.

5.2.3 RELAXATION IN THE ROTATING FRAME

In the presence of an applied rf field (e.g., in a ROESY or TOCSY experiment), the transformation into the interaction frame involves first a transformation into a rotating frame to remove the time dependence of $\mathcal{H}_{rf}(t)$ followed by transformation into the interaction frame of the resulting time-independent Hamiltonian. If $\mathcal{H}_0 \approx \mathcal{H}_z$—that is, if the Zeeman Hamiltonian is dominant (i.e., ignoring the scalar coupling Hamiltonian)—then the interaction frame is equivalent to a doubly rotating tilted frame. For macromolecules with $\omega_1 \tau_c \ll 1$, in which ω_1 is the strength of the applied rf field and τ_c is the rotational correlation time of the molecule, $j^q(\omega + \omega_1) \approx j^q(\omega)$, and approximate values for the relaxation rate constants in the rotating frame can be calculated using [5.57] in which the operators $\mathbf{B}_r$ and $\mathbf{B}_s$ are replaced by the corresponding operators in the tilted frame, $\mathbf{B}_r'$ and $\mathbf{B}_s'$. Thus

$$\Gamma_{rs}' = \langle \mathbf{B}_r' | \hat{\Gamma} \mathbf{B}_s' \rangle / \langle \mathbf{B}_r' | \mathbf{B}_r' \rangle$$

$$= \frac{1}{2} \sum_q \sum_p \{ \langle \mathbf{B}_r' | [\mathbf{A}_{kp}^{-q}, [\mathbf{A}_{kp}^q, \mathbf{B}_s']] \rangle / \langle \mathbf{B}_r' | \mathbf{B}_r' \rangle \} j^q(\omega_p). \qquad [5.62]$$

For an rf field applied with x phase, the operators are transformed as

$$\begin{bmatrix} I_x' \\ I_y' \\ I_z' \end{bmatrix} = \begin{bmatrix} \cos\theta & 0 & -\sin\theta \\ 0 & 1 & 0 \\ \sin\theta & 0 & \cos\theta \end{bmatrix} \begin{bmatrix} I_x \\ I_y \\ I_z \end{bmatrix}, \qquad [5.63]$$

in which θ is defined by [1.21] (if $\mathbf{B}_r$ and $\mathbf{B}_s$ refer to different spins, θ may differ for each spin). The relative orientation of the tilted and untilted reference frames are illustrated in Fig. 5.5. If $\theta = 0$, either because $\omega_1 = 0$ or because $\omega_1 \ll |\omega - \omega_0|$, [5.62] reduces to [5.57]; if the rf field is applied on-resonance ($\omega = \omega_I$), $\theta = \pi/2$. If the rf field is applied midway between the Larmor frequencies of two spins, or if $\omega_1 \gg \omega - \omega_I$ for the spins of interest, the effective frequencies in the rotating frame [1.22] are degenerate, and the relaxation superoperator in the rotating frame is calculated as for like spins (Section 5.2.2).

In general, operators that do not commute with the Hamiltonian in the rotating frame decay rapidly as a consequence of rf inhomogeneity (Section 3.5.3). Thus, if a continuous-wave (CW) rf field is applied, as in

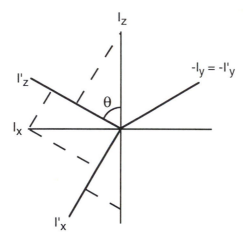

FIGURE 5.5 Relative orientations of the laboratory and tilted reference frames used to determine the transformation [5.63].

a ROESY experiment, only operators with effective frequencies in the rotating frame equal to zero must be considered; such operators are usually limited to longitudinal operators and homonuclear zero-quantum operators. If the rf field is phase-modulated to compensate for resonance offset and rf inhomogeneity, such as by applying the DIPSI-2 or other coherent decoupling scheme, single- and multiple-quantum operators also must be considered (*15*). For operators containing transverse components in the rotating frame, the relaxation rate constant given by [5.62] is an instantaneous rate constant; the effective average rate constant is obtained by averaging the rate constant over the trajectory followed by the operator under the influence of the Hamiltonian in the rotating frame (*16*).

5.3 Spectral Density Functions

A general expression for the spectral density function is given by [5.50]. For relaxation in isotropic liquids in the high-temperature limit (*17*),

$$j^q(\omega) = (-1)^q j^0(\omega) \equiv (-1)^q j(\omega); \qquad [5.64]$$

therefore, only one autospectral density function need be calculated. The relaxation mechanisms of interest in the present context arise from tensor-

TABLE 5.1
Modified Second-Order Spherical Harmonics

q	Y_2^q	$Y_2^{-q} = Y_2^{q*}$
0	$(3\cos^2\theta - 1)/2$	$(3\cos^2\theta - 1)/2$
1	$\sqrt{3/2}\,\sin\theta\,\cos\theta\,e^{i\phi}$	$\sqrt{3/2}\,\sin\theta\,\cos\theta\,e^{-i\phi}$
2	$\sqrt{3/8}\,\sin^2\theta\,e^{i2\phi}$	$\sqrt{3/8}\,\sin^2\theta\,e^{-i2\phi}$

The modified spherical harmonic functions are normalized (to give the conventional spherical harmonic functions) by multiplying by $\sqrt{5/(4\pi)}$.

ial operators of rank $k = 2$. The random functions $F_2^0(t)$ can be written in the form

$$F_2^0(t) = c_0(t)Y_2^0[\Omega(t)], \qquad [5.65]$$

and, consequently,

$$j(\omega) = \mathrm{Re}\left\{\int_{-\infty}^{\infty} \overline{c_0(t)c_0(t+\tau)Y_2^0[\Omega(t)]Y_2^0[\Omega(t+\tau)]}\,\exp(-i\omega\tau)\,d\tau\right\}$$

$$= \mathrm{Re}\left\{\int_{-\infty}^{\infty} C(\tau)\exp(-i\omega\tau)\,d\tau\right\}, \qquad [5.66]$$

in which the *stochastic correlation function* is given by

$$C(\tau) = \overline{c_0(t)c_0(t+\tau)Y_2^0[\Omega(t)]Y_2^0[\Omega(t+\tau)]}, \qquad [5.67]$$

where $c_0(t)$ is a function of physical constants and spatial variables, $Y_2^0[\Omega(t)]$ is a modified second-order spherical harmonic function, and $\Omega(t) = \{\theta(t), \phi(t)\}$ are polar angles in the laboratory reference frame. The polar angles define the orientation of a unit vector that points in the principal direction for the interaction. For the dipolar interaction, the unit vector points along the line between the two nuclei (or between the nucleus and the electron for paramagnetic relaxation). For CSA interaction with an axially symmetrical chemical-shift tensor, the unit vector is collinear with the symmetry axis of the tensor. For the quadrupolar interaction, the unit vector is collinear with the symmetry axis of the electric-field-gradient tensor. The modified spherical harmonics are given in Table 5.1 (*18*). The functions $c_0(t)$ for dipolar, CSA and quadrupolar interactions are given in Table 5.2.

The power spectral density function measures the contribution to

TABLE 5.2
Spatial Functions for Relaxation
Mechanisms

Interaction	$c(t)$
Dipolar	$\sqrt{6}\,(\mu_0/4\pi)\hbar\,\gamma_I\gamma_S r_{IS}(t)^{-3}$
CSA[a]	$\sqrt{2/3}\,(\sigma_\parallel - \sigma_\perp)\,\gamma_I B_0$
Quadrupolar[b]	$e^2qQ/[4\hbar I(2I - 1)]$

[a] The chemical-shift tensor is assumed to be axially symmetrical with principal values $\sigma_{zz} = \sigma_\parallel$ and $\sigma_{xx} = \sigma_{yy} = \sigma_\perp$.

[b] Q is the nuclear quadrupole moment and e is the charge of the electron. The electric field gradient tensor is assumed to be axially symmetric with principal value $V_{zz} = eq$, and $V_{xx} = V_{yy}$.

orientational (rotational) dynamics of the molecule from motions with frequency components in the range ω to $\omega + d\omega$. Not surprisingly, as a molecule rotates stochastically in solution as a result of Brownian motion, the oscillating magnetic fields produced are not distributed uniformly over all frequencies. A small organic molecule tumbles at a greater rate than a biological macromolecule in the same solvent, and the distribution of oscillating magnetic fields resulting from rotational diffusion of the two molecules will be different.

For a rigid spherical molecule undergoing rotational Brownian motion, $c_0(t) = c_0$ is a constant and the autospectral density function is

$$j(\omega) = d_{00}J(\omega), \qquad [5.68]$$

in which the orientational spectral density function is

$$J(\omega) = \mathrm{Re}\left\{\int_{-\infty}^{\infty} C_{00}^2(\tau)\exp\{-i\omega\tau\}\,d\tau\right\}, \qquad [5.69]$$

the orientational correlation function is

$$C_{00}^2(\tau) = \overline{Y_2^0[\Omega(t)]Y_2^0[\Omega(t + \tau)]}, \qquad [5.70]$$

and $d_{00} = c_0^2$. For *isotropic* rotational diffusion of a rigid rotor or spherical top, the correlation function is given by (*12*)

$$C_{00}^2(\tau) = \frac{1}{5}\exp[-\tau/\tau_c], \qquad [5.71]$$

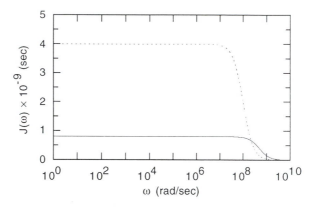

FIGURE 5.6 Spectral density functions for an isotropic rotor. Calculations were performed using [5.72] with (—) τ_c = 2 ns and ($\cdot\cdot\cdot$) τ_c = 10 ns.

in which the *correlation time*, τ_c, is approximately the average time for the molecule to rotate by one radian. The correlation time varies with molecular size, solvent viscosity, and temperature, but generally τ_c is of the order of picoseconds for small molecules and of the order of nanoseconds for biological macromolecules in aqueous solution (Section 1.4). The corresponding spectral density function is

$$J(\omega) = \frac{2}{5}\frac{\tau_c}{(1 + \omega^2\tau_c^2)}.$$ [5.72]

The functional form of the spectral density function for a rigid rotor is Lorentzian; a graph of $J(\omega)$ versus ω is shown in Fig. 5.6. The plot of $J(\omega)$ is relatively constant for $\omega^2\tau_c^2 \ll 1$ and then begins to decrease rapidly at $\omega^2\tau_c^2 \approx 1$. If molecular motion is sufficiently rapid to satisfy $\omega^2\tau_c^2 \ll 1$, the *extreme narrowing* condition obtains and $J(\omega) \approx J(0)$. For sufficiently slow molecular motion, $\omega^2\tau_c^2 \gg 1$, $J(\omega) \propto \omega^{-2}$, and the slow tumbling regime or *spin diffusion* limit is reached.

Local fields are modulated stochastically by relative motions of nuclei in a molecular reference frame as well as by overall rotational Brownian motion. Rigorously for isotropic rotational diffusion and approximately for anisotropic rotational diffusion, the total correlation function is factored as *(19)*

$$C(\tau) = C_O(\tau)\,C_I(\tau).$$ [5.73]

The correlation function for overall motion, $C_O(\tau)$, is given by [5.70] or [5.71]. The correlation function for internal motions, $C_I(\tau)$, is given by

[5.67], in which the orientational variables are defined in a fixed molecular reference frame, rather than the laboratory reference frame. Calculations of $C_1(\tau)$ have been performed for a number of diffusion and lattice jump models for internal motions. N-site lattice jump models assume that the nuclei of the relevant spins jump between N allowed conformations. The jumps are assumed to be instantaneous; therefore, the transition rates reflect the lifetimes of each conformation.

Rather than describing in detail calculations of spectral density functions for diffusion and jump models of intramolecular motions, two useful limiting cases of N-site models are given without proof (see Ref. *10* for a more extensive review). The spectral density function depends on the timescale of the variation in the spatial variables, $c_0(t)$. If the transition rates between sites approaches zero, then

$$j(\omega) = J(\omega) \sum_{\kappa=1}^{N} p_\kappa c_{0\kappa}^2 = J(\omega)\overline{c_0^2}, \qquad [5.74]$$

in which p_κ is the population and $c_{0\kappa}$ is the value of the spatial function for site κ. If the transition rates between sites approaches infinity, then

$$j(\omega) = J(\omega) \sum_{q=-2}^{q} \left| \sum_{\kappa=1}^{N} p_\kappa c_{0\kappa} Y_2^q(\Omega_k) \right|^2 = J(\omega) \sum_{m=-2}^{2} |\overline{c_0 Y_2^q(\Omega)}|^2, \quad [5.75]$$

in which Ω_k are the polar angles for site k.

An extremely useful treatment that incorporates intramolecular motions in addition to overall rotational motion is provided by the Lipari–Szabo model free formalism (*19,20*). In this treatment, the spectral density function is given by

$$j(\omega) = \frac{2}{5}\overline{c_0^2} \left[\frac{S^2 \tau_c}{1 + (\omega\tau_c)^2} + \frac{(1 - S^2)\tau}{1 + (\omega\tau)^2} \right], \qquad [5.76]$$

in which $\tau^{-1} = \tau_c^{-1} + \tau_e^{-1}$, S^2 is the square of the generalized order parameter that characterizes the amplitude of the intramolecular motion in a molecular reference frame and τ_e is the effective correlation time for internal motions. The order parameter is defined by

$$S^2 = [\overline{c_0^2}]^{-1} \sum_{q=-2}^{2} |\overline{c_0 Y_2^q(\Omega)}|^2, \qquad [5.77]$$

in which the overbar indicates an ensemble average performed over the equilibrium distribution of orientations Ω in the molecular reference frame. The order parameter satisfies the inequality, $0 \le S^2 \le 1$, in which lower values indicate larger amplitudes of internal motions. A significant advantage of the Lipari–Szabo formalism is that specification of the microscopic motional model is not required. If τ_e approaches infinity, [5.76] reduces to the same form as [5.74]; if τ_e approaches zero, [5.76] reduces to the same form as [5.75]. Equation [5.76] has been used extensively to analyze spin relaxation in proteins (21,22).

The expressions given in [5.74]–[5.76] are most commonly encountered in discussions of dipolar relaxation between two spins, I and S. Using $c_0(t)$ from Table 5.2 gives

$$j(\omega) = \zeta J(\omega)\overline{r_{IS}^{-6}}; \qquad [5.78]$$

$$j(\omega) = \zeta J(\omega) \sum_{q=-2}^{2} \left| \overline{\frac{Y_2^q(\Omega_k)}{r_{IS}^3}} \right|^2; \qquad [5.79]$$

$$j(\omega) = \frac{2}{5} \zeta \overline{r_{IS}^{-6}} \left[\frac{S^2 \tau_c}{1 + (\omega\tau_c)^2} + \frac{(1 - S^2)\tau}{1 + (\omega\tau)^2} \right]; \qquad [5.80]$$

$$S^2 = [\overline{r_{IS}^{-6}}]^{-1} \sum_{q=-2}^{2} \left| \overline{\frac{Y_2^q(\Omega)}{r_{IS}^3}} \right|^2, \qquad [5.81]$$

in which

$$\zeta = 6[(\mu_0/4\pi)\hbar\gamma_I\gamma_S]^2. \qquad [5.82]$$

Equation [5.78] (slow internal motion) is called r^{-6} averaging and [5.79] (fast internal motion) is called r^{-3} averaging with respect to the conformations of the molecule. The former equation is appropriate for treating the effects of aromatic ring flips, and the latter equation is appropriate for treating methyl group rotations in proteins (23,24).

Other expressions for $j(\omega)$ have been derived for molecules that exhibit anisotropic rotational diffusion or specific internal motional models. Equations [5.78]–[5.81] can be modified in a straightforward manner to incorporate cross-correlation effects, although the resulting expressions are often cumbersome (12).

5.4 Relaxation Mechanisms

A very large number of physical interactions give rise to stochastic Hamiltonians capable of mediating spin relaxation. In the present context, only the intramolecular magnetic dipolar, anisotropic chemical shift (CSA), quadrupolar, and scalar coupling interactions will be discussed. Intramolecular paramagnetic relaxation has the same Hamiltonian as for nuclear dipolar relaxation, except that the interaction occurs between a nucleus and an unpaired electron. Other relaxation mechanisms are of minor importance for macromolecules or are of interest only in very specialized cases. For spin $\frac{1}{2}$ nuclei in diamagnetic biological macromolecules, the dominant relaxation mechanisms are the magnetic dipolar and CSA mechanisms. For nuclei with spin $> \frac{1}{2}$, notably ^{14}N and ^{2}H in proteins, the dominant relaxation mechanism is the quadrupolar interaction.

Relaxation rate constants for nuclei in proteins depend on a large number of factors, including overall rotational correlation times, internal motions, the geometric arrangement of nuclei, and the relative strengths of the applicable relaxation mechanisms. If the overall correlation time and the three-dimensional structural coordinates of the protein are known, relaxation rate constants can be calculated in a relatively straightforward manner using expressions derived in the following sections. In general, ^{1}H relaxation in proteins is dominated by dipolar interactions with other protons (within approximately 5 Å) and by interactions with directly bonded heteronuclei. The latter arise from dipolar interactions with ^{13}C and ^{15}N in labeled proteins or from scalar relaxation of the second kind between the quadrupolar ^{14}N nuclei and amide protons. Relaxation of protonated ^{13}C and ^{15}N heteronuclei is dominated by dipolar interactions with the directly bonded protons, and secondarily by CSA (for ^{15}N spins and aromatic ^{13}C spins). Relaxation of unprotonated heteronuclei, notably carbonyl ^{13}C and unprotonated aromatic ^{13}C spins, is dominated by CSA interactions.

5.4.1 Intramolecular Dipolar Relaxation for *IS* Spin System

Any magnetic nucleus in a molecule generates an instantaneous magnetic dipolar field that is proportional to the magnetic moment of the nucleus. As the molecule tumbles in solution, this field fluctuates and constitutes a mechanism for relaxation of nearby spins. Most importantly for structure elucidation, the efficacy of dipolar relaxation depends on the nuclear moments and on the *inverse sixth power of the distance between the interacting nuclei*. As a result, nuclear spin relaxation can be used to determine distances between nuclei. Protons have a large gyromagnetic ratio; therefore, dipole–dipole interactions cause the most efficient relaxation of proton spins and constitute a sensitive probe for internuclear distances.

Initially, a two-spin system, *IS*, will be considered with $\omega_I \gg \omega_S$ and scalar coupling constant $J_{IS} = 0$. The energy levels of the spin system and the associated transition frequencies are shown in Fig. 5.7. The terms A_{2p}^{q} are given in Table 5.3. The spatial functions for the different interactions are given in Tables 5.1 and 5.2.

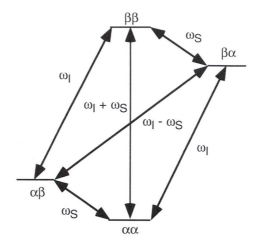

FIGURE 5.7 Transitions and associated eigenfrequencies for a two-spin system.

The relaxation rate constants are calculated using [5.56]. To aid in the calculation of the double commutators, the commutation relations given in Table 5.4 are useful. To begin, the identity operator can be disregarded because it has no effect on the relaxation equations. Next, the block structure of the relaxation matrix can be derived from the coherence orders of the operators and the secular condition. The zero-order block consists of the operators with coherence order equal to zero for both the I and S spins: I_z, S_z, and $2I_zS_z$. Each of the other operators consists of a unique combination of coherence order for the I and S spins; consequently, each of these operators constitutes a block of dimension one and each operator relaxes independently of the others.

The relaxation matrix for the zero-order block has dimensions 3×3, with individual elements, Γ_{rs}, giving the rate constant for relaxation be-

TABLE 5.3
Tensor Operators for the Dipolar Interaction

q	p	A_{2p}^{q}	$A_{2p}^{-q} = A_{2p}^{q}{}^{\dagger}$	ω_p
0	0	$(2/\sqrt{6})I_zS_z$	$(2/\sqrt{6})I_zS_z$	0
0	1	$-1/(2\sqrt{6})I^+S^-$	$-1/(2\sqrt{6})I^-S^+$	$\omega_I - \omega_S$
1	0	$-(1/2)I_zS^+$	$(1/2)I_zS^-$	ω_S
1	1	$-(1/2)I^+S_z$	$(1/2)I^-S_z$	ω_I
2	0	$(1/2)I^+S^+$	$(1/2)I^-S^-$	$\omega_I + \omega_S$

TABLE 5.4

Commutator Relationships[a]

$$[I_x, I_y] = iI_z$$
$$[I_\alpha, 2I_\beta S_\gamma] = 2[I_\alpha, I_\beta]S_\gamma$$
$$[2I_\alpha S_\gamma, 2I_\beta S_\varepsilon] = [I_\alpha, I_\beta]\delta_{\gamma\varepsilon}$$

[a] $I_\alpha = I_x, I_y$, or I_z; $S_\gamma = S_x, S_y$, or S_z. Equivalent expressions for S operators are obtained by exchanging I and S labels. $\delta_{\gamma\varepsilon}$ is the Kronecker delta.

tween operators $\mathbf{B}_r$ and $\mathbf{B}_s$ for $r; s = 1, 2,$ and 3; and $\mathbf{B}_1 = I_z$, $\mathbf{B}_2 = S_z$, and $\mathbf{B}_3 = 2I_z S_z$. The double commutators $[\mathbf{A}_{2p}^{-q}, [\mathbf{A}_{2p}^q, I_z]]$ are calculated as follows for each combination of p and q in Table 5.3:

$$[\mathbf{A}_{20}^0, [\mathbf{A}_{20}^0, I_z]] = \frac{2}{3}[I_z S_z, [I_z S_z, I_z]] = 0;$$

$$[\mathbf{A}_{21}^{-0}, [\mathbf{A}_{21}^0, I_z]] = \frac{1}{24}[I^- S^+, [I^+ S^-, I_z]] = -\frac{1}{24}[I^- S^+, I^+ S^-]$$

$$= -\frac{1}{12}\{I^- I^+ S_z - S^- S^+ I_z\}$$

$$= \frac{1}{24}\{I_z - S_z\};$$

$$[\mathbf{A}_{21}^0, [\mathbf{A}_{21}^{-0}, I_z]] = \frac{1}{24}[I^+ S^-, [I^- S^+, I_z]] = \frac{1}{24}[I^+ S^-, I^- S^+]$$

$$= -\frac{1}{12}\{I^+ I^- S_z - S^+ S^- I_z\}$$

$$= \frac{1}{24}\{I_z - S_z\};$$

$$[\mathbf{A}_{20}^{-1}, [\mathbf{A}_{20}^1, I_z]] = -\frac{1}{4}[I_z S^-, [I_z S^+, I_z]] = 0;$$

$$[\mathbf{A}_{20}^1, [\mathbf{A}_{20}^{-1}, I_z]] = -\frac{1}{4}[I_z S^+, [I_z S^-, I_z]] = 0;$$

$$[\mathbf{A}_{21}^{-1}, [\mathbf{A}_{21}^1, I_z]] = -\frac{1}{4}[I^- S_z, [I^+ S_z, I_z]] = \frac{1}{4}S_z^2[I^-, I^+] \qquad [5.83]$$

$$= -\frac{1}{2}S_z^2 I_z = -\frac{1}{8}I_z;$$

$$[\mathbf{A}_{21}^1, [\mathbf{A}_{21}^{-1}, I_z]] = -\frac{1}{4}[I^+S_z, [I^-S_z, I_z]] = -\frac{1}{24}S_z^2[I^-, I^+]$$

$$= \frac{1}{2}S_z^2 I_z = -\frac{1}{8}I_z;$$

$$[\mathbf{A}_{20}^{-2}, [\mathbf{A}_{20}^2, I_z]] = \frac{1}{4}[I^-S^-, [I^+S^+, I_z]] = -\frac{1}{4}[I^-S^-, I^+S^+]$$

$$= \frac{1}{2}\{I^+I^-S_z + S^-S^+I_z\}$$

$$= \frac{1}{4}\{S_z + I_z\};$$

$$[\mathbf{A}_{20}^2, [\mathbf{A}_{20}^{-2}, I_z]] = \frac{1}{4}[I^+S^+, [I^-S^-, I_z]] = \frac{1}{4}[I^+S^+, I^-S^-]$$

$$= \frac{1}{2}\{I^+I^-S_z + S^-S^+I_z\}$$

$$= \frac{1}{4}\{S_z + I_z\}.$$

For autorelaxation of the I_z operator, the preceding operators are premultiplied by I_z and the trace operation performed:

$$\frac{1}{24}\langle I_z|\{I_z - S_z\}\rangle = \frac{1}{24}\langle I_z^2 - I_z S_z\rangle$$

$$= \frac{1}{24}\{\langle\alpha\alpha|I_z^2 - I_z S_z|\alpha\alpha\rangle + \langle\alpha\beta|I_z^2 - I_z S_z|\alpha\beta\rangle$$

$$+ \langle\beta\alpha|I_z^2 - I_z S_z|\beta\alpha\rangle + \langle\beta\beta|I_z^2 - I_z S_z|\beta\beta\rangle\}$$

$$= \frac{1}{24};$$

$$-\frac{1}{8}\langle I_z|I_z\rangle = -\frac{1}{8}\langle I_z^2\rangle$$

$$= -\frac{1}{8}\{\langle\alpha\alpha|I_z^2|\alpha\alpha\rangle + \langle\alpha\beta|I_z^2|\alpha\beta\rangle \qquad\text{[5.84]}$$

$$+ \langle\beta\alpha|I_z^2|\beta\alpha\rangle + \langle\beta\beta|I_z^2|\beta\beta\rangle\}$$

$$= -\frac{1}{8};$$

$$\frac{1}{4}\langle I_z|\{S_z + I_z\}\rangle = \frac{1}{4}\langle I_z^2 + I_z S_z\rangle$$

$$= \frac{1}{4}\{\langle\alpha\alpha|I_z^2 + I_z S_z|\alpha\alpha\rangle + \langle\alpha\beta|I_z^2 + I_z S_z|\alpha\beta\rangle$$

$$+ \langle\beta\alpha|I_z^2 + I_z S_z|\beta\alpha\rangle + \langle\beta\beta|I_z^2 + I_z S_z|\beta\beta\rangle\}$$

$$= \frac{1}{4}.$$

For cross-relaxation between the S_z and the I_z operator, these operators are premultiplied by S_z and the trace operation performed:

$$\frac{1}{24}\langle S_z|\{I_z - S_z\}\rangle = \frac{1}{24}\langle I_z S_z - S_z^2\rangle$$

$$= \frac{1}{24}\{\langle\alpha\alpha|I_z S_z - S_z^2|\alpha\alpha\rangle + \langle\alpha\beta|I_z S_z - S_z^2|\alpha\beta\rangle$$

$$+ \langle\beta\alpha|I_z S_z - S_z^2|\beta\alpha\rangle + \langle\beta\beta|I_z S_z - S_z^2|\beta\beta\rangle\}$$

$$= -\frac{1}{24};$$

$$-\frac{1}{8}\langle S_z|I_z\rangle = -\frac{1}{8}\langle I_z S_z\rangle$$

$$= -\frac{1}{8}\{\langle\alpha\alpha|I_z S_z|\alpha\alpha\rangle + \langle\alpha\beta|I_z S_z|\alpha\beta\rangle \qquad \text{[5.85]}$$

$$+ \langle\beta\alpha|I_z S_z|\beta\alpha\rangle + \langle\beta\beta|I_z S_z|\beta\beta\rangle\}$$

$$= 0;$$

$$\frac{1}{4}\langle S_z|\{S_z + I_z\}\rangle = \frac{1}{4}\langle S_z^2 + I_z S_z\rangle$$

$$= \frac{1}{4}\{\langle\alpha\alpha|S_z^2 + I_z S_z|\alpha\alpha\rangle + \langle\alpha\beta|S_z^2 + I_z S_z|\alpha\beta\rangle$$

$$+ \langle\beta\alpha|S_z^2 + I_z S_z|\beta\alpha\rangle + \langle\beta\beta|S_z^2 + I_z S_z|\beta\beta\rangle\}$$

$$= \frac{1}{4}.$$

For cross-relaxation between the $2I_z S_z$ operator and the I_z operator, the preceding operators are premultiplied by $2I_z S_z$ and the trace operation performed:

$$\frac{1}{24}\langle 2I_zS_z|\{I_z - S_z\}\rangle = \frac{1}{12}\langle I_z^2S_z - I_zS_z^2\rangle$$

$$= \frac{1}{12}\{\langle\alpha\alpha|I_z^2S_z - I_zS_z^2|\alpha\alpha\rangle + \langle\alpha\beta|I_z^2S_z - I_zS_z^2|\alpha\beta\rangle$$

$$+ \langle\beta\alpha|I_z^2S_z - I_zS_z^2|\beta\alpha\rangle + \langle\beta\beta|I_z^2S_z - I_zS_z^2|\beta\beta\rangle\}$$

$$= 0;$$

$$-\frac{1}{8}\langle 2I_zS_z|I_z\rangle = -\frac{1}{4}\langle I_z^2S_z\rangle$$

$$= -\frac{1}{4}\{\langle\alpha\alpha|I_z^2S_z|\alpha\alpha\rangle + \langle\alpha\beta|I_z^2S_z|\alpha\beta\rangle \qquad [5.86]$$

$$+ \langle\beta\alpha|I_z^2S_z|\beta\alpha\rangle + \langle\beta\beta|I_z^2S_z|\beta\beta\rangle\}$$

$$= 0;$$

$$\frac{1}{4}\langle 2I_zS_z|\{S_z + I_z\}\rangle = \frac{1}{2}\langle I_z^2S_z + I_zS_z^2\rangle$$

$$= \frac{1}{2}\{\langle\alpha\alpha|I_z^2S_z + I_zS_z^2|\alpha\alpha\rangle + \langle\alpha\beta|I_z^2S_z + I_zS_z^2|\alpha\beta\rangle$$

$$+ \langle\beta\alpha|I_z^2S_z + I_zS_z^2|\beta\alpha\rangle + \langle\beta\beta|I_z^2S_z + I_zS_z^2|\beta\beta\rangle\}$$

$$= 0.$$

Autorelaxation and cross-relaxation of the S_z operator can be obtained by exchanging I and S operators in the preceding expressions. Substituting the values of the trace operations above into [5.56] (and using $\langle I_z|I_z\rangle = 1$) yields

$$\Gamma_{11} = \frac{1}{24}\{j(\omega_I - \omega_S) + 3j(\omega_I) + 6j(\omega_I + \omega_S)\};$$

$$\Gamma_{22} = \frac{1}{24}\{j(\omega_I - \omega_S) + 3j(\omega_S) + 6j(\omega_I + \omega_S)\};$$

$$\Gamma_{12} = \frac{1}{24}\{-j(\omega_I - \omega_S) + 6j(\omega_I + \omega_S)\}; \qquad [5.87]$$

$$\Gamma_{13} = 0;$$

$$\Gamma_{23} = 0.$$

If the I and S spins are separated by a constant distance, r_{IS}, then

$$\Gamma_{11} = (d_{00}/4)\{J(\omega_I - \omega_S) + 3J(\omega_I) + 6J(\omega_I + \omega_S)\};$$

$$\Gamma_{22} = (d_{00}/4)\{J(\omega_I - \omega_S) + 3J(\omega_S) + 6J(\omega_I + \omega_S)\}; \qquad [5.88]$$

$$\Gamma_{12} = (d_{00}/4)\{-J(\omega_I - \omega_S) + 6J(\omega_I + \omega_S)\},$$

in which

$$d_{00} = (\mu_0/4\pi)^2 \hbar^2 \gamma_I^2 \gamma_S^2 r_{IS}^{-6}. \qquad [5.89]$$

Dipolar cross-relaxation between the operators $2I_zS_z$ and I_z does not occur; therefore, the $2I_zS_z$ operator relaxes independently of the I_z and S_z operators. This result can be anticipated using symmetry and group theoretical arguments beyond the scope of this text (9,12,14). Cross-relaxation between these operators does arise as a result of interference between dipolar and CSA relaxation mechanisms (11).

Thus, the evolution of the longitudinal operators, I_z and S_z, is governed by

$$d(\langle I_z\rangle(t) - \langle I_z^0\rangle)/dt = -\Gamma_{11}(\langle I_z\rangle(t) - \langle I_z^0\rangle) - \Gamma_{12}(\langle S_z\rangle(t) - \langle S_z^0\rangle);$$
$$\qquad\qquad [5.90]$$
$$d(\langle S_z\rangle(t) - \langle S_z^0\rangle)/dt = -\Gamma_{22}(\langle S_z\rangle(t) - \langle S_z^0\rangle) - \Gamma_{12}(\langle I_z\rangle(t) - \langle I_z^0\rangle).$$

Making the identification $\Gamma_{11} = \rho_I (= R_{1I})$, $\Gamma_{22} = \rho_S (= R_{1S})$, and $\Gamma_{12} = \sigma_{IS}$ puts [5.90] into the form of the Solomon equations [5.9] in which ρ_I and ρ_S are the autorelaxation rate constants and σ_{IS} is the cross-relaxation rate constant. The Solomon transition rate constants (Section 5.1.2) are

$$W_0 = j(\omega_I - \omega_S)/24;$$

$$W_I = j(\omega_I)/16;$$
$$\qquad\qquad [5.91]$$
$$W_S = j(\omega_S)/16;$$

$$W_2 = j(\omega_I + \omega_S)/4.$$

Now consider the relaxation of the transverse I^+ operator; as a consequence of the secular approximation, this operator is immediately seen to relax independently of all other operators except, potentially, for $2I^+S_z$. The double commutators $[A_{2p}^{-q}, [A_{2p}^q, I^+]]$ are calculated as follows for each combination of p and q in Table 5.3:

$$[A_{20}^0, [A_{20}^0, I^+]] = \frac{2}{3}[I_zS_z, [I_zS_z, I^+]] = \frac{2}{3}I^+S_z^2 = \frac{1}{6}I^+;$$

$$[A_{21}^{-0}, [A_{21}^0, I^+]] = \frac{1}{24}[I^-S^+, [I^+S^-, I^+]] = 0;$$

$$[\mathbf{A}_{21}^0, [\mathbf{A}_{21}^{-0}, I^+]] = -\frac{1}{24}[I^+S^-, [I^-S^+, I^+]] = -\frac{1}{12}[I^+S^-, I_zS^+]$$

$$= -\frac{1}{6}I^+I_zS_z + \frac{1}{12}I^+S^+S^- = \frac{1}{24}I^+; \qquad [5.92]$$

$$[\mathbf{A}_{20}^{-1}, [\mathbf{A}_{20}^1, I^+]] = -\frac{1}{4}[I_zS^-, [I_zS^+, I^+]] = \frac{1}{8}[I_zS^-, I^+S^+] = -\frac{1}{8}I^+;$$

$$[\mathbf{A}_{20}^1, [\mathbf{A}_{20}^{-1}, I^+]] = -\frac{1}{4}[I_zS^+, [I_zS^-, I^+]] = -\frac{1}{8}[I_zS^+, I^+S^-] = -\frac{1}{8}I^+;$$

$$[\mathbf{A}_{21}^{-1}, [\mathbf{A}_{21}^1, I^+]] = -\frac{1}{4}[I^-S_z, [I^+S_z, I^+]] = 0;$$

$$[\mathbf{A}_{21}^1, [\mathbf{A}_{21}^{-1}, I^+]] = -\frac{1}{4}[I^+S_z, [I^-S_z, I^+]] = \frac{1}{4}S_z^2[I^+, I_z] = -\frac{1}{8}I^+;$$

$$[\mathbf{A}_{20}^{-2}, [\mathbf{A}_{20}^2, I^+]] = \frac{1}{4}[I^-S^-, [I^+S^+, I^+]] = 0;$$

$$[\mathbf{A}_{20}^2, [\mathbf{A}_{20}^{-2}, I^+]] = \frac{1}{4}[I^+S^+, [I^-S^-, I^+]] = -\frac{1}{4}[I^+S^+, I_zS^-] = \frac{1}{4}I^+.$$

Note that all nonzero results are proportional to I^+; therefore, since the operator basis is orthogonal, no operator cross-relaxes with I^+. For autorelaxation of the I^+ operator, the preceding operators are premultiplied by I^+ and the trace operation performed:

$$\langle I^+ | I^+ \rangle = \langle I^- I^+ \rangle =$$

$$= \{\langle \alpha\alpha | I^- I^+ | \alpha\alpha \rangle + \langle \alpha\beta | I^- I^+ | \alpha\beta \rangle + \langle \beta\alpha | I^- I^+ | \beta\alpha \rangle + \langle \beta\beta | I^- I^+ | \beta\beta \rangle\}$$

$$= 2. \qquad [5.93]$$

This same factor is the normalization in the denominator of [5.56]. Thus

$$R_{2I} = \frac{1}{48}\{4j(0) + j(\omega_I - \omega_S) + 3j(\omega_I) + 6j(\omega_S) + 6j(\omega_I + \omega_S)\}, \qquad [5.94]$$

and

$$d\langle I^+ \rangle / dt = -i\omega_I \langle I^+ \rangle - R_{2I} \langle I^+ \rangle. \qquad [5.95]$$

If r_{IS} is constant, then

$$R_{2I} = (d_{00}/8)\{4J(0) + J(\omega_I - \omega_S) + 3J(\omega_I) + 6J(\omega_S) + 6J(\omega_I + \omega_S)\}. \qquad [5.96]$$

TABLE 5.5

Relaxation Rate Constants for IS Dipolar Interaction

Coherence level	Operator[a]	Relaxation rate constant[b]
Populations	I_z	$(d_{00}/4)\{J(\omega_I - \omega_S) + 3J(\omega_I) + 6J(\omega_I + \omega_S)\}$
	S_z	$(d_{00}/4)\{J(\omega_I - \omega_S) + 3J(\omega_S) + 6J(\omega_I + \omega_S)\}$
	$I_z \leftrightarrow S_z$	$(d_{00}/4)\{-J(\omega_I - \omega_S) + 6J(\omega_I + \omega_S)\}$
0	$2I_zS_z$	$(3d_{00}/4)\{J(\omega_I) + J(\omega_S)\}$
	ZQ_x, ZQ_y	$(d_{00}/8)\{2J(\omega_I - \omega_S) + 3J(\omega_I) + 3J(\omega_S)\}$
± 1	I^+, I^-	$(d_{00}/8)\{4J(0) + J(\omega_I - \omega_S) + 3J(\omega_I) + 6J(\omega_S) + 6J(\omega_I + \omega_S)\}$
	S^+, S^-	$(d_{00}/8)\{4J(0) + J(\omega_I - \omega_S) + 3J(\omega_S) + 6J(\omega_I) + 6J(\omega_I + \omega_S)\}$
	$2I^+S_z, 2I^-S_z$	$(d_{00}/8)\{4J(0) + J(\omega_I - \omega_S) + 3J(\omega_I) + 6J(\omega_I + \omega_S)\}$
	$2I_zS^+, 2I_zS^-$	$(d_{00}/8)\{4J(0) + J(\omega_I - \omega_S) + 3J(\omega_S) + 6J(\omega_I + \omega_S)\}$
± 2	DQ_x, DQ_y	$(d_{00}/8)\{3J(\omega_I) + 3J(\omega_S) + 12J(\omega_I + \omega_S)\}$

[a] Cross-relaxation occurs only between I_z and S_z.
[b] $d_{00} = (\mu_0/4\pi)^2\hbar^2\gamma_I^2\gamma_S^2 r_{IS}^{-6}$.

Analogous equations can be written by inspection for the I^-, S^+, and S^- operators. The complete set of dipolar relaxation rate constants for the basis operators for the two spin system is given in Table 5.5.

The dependence of R_1 and R_2 on τ_c for a rigid molecule is illustrated in Fig. 5.8. Notice that R_1 has a maximum for $\omega_0\tau_c = 1$ whereas R_2 increases monotonically with τ_c.

5.4.2 INTRAMOLECULAR DIPOLAR RELAXATION FOR SCALAR COUPLED IS SPIN SYSTEM

The I_z and S_z operators both commute with the scalar coupling Hamiltonian; consequently, dipolar spin–lattice relaxation is unaffected by the scalar coupling interaction and the expressions given in [5.87] and [5.90] remain valid. The in-phase and antiphase transverse operators, I^+ and $2I^+S_z$ are coupled together by the scalar coupling Hamiltonian. Applying [5.54] yields the following equations:

$$d\langle I^+\rangle(t)/dt = -i\omega_I\langle I^+\rangle(t) - i\pi J_{IS}\langle 2I^+S_z\rangle(t) - R_{2I}\langle I^+\rangle(t);$$
$$d\langle 2I^+S_z\rangle(t)/dt = -i\omega_I\langle 2I^+S_z\rangle(t) - i\pi J_{IS}\langle I^+\rangle(t) - R_{IS}\langle 2I^+S_z\rangle(t),$$

[5.97]

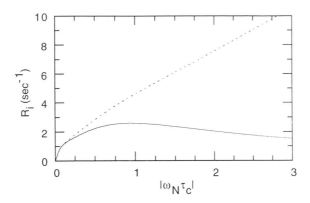

FIGURE 5.8 Relaxation rate constants for ^{1}H–^{15}N dipolar spin system. (—) ^{15}N R_1 spin–lattice rate constants. ($\cdots$) ^{15}N R_2 spin–spin rate constants. Calculations were performed using expressions given in Table 5.5 together with [5.89] and [5.72]. Parameters used were $B_0 = 11.74$ T, $\gamma_I = 2.675 \times 10^8$ T^{-1} s^{-1} (^{1}H), $\gamma_S = -2.712 \times 10^7$ T^{-1} s^{-1} (^{15}N), and $r_{IS} = 1.02$ Å.

in which R_{2I} and R_{IS} are given in Table 5.5. These equations are written in matrix form as

$$
\begin{bmatrix} \langle I^+ \rangle(t) \\ \langle 2I^+ S_z \rangle(t) \end{bmatrix} = - \begin{bmatrix} i\omega_I + R_{2I} & i\pi J_{IS} \\ i\pi J_{IS} & i\omega_I + R_{2IS} \end{bmatrix} \begin{bmatrix} \langle I^+ \rangle(t) \\ \langle 2I^+ S_z \rangle(t) \end{bmatrix}, \qquad [5.98]
$$

and are solved by analogy to [5.13] to yield

$$
\langle I^+ \rangle(t) = \frac{1}{2} \left[\left(1 - \frac{R_{2I} - R_{2IS}}{(\lambda_+ - \lambda_-)} \right) \exp(-\lambda_- t) \right.
$$

$$
\left. + \left(1 + \frac{R_{2I} - R_{2IS}}{(\lambda_+ - \lambda_-)} \right) \exp(-\lambda_+ t) \right] \langle I^+ \rangle(0)
$$

$$
- \frac{i\pi J_{IS}}{(\lambda_+ - \lambda_-)} [\exp(-\lambda_- t) - \exp(-\lambda_+ t)] \langle 2I^+ S_z \rangle(0); \qquad [5.99]
$$

$$
\langle 2I^+ S_z \rangle(t) = \frac{1}{2} \left[\left(1 + \frac{R_{2I} - R_{2IS}}{(\lambda_+ - \lambda_-)} \right) \exp(-\lambda_- t) \right.
$$

$$
\left. + \left(1 - \frac{R_{2I} - R_{2IS}}{(\lambda_+ - \lambda_-)} \right) \exp(-\lambda_+ t) \right] \langle 2I^+ S_z \rangle(0)
$$

$$
- \frac{i\pi J_{IS}}{(\lambda_+ - \lambda_-)} [\exp(-\lambda_- t) - \exp(-\lambda_+ t)] \langle I^+ \rangle(0),
$$

in which

$$\lambda_\pm = \{i\omega + (R_{2I} + R_{2IS})/2 \pm [((R_{2I} - R_{2IS})/2)^2 - (\pi J_{IS})^2]^{1/2}\}. \quad [5.100]$$

If $(2\pi J_{IS})^2 \gg (R_{2I} - R_{2IS})^2$, then

$$\langle I^+\rangle(t) = \frac{1}{2}[\exp\{-(i\omega - i\pi J_{IS} + R_{\mathrm{av}})t\}$$

$$+ \exp\{-(i\omega + i\pi J_{IS} + R_{\mathrm{av}})t\}]\langle I^+\rangle(0)$$

$$-[\exp\{-(i\omega - i\pi J_{IS} + R_{\mathrm{av}})t\}$$

$$-\exp\{-(i\omega + i\pi J_{IS} + R_{\mathrm{av}})t\}]\langle 2I^+S_z\rangle(0); \quad [5.101]$$

$$\langle 2I^+S_z\rangle(t) = \frac{1}{2}[\exp\{-(i\omega - i\pi J_{IS} + R_{\mathrm{av}})t\}$$

$$+\exp\{-(i\omega + i\pi J_{IS} + R_{\mathrm{av}})t\}]\langle 2I^+S_z\rangle(0)$$

$$-[\exp\{-(i\omega - i\pi J_{IS} + R_{\mathrm{av}})t\}$$

$$-\exp\{-(i\omega + i\pi J_{IS} + R_{\mathrm{av}})t\}]\langle I^+\rangle(0),$$

with

$$R_{\mathrm{av}} = (R_{2I} + R_{2IS})/2$$

$$= (1/48)\{4j(0) + j(\omega_I - \omega_S) + 3j(\omega_I) + 3j(\omega_S) + 6j(\omega_I + \omega_S)\}. \quad [5.102]$$

Equation [5.101] predicts that the signal arising from I^+ has the expected form of a doublet with linewidth R_{av}/π. The doublet is in-phase if $\langle 2I^+S_z\rangle(0) = 0$ and antiphase if $\langle I^+\rangle(0) = 0$. Evolution of the scalar coupling interaction on a faster time scale than the relaxation processes averages the two relaxation rate constants because coherence is rapidly exchanged between the I^+ and $2I^+S_z$ operators.

For the purely dipolar IS interaction in the spin diffusion limit,

$$R_{2I} - R_{2IS} = 3d_{00}J(\omega_S)/8 = \frac{3\mu_0^2\hbar^2\gamma_I^2}{320\,\pi^2B_0^2r_{IS}^2\tau_c} \quad [5.103]$$

normally is quite small. For example, if $I = {}^{15}N$, $S = {}^{1}H$, and $\tau_c = 5$ ns, then $R_{2I} - R_{2IS} = 0.016$ s^{-1}, compared with $J_{IS} = 92$ Hz. However, the S_z operator may have relaxation pathways other than the IS dipolar interaction. In the cited example, the proton S_z operator would be dipolar coupled to other protons, and the relaxation rate constant for the $2I^+S_z$ operator contains a contribution, R_{ext}, from proton dipolar longitudinal relaxation. Ignoring cross-correlation and cross-relaxation effects, R_{ext} is simply additive to R_{2IS}. The additional contribution from R_{ext} has two important effects. First, R_{av} is increased by $R_{ext}/2$, as seen from [5.102]. Practical consequences of the increased linewidth in heteronuclear NMR spectra are discussed in Section 7.1.2.4. Second, if R_{ext} is sufficiently large, then $(R_{2I} - R_{2IS} - R_{ext})^2 \approx R_{ext}^2 \gg (2\pi J_{IS})^2$, $\lambda_+ = i\omega + R_{2I}$, $\lambda_- = i\omega + R_{2IS} + R_{ext}$, and [5.99] reduces to

$$\langle I^+ \rangle(t) = \langle I^+ \rangle(0)\exp[-(i\omega + R_{2I})t];$$

$$\langle 2I^+ S_z \rangle(t) = \langle 2I^+ S_z \rangle(0)\exp[-(i\omega + R_{2IS} + R_{ext})t].$$

[5.104]

The expected doublet has been reduced to a singlet resonance in a process commonly called *self-decoupling*, which is similar both to scalar relaxation of the second kind (Section 5.4.5) and chemical exchange (Section 5.6.2). For $(R_{2I} - R_{2IS} - R_{ext})^2 \approx (2\pi J_{IS})^2$, the doublet is partially decoupled and broadened as for intermediate chemical exchange (Section 5.6.1). Self-decoupling can complicate the measurement of scalar coupling constants (Sections 6.2.1.5, 6.3.3, 7.5) (25).

A similar set of equations can be obtained for the S^+ and $2S^+ I_z$ coherences by interchanging the I and S labels. Notice that for an uncoupled IS spin system, $R_{2I} \neq R_{2S}$, but for a scalar-coupled spin system undergoing free precession, R_{av} is identical for the I and S spins.

5.4.3 INTRAMOLECULAR DIPOLAR RELAXATION FOR *IS* SPIN SYSTEM IN THE ROTATING FRAME

An *IS* homonuclear spin system, in which the two spins interact through the dipolar interaction but are not scalar-coupled, will be treated. The spin-lock field is assumed to be applied with x phase. The autorelaxation rate constant of the I'_z operator and the cross-relaxation rate constant between the I'_z and S'_z operators will be calculated in the tilted rotating frame. As discussed in Section 5.2.3, in the presence of the spin-lock field the I and S spins are degenerate and are treated as like spins; thus, the components of the dipolar interaction listed in Table 5.3 must be redefined according to [5.61] as

$$\mathbf{A}_2^0 = \mathbf{A}_{20}^0 + \mathbf{A}_{21}^{-0} + \mathbf{A}_{21}^0;$$

$$\mathbf{A}_2^{\pm 1} = \mathbf{A}_{20}^{\pm 1} + \mathbf{A}_{21}^{\pm 1};$$

[5.105]

$$\mathbf{A}_2^{\pm 2} = \mathbf{A}_{20}^{\pm 2}.$$

From [5.63],

$$I'_z = \sin \theta_I I_x + \cos \theta_I I_z;$$

$$S'_z = \sin \theta_S S_x + \cos \theta_S S_z. \qquad [5.106]$$

Applying [5.62], the double commutators $[A_2^{-q}, [A_2^q, I'_z]]$ are calculated first. Straightforward, but tedious, calculations yield

$$[A_2^0, [A_2^0, I'_z]] = \sin \theta_I (5 I_x + 4 S_x)/24 + \cos \theta_I (I_z - S_z)/6$$

$$[A_2^{-1}, [A_2^1, I'_z]] = [A_2^1, [A_2^{-1}, I'_z]]$$

$$= -\sin \theta_I (2 I_x + 2 S_x + 2 I^-)/8 - \cos \theta_I I_z/8; \quad [5.107]$$

$$[A_2^{-2}, [A_2^2, I'_z]] = [A_2^2, [A_2^{-2}, I'_z]]$$

$$= -\sin \theta_I I^-/8 - \cos \theta_I (I_z + S_z)/8.$$

The autorelaxation rate constant is determined by premultiplying the preceding expressions by I'_z and forming the trace:

$$\langle \sin \theta_I I_x + \cos \theta_I I_z | \sin \theta_I (5 I_x + 4 S_x)/24 + \cos \theta_I (I_z - S_z)/6 \rangle$$

$$= \left(\frac{5}{24} \right) \sin^2 \theta_I + \left(\frac{1}{12} \right) \cos^2 \theta_I;$$

$$\langle \sin \theta_I I_x + \cos \theta_I I_z | -\sin \theta_I (2 I_x + 2 S_x + 2 I^-)/8 - \cos \theta_I I_z/8 \rangle \quad [5.108]$$

$$= -\left(\frac{3}{16} \right) \sin^2 \theta_I - \left(\frac{1}{8} \right) \cos^2 \theta_I;$$

$$\langle \sin \theta_I I_x + \cos \theta_I I_z | -\sin \theta_I I^-/8 - \cos \theta_I (I_z + S_z)/8 \rangle$$

$$= \frac{1}{8} \sin^2 \theta_I + \frac{1}{4} \cos^2 \theta_I.$$

Thus, the autorelaxation rate, $R_1(\theta_I)$ (which commonly is called $R_{1\rho}$), is given by

$$R_1(\theta_I) = \frac{1}{48} \{ (2 \cos^2 \theta_I + 5 \sin^2 \theta_I) j(0) + (6 \cos^2 \theta_I + 9 \sin^2 \theta_I) J(\omega_0)$$

$$+ (12 \cos^2 \theta_I + 6 \sin^2 \theta_I) j(2 \omega_0) \} \qquad [5.109]$$

$$= R_{1I} \cos^2 \theta_I + R_{2I} \sin^2 \theta_I.$$

Similarly, the cross-relaxation rate constant is found by premultiplying the expressions in [5.107] by S'_z and forming the trace:

$$\langle \sin \theta_S S_x + \cos \theta_S S_z | \sin \theta_I (5 I_x + 4 S_x)/24 + \cos \theta_I (I_z - S_z)/6 \rangle$$

$$= \frac{1}{6} \sin \theta_S \sin \theta_I - \left(\frac{1}{12} \right) \cos \theta_S \cos \theta_I;$$

$$\langle \sin \theta_S\, S_x + \cos \theta_S\, S_z | -\sin \theta_I (2I_x + 2S_x + 2I^-)/8 - \cos \theta_I I_z/8 \rangle$$

$$= -\frac{1}{8} \sin \theta_S \sin \theta_I\,;$$

$$\langle \sin \theta_S\, S_x + \cos \theta_S\, S_z | -\sin \theta_I I^-/8 - \cos \theta_I (I_z + S_z)/8 \rangle \qquad [5.110]$$

$$= \frac{1}{4} \cos \theta_S \cos \theta_I\,.$$

Thus, the cross-relaxation rate, $R_{IS}(\theta_I, \theta_S)$ is given by

$$R_{IS}(\theta_I, \theta_S) = \frac{1}{24} \{ (-\cos \theta_S \cos \theta_I + 2 \sin \theta_S \sin \theta_I)\, j(0)$$

$$+ 3 \sin \theta_S \sin \theta_I\, j(\omega_0) + 6 \cos \theta_S \cos \theta_I\, j(2\omega_0) \} \quad [5.111]$$

$$= \cos \theta_I \cos \theta_S\, \sigma_{IS}^{\mathrm{NOE}} + \sin \theta_I \sin \theta_S\, \sigma_{IS}^{\mathrm{ROE}}\,,$$

in which pure laboratory-frame cross-relaxation rate constant, $\sigma_{IS}^{\mathrm{NOE}}$, is given in [5.87] and the pure rotating-frame cross-relaxation rate constant is given by (26):

$$\sigma_{IS}^{\mathrm{ROE}} = \left(\frac{1}{24}\right) \{ 2j(0) + 3j(\omega_0) \}\,. \qquad [5.112]$$

For both autorelaxation and cross-relaxation, the effect of the tilted field is to average laboratory (longitudinal) and rotating-frame (transverse) relaxation rate constants by the projection of the spin operators onto the tilted reference frame.

5.4.4 CHEMICAL-SHIFT ANISOTROPY AND QUADRUPOLAR RELAXATION

Chemical shifts are reflections of the electronic environments that modify the local magnetic fields experienced by different nuclei (Section 1.5). These local fields are anisotropic; consequently, the components of the local fields in the laboratory reference frame vary as the molecule reorients as a result of molecular motion. These varying magnetic fields are a source of relaxation. Very approximately, the maximum CSA for a particular nucleus is of the order of the chemical-shift range for the nucleus. CSA is important as a relaxation mechanism only for nuclei with a large chemical-shift range. In the NMR spectroscopy of biological molecules, ^{13}C, ^{15}N, and ^{31}P nuclei have significant CSA contributions to relaxation. CSA is generally a negligible effect for proton relaxation. CSA rate constants have a quadratic dependence on the applied magnetic field strength. Thus, use of higher magnetic field strengths does not always increase the achievable signal-to-noise ratio as much as expected theoretically (Section 3.3.3), because increased CSA relaxation broadens the resonance linewidths.

Nuclei with $I > \frac{1}{2}$ also possess nuclear electric quadrupole moments. The quadrupole moment is a characteristic of the particular nucleus and represents a departure of the nuclear charge distribution from spherical symmetry. The interaction of the quadrupole moment with local oscillating electric field gradients (due to electrons) provide a relaxation mechanism. Quadrupolar interactions can be very large and efficient for promoting relaxation. Quadrupolar nuclei display broad resonance lines in NMR spectra, unless the nuclei are in highly symmetric electronic environment (which reduces the magnitudes of the electric field gradients at the locations of the nuclei). As discussed in more detail elsewhere, Bloch spin–lattice and spin–spin relaxation rate constants can only be defined for quadrupolar nuclei under extreme narrowing conditions or for quadrupolar nuclei with $I = 1$ (2).

The terms $\mathbf{A}_{2p}^q$ for the CSA and quadrupolar interactions are given in Table 5.6. The spherical harmonic and spatial functions for the different interactions are given in Tables 5.1 and 5.2. Relaxation rate constants are calculated for a single spin I by using the basis operators, I_z, I^-, I^+. Spin–lattice and spin–spin relaxation rate constants for the CSA and quadrupolar interactions are calculated by the same procedure as for the dipolar interactions and are given in Table 5.7. The results are calculated for axially symmetrical chemical-shift and electric field gradient tensors (i.e., $\sigma_{xx} = \sigma_{yy} \neq \sigma_{zz}$ and $V_{xx} = V_{yy} \neq V_{zz}$). Extensions to these results for anisotropic tensors are given elsewhere (2).

5.4.5 SCALAR RELAXATION

As discussed in Sections 1.6 and 2.5.2, the isotropic scalar coupling Hamiltonian, $\mathcal{H}_J = 2\pi J_{IS}\, \mathbf{I} \cdot \mathbf{S}$, slightly perturbs the Zeeman energy levels of the coupled spins; the resonances thereby are split into characteristic multiplet patterns. Spin I experiences a local magnetic field that depends on the value of the coupling constant and the state of spin S. The local magnetic field becomes time-dependent if the value of J_{IS} is time-dependent or if state of the S spin varies rapidly. The former relaxation mechanism is termed *scalar relaxation of the first kind*; the latter mechanism is termed *scalar relaxation of the second kind*. Scalar relaxation of the first kind results from transitions of the spin system between environments with different values of J_{IS}. For example, the three-bond scalar coupling constant for a pair of protons depends on the intervening dihedral angle according to the Karplus relationship [8.2]. If the dihedral angle is time-

TABLE 5.6
Tensor Operators for the CSA and Quadrupolar Interactions

q	p	CSA		Quadrupolar		ω_p
		A_{2p}^q	$A_{2p}^{-q} = A_{2p}^{q\dagger}$	A_{2p}^q	$A_{2p}^{-q} = A_{2p}^{q\dagger}$	
0	0	$(2/\sqrt{6})I_z$	$(2/\sqrt{6})I_z$	$(1/2\sqrt{6})[4I_z^2 - I^+I^- - I^-I^+]$	$(1/2\sqrt{6})[4I_z^2 - I^+I^- - I^-I^+]$	0
1	0	$-(1/2)I^+$	$(1/2)I^-$	$-(1/2)(I_zI^+ + I^+I_z)$	$-(1/2)(I_zI^- + I^-I_z)$	ω_I
2	0	—	—	$(1/2)I^+I^+$	$(1/2)I^-I^-$	$2\omega_I$

TABLE 5.7

CSA and Quadrupolar Relaxation Rate Constants

Rate Constant	CSA[a]	Quadrupolar[b]
R_1	$d_{00} J(\omega_I)$	$3d_{00}\{J(\omega_I) + 2J(2\omega_I)\}$
R_2	$(d_{00}/6)\{4J(0) + 3J(\omega_I)\}$	$(3d_{00}/2)\{3J(0) + 5J(\omega_I) + 2J(2\omega_I)\}$

[a] $d_{00} = (\sigma_\| - \sigma_\perp)^2 \gamma_I^2 B_0^2/3 = (\sigma_\| - \sigma_\perp)^2 \omega_I^2/3$.
[b] A spin 1 quadrupolar nucleus is assumed; $d_{00} = [e^2 qQ/(4\hbar)]^2$.

dependent, the consequent time dependence of J_{IS} can lead to scalar relaxation. Scalar relaxation of the second kind results if the S spin relaxes rapidly (e.g., S is a quadrupolar nucleus) or is involved in rapid chemical exchange. Scalar relaxation of the second kind also can be significant if the S spin is a proton nucleus in a macromolecule, in which case the homonuclear relaxation rate constants (reflecting the dipolar interaction of the S spin with protons other than the I spin) can be large. Normally field fluctuations produced by this mechanism are not fast enough for effective longitudinal relaxation, but transverse relaxation may be induced.

In contrast to earlier sections, the relaxation rate constants for scalar relaxation will not be explicitly calculated; instead, the appropriate expressions for R_1^{sc} and R_2^{sc} are given by (2)

$$R_1^{sc} = \frac{2A^2}{3} S(S+1) \frac{\tau_2}{1 + (\omega_I - \omega_S)^2 \tau_2^2};$$

$$R_2^{sc} = \frac{A^2}{3} S(S+1) \left[\frac{\tau_2}{1 + (\omega_I - \omega_S)^2 \tau_2^2} + \tau_1 \right].$$

[5.113]

For scalar relaxation of the first kind, $A = 2\pi(p_1 p_2)^{1/2}(J_1 - J_2)$, in which J_1 and J_2 are the scalar coupling constants, p_1 and p_2 are the relative populations ($p_1 + p_2 = 1$), and $\tau_1 = \tau_2 = \tau_e$ are the exchange time constants for the two environments. For scalar relaxation of the second kind, $A = 2\pi J_{IS}$; τ_1 and τ_2 are the spin–lattice and spin–spin relaxation time constants for the S spin, respectively. If the S spin is a quadrupolar nucleus, the relaxation time constants can be calculated using the expressions given in Table 5.7.

5.5 Nuclear Overhauser Effect

By far the most important manifestation of the prediction [5.90] that dipolar-coupled spins do not relax independently is the nuclear Overhauser effect (NOE). The Solomon equations [5.9] are extremely useful for explication of NOE experiments. The NOE is characterized by the cross relaxation rate constant, σ_{IS}^{NOE}, defined by [5.87], or the steady-state *NOE enhancement*, η_{IS}, which is defined below. These two quantities naturally arise in transient or steady-state NOE experiments, respectively. The NOE is without doubt one of the most important effects in NMR spectroscopy; more detailed discussions can be found in monographs devoted to the subject (*27,28*).

The steady-state NOE experiment will be illustrated by using a two spin system as an example. If the S spin is irradiated by a weak rf field (so as not to perturb the I spin) for a lengthy period of time $t \gg 1/\rho_S$, $1/\rho_I$, then the populations across the S spin transitions are equalized and the I spin magnetization evolves to a steady-state value, $\langle I_z^{ss} \rangle$. In this situation, the S spins are said to be saturated. Setting $d\Delta I_z(t)/dt = 0$ and $\langle S_z \rangle(t) = 0$ in [5.9] and solving for $\langle I_z^{ss} \rangle / \langle I_z^0 \rangle$ yields

$$\frac{d\langle I_z^{ss} \rangle}{dt} = -\rho_I(\langle I_z^{ss} \rangle - \langle I_z^0 \rangle) + \sigma_{IS}^{NOE}\langle S_z^0 \rangle = 0;$$

$$\langle I_z^{ss} \rangle / \langle I_z^0 \rangle = 1 + \sigma_{IS}^{NOE}\langle S_z^0 \rangle / (\rho_I \langle I_z^0 \rangle); \qquad [5.114]$$

$$\langle I_z^{ss} \rangle / \langle I_z^0 \rangle = 1 + \frac{\sigma_{IS}^{NOE}\gamma_S}{\rho_I \gamma_I} = 1 + \eta_{IS},$$

in which

$$\eta_{IS} \equiv \frac{\sigma_{IS}^{NOE}\gamma_S}{\rho_I \gamma_I}. \qquad [5.115]$$

As shown, the value of the longitudinal magnetization (or population difference) for the I spin is altered by saturating (equalizing the population difference) the S spin. If η_{IS} is positive, the population differences across the I spin transitions are increased by reducing the population differences across the S spin transitions. Since the equilibrium population differences are inversely proportional to temperature [1.7], this result appears to indicate that heating the S spins (reducing the population difference) has the effect of cooling the I spins (increasing the population difference). This conclusion would appear to violate the Second Law of Thermodynamics; however, if coupling between the spin system and the lattice is properly taken into account, then no inconsistency with thermodynamics exists.

The value of the NOE enhancement, η_{IS}, can be measured using the *steady-state NOE difference experiment*. In this experiment, two spectra are recorded. In the first spectrum, the S spin is saturated for a period of time sufficient to establish the NOE enhancement of the I spin, a 90° pulse is applied to the system, and the FID recorded. The intensity of the I spin resonance in the spectrum is proportional to $\langle I_z^{ss} \rangle$. In the second experiment, the S spin is not saturated. Instead, a 90° pulse is applied to the system at equilibrium and the FID is recorded. The intensity of the I spin resonance in this spectrum is proportional to $\langle I_z^0 \rangle$. The value of η_{IS} can then be calculated from [5.115]. In practice, the steady-state NOE difference experiment is performed somewhat differently than described in order to maximize the accuracy of the results; such complications are not relevant to the present discussion (28).

Measurements of σ_{IS}^{NOE} can be made by use of the one-dimensional *transient NOE experiment*, discussed in Section 5.1.2 or the two-dimensional *NOESY experiment* (Section 6.6.1). These laboratory-frame relaxation transient NOE experiments have rotating frame analogs; the *transient ROE experiment* and the two dimensional *ROESY experiment* (Section 6.6.2) in which the rotating-frame cross relaxation rate constant, σ_{IS}^{ROE}, is given by [5.112].

Using the isotropic rotor spectral density function [5.72], the cross-relaxation rate constants, for a homonuclear spin system ($\gamma_I = \gamma_S = \gamma$) are given by

$$\sigma_{IS}^{NOE} = \frac{\hbar^2 \mu_0^2 \gamma^4 \tau_c}{40\pi^2 r_{IS}^6} \left\{ -1 + \frac{6}{1 + 4\omega_0^2 \tau_c^2} \right\};$$

[5.116]

$$\sigma_{IS}^{ROE} = \frac{\hbar^2 \mu_0^2 \gamma^4 \tau_c}{40\pi^2 r_{IS}^6} \left\{ 2 + \frac{3}{1 + \omega_0^2 \tau_c^2} \right\},$$

and the NOE enhancement is given by

$$\eta_{IS} = \left\{ -1 + \frac{6}{1 + 4\omega_0^2 \tau_c^2} \right\} \bigg/ \left\{ 1 + \frac{3}{1 + \omega_0^2 \tau_c^2} + \frac{6}{1 + 4\omega_0^2 \tau_c^2} \right\}. \quad [5.117]$$

The cross-relaxation rate constants are proportional to the inverse sixth power of the distance between the two dipolar interacting spins, but η_{IS} does not depend on the distance r_{IS} between the two spins. Thus, a measurement of η_{IS} can indicate that two spins are close enough in space to experience dipolar cross-relaxation, but a quantitative estimate of the distance separating the spins cannot be obtained. To estimate the distance between two nuclei, σ_{IS}^{NOE} or σ_{IS}^{ROE} must be measured directly (or η_{IS} measured in one experiment and ρ_I in a second experiment).

In the extreme narrowing limit ($\omega_0 \tau_c \ll 1$), [5.116] and [5.117] reduce to

$$\sigma_{IS}^{NOE} = \sigma_{IS}^{ROE} = \frac{\hbar^2 \mu_0^2 \gamma^4 \tau_c}{8\pi^2 r_{IS}^6};$$

[5.118]

$$\eta_{IS} = \frac{1}{2},$$

and in the spin diffusion limit ($\omega_0\tau_c \gg 1$),

$$\sigma_{IS}^{NOE} = -\frac{\hbar^2\mu_0^2\gamma^4\tau_c}{40\pi^2 r_{IS}^6};$$

$$\sigma_{IS}^{ROE} = \frac{\hbar^2\mu_0^2\gamma^4\tau_c}{20\pi^2 r_{IS}^6};$$ \[5.119\]

$$\eta_{IS} = -1.$$

In the slow tumbling regime, the laboratory- and rotating-frame cross relaxation rate constants are related by

$$\sigma_{ROE} = -2\sigma_{NOE}.$$ \[5.120\]

This relationship has been used to compensate approximately for cross-relaxation effects in NMR spectra (*29,30*). The values of σ_{IS}^{NOE} and η_{IS} are zero if $\omega\tau_c = 1.12$, whereas $\sigma_{IS}^{ROE} > 0$ for all τ_c.

The principal use of the NOE in biological NMR spectroscopy is the determination of distances between pairs of protons (*31*). The NOE enhancements of interest arise from slowly tumbling biological macromolecules in the spin diffusion limit. For such molecules, relatively large transient homonuclear proton NOE (or ROE) enhancements build up quickly and are detected most effectively by transient NOE and NOESY (or transient ROE and ROESY) methods (Section 6.6).

5.6 Chemical-Exchange Effects in NMR Spectroscopy

NMR spectroscopy provides an extremely powerful and convenient method for monitoring the *exchange* of a nucleus between environments due to chemical reactions or conformational transitions. In the first instance, the nucleus exchanges intermolecularly between sites in different molecules; in the second, the nucleus exchanges intramolecularly between conformations. The exchange process can be monitored by NMR spectroscopy even if the sites are chemically equivalent provided that the sites are magnetically distinct. Nuclear spins can be manipulated during the NMR experiment without affecting the chemical states of the system, because of the weak coupling between the spin system and the lattice. Thus, chemical reactions and conformational exchange processes can be studied by NMR spectroscopy while the system remains in chemical equilibrium.

To establish a qualitative picture of the effects of exchange on an NMR spectrum, suppose that a given nucleus exchanges with rate constant

k between two magnetically distinct sites with resonance frequencies that differ by $\Delta\nu$. On average, the resonance frequency of the spin in each site can only be observed for a time of the order of $1/k$ before the spin jumps to the other site and begins to precess with a different frequency. The finite observation time places a lower limit on the magnitude of $\Delta\nu$ required to distinguish the two sites. If the exchange rate is *slow* ($k \ll \Delta\nu$), distinct signals are observed from the nuclei in the two sites; in contrast, if the exchange rate is fast ($k \gg \Delta\nu$), a single resonance is observed at the population-weighted average chemical shift of the nuclei in the two sites. *The NMR chemical-shift timescale is defined by the difference between the frequencies of the two exchanging resonances.*

5.6.1 CHEMICAL EXCHANGE FOR ISOLATED SPINS

For simplicity, only the case of chemical exchange in spin systems without scalar coupling interactions will be treated. In this situation, the exchange process can be treated using an extension of the Bloch equations (Section 1.2). A first-order chemical reaction (or two-site chemical exchange) between two chemical species, A_1 and A_2, is described by the reaction

$$A_1 \underset{k_{-1}}{\overset{k_1}{\rightleftharpoons}} A_2, \qquad [5.121]$$

in which k_1 is the reaction rate constant for the forward reaction and k_{-1} is the reaction rate constant for the reverse reaction. The chemical kinetic rate laws for this system can be written in matrix form as

$$\frac{d}{dt}\begin{bmatrix} [A_1](t) \\ [A_2](t) \end{bmatrix} = \begin{bmatrix} -k_1 & k_{-1} \\ k_1 & -k_{-1} \end{bmatrix}\begin{bmatrix} [A_1](t) \\ [A_2](t) \end{bmatrix}. \qquad [5.122]$$

For a coupled set of N first-order chemical reactions between N chemical species, this equation can be generalized to

$$\frac{d\mathbf{A}(t)}{dt} = \mathbf{K}\mathbf{A}(t), \qquad [5.123]$$

in which the matrix elements of the rate matrix, $\mathbf{K}$, are given by

$$K_{ij} = k_{ji}, \qquad (i \neq j)$$

$$K_{ii} = -\sum_{\substack{j=1 \\ j\neq i}}^{N} k_{ij}, \qquad [5.124]$$

and the chemical reaction between the ith and jth species is

$$A_i \underset{k_{ji}}{\overset{k_{ij}}{\rightleftarrows}} A_j. \qquad [5.125]$$

The modified Bloch equations can be written in matrix form for the jth chemical species as

$$\frac{d\mathbf{M}_{jz}(t)}{dt} = \gamma(1 - \sigma_j)[\mathbf{M}_j(t) \times \mathbf{B}(t)]_z - \mathbf{R}_{1j}\{\mathbf{M}_{jz}(t) - \mathbf{M}_{j0}(t)\} + \sum_{k=1}^{N} K_{jk}\mathbf{M}_{kz}(t);$$

$$\frac{d\mathbf{M}_{jx}(t)}{dt} = \gamma(1 - \sigma_j)[\mathbf{M}_j(t) \times \mathbf{B}(t)]_x - \mathbf{R}_{2j}\mathbf{M}_{jx}(t) + \sum_{k=1}^{N} K_{jk}\mathbf{M}_{kx}(t); \qquad [5.126]$$

$$\frac{d\mathbf{M}_{jy}(t)}{dt} = \gamma(1 - \sigma_j)[\mathbf{M}_j(t) \times \mathbf{B}(t)]_y - \mathbf{R}_{2j}\mathbf{M}_{jy}(t) + \sum_{k=1}^{N} K_{jk}\mathbf{M}_{ky}(t),$$

with

$$\mathbf{M}_{j0}(t) = \mathbf{M}_0[A_j](t) \bigg/ \sum_{j=1}^{N} [A_j](t). \qquad [5.127]$$

The modified Bloch equations for chemical reactions are called the McConnell equations (32). If the system is in chemical equilibrium, then $[A_j](t) = [A_j]$.

The preceding equations can be generalized to the case of higher-order chemical reactions by defining the pseudo-first order rate constants:

$$k_{ij} = \frac{\dot{\zeta}_{ij}(t)}{[A_i](t)}, \qquad [5.128]$$

in which $\dot{\zeta}_{ij}(t)$ is the rate law for conversion of the ith species containing the nuclear spin of interest into the jth species containing the nuclear spin of interest. The effect of the chemical reactions is to shift the spin of interest between molecular environments. For example, consider the elementary reaction

$$A_1 + B \underset{k_{-1}}{\overset{k_1}{\rightleftarrows}} A_2 + C, \qquad [5.129]$$

in which a spin in species A_1 is transferred to species A_2 as a result of the chemical reaction. The chemical kinetic rate laws for this system can be written in matrix form as

$$\frac{d}{dt}\begin{bmatrix} [A_1](t) \\ [A_2](t) \end{bmatrix} = \begin{bmatrix} -k_1[B](t) & k_{-1}[C](t) \\ k_1[B](t) & -k_{-1}[C](t) \end{bmatrix}\begin{bmatrix} [A_1](t) \\ [A_2](t) \end{bmatrix}, \qquad [5.130]$$

which has the same form as [5.123] in which the elements of $\mathbf{K}$ are defined using [5.124] and [5.128]. Notice that the rate expressions for $[B](t)$ and $[C](t)$ are not included in [5.130] because the spin of interest is not contained in either species.

In the absence of applied rf fields, the equation for longitudinal magnetization becomes

$$\frac{d\mathbf{M}_{jz}(t)}{dt} = -\mathbf{R}_{1j}\{\mathbf{M}_{jz}(t) - \mathbf{M}_{j0}(t)\} + \sum_{k=1}^{N} K_{jk}\mathbf{M}_{kz}(t). \qquad [5.131]$$

Defining

$$\mathbf{M}_z(t) = \begin{bmatrix} \mathbf{M}_{1z}(t) \\ \vdots \\ \mathbf{M}_{Nz}(t) \end{bmatrix} \qquad [5.132]$$

yields the compact expression

$$\frac{d\mathbf{M}_z(t)}{dt} = (-\mathbf{R} + \mathbf{K})\{\mathbf{M}_z(t) - \mathbf{M}_0(t)\} + \mathbf{K}\mathbf{M}_0(t), \qquad [5.133]$$

in which the elements of $\mathbf{R}$ are given by $R_{ij} = \delta_{ij} R_{1j}$ (for simplicity, the possibility of simultaneous dipolar cross-relaxation and chemical exchange is not considered herein). If the system is in chemical equilibrium, $\mathbf{K}\mathbf{M}_0(t) = \mathbf{K}\mathbf{M}_0 = 0$ and

$$\frac{d\,\Delta\mathbf{M}_z(t)}{dt} = (-\mathbf{R} + \mathbf{K})\,\Delta\mathbf{M}_z(t). \qquad [5.134]$$

The equation of motion for the transverse magnetization can be written in the rotating frame as

$$\frac{d\mathbf{M}^+(t)}{dt} = (i\mathbf{\Omega} - \mathbf{R} + \mathbf{K})\mathbf{M}^+(t), \qquad [5.135]$$

in which the elements of $\mathbf{\Omega}$ are given by $\Omega_{ij} = \delta_{ij}\Omega_j$, and the elements of $\mathbf{R}$ are given by $R_{ij} = \delta_{ij}R_{2j}$.

Equations [5.134] and [5.135] have the same functional form as [5.12] and can be solved by the same methods [5.13]. For example, the rate matrix for longitudinal relaxation in a system undergoing two-site exchange is given by

$$\mathbf{R} - \mathbf{K} = \begin{bmatrix} \rho_1 + k_1 & -k_{-1} \\ -k_1 & \rho_2 + k_{-1} \end{bmatrix}, \qquad [5.136]$$

with eigenvalues

$$\lambda_{\pm} = \tfrac{1}{2}\{(\rho_1 + \rho_2 + k_1 + k_{-1}) \pm [(\rho_1 - \rho_2 + k_1 - k_{-1})^2 + 4k_1 k_{-1}]^{1/2}\}. \quad [5.137]$$

The time course of the magnetization is given by

$$\begin{bmatrix} \Delta M_1(t) \\ \Delta M_2(t) \end{bmatrix} = \begin{bmatrix} a_{11}(t) & a_{12}(t) \\ a_{21}(t) & a_{22}(t) \end{bmatrix} \begin{bmatrix} \Delta M_1(0) \\ \Delta M_2(0) \end{bmatrix}, \qquad [5.138]$$

in which

$$a_{11}(t) = \frac{1}{2}\left[\left(1 - \frac{\rho_1 - \rho_2 + k_1 - k_{-1}}{(\lambda_+ - \lambda_-)}\right)\exp(-\lambda_- t)\right.$$
$$\left. + \left(1 + \frac{\rho_1 - \rho_2 + k_1 - k_{-1}}{(\lambda_+ - \lambda_-)}\right)\exp(-\lambda_+ t)\right];$$

$$a_{22}(t) = \frac{1}{2}\left[\left(1 + \frac{\rho_1 - \rho_2 + k_1 - k_{-1}}{(\lambda_+ - \lambda_-)}\right)\exp(-\lambda_- t)\right.$$
$$\left. + \left(1 - \frac{\rho_1 - \rho_2 + k_1 - k_{-1}}{(\lambda_+ - \lambda_-)}\right)\exp(-\lambda_+ t)\right]; \qquad [5.139]$$

$$a_{12}(t) = \frac{k_{-1}}{(\lambda_+ - \lambda_-)}[\exp(-\lambda_- t) - \exp(-\lambda_+ t)];$$

$$\alpha_{21}(t) = \frac{k_1}{(\lambda_+ - \lambda_-)}[\exp(-\lambda_- t) - \exp(-\lambda_+ t)].$$

To obtain some insight in to the form of these equations, assume that $\rho_1 = \rho_2 = \rho$, and that exchange is symmetrical with $k_1 = k_{-1} = k$. Under these conditions, the time dependence of the longitudinal magnetization is given by

$$a_{11}(t) = a_{22}(t) = \tfrac{1}{2}[1 + \exp(-2kt)]\exp(-\rho t); \qquad [5.140]$$

$$a_{12}(t) = a_{21}(t) = \tfrac{1}{2}[1 - \exp(-2kt)]\exp(-\rho t).$$

The homology between [5.20] and [5.140] illustrates the fundamental similarity between the effects of cross-relaxation and chemical exchange on longitudinal magnetization. Indeed, similar experimental techniques are utilized to study both phenomena (such as NOESY and ROESY experiments; Section 6.6).

The rate matrix for transverse relaxation in a system undergoing two-site exchange is given by

$$-i\Omega + \mathbf{R} - \mathbf{K} = \begin{bmatrix} -i\Omega_1 + \rho_1 + k_1 & -k_{-1} \\ -k_1 & -i\Omega_2 + \rho_2 + k_{-1} \end{bmatrix}, \quad [5.141]$$

with eigenvalues

$$\lambda_{\pm} = \tfrac{1}{2}\{(-i\Omega_1 - i\Omega_2 + \rho_1 + \rho_2 + k_1 + k_{-1})$$
$$\pm [(-i\Omega_1 + i\Omega_2 + \rho_1 - \rho_2 + k_1 - k_{-1})^2 + 4k_1 k_{-1}]^{1/2}\}. \quad [5.142]$$

The time course of the magnetization is given by

$$\begin{bmatrix} M_1^+(t) \\ M_2^+(t) \end{bmatrix} = \begin{bmatrix} a_{11}(t) & a_{12}(t) \\ a_{21}(t) & a_{22}(t) \end{bmatrix} \begin{bmatrix} M_1^+(0) \\ M_2^+(0) \end{bmatrix}, \quad [5.143]$$

in which

$$a_{11}(t) = \frac{1}{2}\left[\left(1 - \frac{-i\Omega_1 + i\Omega_2 + \rho_1 - \rho_2 + k_1 - k_{-1}}{(\lambda_+ - \lambda_-)}\right)\exp(-\lambda_- t)\right.$$
$$\left. + \left(1 + \frac{-i\Omega_1 + i\Omega_2 + \rho_1 - \rho_2 + k_1 - k_{-1}}{(\lambda_+ - \lambda_-)}\right)\exp(-\lambda_+ t)\right];$$

$$a_{22}(t) = \frac{1}{2}\left[\left(1 + \frac{-i\Omega_1 + i\Omega_2 + \rho_1 - \rho_2 + k_1 - k_{-1}}{(\lambda_+ - \lambda_-)}\right)\exp(-\lambda_- t)\right.$$
$$\left. + \left(1 - \frac{-i\Omega_1 + i\Omega_2 + \rho_1 - \rho_2 + k_1 - k_{-1}}{(\lambda_+ - \lambda_-)}\right)\exp(-\lambda_+ t)\right]; \quad [5.144]$$

$$a_{12}(t) = \frac{k_{-1}}{(\lambda_+ - \lambda_-)}[\exp(-\lambda_- t) - \exp(-\lambda_+ t)]$$

$$a_{21}(t) = \frac{k_1}{(\lambda_+ - \lambda_-)}[\exp(-\lambda_- t) - \exp(-\lambda_+ t)].$$

To obtain some insight in to the form of these equations, assume that $\Omega_1 = -\Omega_2 = \Omega$ (i.e., the reference frequency in the rotating frame is midway between the frequencies of the two sites), that $\rho_1 = \rho_2 = \rho$, and that exchange is symmetrical with $k_1 = k_{-1} = k$. Under these conditions, the time dependence of the transverse magnetization is given by

$$a_{11}(t) = \frac{1}{2}\left[\left(1 + \frac{i\Omega}{\Delta}\right)\exp\{-(\rho + k - \Delta)t\} + \left(1 - \frac{i\Omega}{\Delta}\right)\exp\{-(\rho + k + \Delta)t\}\right];$$

$$a_{22}(t) = \frac{1}{2}\left[\left(1 - \frac{i\Omega}{\Delta}\right)\exp\{-(\rho + k - \Delta)t\} + \left(1 + \frac{i\Omega}{\Delta}\right)\exp\{-(\rho + k + \Delta)t\}\right];$$

$$a_{12}(t) = a_{21}(t) = \frac{k}{2\Delta}[\exp\{-(\rho + k - \Delta)t\} - \exp\{-(\rho + k + \Delta)t\}], \qquad [5.145]$$

in which $\Delta = (k^2 - \Omega^2)^{1/2}$. In the slow exchange limit, $\Omega \gg k$ and

$$a_{11}(t) = \exp\{-(\rho + k - i\Omega)t\};$$

$$a_{22}(t) = \exp\{-(\rho + k + i\Omega)t\}; \qquad [5.146]$$

$$a_{12}(t) = a_{21}(t) = 0.$$

Two resonances are observed at $\Omega_1 = \Omega$ and $\Omega_2 = -\Omega$ with linewidths equal to $(\rho + k)/\pi$. In the fast exchange limit, $\Omega \ll k$ and

$$a_{11}(t) = a_{22}(t) = \tfrac{1}{2}[1 + \exp(-2kt)]\exp(-\rho t);$$
$$\qquad\qquad [5.147]$$
$$a_{12}(t) = a_{21}(t) = \tfrac{1}{2}[1 - \exp(-2kt)]\exp(-\rho t).$$

In this case, [5.147] are purely real. Consequently, a single resonance line is observed at a shift of $(\Omega_1 + \Omega_2)/2 = 0$. The observed signal is equal to

$$M_1^+(t) + M_2^+(t) = a_{11}(t)M_1^+(0) + a_{12}(t)M_2^+(0) + a_{22}(t)M_2^+(0) + a_{21}(t)M_1^+(0)$$

$$= [a_{11}(t) + a_{12}(t) + a_{22}(t) + a_{21}(t)]M^+(0)/2$$

$$= M^+(0)\exp(-\rho t), \qquad [5.148]$$

and has a linewidth equal to ρ/π. Equations [5.146] and [5.148] confirm the qualitative conclusions about the slow and fast exchange regimes stated above. In addition, these equations indicate that the linewidths of the resonance signals are perturbed in the slow exchange limit but that the natural linewidths are restored in the fast exchange limit.

For two-site exchange, lineshapes expected for various exchange rates can be calculated by Fourier transformation of [5.144] or [5.145]. Figure 5.9 shows calculated spectra in several exchange regimes. Fig. 5.9a shows the case of slow exchange. As expected from [5.146], two resonance lines are observed. As the exchange rate increases, the resonance lines broaden as shown in Figure 5.9b. When the exchange rate is of the order of the

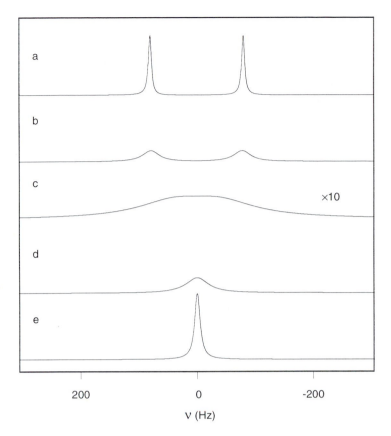

FIGURE 5.9 Chemical exchange for a two-site system. Shown are the Fourier transformations of FIDs calculated by using [5.145]. The calculations used $\Omega = 80$ Hz and $\rho = 10$ s^{-1}. Calculations were performed for values of the exchange rate, k, equal to (a) 10 s^{-1}, (b) 100 s^{-1}, (c) 450 s^{-1}, (d) 1,000 s^{-1}, and (e) 5,000 s^{-1}. The spectrum for (c) has been expanded vertically by a factor of 10 for clarity.

chemical shift separation between the two sites, the lines become very broad and begin to coalesce (Fig. 5.9c). This is known as the *intermediate exchange* regime or *coalescence*. Intermediate exchange processes can cause peaks to disappear in spectra because the broadening becomes so great that the resonance line becomes indistinguishable from the baseline noise. Increasing the exchange rate for the system above the coalescence point pushes the system into fast exchange (Fig. 5.9d,e). As expected from [5.148], a single resonance is observed at the average resonance frequency for the two sites.

5.6.2 QUALITATIVE EFFECTS OF CHEMICAL EXCHANGE IN SCALAR-COUPLED SYSTEMS

Multiplet structure due to scalar couplings is affected by chemical exchange. Detailed theoretical treatment using the density matrix formalism (33) is beyond the subject matter of this text; instead, the present discussion will present qualitatively the most important effects. Formally, scalar relaxation (Section 5.4.5) and chemical exchange in scalar-coupled systems are homologous. Two different cases must be considered: intermolecular (homologous to scalar relaxation of the second kind) and intramolecular exchange (homologous to scalar relaxation of the first kind).

Intermolecular chemical exchange in scalar coupled systems is encountered frequently in biological NMR applications. For example, exchange between labile amide protons and solvent protons perturbs the NH to Hα scalar coupling interaction. In an IS spin system, the I spin resonance is a doublet, with the lines separated by J_{IS}. One line of the doublet is associated with the S spin in the α state, and the other line is associated with the S spin in the β state. Suppose that a given I spin is coupled to an S spin in the α state. If the S spin exchanges with another S spin originating from the solvent (intermolecular exchange), then after the exchange the I spin has equal probability of being coupled to an S spin in the α and β states because the incoming spin has a 50% chance of either being in its α state or in its β state. Similar considerations hold for an I spin initially coupled to an S spin in the β state. Consequently, the I spin sees the S spin state constantly changing, due to exchange, and thus the frequency of the I spin resonance constantly changes between the frequencies of the two lines of the doublet. This phenomenon constitutes a two-site exchange process and exhibits properties of slow, intermediate, and fast exchange. If the exchange is fast compared to the difference in frequency between the two lines (i.e., compared to the scalar coupling constant), a single line is observed at the mean frequency (the Larmor frequency of the I spin). Since homonuclear scalar coupling constants tend to be small, relatively slow exchange processes, which would minimally perturb the chemical shifts of the exchanging spins, can result in collapse of multiplet structure. Indeed, the broadening of multiplets and the disappearance of multiplet structure are the first clues to the existence of exchange phenomena in NMR spectra.

Intramolecular exchange constitutes a slightly different situation. Consider a system in which spin I is scalar coupled to spin S, but because of the presence of multiple conformers, spin S can be in n environments, $S^1, S^2, \ldots, S^n$, with different scalar coupling constants. For simplicity, the chemical shift of the I spin is assumed to be identical in all conformers.

If the conformers interconvert on a timescale long compared to the scalar coupling constants, the I spin multiplet is a superposition of n doublets arising from the $IS^1, IS^1, \ldots, IS^n$ scalar coupling interactions. On the other hand, if the conformers interconvert at a rate much larger than the scalar coupling constants, the I spin resonance is a doublet with an effective scalar coupling constant that is a population-weighted average of the n scalar coupling constants. An example of this effect arises for the scalar coupling between H^α and H^β protons in amino acids. If the conformations of the H^β protons are fixed relative to the H^α proton, the H^α multiplet is split by two coupling constants, one from each of the H^β protons to the H^α proton (e.g., 12 Hz and 3 Hz for a *trans,gauche* conformation). On the other hand, if the H^β protons exchange between *trans,gauche*$^+$ and *gauche*$^-$ rotameric sites as a result of free rotation about the C^α–C^β bond, then the H^α multiplet is split by a single average coupling constant [with a value $(12 + 3 + 3)/3 = 6$ Hz] due to the H^β protons.

References

1. J. McConnell, "The Theory of Nuclear Magnetic Relaxation in Liquids," pp. 1–196. Cambridge University Press, New York, 1987.
2. A. Abragam, "Principles of Nuclear Magnetism," pp. 1–599, Clarendon Press, Oxford, 1961.
3. M. Goldman, "Quantum Description of High-Resolution NMR in Liquids," pp. 1–268. Clarendon Press, New York, 1988.
4. F. Bloch, *Phys. Rev.* **70**, 460–474 (1946).
5. N. Bloembergen, E. M. Purcell, and R. V. Pound, Phys. Rev. **73**, 679–712 (1948).
6. I. Solomon, *Phys. Rev.* **99**, 559–565 (1955).
7. R. K. Wangsness and F. Bloch, *Phys. Rev.* **89**, 728–739 (1953).
8. A. G. Redfield, *Adv. Magn. Reson.* **1**, 1–32 (1965).
9. N. C. Pyper, *Mol. Phys.* **22**, 433–458 (1971).
10. R. Brüschweiler and D. A. Case, *Prog. NMR Spectrosc.* **26**, 27–58 (1994).
11. M. Goldman, *J. Magn. Reson.* **60**, 437–452 (1984).
12. L. G. Werbelow and D. M. Grant, *Adv. Magn. Reson.* **9**, 189–299 (1977).
13. N. Müller, G. Bodenhausen, K. Wüthrich, and R. R. Ernst, *J. Magn. Reson.* **65**, 531–534 (1985).
14. N. C. Pyper, *Mol. Phys.* **21**, 1–33 (1971).
15. M. H. Levitt, G. Bodenhausen, and R. R. Ernst, *J. Magn. Reson.* **53**, 443–461 (1983).
16. C. Griesinger and R. R. Ernst, *Chem. Phys. Lett.* **152**, 239–247 (1988).
17. P. S. Hubbard, *Phys. Rev.* **180**, 319–326 (1969).
18. D. M. Brink and G. R. Satchler, "Angular Momentum," pp. 1–170. Clarendon Press, Oxford, 1993.
19. G. Lipari and A. Szabo, *J. Am. Chem. Soc.* **104**, 4546–4559 (1982).
20. G. Lipari and A. Szabo, *J. Am. Chem. Soc.* **104**, 4559–4570 (1982).
21. A. G. Palmer, *Curr. Opin. Biotechnol.* **4**, 385–391 (1993).
22. G. Wagner, *Curr. Opin. Struc. Biol.* **3**, 748–753 (1993).

23. T. M. G. Koning, R. Boelens, and R. Kaptein, *J. Magn. Res.* **90,** 111–123 (1990).
24. H. Liu, P. D. Thomas, and T. L. James, *J. Magn. Reson.* **98,** 163–175 (1992).
25. G. S. Harbison, *J. Am. Chem. Soc.* **115,** 3026–3027 (1993).
26. A. A. Bothner-By, R. L. Stephens, and J. Lee, *J. Am. Chem. Soc.* **106,** 811–813 (1984).
27. J. H. Noggle and R. E. Shirmer, "The Nuclear Overhauser Effect: Chemical Applications," pp. 1–259. Academic Press, New York, 1971.
28. D. Neuhaus and M. Williamson, "The Nuclear Overhauser Effect in Structural and Conformational Analysis," pp. 1–522. VCH Publishers, New York, 1989.
29. C. Griesinger, G. Otting, K. Wüthrich, and R. R. Ernst, *J. Am. Chem. Soc.* **110,** 7870–7872 (1988).
30. J. Cavanagh and M. Rance, *J. Magn. Reson.* **96,** 670–678 (1992).
31. K. Wüthrich, "NMR of Proteins and Nucleic Acids," pp. 1–292. Wiley, New York, 1986.
32. H. M. McConnell, *J. Chem. Phys.* **28,** 430–431 (1958).
33. R. R. Ernst, G. Bodenhausen, and A. Wokaun, "Principles of Nuclear Magnetic Resonance in One and Two Dimensions," pp. 1–610. Clarendon Press, Oxford, 1987.

EXPERIMENTAL ^{1}H NMR METHODS

Because of the high natural abundance and large gyromagnetic ratio of the ^{1}H nucleus, protein NMR studies traditionally utilized predominantly homonuclear ^{1}H spectroscopic techniques. This chapter describes the homonuclear ^{1}H NMR experiments required to obtain complete ^{1}H resonance assignments and to ascertain structural and dynamical features of proteins with molecular masses of up to 10–12 kDa, provided that the proteins are well-behaved in solution and display reasonable chemical-shift dispersion (as a consequence of differences in chemical-shift dispersion, relatively larger β-sheet proteins than α-helical proteins are amenable to investigation). Over the last decade, hundreds of two- and three-dimensional ^{1}H NMR experiments have been described in the literature. However, many of these are not generally applicable, or have been superseded by superior techniques. This chapter provides a concise compendium of useful NMR experiments from which resonance assignments and subsequent structural and dynamical investigations can be performed by using a minimum of spectrometer time. Heteronuclear NMR spectroscopy, utilizing ^{13}C and ^{15}N spins as well as ^{1}H spins, is described in Chapter 7.

Throughout this chapter, phase-sensitive, rather than magnitude-mode, spectra have been recorded because resolution of the resonances is superior and analysis of the cross-peak fine structure is facilitated.

Sections of the correlation spectra with antiphase lineshapes are shown with multiple contours for the positive peak components, and a single contour for negative components. Generally, spectra with in-phase line-shapes are depicted with only positive or negative levels displayed; spectra in which both sets of peaks are displayed are discussed in the appropriate figure captions.

6.1 Assessment of the 1D ¹H Spectrum

Although the majority of spectroscopic analyses will depend on 2D and 3D NMR spectroscopy, important preparatory work can be performed by using one-dimensional (1D) ¹H NMR spectroscopy. To begin, a 1D ¹H NMR spectrum of the protein in H_2O solution (containing 5–10% D_2O as a lock reference) is acquired using the one-dimensional Hahn echo experiment (Section 3.6.4.2).

The first spectral feature of interest is the signal-to-noise (S/N) ratio. If more than the most basic correlation experiments (i.e., COSY for homonuclear spectroscopy, HSQC/HMQC for heteronuclear spectroscopy) are to be feasible, then the 1D ¹H spectrum must contain a reasonable signal amplitude after coaddition of 16 or 32 transients. The standard sample used for ¹H experiments in this text is a 2-mM ubiquitin solution (see Preface). The ¹H NMR spectrum shown in Fig. 6.1a was collected with 32 transients, and has a S/N ratio of 243 for a resolved upfield-shifted methyl group, and 46 for a resolved downfield-shifted amide group. S/N ratios are measured as the resonance peak height divided by the root-mean-square baseline noise in the spectrum. Rapid rotation narrows the resonance linewidths for methyl groups (relative to ¹Hᵅ resonances); in contrast, amide proton linewidths are broadened by amide proton solvent exchange, scalar relaxation by the ¹⁴N nucleus, and partially resolved scalar coupling to the ¹Hᵅ spin. In addition the intensity of the amide resonance is reduced by the fraction of D_2O present in the sample. Thus, the value of the S/N ratio for the amide is less than one-third of the value for the methyl group. The S/N ratio depends on the concentration of protein and is also affected by the linewidth of the signals. For ubiquitin (M_r = 8565 Da), the half-height linewidths of a resolved upfield methyl group and downfield amide proton are 6.4 and 8.6 Hz, respectively, at 300 K. Linewidths were obtained by curve fitting to an in-phase Lorentzian doublet because the observed half-height width of both resonances has a contribution from a partially resolved scalar coupling. The linewidths are related to the rotational correlation time of the protein, which in turn depends on the solution viscosity and temperature (Section 1.4). Larger

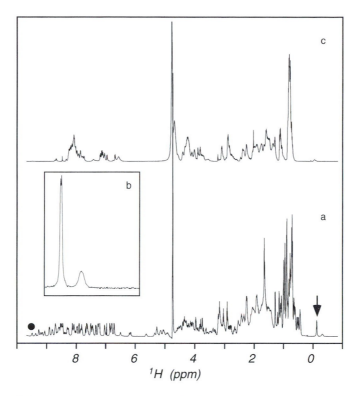

FIGURE 6.1 (a) 1D spectrum of a 2-mM ubiquitin in H_2O acquired at 500 MHz and 27°C. The $\pi/2$ acquisition pulse was preceded by 1.5 s of solvent presaturation and followed by a Hahn echo (Section 3.6.4.2). The spectrum is the result of 32 transients collected with a spectral width of 12.5 kHz and a 30-kHz filter width. (b) The region around the upfield-shifted methyl group of Leu50 (arrow) is enlarged in the inset. The RMS noise at the edges of the spectrum is 0.0023 (arbitrary units), whereas the Leu50 methyl ¹H signal has a height of 0.56, giving a S/N ratio of 243. A S/N ratio of 46 is obtained for the resolved downfield-shifted ¹H^N resonance of Ile13 (filled circle). (c) 1D spectrum of a 2-mM ubiquitin sample in 8-M urea/H_2O solution. The reduced chemical-shift dispersion is characteristic of unfolded or denatured proteins.

proteins have larger linewidths and lower apparent S/N ratios even at similar concentrations. Aggregation causes line broadening, and in unfortunate cases, proteins may aggregate to such an extent that no peaks are observed in the 1D ¹H spectrum (the linewidth becomes so large that the peaks merge into the baseline), even though other methods of detection indicate that the actual protein concentration in solution is in the millimolar range.

The majority of multidimensional homonuclear and heteronuclear NMR experiments used for protein NMR spectroscopy initially excite equilibrium ^{1}H magnetization. The recovery of an equilibrium magnetization state, and therefore the repetition rate of the experiment, depends on the relaxation properties of the ^{1}H nuclei. The (nonselective) ^{1}H R_1 relaxation rate constant can be estimated from a 1D inversion recovery experiment (Section 3.6.2.5). In order to avoid steady-state artifacts and low sensitivity per unit of measuring time, the sum of acquisition time and recycle delay should be greater than $1.3/R_1$ (Fig. 3.34). In indirectly detected ^{1}H dimensions, the transverse relaxation rate constant, R_2, of the coherence of interest is important in deciding an appropriate value for $t_{1,max}$. Extending t_1 beyond $2/R_2$ or $3/R_2$ is undesirable because the later increments of t_1 only contribute noise to the processed spectrum. The phenomenological ^{1}H single-quantum relaxation rate constant, $R_2^* = R_2 + R_{inhom}$, in which R_{inhom} represents the contribution from magnetic field inhomogeneity, can be estimated from the full-width-at-half-height linewidth (Δv_{FWHH}) of resolved peaks as $R_2^* \approx (\pi \Delta v_{FWHH})$.

Resonance dispersion in the 1D spectrum indicates the integrity of the protein under the particular experimental conditions chosen. Denatured proteins have chemical shifts close to those found in short linear peptides (the so-called random-coil shifts, Section 1.5), whereas folded proteins will exhibit a range of chemical shifts due to the anisotropic magnetic fields of proximal aromatic or carbonyl groups. Thus, if very little chemical shift dispersion is observed, the protein may be unfolded, or may have very little stable structure. As an example, a spectrum of ubiquitin denatured in 8-M urea is shown in Fig. 6.1b. The key resonances to examine arise from the amide protons (random-coil shifts 8.5–8.0 ppm), α-protons (random-coil shifts 4.4–4.1 ppm), and methyl groups of valine, isoleucine, and leucine (random-coil shifts 1.1–0.8 ppm.). Examination of the chemical-shift dispersion also indicates the ease with which resonance assignments can be made by the sequential spin system method described in Section 8.1.1. If overlap in the important ^{1}H^N region is significant (and to a lesser extent in the ^{1}H$^\alpha$ and methyl ^{1}H regions), sequential assignments will be hard-won, and the number of NOEs that can be assigned unambiguously will be low (leading to poor structural definition). Recently, schemes have been proposed that use the observed dependence of ^{1}H^N and ^{1}H$^\alpha$ chemical shift on secondary structure to estimate the number of residues in different types of regular secondary structure (1,2).

Finally, the purity of the sample can be gauged from the 1D spectrum. Low-molecular-weight impurities are apparent as sharp peaks amid the broader envelope of protein resonances. Of course, a pure protein sample also can exhibit linewidth variations due to differential internal mobility, particularly at side-chain termini, or in flexible-loop regions, so sharp lines are not necessarily proof of contamination. Observation of variations in the relative peak heights of resolved resonances may indicate inhomogeneous protein preparations, although such information usually is better gauged from COSY (Section 6.2.1) or HSQC/HMQC spectra (Section 7.1.1). Low-molecular-weight impurities also can be identified from a TOCSY experiment (Section 6.5) recorded with a long (200-ms) mixing time because protein resonances are attenuated preferentially by relaxation.

If any of the basic attributes of the 1D ^{1}H NMR spectrum (S/N ratio, linewidth, chemical-shift dispersion) are less than ideal, the 1D spectrum provides an efficient way to probe the dependence of these aspects on sample conditions, including concentration, temperature, pH, and ionic strength. As a word of caution, care should be taken when performing such studies, as extremes of temperature and pH may lead to irreversible denaturation or loss of protein integrity. Furthermore, unless spectra are recorded without presaturation of the solvent resonance (Section 3.5.3), spectra may suffer from a loss of amide proton signal intensity from saturation transfer via exchange with solvent at very high temperature or pH.

6.2 COSY-Type Experiments

6.2.1 COSY

COSY, or correlated spectroscopy, was the first 2D NMR experiment to be devised (3,4), and remains useful for NMR studies of proteins. COSY cross-peaks arise through coherence transfer between coupled spins; in practice, for protein studies, this limits transfer to protons separated by two or three bonds. The pulse sequence simply consists of two pulses separated by an incrementable delay (t_1). The recycle delay precedes the first pulse, and the acquisition period (t_2) follows the second pulse, as shown in Fig. 6.2. The basic phase cycle consists of eight steps: the phases of both pulses and the receiver are cycled together using the CYCLOPS scheme, and the phases of the first pulse and receiver are inverted to reduce axial peak intensity.

6.2.1.1 Product Operator Analysis Most of the homonuclear NMR experiments discussed in this chapter begin with a $90°_x$–t_1–$90°_x$ pulse sequence element (multiple-quantum experiments are the principal excep-

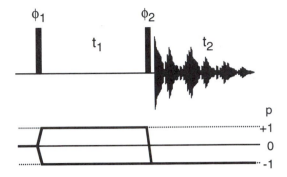

FIGURE 6.2 Pulse sequence and coherence-level diagram for the COSY experiment. Narrow bars represent 90° pulses. The basic phase cycle is $\phi_1 = x$; $\phi_2 = x$; and receiver $= x$. Axial peak suppression and CYCLOPS phase cycling are performed to obtain an eight-step phase cycle.

tions). In the following, spins will be designated I_k for $k = 1, 2, ..., K$ (for a K-spin system). The chemical shift of the kth spin is Ω_k, and the scalar coupling constant between the jth and kth spins is J_{jk} (assumed to represent a three-bond scalar interaction). For a two-spin system, initial I_{1z} magnetization evolves through the pulse sequence element as

$$
I_{1z} \xrightarrow{\left(\frac{\pi}{2}\right)_x - t_1 - \left(\frac{\pi}{2}\right)_x} - I_{1z} \cos(\Omega_1 t_1) \cos(\pi J_{12} t_1)
$$

$$
- 2I_{1x}I_{2y} \cos(\Omega_1 t_1) \sin(\pi J_{12} t_1)
$$

$$
+ I_{1x} \sin(\Omega_1 t_1) \cos(\pi J_{12} t_1)
$$

$$
- 2I_{1z}I_{2y} \sin(\Omega_1 t_1) \sin(\pi J_{12} t_1), \qquad [6.1]
$$

in which $(\pi/2)_x$ represent nonselective rf pulses with x phase applied to the I spins. Parallel evolution beginning with I_{2z} magnetization is exhibited by exchanging I_1 and I_2 labels.

A product operator analysis of the COSY experiment reveals important features that must be considered while acquiring, processing, and analyzing COSY spectra. The essence of the product operator analysis for a two-spin system has already been presented (Section 4.2.1). In summary, the first two terms of [6.1] do not lead to observable magnetization and can be ignored. The third term gives rise to a diagonal peak, and the fourth term leads to a cross-peak modulated by Ω_1 in t_1 and Ω_2 in t_2.

Manipulation of the trigonometric terms of [6.1] leads to a clearer understanding of the multiplet fine structure:

$$\sin(\Omega_1 t_1)\cos(\pi J_{12} t_1) = \frac{1}{2}[\sin(\Omega_1 t_1 - \pi J_{12} t_1) + \sin(\Omega_1 t_1 + \pi J_{12} t_1)]; \quad [6.2]$$

$$\sin(\Omega_1 t_1)\sin(\pi J_{12} t_1) = \frac{1}{2}[\cos(\Omega_1 t_1 - \pi J_{12} t_1) - \cos(\Omega_1 t_1 + \pi J_{12} t_1)]. \quad [6.3]$$

As indicated by [6.2], the diagonal peak has an in-phase lineshape in F_1 with the two multiplet components centered at Ω_1 and separated by $2\pi J_{12}$; in contrast, [6.3] indicates that the cross-peak has an antiphase lineshape with the two components centered at Ω_1 and separated by $2\pi J_{12}$ (if Ω is given in units of angular frequency). The sinusoidal modulation of [6.2] and cosinusoidal modulation [6.3] means that the diagonal and cross-peaks differ in phase by 90° and cannot both be phased to absorption simultaneously. Consideration of the evolution of I_{1x} and $2I_{1z}I_{2y}$ during t_2 indicates that the F_2 lineshapes of the diagonal and cross-peaks are the same as in the F_1 dimension.

The antiphase lineshapes of COSY cross-peaks have important implications for the way in which these data are collected and processed (Sections 6.2.1.2 and 6.2.1.3). In addition, the differences in the relative phase of diagonal and cross-peaks is one of the main shortcomings of the COSY experiment: namely, when the cross-peaks are phased to absorption, the in-phase dispersive tails of the diagonal peaks obscure cross-peaks near the diagonal (arising from scalar-coupled protons close in chemical shift).

COSY cross-peaks between spins in more complex spin systems have fine structure in addition to the simple antiphase splitting of [6.3] (5). The product operator formalism can be used to elucidate the nature of this fine structure, although the calculations quickly become impractical to perform by hand if more than a few coupled spins are involved. For a cross-peak between spins I_1 and I_2, the component of the time-domain signal at the chemical shift of spin I_1 is proportional to (6)

$$\sin(\Omega_1 t)\sin(\pi J_{12} t)\prod_{k=3}^{K}\cos(\pi J_{1k} t)$$

$$= \frac{1}{2}[\cos(\Omega_1 t - \pi J_{12} t) - \cos(\Omega_1 t + \pi J_{12} t)]\prod_{k=3}^{K}\cos(\pi J_{1k} t), \quad [6.4]$$

in which t represents either the t_1 or t_2 evolution period (depending on whether the cross-peak represents $I_1 \rightarrow I_2$ or $I_2 \rightarrow I_1$ coherence transfer), J_{12} is the active scalar coupling between spin I_1 and spin I_2, and J_{1k} is the magnitude of the passive scalar coupling between spin I_1 and spin

I_k ($k > 2$). A similar equation would represent the signal at the chemical shift of the I_2 spin, except that the product would extend over the passive scalar-coupled partners of the I_2 spin. The product of the two sine terms on the left-hand side of [6.4] gives rise to the antiphase splitting by the active coupling J_{12}. Because the product of two cosine terms can be decomposed into a sum of cosine terms, after Fourier transformation, [6.4] yields in-phase absorption components for the passive scalar coupling interactions. The appearance of cross-peaks for amino acid spin systems for a variety of coupling constants and linewidths have been described (7). In the special case of scalar coupling between spin I_1 and a I_n group, in which n is the number of magnetically equivalent spins, the cross-peak is considered to have one active coupling and ($n - 1$) passive couplings; as a result, the relative intensities of the fine-structure components are described by the antiphase Pascal triangle (6).

6.2.1.2 Experimental Protocol Aside from the details of experimental protocols common to all 2D experiments (setting the rf transmitter frequency, calibrating the 90° pulse length, choosing spectral widths, determining the recycle delay, etc.), the nature of coherence transfer and lineshapes in a COSY experiment require consideration of two additional aspects. First, digital resolution has a profound influence on the relative cross-peak intensity of antiphase lineshapes. If the digital resolution in F_1 or F_2 is too low, the positive and negative lobes of the cross-peak will cancel partially, and the intensity of the cross-peak will be reduced as indicated in Fig. 6.3. Second, the cross-peak product operators contain $\sin(\pi J_{12} t_1)$ and $\sin(\pi J_{12} t_2)$ trigonometric terms arising from the active scalar coupling interaction [6.1] that are superposed on the terms reflecting chemical-shift evolution; in this respect, the COSY experiment is said to generate sine-modulated data. Consequently, evolution in t_1 and t_2 must occur for times comparable to or greater than $1/(4J_{12})$ to $1/(2J_{12})$ (approximately 62–125 ms for a 4-Hz coupling constant) if observable cross-peaks are to be obtained. The parameter $t = 1/(2J_{12})$ occurs many times in discussions of scalar correlated experiments, because coherence transfer occurs via antiphase terms with magnitudes proportional to $\sin(\pi J_{12} t)$. Acquisition for a sufficient length of time in t_2 is rarely a problem as a greater proportion of the recycle delay can be spent recording each FID without lengthening the total acquisition time of the experiment. Commonly, each FID is collected for $t_{2,max} \geqslant 1/(2J_{12})$, and truncated at the processing stage to achieve the desired amount of sensitivity or resolution enhancement. In contrast, increasing $t_{1,max}$ by increasing the number of data points in each interferogram adds linearly to the total acquisition time because a new FID must be collected for each additional t_1 value.

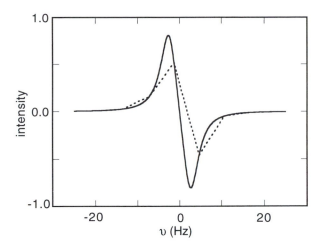

FIGURE 6.3 Variation in the peak height of an antiphase absorptive doublet as the digital resolution is decreased to approach the size of the peak splitting. Both curves represent data for an antiphase pair of Lorentzian lines with half-height width 5 Hz and a separation of 5 Hz. The solid curve has a digital resolution of 0.2 Hz per point and accurately traces the lineshape. The broken line has a digital resolution of 6 Hz per point and clearly has a smaller vertical separation of the positive and negative extrema. Note that the exact decrease in peak height will depend on where the low-digital-resolution points fall on the curve.

For a protein in the 8–10-kDa range, COSY is one of the more sensitive proton 2D experiments. With a protein concentration of around 2–4 mM, a reasonable S/N ratio can be obtained in as short a time as a few hours with a 500-MHz NMR spectrometer. For example, assuming a 10-ppm spectral width in F_1, and a recycle delay of 2.0 s (t_2 acquisition for 0.3 s and equilibrium recovery and/or solvent presaturation for 1.7 s), a COSY experiment with 8 transients for each of 512 t_1 increments could be recorded in just over 2 h and would have a $t_{1,max}$ of 50 ms. If cross-peaks arising from small coupling constants are to be observed (e.g., $^3J_{H^NH^\alpha}$ couplings in residues adopting helical conformations), then additional t_1 increments can be recorded so as to bring $t_{1,max}$ into the 80-ms range. Alternatively, the number of t_1 increments can be kept constant and $t_{1,max}$ increased by decreasing the F_1 spectral width; however, care must be taken to avoid folding or aliasing diagonal peaks on top of cross-peaks. Increasing $t_{1,max}$ beyond 100 ms is unlikely to bring significant improvements in S/N because of extensive relaxation during the longer values of t_1. For more dilute samples, the number of transients collected for

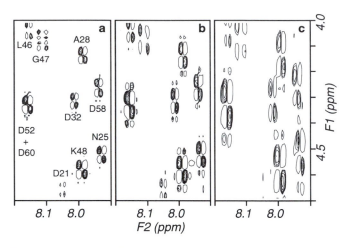

FIGURE 6.4 Sections of the ^{1}H^N–^{1}H$^\alpha$ region of COSY spectra acquired with the same total acquisition time (~7 hs) but with different values of $t_{1,max}$. (a) 16 transients per increment, $t_{1,max}$ = 69 ms; (b) 32 transients per increment, $t_{1,max}$ = 34.5 ms; (c) 64 transients per increment, $t_{1,max}$ = 17.2 ms. All spectra were processed with an unshifted sine bell in t_2 and a sine bell shifted 20° over the available data and zero-filled to 1024 points in t_1. As $t_{1,max}$ decreases, many of the peaks become weaker and that of Asp21 is lost completely. Also note that the characteristic glycine fine structure is not present at the shorter values of $t_{1,max}$ (see cross-peak of Gly47), making identification of these residues difficult.

each t_1 increment must be increased. Useful information can usually be obtained with as little as 0.25-mM protein samples with total acquisition times of 24 h.

Except as noted in the figure captions, the COSY spectra of ubiquitin were acquired with 16 transients for each of 800 t_1 values with an F_1 spectral width of 5800 Hz; TPPI was used for frequency discrimination in t_1; and $t_{1,max}$ = 69 ms. The acquisition time for each FID was 330 ms (spectral width 6250 Hz over 2048 complex points), although $t_{2,max}$ was usually reduced at the processing stage (see Section 6.2.1.3). The water resonance was suppressed by presaturation for 1.5 s. The total acquisition time was just over 7 h.

The effect of $t_{1,max}$ on COSY spectra of ubiquitin is illustrated in Fig. 6.4. As $t_{1,max}$ decreases from 69 to 17.2 ms, the cross-peaks become less intense because of self-cancellation of the antiphase multiplet components. At the shortest values of $t_{1,max}$, the cross-peak of Asp21 is not observed and the characteristic glycine fine structure of the Gly47 cross-peak disappears.

6.2.1.3 Processing   As surmised from [6.1], the cross- and diagonal peaks are 90° out of phase in COSY spectra. In the normal mode of display, the cross-peaks are phased to have antiphase absorptive lineshapes in both dimensions, and the diagonal peaks are phased to have in-phase dispersive lineshapes in both dimensions. Because the tails of the diagonal peaks can obscure information-containing cross-peaks, processing procedures for COSY spectra are designed to maximize cross-peak intensity and minimize diagonal peak intensity. Because cross-peak lineshapes present in a COSY spectrum are antiphase, apodization functions emphasizing the initial parts of each FID do not improve the S/N ratio; instead, window functions must include data points up to $1/(2J_{12})$, where J_{12} is the magnitude of the active coupling constant leading to the cross-peak. Strongly resolution-enhancing window functions such as unshifted or slightly phase shifted sine bells are used for COSY spectra. As an added bonus, the use of an unshifted sine bell in t_2 severely attenuates any residual H_2O peak, leading to spectra with very flat baselines and no significant ridges emanating from the F_2 water stripe. The t_1 window function is a compromise between reduction of the diagonal tails (unshifted or slightly shifted sine bell) and sensitivity (increasingly shifted sine bell).

An unshifted sine bell applied over 150 ms of the acquired data in t_2, and a 15–30° phase-shifted sine bell applied over all data points in t_1 will usually give adequate results for cross-peaks far from the diagonal such as those arising from $^1H^N$–$^1H^\alpha$ correlations. Enhanced resolution, and a concomitant reduction in diagonal peak intensity, can be achieved by increasing $t_{2,max}$ to 200–300 ms, and by shifting the F_1 sine bell by 0–10°. Such processing is used for the observation of cross-peaks close to the diagonal, such as in those involving leucine, valine, or isoleucine methyl resonances. The exact processing parameters used for the example ubiquitin spectra are described in the appropriate figure captions. Figure 6.5 illustrates the use of strong resolution enhancement to aid in the observation of the correlations involving the methyl groups of the leucine, isoleucine, and valine.

The choice of correct phase parameters in F_2 is not readily apparent from the Fourier transform of the row of data collected with the smallest value of t_1. In order to determine the phase parameters, a 1D pulse–acquire spectrum is acquired with the same pulse length, carrier position, and spectral width as the 2D COSY experiment. The phase corrections determined for the 1D spectrum can then be applied during the F_2 processing of the COSY, possibly with the zero-order phase correction adjusted by 90° depending on the exact phase cycles used in the two experiments. With the high speed of modern computer workstations, an alternative is to process the spectrum in both dimensions, determine the required phase

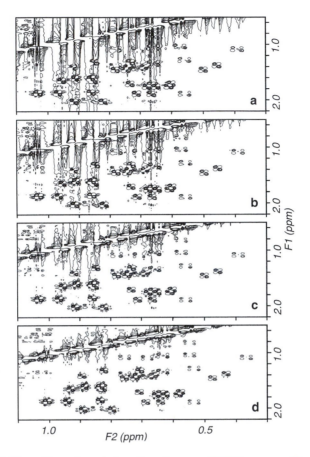

FIGURE 6.5 The effect of t_1 window functions on COSY spectra. The same data set was reprocessed with the sine bells applied in t_1 shifted by the following amounts: (a) 30°, (b) 20°, (c) 10°, (d) 5°. As the window function becomes more resolution-enhancing, the peaks close to the diagonal are less obscured by the dispersive tails from the diagonal.

parameters by examining several rows of the data, and then reprocess the entire spectrum with phase corrections added. The phase correction required in F_1 will result from precession of the spins during the finite pulse lengths of the pulses on either side of t_1, and during the initial value of t_1. The amount of precession can be calculated, thus, the F_1 phase parameters may be calculated and applied during processing (Section 3.3.2.3).

During the extraction of coupling constants from the cross-peak fine structure (Section 6.2.1.5), analysis should be performed on F_2 sections of spectra that have been substantially zero-filled in this dimension. Obtaining a digital resolution of about 0.5 Hz per point is usually adequate. In order to reduce the data storage requirements and processing time of such a large spectrum, F_2 regions that do not contain the peaks of interest may be discarded after the F_2 Fourier transformation.

6.2.1.4 Information Content The power of the COSY experiment lies in the ability to provide correlations between pairs of protons separated by three bonds. However, assigning entire spin systems in the COSY spectrum is rarely possible because of chemical-shift degeneracy in the upfield region of the spectrum. Instead, the COSY spectrum is best used to identify correlations in the so-called fingerprint regions. These regions are well separated from each other and usually contain well-resolved cross-peaks, the number of which reflect the amino acid composition of the protein. The usual regions of interest are the $^1\text{H}^N$–$^1\text{H}^\alpha$ cross-peaks (the backbone fingerprint); the $^1\text{H}^\alpha$–$^1\text{H}^\beta$ cross-peaks (this region can get very crowded and is therefore of limited usefulness); cross-peaks between the aromatic resonances of phenylalanine, tyrosine, tryptophan, and histidine side chains; cross-peaks involving the isoleucine, valine, and leucine methyl groups; and the cross-peaks involving alanine and threonine methyl groups. Other correlation experiments such as TOCSY (Section 6.5) or relayed-COSY (Section 6.2.2) can then be used to connect these fragments together to form complete spin-system assignments. Figure 6.6 indicates these regions of the COSY spectrum, and Figs. 6.7 and 6.8 show details of the two methyl fingerprint regions of ubiquitin. COSY is a relatively sensitive experiment (with the provisos given below for resonances with large linewidths or small coupling constants), and can be used to check sample purity or homogeneity. Although the absence of some expected correlations in the fingerprint regions usually is indicative of experimental shortcomings (e.g., resonance overlap), the presence of extra resonances indicates deficiencies in the purity or conformational homogeneity of the sample.

The COSY spectrum of a protein of N residues, containing P proline and G glycine residues, should display $N - P + G - 1$ correlations in the $^1\text{H}^N$–$^1\text{H}^\alpha$ fingerprint region. At low pH, arginine $^1\text{H}^\delta$–$^1\text{H}^\varepsilon$ cross peaks can be observed in the $^1\text{H}^N$–$^1\text{H}^\alpha$ fingerprint region. The number of observed correlations is often less than expected because of rapid amide proton exchange with solvent, coincidence of the $^1\text{H}^\alpha$ and water resonances (and therefore attenuation by the solvent presaturation process), or degeneracy of both $^1\text{H}^\alpha$ protons of a glycine residue. Double-quantum (2Q) spectroscopy (Section 6.4.1) can provide a useful means of observing correlations absent for the two latter reasons, while the preTOCSY-COSY

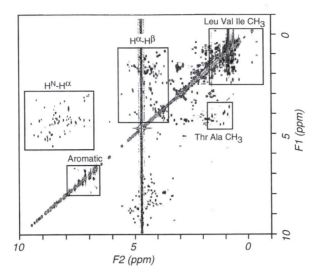

FIGURE 6.6 The five regions of the COSY spectrum containing the fingerprint cross-peaks.

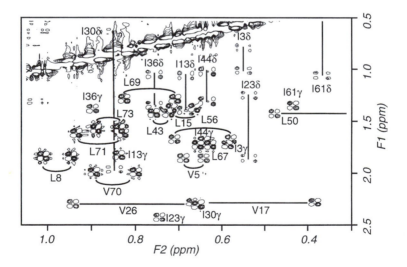

FIGURE 6.7 Section of the H₂O COSY spectrum showing the cross-peaks in the leucine, valine, and isoleucine fingerprint region. The cross-peaks are labeled with the corresponding resonance assignment. This spectrum is the same as in Fig. 6.5d.

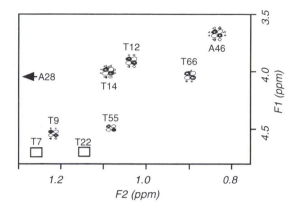

FIGURE 6.8 Section of the COSY spectrum showing the cross-peaks in the alanine and threonine methyl fingerprint region. The cross-peaks are labeled with the corresponding resonance assignment, and the arrow indicates a cross-peak falling outside the spectral region shown. For Thr7 and Thr22, the ^{1}H$^\beta$ protons are coincident with the water resonance and are attenuated by presaturation. These correlations are observed in D$_2$O solution and in a pre-TOCSY COSY (Section 6.2.1.5 and Fig. 6.15).

experiment (Sections 6.2.1.6 and 6.5.5) circumvents problems of ^{1}H$^\alpha$ coincidence with the water resonance.

Ubiquitin contains 76 residues, including 3 prolines and 6 glycines; therefore, 78 ^{1}H^N–^{1}H$^\alpha$ resonances are expected. Seventy peaks are plainly visible in Fig. 6.9. A more detailed analysis of the COSY and other homonuclear correlation spectra indicates that eight cross-peaks are not observed for the following reasons: (1) the ^{1}H$^\alpha$ resonances of Val5, Leu15, and Arg54 are coincident with the H$_2$O signal; (2) the amide protons of Glu24 and Gly53 are very broad; (3) ^{1}H^N and ^{1}H$^\alpha$ resonances of Asp52 and Asp60 have identical shifts; and (4) ^{1}H$^\alpha$ and ^{1}H$^{\alpha\prime}$ of Gly75 are degenerate. Thus, the backbone resonances of all residues can be identified.

6.2.1.5 Quantitation of Scalar Coupling Constants in COSY Spectra The value of the ^3J scalar coupling constant can be determined from the fine structure in a COSY cross-peak for spins without passive coupling partners, such as the $^3J_{H^N H^\alpha}$ of residues other than glycine in H$_2$O solution, and the $^3J_{H^\alpha H^\beta}$ for residues with a single ^{1}H$^\beta$ in D$_2$O solution. The presence of a passive coupling in glycine ^{1}H^N–^{1}H$^\alpha$ cross-peaks and ^{1}H$^\alpha$–^{1}H$^\beta$ cross-peaks within residues with two ^{1}H$^\beta$ protons complicates the analysis and the scalar coupling constants are more conveniently obtained from E-COSY spectra (Section 6.3.3).

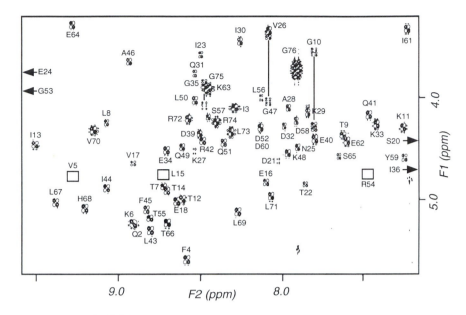

FIGURE 6.9 ^{1}H^N–^{1}H$^\alpha$ fingerprint region of ubiquitin. The cross-peaks are labeled by the one-letter amino acid code and the residue number. The boxes indicate residues whose H$^\alpha$ protons are coincident with the H$_2$O resonance; hence the cross-peak is suppressed by presaturation. These correlations are observed in a pre-TOCSY COSY (Section 6.2.1.6 and Fig. 6.15). See text for a more complete account of the peaks present in this spectrum.

Although the product operator analysis of COSY indicates that the lobes of a cross-peak in a two-spin system are separated by J_{12}, the effect of the linewidth on the anti-phase lineshape has not been considered. Qualitatively, as the linewidth approaches or exceeds the size of the coupling constant, cancellation of the positive and negative multiplet components reduces the intensity and increases the apparent separation of the multiplet as shown in Fig. 6.10. Quantitatively, the antiphase absorptive lineshape is given by [3.24]:

$$A(\nu) = \frac{1}{1 + (J_{12} - 2\nu)^2/\Delta\nu^2_{\text{FWHH}}} - \frac{1}{1 + (J_{12} + 2\nu)^2/\Delta\nu^2_{\text{FWHH}}}$$
$$= \frac{1}{1 + p^2(1 - 2\nu/J_{12})^2} - \frac{1}{1 + p^2(1 + 2\nu/J_{12})^2}, \qquad [6.5]$$

in which the individual multiplet components are assumed to have Lorentzian lineshapes with linewidth $\Delta\nu_{\text{FWHH}}$, $p = J_{12}/\Delta\nu_{\text{FWHH}}$, and the line-

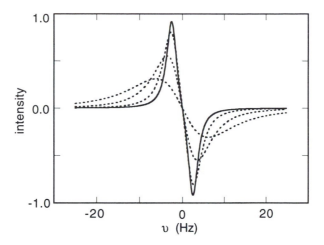

FIGURE 6.10 Changes observed for a Lorentzian antiphase absorptive doublet as the peak separation varies as a function of linewidth. The Lorentzian lines are separated by 5 Hz and have a half-height width of 3 (solid curve), 5, 10, or 20 Hz. All curves are plotted such that a single Lorentzian line would have a height of unity.

shape is normalized to unit amplitude. The extrema are obtained by setting the derivative of [6.5] with respect to ν equal to zero, and solving the resulting equation for ν:

$$\nu = \pm \frac{J_{12}}{2\sqrt{3}p}\left[p^2 + 2\sqrt{p^4 + p^2 + 1} - 1\right]^{1/2}. \qquad [6.6]$$

Since this equation provides the values of ν at the maximum and minimum amplitude of the lineshape, substitution of [6.6] back into [6.5] allows a determination of the apparent peak heights of the multiplet lines. The resulting equation is shown graphically in Fig. 6.11. This confirms the qualitative view of Fig. 6.10: As J_{12} is reduced with respect to $\Delta\nu_{FWHH}$, the intensity of a COSY cross-peak decreases. Thus, in the case of equivalent linewidth and coupling constant ($p = 1$), the resultant COSY peak intensity will be $\approx 80\%$ of the value of a single Lorentzian line; linewidths in the range 6–8 Hz are expected for ubiquitin, and the intensity of many cross-peaks will be reduced by this amount or more.

The decrease in cross-peak intensity caused by broad lines can severely limit the information content of a COSY spectrum, but more insidious is the effect on the separation of the peak maxima. Since the values of ν given by Eq. [6.6] represent the frequency at which the amplitude is

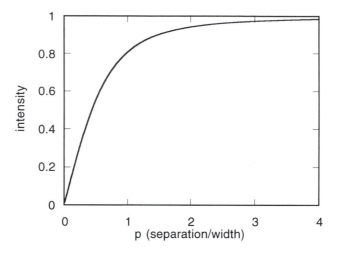

FIGURE 6.11 Peak height of a Lorentzian antiphase absorptive doublet as the peak separation varies as a function of linewidth. The variable p is defined as the ratio of the scalar coupling constant to the half-height linewidth.

a maximum or a minimum (given by the "+" or "−" solutions, respectively), the total separation of the peak maxima is given by

$$\nu_a = \frac{J_{12}}{\sqrt{3}p}\left[p^2 + 2\sqrt{p^4 + p^2 + 1} - 1\right]^{1/2}. \qquad [6.7]$$

This equation was first described by Neuhaus *et al.* (*6*), and is most useful in the form shown in Fig. 6.12. The graph indicates that the observed separation is always larger than the actual value of J_{12}. Moreover, Fig. 6.12 indicates that the smaller the value of p (the larger the linewidth with respect to the multiplet separation), the larger the difference between the actual separation of the initial Lorentzian lines (J_{12}) and the observed separation of the peak maxima. Thus, to continue the example described above, when the linewidth is equal to the J_{12} coupling ($p = 1$), the observed separation of peak maxima will be 7% higher than J_{12}. The error in J_{12} increases rapidly as p decreases below 1. Self-cancellation is especially problematic for residues in a helical environment; the $^3J_{H^NH^\alpha}$ coupling constant calculated for an ideal helix (with $\phi = -60°$ and $\theta = |\phi - 60°|$) using the Karplus equation [8.2] is 4.2 Hz.

A method of determining scalar coupling constants that takes into account the natural properties of a Lorentzian line to overcome the problems of self-cancellation has been proposed by Kim and Prestegard (*8*).

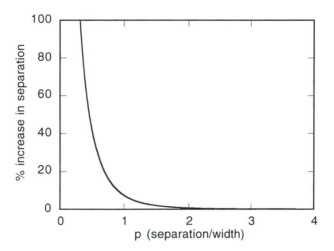

FIGURE 6.12 Apparent separation of peak maxima relative to the actual separation of a pair of absorptive antiphase Lorentzian lines as a function of linewidth. The variable p is defined as the ratio of the scalar coupling constant to the half-height linewidth. As p approaches infinity, the separation approaches J_{12}. As p approaches to zero, the separation tends to a limit of $\Delta\nu_{FWHH}/\sqrt{3}$.

The method takes advantage of the differences in lineshapes of absorptive and dispersive antiphase Lorentzian multiplets. Analogously to [6.7], a pair of simultaneous equations describing the coupling constant and the linewidth, J_{12} and $\Delta\nu_{FWHH}$, in terms of the absorptive and dispersive peak separations, ν_a and ν_d, can be solved to give (Fig. 6.13):

$$64v_a^2 J_{12}^6 - 64v_a^2 v_d^2 J_{12}^4 + (-144v_a^6 + 96v_a^4 v_d^2 + 36v_a^2 v_d^4) J_{12}^2$$

$$+ 81v_a^8 - 36v_a^6 v_d^2 - 42v_a^4 v_d^4 - 4v_a^2 v_d^6 + v_d^8 = 0 \qquad [6.8]$$

The real root of this cubic equation in J_{12}^2 provides the coupling constant. The method is performed most easily by processing the COSY with anti-phase absorptive cross-peaks (i.e., in the normal way, although with high digital resolution). The absorptive peak separation, ν_a, is measured from a row (F_2 cross section) through a given cross-peak. Adding a zero-order phase correction of 90° to the row allows measurement of the dispersive peak separation, ν_d, for the same cross-peak. This method assumes that the lineshape is Lorentzian and is not applicable if window functions such as shifted sine bells or Lorentzian–Gaussian transformations have been employed in t_2. The method works reasonably well for cross-peaks with high S/N ratio; however, for noisy data,

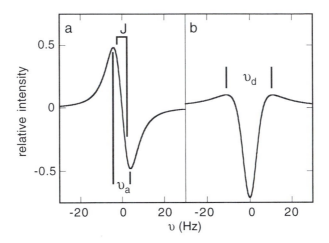

FIGURE 6.13 Measurement of the maxima separation for a pair of antiphase Lorentzian lines in absorption (v_a) and dispersion (v_d). In both panels, the real separation is 5.0 Hz (denoted by J in the left-hand panel) and the half-height linewidth is 10 Hz ($p = 0.5$). Measurement of v_d is difficult when $p < 1.0$ because of the low intensity of the broad positive lobes.

accurate measurement of v_d is difficult because the tops of the peaks are broad.

Alternatively, [6.5] can be fit directly to the F_2 cross section) through a cross-peak (or several rows coadded to improve the S/N ratio) by a nonlinear least-squares algorithm to determine the values of the scalar coupling constant and linewidth most consistent with the lineshape (9,10). Once again, the spectrum should be processed to maintain Lorentzian lineshapes in F_2. The accuracy of this approach also is compromised as p decreases owing to the lower S/N ratios. Several methods of analysis of ¹H spectra have been reported to be of use in situations where the linewidth is greater than J_{12} (11–13).

Practically, scalar coupling constants that are less than about half of the linewidth are difficult to measure by analysis of antiphase splittings in the COSY experiment. Even the "best" method of direct line fitting is rarely useful for the larger proteins studied by NMR (>15 kDa) and scalar coupling constants are better obtained using heteronuclear experiments (Section 7.5).

6.2.1.6 Experimental Variants   The basic COSY pulse sequence has been extensively modified since its initial conception. Three examples will be briefly mentioned here: COSY-β, purged COSY or P-COSY, and

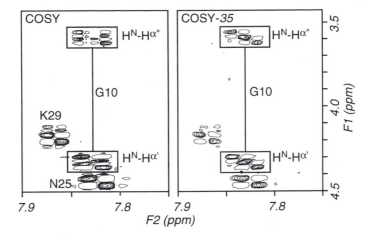

FIGURE 6.14 Cross-peaks of Gly10 in the COSY-35 experiment. Compared to the normal COSY, contributions from the passive coupling $^2J_{H^{\alpha'}H^{\alpha''}}$ are removed from the $^1H^N$–$^1H^\alpha$ cross-peaks in F_1 while the passive coupling of $^1H^N$ to the other $^1H^\alpha$ is removed in F_2. The fine structure of $^1H^N$–$^1H^\alpha$ cross-peaks to nonglycine residues is not affected by the reduced flip angle.

pre-TOCSY COSY. The first of these provides a means to simplify cross-peak fine structure while the latter two address deficiencies in the basic experiment.

In the COSY-β experiment, the rotation angle of the second pulse is $\beta < 90°$. Reduction of the length of the final pulse in the COSY-β experiment has two effects on the resulting spectrum: (1) the diagonal peaks are more intense relative to the cross-peaks because coherence transfer scales as $\sin^2(\beta)$ and (2) the fine structure of the cross-peaks changes because connected and unconnected transitions have different intensities. In spin systems with three or more mutually coupled spins, certain components of a given cross-peak arising from passive coupling will be reduced in intensity. Acquisition with $\beta = 35°$ provides reasonable sensitivity while suppressing cross-peak components arising from unconnected transitions by 10-fold compared to those from connected transitions (4). An example of the simplification obtained is shown in Fig. 6.14 for the cross-peaks arising from Gly10. The multiplet fine structure in COSY-β is similar to that observed in E-COSY (14), and will be discussed in more detail in Section 6.3.3. Although the removal of some elements of fine structure can facilitate identification of peaks in crowded regions of the spectrum, the COSY-β experiment offers little benefit for spin-system identification and principally permits scalar coupling constants to be measured without errors of the type described in Fig. 6.12.

The P-COSY experiment provides a way to remove the dispersive diagonal tails of the COSY experiment without a significant loss in sensitivity *(15)*. The central idea in the technique is similar to that of the primitive E-COSY (Section 6.3.3) or PE-COSY. In the PE-COSY, a COSY-35 and a COSY-0 are combined with the result that the diagonal is severely attenuated *(16)*. This experiment suffers from a lack of sensitivity due to the low inherent magnitude of cross-peaks in COSY-β, the need to acquire two complete 2D spectra, and the losses incurred by subtracting two spectra of comparable S/N ratio. These issues are rectified in the P-COSY by generating cross-peaks in a normal COSY experiment and preparing a "diagonal only" spectrum by repeated left shifting of a 1D spectrum acquired with high S/N *(15)*. The resulting spectrum is comparable in quality to a 2QF-COSY (Section 6.3.1) but with 2-fold greater sensitivity.

Cross-peaks in the ^{1}H^N–^{1}H$^\alpha$ fingerprint region of COSY spectra are attenuated by presaturation of the solvent resonance. The simple expedient of including a short isotropic mixing period (Section 4.2.2) between the presaturation period and the first 90° pulse restores intensity to these cross-peaks. The mixing period transfers magnetization from ^{1}H^N to ^{1}H$^\alpha$ via the $^3J_{\text{H}^N\text{H}^\alpha}$ scalar coupling interaction; therefore, ^{1}H$^\alpha$ magnetization is present at the start of the COSY sequence even for the protons coincident with the water resonance. This variant is referred to as the pre-TOCSY COSY *(17)*. An example of such a spectrum is shown in Fig. 6.15; the cross-peaks from Val5, Leu15, and Arg54 are plainly visible in this spectrum, but are missing from Fig. 6.9. Further discussion of pre-TOCSY sequences can be found in Section 6.5.5.

6.2.2 RELAYED COSY

Conceptually, the relayed coherence transfer experiments (RCT-COSY) or relayed-COSY (R-COSY) are simple extensions of the COSY experiment *(18)*. Instead of acquiring the spectrum after a single coherence transfer step from spin I_1 to spin I_2, a delay is introduced prior to acquisition to allow antiphase magnetization to develop for a second time. If spin I_2 has coupling partners other than I_1 (e.g., spin I_3), antiphase coherence will develop between I_2 and I_3 and a third 90° pulse will transfer coherence from I_2 to I_3. Since chemical shifts are monitored only during t_1 (prior to the first coherence transfer step) and t_2 (after the second coherence transfer step), the result will be a cross-peak between spins I_1

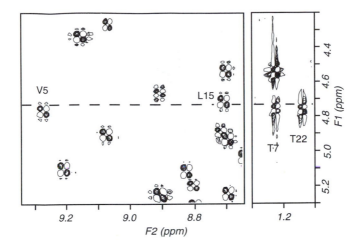

FIGURE 6.15 Two sections of the pre-TOCSY COSY spectrum of ubiquitin. The experiment was recorded with 1.5 s of presaturation followed by 27 ms of DIPSI-2rc isotropic mixing (*47*). Sixteen transients were collected for each of 800 increments of t_1 ($t_{1,\max}$ = 69 ms). The labeled cross-peaks were not present in the standard COSY experiment (Figs. 6.8 and 6.9) because of presaturation.

and I_3 even if I_1 and I_3 are not directly coupled. For simple molecules containing only a few resonances, such spin gymnastics are redundant, since in the COSY experiment both the $I_1 \rightarrow I_2$ and $I_2 \rightarrow I_3$ cross-peaks are readily observed. However, in the complex spectra of proteins, chemical-shift degeneracy of spin I_2 with resonances of other spin systems can hinder the assignment of I_1 and I_3 to the same spin system. In such cases the cross-peak between I_1 and I_3 in the R-COSY experiment places I_1 and I_3 unequivocally in the same amino acid spin system. The pulse sequence used to observe relayed coherence transfer is shown in Fig. 6.16.

6.2.2.1 Product Operator Analysis

For this analysis, the linear spin system I_1–I_2–I_3 is considered, in which J_{12} and J_{23} are resolved and $J_{13} = 0$. Considering initial I_1 magnetization, the operator terms present after the 90°_x–t_1–90°_x sequence are given by [6.1]. The $2I_{1z}I_{2y}$ term of this expression represents transfer of coherence from I_1 to I_2 and is the first step in the generation of an $I_1 \rightarrow I_3$ cross-peak. Appropriate phase cycling preserves this term (and also the I_{1x} single quantum coherence) and suppresses other coherence orders. During the spin-echo $\tau/2$–180°_x–$\tau/2$ sequence,

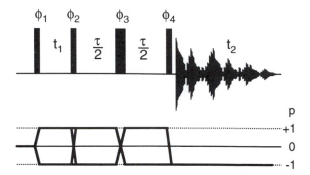

FIGURE 6.16 The pulse sequence and coherence level diagram for the R-COSY experiment. The phases are cycled as follows: $\phi_1 = 4(x, -x)$; $\phi_2 = 4(x)\ 4(-x)$; $\phi_3 = 8(x)$; $\phi_4 = 2(x, x, -x, -x)$; and receiver $= 2(x, -x)$. CYCLOPS is performed on all pulses and the receiver to produce a 32-step cycle. The 180° pulse during the mixing period can be a composite $90°_{\phi2} - 180°_{\phi2+\pi/2} - 90°_{\phi2}$ pulse.

chemical shifts are refocused and evolution under the J_{12} and J_{23} scalar coupling Hamiltonians yields the following terms:

$$I_{1x} \sin(\Omega_1 t_1) \cos(\pi J_{12} t_1) - 2I_{1z}I_{2y} \sin(\Omega_1 t_1) \sin(\pi J_{12} t_1) \xrightarrow{\tau/2 - \pi_x - \tau/2}$$

$$[I_{1x} \cos(\pi J_{12}\tau) + 2I_{1y}I_{2z} \sin(\pi J_{12}\tau)] \sin(\Omega_1 t_1) \cos(\pi J_{12} t_1)$$

$$+ [-2I_{1z}I_{2y} \cos(\pi J_{23}\tau) + 4I_{1z}I_{2x}I_{3z} \sin(\pi J_{23}\tau)]$$

$$\times \sin(\Omega_1 t_1) \sin(\pi J_{12} t_1) \cos(\pi J_{12}\tau)$$

$$+ [I_{2x} \cos(\pi J_{23}\tau) + 2I_{2y}I_{3z} \sin(\pi J_{23}\tau)]$$

$$\times \sin(\Omega_1 t_1) \sin(\pi J_{12} t_1) \sin(\pi J_{12}\tau). \tag{6.9}$$

Application of the final $90°_x$ pulse creates the following terms immediately prior to t_2:

$$[I_{1x} \cos(\pi J_{12}\tau) - 2I_{1z}I_{2y} \sin(\pi J_{12}\tau)] \sin(\Omega_1 t_1) \cos(\pi J_{12} t_1)$$

$$+ [2I_{1y}I_{2z} \cos(\pi J_{23}\tau) + 4I_{1y}I_{2x}I_{3y} \sin(\pi J_{23}\tau)]$$

$$\times \sin(\Omega_1 t_1) \sin(\pi J_{12} t_1) \cos(\pi J_{12}\tau)$$

$$+ [I_{2x} \cos(\pi J_{23}\tau) - 2I_{2z}I_{3y} \sin(\pi J_{23}\tau)]$$

$$\times \sin(\Omega_1 t_1) \sin(\pi J_{12} t_1) \sin(\pi J_{12}\tau). \tag{6.10}$$

Of the six components in [6.10], only the three-spin coherence $4I_{1y}I_{2x}I_{3y}$ is unobservable. The most interesting peak arises by relay from

I_1 via I_2 to I_3, and is described by the $2I_{2z}I_{3y}$ term. This peak appears at the I_3 chemical shift and is antiphase with respect to spin I_2 in F_2. It appears at the chemical shift of I_1 in F_1, and the $\sin(\Omega_1 t_1)\sin(\pi J_{12}t_1)$ modulation indicates antiphase lineshape with respect to I_2. In the normal mode of display, relay peaks are phased to be absorptive in both dimensions, and this will be assumed in the following discussion of the phases of other peaks.

The I_{1x} and $2I_{1y}I_{2z}$ terms contribute to the I_1 diagonal peak. The former is analogous to the diagonal peak in a normal COSY experiment and has a dispersive in-phase lineshape in both dimensions. The second diagonal component is an absorptive antiphase signal in both dimensions. Thus, as in the regular COSY experiment, the diagonal is dominated by dispersive in-phase terms that must be attenuated by severe window functions.

Finally, the R-COSY spectrum contains cross-peaks between spins that are directly coupled ("COSY"-type peaks). The $2I_{1z}I_{2y}$ and I_{2x} terms of [6.10] contribute to these peaks. The former possesses absorptive antiphase character in F_2 and dispersive in-phase character in F_1; the latter possesses dispersive in-phase character in F_2 and absorptive antiphase character in F_1. The contributions of each component to the overall lineshape of these peaks is dependent on the length of the mixing period as described by the $\sin(\pi J_{12}\tau)$ and $\cos(\pi J_{23}\tau)\sin(\pi J_{12}\tau)$ trigonometric functions, respectively. These peaks can never be phased to pure absorption lineshapes, and dispersive tails emanate from the COSY peaks in the R-COSY spectrum.

Although the results observed in R-COSY spectra of the larger spin systems found in real amino acids are more complex than the equations described above, the general trends are similar. A more complete description of the transfer functions in particular amino acids has been presented by Bax and Drobny (*19*).

6.2.2.2 Experimental Protocol From the preceding discussion, cancellation of antiphase cross-peaks also must be avoided in R-COSY experiments. In addition to the standard COSY acquisition parameters, a value must be chosen for the mixing delay, τ, during which the second antiphase state develops. In H_2O solution, the most useful peaks appear at the chemical shift of $^1H^\beta$ in F_1 and at the chemical shift of $^1H^N$ in F_2 (Section 8.1.1). For such cross-peaks, coherence will be transferred from $^1H^\beta$ to $^1H^\alpha$ during t_1, and from $^1H^\alpha$ to $^1H^N$ during the relay period, i.e., $I_1 = {}^1H^\beta$, $I_2 = {}^1H^\alpha$ and $I_3 = {}^1H^N$.

Given the preceding discussion, $t_{1,\max}$ must be long enough to allow transfer via the $^1H^\alpha$–$^1H^\beta$ coupling (i.e., $t_{1,\max} \geq 1/(2\,{}^3J_{H^\alpha H^\beta})$). For many residues in folded proteins, at least one $^3J_{H^\alpha H^\beta}$ coupling will be in the 8–

12-Hz range. Thus, R-COSY experiments utilize $t_{1,max}$ in the range 40–60 ms. For an F_1 spectral width of 5 kHz, 400–600 t_1 increments are recorded (assuming quadrature detection with TPPI).

During the mixing time τ, the anti-phase coherence transferred from $^1H^\beta$ to $^1H^\alpha$ during t_1 must refocus with respect to $^1H^\beta$ and defocus with respect to $^1H^N$. Thus, the magnitude of the transfer from $^1H^\beta$ to $^1H^N$ depends on both $^3J_{H^NH^\alpha}$ and $^3J_{H^\alpha H^\beta}$. Analytical functions describing the transfer have been calculated for a variety of spin-system types using values of coupling constants commonly found in proteins (19,20). Maxima in the transfer functions occur for τ = 40–60 ms; however, relaxation during τ reduces the intensity of all peaks. Therefore, compromise values of τ = 20–40 ms provide adequate coherence transfer without extensive relaxation losses. Usually, with a 2–4 mM sample, adequate $^1H^N$–$^1H^\beta$ cross-peak intensities are obtained by acquiring 32–64 transients per t_1 increment.

The R-COSY spectrum analyzed in Sections 6.2.2.3 and 6.2.2.4 was acquired from a 2-mM ubiquitin solution in H_2O. In this study 32 transients were collected for each of 800 increments ($t_{1,max}$ = 69 ms). A mixing time of 22 ms was used during the relay portion of the experiment (τ). A composite 180° pulse was used to refocus chemical shifts in the middle of τ (Section 3.4.2). The total acquisition time was 14 hs.

6.2.2.3 Processing   Given the results of the product operator analysis, processing of R-COSY data is very similar to that used for COSY data. Harsh window functions (unshifted sine bells in F_2 and slightly shifted sine bells in F_1) are required to minimize the dispersive diagonal peaks and COSY-type cross-peaks.

6.2.2.4 Information Content   The most interesting cross-peaks in the R-COSY experiment result from two coherence transfer steps. As a rule, all possible $^1H^\beta$–$^1H^N$ cross-peaks in the R-COSY spectrum are not observed because of the dependence on two coupling constants (either or both of which could be small). However, in side chains containing $C^\beta H_2$ groups, at least one of the $^1H^\alpha$–$^1H^\beta$ couplings is commonly greater than 8 Hz and provides an efficient means of generating one $^1H^\beta$–$^1H^N$ cross-peak.

These generalizations are exemplified by the peaks observed in the R-COSY spectrum of ubiquitin. Of the 45 side chains that have an observable backbone amide proton and a β-methylene group, six residues have no $^1H^N$–$^1H^\beta$ correlations, 30 (67%) have a single $^1H^N$–$^1H^\beta$ correlation, and in only 9 cases (20%) are both $^1H^N$–$^1H^\beta$ correlations observed. Also, 11 of the 18 side chains with an amide proton and a single β-proton exhibit

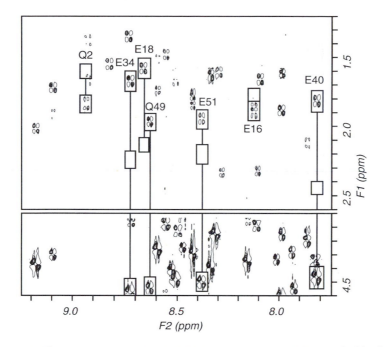

FIGURE 6.17 Two sections of a R-COSY spectrum (τ = 22 ms) of ubiquitin in
H_2O solution showing $^1H^N$–$^1H^\beta$ (top) and $^1H^N$–$^1H^\alpha$ (bottom) correlations. The
glutamate and glutamine resonances in the depicted region are surrounded by
rectangles. Note that only one of the possible $^1H^N$–$^1H^\beta$ correlations has appreciable
intensity (top) and that many of the $^1H^N$–$^1H^\alpha$ COSY-type peaks have distinct
dispersive tails.

an $^1H^N$–$^1H^\beta$ cross-peak. Although acquisition of R-COSY spectra with
longer mixing times can permit identification of additional $^1H^N$–$^1H^\beta$ correlations arising from small $^3J_{H^NH^\alpha}$ coupling constants, DR-COSY (Section
6.2.3) and TOCSY (Section 6.5) experiments are better suited to detecting
such correlations. The section of the R-COSY spectrum shown in Fig.
6.17 (top) depicts several of the relayed $^1H^N$–$^1H^\beta$ cross-peaks. The
$^1H^N$–$^1H^\alpha$ section is also shown in Fig. 6.17 (bottom) to emphasize the
dispersive nature of the COSY-type peaks in the R-COSY spectrum.

6.2.3 DOUBLE-RELAYED COSY

Additional coherence transfer steps can be obtained by inserting a
second $\tau/2$–$180°$–$\tau/2$–$90°$ sequence prior to t_2 in the R-COSY experiment
(21). This variant is termed the double-relayed-COSY (DR-COSY), and

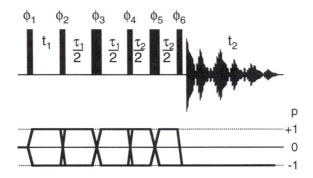

FIGURE 6.18 Pulse sequence and coherence-level diagram for the DR-COSY experiment. The 180° pulses can be replaced with composite pulses as described in Fig. 6.16. The phases are cycled as follows: $\phi_1 = 8(x, -x)$; $\phi_2 = 8(x)\,8(-x)$; $\phi_3 = 16(x)$; $\phi_4 = 4(x)\,4(-x)\,4(x)\,4(-x)$; $\phi_5 = 16(y)$; $\phi_6 = 4(x, x, -x, -x)$; and receiver $= x, -x, -x, x, 2(-x, x, x, -x), x, -x, -x, x$. CYCLOPS is performed on all pulses to yield a 64-step phase cycle.

the pulse sequence is shown in Fig. 6.18. The DR-COSY correlates protons through a network of three scalar couplings. In principle, the COSY experiment could be extended indefinitely until correlations were observed between all spins in a particular spin system. However, inefficient coherence transfer and relaxation during the periods of free precession reduce the sensitivity to such an extent that the DR-COSY represents the practical limit of such experiments.

The protocol for DR-COSY contains essentially the same considerations as for the R-COSY, except that values must be evaluated for both τ_2 and τ_1. The most useful peaks will be transferred through three couplings to ^{1}H^N. Instead of optimizing transfer from ^{1}H$^\gamma$ to ^{1}H^N, the most effective results are obtained by transfer from ^{1}H$^{\beta'}$ to ^{1}H$^{\beta''}$ during t_1 (via the large ^{1}H$^{\beta'}$–^{1}H$^{\beta''}$ coupling), from ^{1}H$^{\beta''}$ to ^{1}H$^\alpha$ during τ_1 (via the largest of the two possible ^{1}H$^\beta$–^{1}H$^\alpha$ couplings), and from ^{1}H$^\alpha$ to ^{1}H^N during τ_2. In this way, relatively intense cross-peaks are obtained to both ^{1}H$^\beta$ protons, and DR-COSY represents a significant improvement over R-COSY in which correlations usually are observed to only one ^{1}H$^\beta$.

Although a product operator analysis is not presented here for the DR-COSY, important results are obtained by analogy with the R-COSY sequence. Double-relay cross-peaks are absorptive antiphase in both dimensions, whereas COSY and single-relay peaks contain some in-phase dispersive contributions. Choices of mixing time to optimize the intensity of various correlations have been determined by calculating the cross-peak intensities as a function of τ_2 and τ_1 for a variety of possible coupling constants and spin systems (20). Setting the delays $\tau_1 = 28$ ms and $\tau_2 = 35$ ms maximizes the $^1H^{\beta'}$–$^1H^{\beta''}$–$^1H^\alpha$–$^1H^N$ transfer. Transfer during t_1 is via the large 15-Hz $^1H^{\beta'}$–$^1H^{\beta''}$ coupling, so $t_{1,max}$ can be slightly shorter than in the R-COSY, although the cross-peaks are still antiphase, and subject to self-cancellation if $t_{1,max}$ is too short. Alternatively, setting the delays $\tau_1 = 20$ ms and $\tau_2 = 31$ ms will maximize transfer from $^1H^\gamma$ to $^1H^N$ (20,22). Total acquisition times usually need to be longer than for R-COSY because the most interesting correlations result from three transfer steps. The literature indicates that 48-h acquisitions are not uncommon for this experiment. Processing is analogous to that used in COSY and R-COSY experiments (Section 6.2.1.3).

A DR-COSY spectrum was acquired from an H_2O solution of ubiquitin. In this study 64 transients were collected for each of 576 increments in t_1 with a $t_{1,max} = 50$ ms and a total acquisition time of 21 hs. The mixing times τ_1 and τ_2 were set to 28 ms and 35 ms, respectively, to emphasize H^β–H^N correlations, and composite 180° pulses were used in the middle of both delays (Section 3.4.2).

Figure 6.19 depicts a region of the DR-COSY in which several correlations from both protons of β-methylene groups are observed to the backbone amide proton. In the equivalent region of the R-COSY only one of the correlations is observed (compare with Fig. 6.17). Correlations from $^1H^\gamma$ to $^1H^N$ are absent for the glutamine and glutamate side chains identified in Fig. 6.19. In the spectrum as a whole, with mixing times $\tau_1 = 28$ ms and $\tau_2 = 35$ ms, correlations from $^1H^{\beta''}$ and $^1H^{\beta'}$ to $^1H^N$ are observed in 39 (87%) of the 45 residues with a β-methylene group. Of these side chains, 31 also contained γ-protons, but in only three cases were $^1H^\gamma$–$^1H^N$ correlations observed (both $^1H^\beta$–$^1H^N$ correlations were also observed for these three cases). For side chains containing a single β-proton, more $^1H^N$–$^1H^\beta$ correlations were observed in the DR-COSY than in the R-COSY experiment (14 out of 18 compared to 11 out of 18), and 13 correlations from $^1H^N$ to at least one $^1H^\gamma$ were observed.

6.3 Multiple-Quantum-Filtered COSY

The double-quantum-filtered (2QF) COSY experiment was developed by Ernst and coworkers as an alternative to COSY (23,24). The 2QF-COSY is the simplest version of a family of experiments based on filtration through a p-quantum state. Not all amino acid spin systems are capable of achieving some of the higher quantum states; hence, experiments with $p > 2$ provide useful spectral simplification (25).

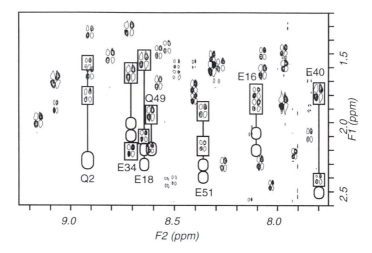

FIGURE 6.19 Section of a DR-COSY spectrum ($\tau_1 = 28$ ms, $\tau_2 = 35$ ms) of ubiquitin acquired from H_2O solution. The same spectral region as the top of Fig. 6.17 is shown, with glutamine and glutamate $^1H^\beta$ and $^1H^\gamma$ resonance positions denoted by the rectangles and ellipses, respectively. The particular delays used during the relay periods have optimized transfer of both $^1H^\beta$ to $^1H^N$ at the expense of $^1H^\gamma$–$^1H^N$ correlations.

All the pQF COSY experiments have the same basic pulse sequence consisting of three 90° pulses, with the first two separated by t_1 and the last two separated by a short delay as shown in Fig. 6.20. As described in Section 4.3.2.1, phase cycling is used to select the particular coherence level of interest. Selection of a desired order of multiple-quantum filtration is achieved by incrementing the phases of the pulses as follows:

$$\phi_1 \text{ and } \phi_2: \quad n\pi/p;$$

$$\phi_3: \qquad\quad 0; \tag{6.11}$$

$$\text{Receiver:} \quad -n\pi,$$

in which the integer n is incremented from 0 to $2p - 1$. Thus, the basic phase cycle consists of $2p$ steps. Note that for $p > 2$, appropriate selection involves phase shift increments of less than 90°; hence the x, y, $-x$, $-y$ notation is no longer appropriate to describe the phases. Such shifts are readily accomplished with the digital phase-shifting hardware present on modern spectrometers. Alternative cycles involving phase shifting of ϕ_3 also can be used to obtain the desired coherence selection. Formally,

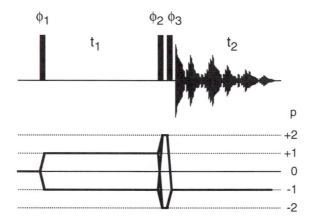

FIGURE 6.20 Pulse sequence and coherence-level diagram for 2QF-COSY. The basic 4-step phase cycle for 2QF-COSY is $\phi_1 = \phi_2 = x, y, -x, -y$; $\phi_3 = x$; and receiver $= 2(x, -x)$. The same pulse sequence is used for 3QF-COSY with the following 6-step phase cycle: $\phi_1 = \phi_2 = 0°, 60°, 120°, 180°, 240°, 300°$; $\phi_3 = 0°$; and receiver $= 3(0°, 180°)$. Axial peak suppression and CYCLOPS phase cycling yields a 32-step phase cycle for 2QF-COSY and 48-step phase cycle for 3QF-COSY.

these phase cycles retain coherence orders kp, where $k = \pm1, \pm3, \pm5$ ..., but the higher-order coherences have low intensity (see Section 6.3.1). In addition to the basic $2p$ cycle, artifacts in the spectra are reduced by performing CYCLOPS on all pulses and axial peak suppression on the first pulse (Section 4.3.2.3), leading to $16p$ steps in the cycle.

6.3.1 2QF-COSY

As noted above, one of the major shortcomings of the COSY experiment is the 90° phase difference between the diagonal and cross-peaks, leading to a dispersive diagonal in the most usual mode of presentation. The intense, sprawling nature of the diagonal, particularly for uncoupled singlet resonances, distorts and obscures cross-peaks close to the diagonal. The double-quantum-filtered (2QF) COSY experiment overcomes some of the drawbacks of COSY (23,24). The multiplet structures of cross-peaks in the COSY and 2QF-COSY are identical. The benefits of the 2QF-COSY are the absorptive antiphase lineshapes of the diagonal peaks of coupled spins, and the severe attenuation of diagonal peaks from uncoupled resonances. The drawbacks of 2QF-COSY are a 2-fold decrease in sensitivity and a longer phase cycle.

6.3.1.1 Product Operator Analysis   To begin, a scalar-coupled two-spin system will be treated. The product operators present following the pulse sequence segment $90^\circ_x-t_1-90^\circ_x$ are given by [6.1]. The phase cycle selects only double quantum coherence present after the second pulse; therefore, all terms can be disregarded except the multiple-quantum term $-2I_{1x}I_{2y} \cos(\Omega_1 t_1) \sin(\pi J_{12} t_1)$. This product operator is a superposition of double-quantum and zero-quantum coherence and can be expanded accordingly as (see Section 2.7.5),

$$-\tfrac{1}{2}[2I_{1x}I_{2y} + 2I_{1y}I_{2x} + 2I_{1x}I_{2y} - 2I_{1y}I_{2x}] \cos(\Omega_1 t_1) \sin(\pi J_{12} t_1)$$
$$= -\tfrac{1}{2}[DQ_y - ZQ_y] \cos(\Omega_1 t_1) \sin(\pi J_{12} t_1). \qquad [6.12]$$

The last two terms in brackets, $2I_{1x}I_{2y} - 2I_{1y}I_{2x}$, represent zero-quantum coherence and are rejected by the phase cycle. The first two terms in the brackets, $2I_{1x}I_{2y} + 2I_{1y}I_{2x}$, represent double-quantum coherence and are selected by the phase cycle. The final 90°_x pulse generates the following observable single quantum coherence terms:

$$-\tfrac{1}{2}[2I_{1x}I_{2y} + 2I_{1y}I_{2x}] \cos(\Omega_1 t_1) \sin(\pi J_{12} t_1) \xrightarrow{\left(\tfrac{\pi}{2}\right)_x}$$
$$-\tfrac{1}{2}[2I_{1x}I_{2z} + 2I_{1z}I_{2x}] \cos(\Omega_1 t_1) \sin(\pi J_{12} t_1). \qquad [6.13]$$

Simple inspection shows that the $2I_{1x}I_{2z}$ product operator evolves at the chemical shift of the I_1 spin and is antiphase with respect to the scalar coupling during t_2. The $2I_{1z}I_{2x}$ product operator evolves at the chemical shift of the I_2 spin and is antiphase with respect to the scalar coupling during t_2. Because both terms evolve with a frequency of Ω_1 during t_1, the former represents an I_1 spin diagonal peak, and the latter represents an $I_1 \rightarrow I_2$ cross-peak. The diagonal and cross-peak terms have the same phase in F_2 (both terms contain x operators) and are modulated by the same t_1 trigonometric terms; expansion of the trigonometric coefficients indicates that both the diagonal and cross-peaks are antiphase in the F_1 dimension. Consequently, the diagonal and cross-peaks in a 2QF-COSY spectrum can be phased to pure absorption in both dimensions. Thereby, the 2QF-COSY experiment eliminates problems associated with dispersive tails emanating from the diagonal resonances in COSY spectra. This attribute influences the processing and extraction of information from 2QF-COSY spectra.

Consideration of the evolution of a three-spin system during a 2QF-COSY is useful at this point to reveal further features of the spectrum. The system now contains three spins—I_1, I_2, and I_3—with couplings J_{12}, J_{13}, and J_{23}. Evolution of initial I_{1z} magnetization under the chemical shift,

J_{12} scalar coupling, and J_{13} scalar coupling Hamiltonians leads to the following antiphase terms:

$$I_z \xrightarrow{\left(\frac{\pi}{2}\right)_x} -I_{1y}$$

$$\xrightarrow{t_1} 2I_{1x}I_{2z}\cos(\Omega_1 t_1)\sin(\pi J_{12}t_1)\cos(\pi J_{13}t_1)$$
$$+2I_{1y}I_{2z}\sin(\Omega_1 t_1)\sin(\pi J_{12}t_1)\cos(\pi J_{13}t_1)$$
$$+2I_{1x}I_{3z}\cos(\Omega_1 t_1)\cos(\pi J_{12}t_1)\sin(\pi J_{13}t_1)$$
$$+2I_{1y}I_{3z}\sin(\Omega_1 t_1)\cos(\pi J_{12}t_1)\sin(\pi J_{13}t_1)$$
$$+4I_{1y}I_{2z}I_{3z}\cos(\Omega_1 t_1)\sin(\pi J_{12}t_1)\sin(\pi J_{13}t_1)$$
$$-4I_{1x}I_{2z}I_{3z}\sin(\Omega_1 t_1)\sin(\pi J_{12}t_1)\sin(\pi J_{13}t_1)$$

$$\xrightarrow{\left(\frac{\pi}{2}\right)_x} -2I_{1x}I_{2y}\cos(\Omega_1 t_1)\sin(\pi J_{12}t_1)\cos(\pi J_{13}t_1)$$
$$-2I_{1z}I_{2y}\sin(\Omega_1 t_1)\sin(\pi J_{12}t_1)\cos(\pi J_{13}t_1)$$
$$-2I_{1x}I_{3y}\cos(\Omega_1 t_1)\cos(\pi J_{12}t_1)\sin(\pi J_{13}t_1)$$
$$-2I_{1z}I_{3y}\sin(\Omega_1 t_1)\cos(\pi J_{12}t_1)\sin(\pi J_{13}t_1)$$
$$+4I_{1z}I_{2y}I_{3y}\cos(\Omega_1 t_1)\sin(\pi J_{12}t_1)\sin(\pi J_{13}t_1)$$
$$-4I_{1x}I_{2y}I_{3y}\sin(\Omega_1 t_1)\sin(\pi J_{12}t_1)\sin(\pi J_{13}t_1). \qquad [6.14]$$

The $2I_{1z}I_{2y}$ and $2I_{1z}I_{3y}$ terms on lines 2 and 4 of [6.14] describe antiphase single-quantum coherence and will be rejected by the phase cycling. Similarly, the $4I_{1x}I_{2y}I_{3y}$ term on line 6 describes a mixture of 3Q and three-spin single-quantum coherences and will also be rejected (this term is discussed in more detail with respect to the 3QF-COSY in Section 6.3.2.1). The terms on lines 1, 3, and 5 of [6.14] contain mixtures of ZQ and DQ coherence (Section 2.7.5) and can be rewritten as

$$-2I_{1x}I_{2y} = -\frac{1}{2}[2I_{1x}I_{2y} + 2I_{1y}I_{2x} - 2I_{1y}I_{2x} + 2I_{1x}I_{2y}]$$

$$= -[DQ_y^{12} - ZQ_y^{12}]; \qquad [6.15]$$

$$-2I_{1x}I_{3y} = -\frac{1}{2}[2I_{1x}I_{3y} + 2I_{1y}I_{3x} - 2I_{1y}I_{3x} + 2I_{1x}I_{3y}]$$

$$= -[DQ_y^{13} - ZQ_y^{13}]; \qquad [6.16]$$

$$4I_{1z}I_{2y}I_{3y} = I_{1z}[2I_{2y}I_{3y} - 2I_{2x}I_{3x} + 2I_{2x}I_{3x} + 2I_{2y}I_{3y}]$$

$$= -2I_{1z}[DQ_x^{23} - ZQ_x^{23}], \qquad [6.17]$$

in which ZQ_p^{ij} and DQ_p^{ij} indicate zero-quantum coherence and two-quantum coherence, respectively, of phase p (x or y) between spins I_i and I_j. The DQ_x^{23} operator is antiphase with respect to I_1. Only the double-quantum components of [6.15] to [6.17] are retained by the phase cycle.

The final 90_x° pulse will transfer the DQ terms back to observable magnetization:

$$-\frac{1}{2}\cos(\Omega_1 t_1)\,\sin(\pi J_{12}t_1)\,\cos(\pi J_{13}t_1)\,[2I_{1x}I_{2z} + 2I_{1z}I_{2x}];$$

$$-\frac{1}{2}\cos(\Omega_1 t_1)\,\cos(\pi J_{12}t_1)\,\sin(\pi J_{13}t_1)\,[2I_{1x}I_{3z} + 2I_{1z}I_{3x}]; \qquad [6.18]$$

$$+\frac{1}{2}\cos(\Omega_1 t_1)\,\sin(\pi J_{12}t_1)\,\sin(\pi J_{13}t_1)\,[4I_{1y}I_{2x}I_{3x} - 4I_{1y}I_{2z}I_{3z}].$$

The two terms in the first line of [6.18] describe an I_1 diagonal peak, and an $I_1 \rightarrow I_2$ cross-peak; both have antiphase character and are of the same relative phase, as discussed above for [6.13]. Similarly, line 2 describes an I_1 diagonal peak, and an $I_1 \rightarrow I_3$ cross-peak. The $\cos(\pi J t_1)$ trigonometric terms in the first two lines indicate additional in-phase splittings in F_1 arising from passive coupling (see also Section 6.2.1.1); similar splittings will also be present in F_2. The first term on line 3 is unobservable, and the second term is a third I_1 diagonal component doubly antiphase with respect to I_2 and I_3. When the cross-peaks are phased to absorption, this later term will be in dispersion. Fortunately, the antiphase nature of this term leads to pronounced self-cancellation; hence the dispersive tails will not extend far from the diagonal. In conclusion, this analysis has demonstrated that, even in more complex spin systems, the diagonal resonances of the 2QF-COSY experiment have predominantly in-phase absorption lineshapes and do not obscure cross-peaks between protons close in chemical shift.

6.3.1.2 Experimental Protocol   The cross-peaks in the COSY and 2QF-COSY experiments have similar lineshape properties. Thus, many experimental requirements for the 2QF-COSY, notably high digital resolution in F_1 and F_2, are similar to those described above for the COSY experiment. The only extra parameter required in the experimental protocol is the delay between the two final 90° pulses. This delay should be on

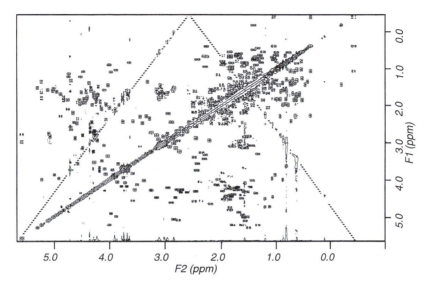

FIGURE 6.21 2QF COSY spectrum acquired with a short recycle delay. The "double-diagonal" artifacts lie along the dotted line. Quadrature detection in F_1 was achieved with TPPI; consequently, upfield of 2.6 ppm, the artifact peaks fold to $F_1 = 4.8 - 2F_2$. The spectrum was acquired using the pulse sequence of Fig. 6.20 (16-step phase cycle) with 32 scans for each of 512 increments with a total recycle delay of 1.33 s (0.33-s acquisition time and 1-s weak presaturation).

the order of a few microseconds, i.e., as short as possible, while still allowing the rf hardware to make accurate phase shifts of the pulses.

When 2QF-COSY experiments are acquired from D_2O solution, the digital resolution in F_1 is improved by reducing the spectral width to span only the aliphatic resonances. Because correlations between the aromatic and aliphatic portions of the spectrum are not observed, folding of the diagonal and cross-peaks of the aromatic resonances do not confuse analysis of the F_1 chemical shifts (Section 4.3.4.3). By reduction of the spectral width, a high $t_{1,max}$ is obtained with fewer increments of t_1.

One artifact common in 2QF-COSY spectra results from incomplete recovery of the spins during the recycle delay and gives rise to extra diagonal peaks at $F_1 = 2F_2$ (in this case F_1 and F_2 refer to absolute frequencies from the carrier position, not chemical shifts). Figure 6.21 depicts a region of a spectrum acquired with a short recycle delay (1.33 s) where such artifacts are particularly prominent. Because of the particular spectral widths and carrier position in this spectrum, the artifact peaks occur at $F_1(ppm) = 2 \times F_2(ppm) - 5.6$ for F_2 between 5.6 and 2.6 ppm;

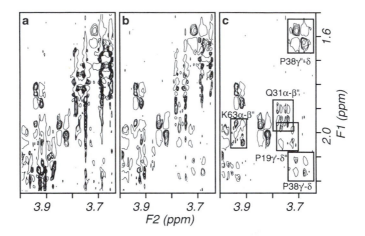

FIGURE 6.22 Cross-peaks obscured by the double-diagonal artifacts. (a) Enlargement of the spectrum shown in Fig. 6.21. The spectra in (b) and (c) were obtained with a longer total recycle delay (2.3 s), and the phase cycle for (c) included axial peak suppression. Other acquisition and processing parameters were the same for all three spectra. Assignments for the obscured cross-peaks are included in panel (c).

as a result of folding, above F_2(ppm) = 2.6 they occur at F_1(ppm) = 4.8 − 2 × F_2(ppm). In particular, the artifacts obscure weak peaks in the ^{1}H$^\alpha$–^{1}H$^\beta$ fingerprint region (Fig. 6.22, panel a). Lengthening the recycle delay (panel b) and the incorporation of axial peak suppression in the phase cycle (panel c) reduces the size of these artifacts and allows several weak proline ^{1}H$^\delta$–^{1}H$^\gamma$ cross-peaks to be identified.

6.3.1.3 Processing   Although the cross-peaks of the 2QF-COSY are similar to those of the COSY experiment, processing requirements are usually quite different. Strongly resolution-enhancing window functions are not required to reduce the intensity of the predominantly antiphase absorptive diagonal peaks. Nonetheless, the appearance of the spectrum is improved by using mild resolution enhancement to minimize overlap of neighboring cross-peaks, and reduce self-cancellation of the multiplet components. Sine bells shifted by 30–90° in both t_1 and t_2 are appropriate window functions. The sine bells should span between 150 and 300 ms of each FID (depending on the degree of resolution required and the S/N ratio of the data) and over the entire t_1 interferogram. Phase parameters are determined by the methods described above for the COSY experiment. All the D$_2$O 2QF-COSY spectra depicted in this chapter were processed with a sine-bell-shifted 60° in t_2 for a $t_{2,max} = 256$ ms, and a sine-bell-shifted 45° over all 512 t_1 points for $t_{1,max} = 82.5$ ms.

6.3.1.4 Information Content   The 2QF-COSY contains essentially the same information as the normal COSY experiment. The regions of interest in the 2QF-COSY experiment are still the fingerprint regions described in Fig. 6.6. A disadvantage of using the double-quantum filter is a decrease in S/N ratio by a theoretical factor of 2 relative to the COSY experiment (23). The origin of the decrease can be appreciated by considering the first four steps of the phase cycle of the 2QF-COSY experiment.

$$\phi_1: \qquad x, \quad y, -x, -y;$$

$$\phi_2: \qquad x, \quad y, -x, -y;$$

$$\phi_3: \qquad x, \quad x, \quad x, \quad x;$$

$$\text{Receiver:} \quad x, -x, \quad x, -x.$$

Because the final two pulses of the experiment are contiguous, the first and third transients have a 180° pulse, or 0° pulse, respectively, after t_1; these transients suppress the in-phase dispersive diagonal components and add noise to the spectrum but do not contribute any intensity to the cross-peaks (*15*). Thus, only half of the transients in the 2QF-COSY experiment contribute to cross-peak intensity, leading to a 2-fold decrease in sensitivity relative to a COSY experiment acquired in the same amount of time. The difference in sensitivity is demonstrated in Fig. 6.23, which shows cross sections through H_2O COSY spectra at the F_1 chemical shift of the $^1H^\alpha$ resonance of Ile13. The absolute magnitude of the cross-peaks are similar as the H_2O COSY was acquired with 16 scans per t_1 increment while the 2QF-COSY spectra was acquired with 32 scans per t_1 increment; all other acquisition and processing parameters are identical, with a sine bell shifted by 20° being necessary to prevent the tails of diagonal peaks interfering with the COSY cross-peaks. Although doubling the number of transients in the 2QF-COSY experiment equalizes the signal amplitude, the root-mean-square (RMS) noise also increases by a factor of $\sqrt{2}$; hence the actual S/N ratio is lower than for the COSY.

The twofold difference in sensitivity is observed if both spectra are recorded and processed in exactly the same way. In practice, however, the more severe window functions used for processing COSY spectra reduce the sensitivity of the experiment. Thus, in Fig. 6.23 the top trace depicts a row of data from the 2QF-COSY reprocessed with sine bells shifted 60° and 90° in t_2 and t_1, respectively. The S/N ratio is now 1.3-fold higher than in the COSY spectrum. As a caveat, the severe window

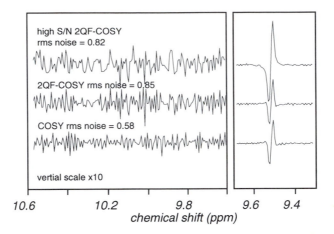

FIGURE 6.23 Comparison of the S/N ratio in COSY and 2QF-COSY spectra for the ^{1}H^N–^{1}H$^\alpha$ cross-peak of Ile13. Except for the number of transients (COSY = 16, 2QF-COSY = 32), both spectra were acquired under identical conditions. The two lower traces were also processed identically (sine bell shifted 20° in t_1 and an unshifted sine bell in t_2), whereas the top trace was processed with less resolution-enhancing window functions (a cosine bell in t_1 and a sine bell shifted 60° in t_2). See text for a more detailed description.

functions used with COSY spectra also are resolution-enhancing, so although the final COSY and 2QF-COSY experiments can have similar S/N ratios, the resolution in the COSY will be higher.

A useful strategy for the study of proteins employs the COSY experiment in H$_2$O solution to observe ^{1}H^N–^{1}H$^\alpha$ cross-peaks that are far from the diagonal, and uses the 2QF-COSY in D$_2$O solution to identify cross-peaks in the vicinity of the diagonal. Even though harsh window functions can be used to observe peaks close to the diagonal in COSY spectra, distortions arise and the peak are more readily observed in the 2QF-COSY experiment (Fig. 6.24).

6.3.2 3QF-COSY

As described in the Section 6.3, the 2QF-COSY is just one member of a family of experiments involving filtration through a p-quantum state; experiments with $p > 2$ offer spectral simplification because some resonances cannot participate in a p-quantum coherence (see discussion of selection rules below). As a practical matter, pQF-COSY experiments with $p > 3$ are applied rarely to proteins.

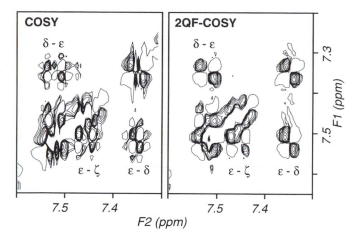

FIGURE 6.24 Comparison of cross-peaks close to the diagonal in COSY and 2QF-COSY. The COSY spectrum was acquired with 16 transients for each of 800 t_1 increments and processed with a sine bell shifted 5° in t_1 and an unshifted sine bell in t_2. The 2QF-COSY was acquired with 32 transients for 800 t_1 increments and processed with a cosine bell in t_1 and a sine bell shifted 60° in t_2. The severe window functions required to reduce the streaks emanating from the diagonal of the COSY spectrum distort the aromatic cross-peaks of Phe45, especially those between $^1H^\varepsilon$ and $^1H^\zeta$; these cross-peaks are clearly observable in the 2QF-COSY.

6.3.2.1 Product Operator Analysis   A product operator analysis of the 3QF-COSY experiment initially proceeds as described above for the 2QF-COSY, and [6.14] describes the terms present after the second 90° pulse in the sequence. The phase cycling for the 3QF-COSY will eliminate all terms in [6.14] except $4I_{1x}I_{2y}I_{3y}$. Just as the product operator $2I_{1x}I_{2y}$ term in the 2QF-COSY experiment was expressed as the combination of DQ and ZQ coherences [6.15], the $4I_{1x}I_{2y}I_{3y}$ operator is expanded as the combination of 3Q coherences and three-spin single quantum coherences. Using the single-element basis set of Section 2.7.2, and referring to the eigenstates for a three-spin system (Fig. 2.4), a $3Q_x$ coherence is formally described by

$$3Q_x = \frac{1}{2}[I_1^+ I_2^+ I_3^+ + I_1^- I_2^- I_3^-]$$

$$= \frac{1}{2}[(I_{1x} + iI_{1y})(I_{2x} + iI_{2y})(I_{3x} + iI_{3y}) + (I_{1x} - iI_{1y})(I_{2x} - iI_{2y})(I_{3x} - iI_{3y})]$$

$$= \frac{1}{4}[4I_{1x}I_{2x}I_{3x} - 4I_{1y}I_{2y}I_{3x} - 4I_{1x}I_{2y}I_{3y} - 4I_{1y}I_{2x}I_{3y}]. \qquad [6.19]$$

Three-spin single-quantum coherences connect eigenstates for which two spins change quantum number by $+1$ and the third changes by -1, or for which two spins change quantum number by -1 and the third changes it by $+1$; formally, the net coherence order is $+1$ or -1. For example,

$$1Q_{12\bar{3}}$$

$$= \frac{1}{2}[I_1^+ I_2^+ I_3^- + I_1^- I_2^- I_3^+]$$

$$= \frac{1}{2}[(I_{1x} + iI_{1y})(I_{2x} + iI_{2y})(I_{3x} - iI_{3y}) + (I_{1x} - iI_{1y})(I_{2x} - iI_{2y})(I_{3x} + iI_{3y})]$$

$$= \frac{1}{4}[4I_{1x}I_{2x}I_{3x} - 4I_{1y}I_{2y}I_{3x} + 4I_{1x}I_{2y}I_{3y} + 4I_{1y}I_{2x}I_{3y}], \qquad [6.20]$$

in which the overbar indicates the spin-changing quantum number in the sense opposite that for the other two spins. Taking the appropriate combinations of the three-spin single-quantum terms and the $3Q_x$ term yields the following result:

$$4I_{1x}I_{2y}I_{3y} = -3Q_x + 1Q_{12\bar{3}} + 1Q_{1\bar{2}3} - 1Q_{\bar{1}23}. \qquad [6.21]$$

Thus, from [6.19], the action of the final 90° pulse acting on the $3Q_x$ term is

$$-3Q_x \sin(\pi\Omega_1 t_1) \sin(\pi J_{12}t_1) \sin(\pi J_{13}t_1) \xrightarrow{\left(\frac{\pi}{2}\right)_x}$$

$$\frac{1}{4}[-4I_{1x}I_{2x}I_{3x} + 4I_{1z}I_{2z}I_{3x} + 4I_{1x}I_{2z}I_{3z} + 4I_{1z}I_{2x}I_{3z}] \qquad [6.22]$$

$$\times \sin(\pi\Omega_1 t_1) \sin(\pi J_{12}t_1) \sin(\pi J_{13}t_1).$$

The first term of [6.22] is unobservable, whereas the last three correspond to an $I_1 \rightarrow I_3$ cross-peak, an I_1 diagonal peak, and an $I_1 \rightarrow I_2$ cross-peak, respectively. The three observable terms are modulated by the same trigonometric functions; hence the resonances have the same relative phase in F_1. The combination of three sine terms indicates that all three peaks are antiphase with respect to J_{12} and J_{13}. Likewise in F_2, all three peaks have the same relative phase (x) and are in antiphase with respect to both couplings. The appearance of a double antiphase Lorentzian lineshape is discussed in more detail in Section 6.4.1.1 and shown in Fig. 6.34 (later).

6.3.2.2 Experimental Protocol and Processing The relative sensitivity of a cross-peak arising from a p-quantum coherence in a pQF-COSY decreases by a factor of 2^{p-1} (cf. the factors of $\frac{1}{2}$ in [6.18] and $\frac{1}{4}$ in [6.22]) (25); consequently, more transients must be coadded to obtain a suitable S/N ratio. For this reason, the increased number of steps in the phase cycle to achieve p-quantum selection is rarely a problem. Other practical considerations in implementing 3QF-COSY experiments parallel those described above for the 2QF-COSY (Section 6.3.1.2).

The 3QF-COSY spectra of ubiquitin were acquired from D_2O solution with 48 transients for each of 512 increments; the spectral width was 3100 Hz, and $t_{1,max}$ was 82.6 ms. The spectrum was folded in F_1. The complete phase cycle used in this experiment is listed in the caption to Fig. 6.20.

Considerations for processing pQF-COSY spectra are similar to those of the 2QF-COSY described in Section 6.3.1.3. The cross-peak fine structure will be antisymmetrical with respect to the chemical-shift axes for even p, and symmetrical for odd p (e.g., see Figs. 6.25 or 6.28). This simple observation aids in determining appropriate phase corrections for higher order pQF-COSY spectra.

6.3.2.3 Information Content

The main reason for filtration via higher coherence orders is the spectral simplification achieved. The following selection rules provide the basis for this simplification: (1) diagonal peaks will be observed when the active spin has *resolved scalar couplings* to p-1 spins and (2) a cross-peak between two spins will be observed when the two active spins have p-2 *mutual* resolved coupling partners (25). A resolved scalar coupling is a scalar interaction that would give rise to an observable multiplet splitting in the 1D spectrum (assuming no overlap with resonances of other spin systems); thus, degenerate protons can never have a mutual resolved coupling. The cross-peaks expected from these definitions are strictly valid only in the weak coupling limit and are summarized in Figure 2 of Müller et al. (25). Experiments with $p > 3$ have been reported in the literature (25,26), but the lack of sensitivity and modest gains in spectral simplification preclude their common application to proteins.

Some of the specific advantages of D_2O 3QF-COSY over the 2QF counterpart are a removal of $^1H^\alpha$ diagonal peaks and $^1H^\alpha$–$^1H^\beta$ cross-peaks of threonine, alanine, valine, and isoleucine residues; complete removal of all glycine peaks; removal of all diagonal and cross-peaks involving methyl groups except $^1H^\gamma$–$^1H^\delta$ of isoleucine; removal of tyrosine aromatic-ring resonances (unless the four bond $^1H^{\delta 1}$–$^1H^{\delta 2}$ or $^1H^{\varepsilon 1}$–$^1H^{\varepsilon 2}$ couplings are resolved) and removal of $^1H^\alpha$–$^1H^\beta$ cross-peaks of spin systems with degenerate β-methylene protons.

The effect on the methyl fingerprint region is shown in Fig. 6.25. The methyl groups of alanine, threonine, valine, and leucine residues have a

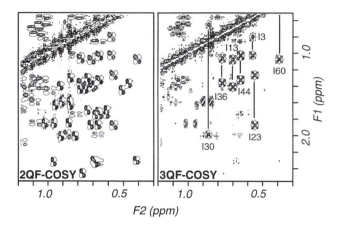

FIGURE 6.25 Comparison of the methyl fingerprint region in 2QF- and 3QF-COSY experiments. Assignments for the isoleucine ^{1}H$^{\gamma 1}$–^{1}H$^{\delta}$ cross-peaks are labeled in the 3QF-COSY. The single resolved couplings to the methyl groups of leucine and valine and the $C^{\gamma 2}H_3$ group of isoleucine are insufficient to allow cross-peaks in the 3QF spectrum.

single resolved coupling, and formally cannot produce diagonal or cross-peaks in pQF-COSY experiments with $p > 2$. This simplification allows ready identification of isoleucine ^{1}H$^{\gamma}$–^{1}H$^{\delta}$ cross-peaks. Multiexponential relaxation of methyl group coherences results from cross-correlation (Section 5.2.1) and causes violation of the selection rules. Weak *forbidden* cross-peaks and diagonal peaks are sometimes observed from methyl-containing spin systems in 3QF- and 4QF-COSY experiments (27,28).

In the ^{1}H$^{\alpha}$–^{1}H$^{\beta}$ fingerprint region, serine ^{1}H$^{\alpha}$–^{1}H$^{\beta}$ and proline ^{1}H$^{\gamma}$–^{1}H$^{\delta}$ cross-peaks are often more readily observed because of the removal of glycine and threonine cross-peaks in this region. In opposition to the general trend in sensitivity for pQF-COSY experiments, the double anti-phase absorptive fine structure in the 3QF-COSY leads to less self-cancellation than do the absorptive single antiphase peaks in the 2QF-COSY, and a higher apparent S/N ratio sometimes is observed for particular resonances in 3QF-COSY spectra. An example of an ideal absorptive double antiphase Lorentzian line is shown in Fig. 6.34. An experimental example is illustrated in Fig. 6.26. Although ^{1}H$^{\alpha}$–^{1}H$^{\beta'}$–^{1}H$^{\beta''}$ fragments with identical ^{1}H$^{\beta}$ shifts cannot participate in a three-quantum coherence, even slight deviations from degeneracy can lead to peaks in the 3QF-COSY. In the case of Glu24, other correlation spectra were unable to resolve the slight difference in chemical shift of ^{1}H$^{\beta'}$ and ^{1}H$^{\beta''}$, but a peak

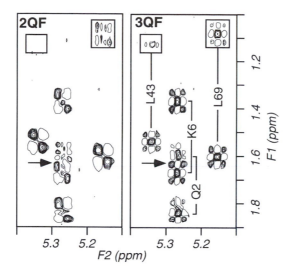

FIGURE 6.26 Comparison of several $^1H^\alpha$–$^1H^\beta$ cross-peaks in 2QF-COSY and 3QF-COSY spectra. For the upfield $^1H^\beta$ of Leu43 and Leu69 (boxed cross-peaks), the active coupling is small and the cancellation of the antiphase components severely attenuates the cross-peaks in the 2QF-COSY. The components of the 3QF-COSY cross-peaks are not separated by the sum of the individual $^3J_{H^\alpha H^\beta}$; hence the cancellation is not as pronounced. The arrows indicate cancellation (2QF-COSY) and reinforcement (3QF-COSY) of overlapping parts of the downfield Lys6 and the upfield Gln2 $^1H^\alpha$–$^1H^\beta$ cross-peaks as shown schematically in Fig. 6.28.

was observed in the 3QF-COSY (Fig. 6.27). The fine structure clearly indicates that $^1H^{\beta'}$ and $^1H^{\beta''}$ are separated by approximately 0.02 ppm.

A detailed catalog of the fine structure observed in cross-peaks in pQF-COSY has been presented elsewhere (25). In a system of n mutually coupled spins, cross-peaks in a pQF-COSY (where $n > p$) will contain contributions from all possible p-quantum states (including combination bands if $n - p = 2, 4, ...$). Symmetry properties of the cross-peak multiplets indicate that partially overlapping cross-peaks will be more subject to cancellation in spectra filtered with an even value of p than an odd value of p (Fig. 6.28) (29).

6.3.3 E-COSY

Values of ^3J can provide dihedral angle information that can be used as experimental restraints during protein structure determination (Section 8.2.1). For spins that possess a single active coupling, the scalar coupling

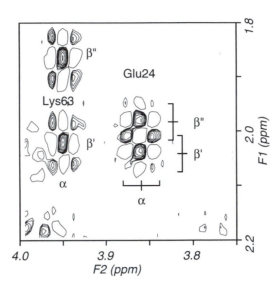

FIGURE 6.27 ^{1}H$^\alpha$–^{1}H$^{\beta'}$ and ^{1}H$^\alpha$–^{1}H$^{\beta''}$ cross-peaks of Glu24 in the 3QF-COSY spectrum. Although the β-protons appear degenerate in other spectra (COSY, 2QF-COSY, and 2Q), the fine structure in the 3QF-COSY clearly shows that the chemical shifts are slightly different (approximately 0.02 ppm).

constant can be extracted from the fine structure of a peak in the COSY spectrum (Section 6.2.1.5). If both actively coupled spins have additional passive coupling partners, the COSY cross-peak fine structure becomes more complicated, and in general the individual multiplet component positions (and hence the scalar coupling constants) are not well resolved for the broad resonance lines observed in protein spectra. A number of methods have been developed that excite only connected transitions (Section 2.6 and Fig. 2.4); specifically, the passive spins remain unperturbed and the cross-peaks contain only active coupling components. These methods have been given the generic name of "Exclusive" or E-COSY (30). In ^{1}H NMR of proteins, E-COSY spectra provide a useful means of measuring $^3J_{\mathrm{H}^\alpha\mathrm{H}^\beta}$ in side chains containing β-methylene groups; limitations of sensitivity and linewidth have precluded widespread use in measuring other scalar coupling interactions. However, the small linewidths and high concentrations available in peptide studies have allowed the measurement of virtually all ^{1}H–^{1}H couplings in some peptide systems using E-COSY spectra (31).

6.3.3.1 Product Operator Analysis   The most common approach to obtaining E-COSY data is construction of a linear combination of pQF-

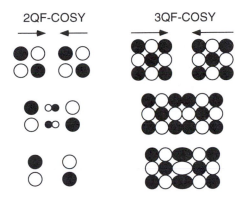

FIGURE 6.28 Schematic view of partially overlapped cross-peaks in 2QF- and 3QF-COSY spectra. Because of the symmetry of the cross-peaks, those in the 2QF-COSY cancel, whereas those in the 3QF-COSY reinforce. A real example is observed in Fig. 6.26.

COSY spectra where the level of filtration includes the orders $p = 0, 1, 2 \ldots$ (*30,32*). The weight factor given to a particular COSY of multiple quantum order p in this combination is given by

$$
W_p = \begin{cases} \dfrac{1}{4}p^2 + W_0 & (p \text{ even}) \\[2ex] \dfrac{1}{4}(p^2 - 1) + W_1 & (p \text{ odd}) \end{cases} \tag{6.23}
$$

where the first and second lines of [6.23] are applicable to even and odd p, respectively. Because spectra acquired with $p = 0$ and $p = 1$ contribute only to the diagonal, W_0 and W_1 are customarily set to zero to minimize interference from intense diagonal peaks. Thus, the appropriate weights become $W_2 = 1$, $W_3 = 2$, $W_4 = 4$, etc. For a given cross-peak, pQF-COSY is included in the combination if p-2 passive spins are common to the two active spins (*30*). Given the pQF-COSY selection rules (Section 6.3.2.3), if a cross-peak appears in the pQF-COSY, the spectrum should be included in the linear combination in the E-COSY experiment. For a spin system containing a β-methylene group, a single common passive spin $^1H^\beta$ contributes to both $^1H^\alpha$–$^1H^\beta$ cross-peaks; therefore, the maximum level of multiple quantum coherence required in the linear combination to remove the passive spin components is 3.

The results of the product operator analyses performed for the 2QF-COSY (Section 6.3.1.1) and 3QF-COSY (Section 6.3.2.1) demonstrate the appearance of cross-peaks in the E-COSY of a three-spin system. From

[6.18], the observable term describing the $I_1 \rightarrow I_2$ cross-peak in the 2QF-COSY spectrum of the I_1, I_2, I_3 spin system is proportional to

$$\sigma(2QF)_{12} = -\frac{1}{2}\cos(\Omega_1 t_1)\sin(\pi J_{12} t_1)\cos(\pi J_{13} t_1)\, 2I_{1z}I_{2x}. \quad [6.24]$$

Similarly from [6.22], the equivalent operator in the 3QF-COSY is

$$\sigma(3QF)_{12} = +\frac{1}{4}\sin(\pi \Omega_1 t_1)\sin(\pi J_{12} t_1)\sin(\pi J_{13} t_1)\, 4I_{1z}I_{2x}I_{3z}. \quad [6.25]$$

Considering quadrature detection in t_1 and evolution and detection during t_2 gives the following signals for the 2QF-COSY and 3QF-COSY experiments, respectively:

$s(2QF)_{12}$

$$= -\frac{i}{4}\exp(\Omega_1 t_1)\sin(\pi J_{12} t_1)\cos(\pi J_{13} t_1)\exp(\Omega_2 t_2)\sin(\pi J_{12} t_2)\cos(\pi J_{23} t_2)$$

$$= -\frac{i}{16}\exp(\Omega_1 t_1)\sin(\pi J_{12} t_1)\exp(\Omega_2 t_2)\sin(\pi J_{12} t_2)$$

$$\times\{\exp(\pi J_{13} t_1)\exp(\pi J_{23} t_2) + \exp(\pi J_{13} t_1)\exp(-\pi J_{23} t_2)$$

$$+ \exp(-\pi J_{13} t_1)\exp(\pi J_{23} t_2) + \exp(-\pi J_{13} t_1)\exp(-\pi J_{23} t_2)\}; \quad [6.26]$$

$s(3QF)_{12}$

$$= -\frac{i}{8}\exp(\Omega_1 t_1)\sin(\pi J_{12} t_1)\sin(\pi J_{13} t_1)\exp(\Omega_2 t_2)\sin(\pi J_{12} t_2)\sin(\pi J_{23} t_2)$$

$$= -\frac{i}{32}\exp(\Omega_1 t_1)\sin(\pi J_{12} t_1)\exp(\Omega_2 t_2)\sin(\pi J_{12} t_2)$$

$$\times\{\exp(\pi J_{13} t_1)\exp(\pi J_{23} t_2) + \exp(\pi J_{13} t_1)\exp(-\pi J_{23} t_2)$$

$$- \exp(-\pi J_{13} t_1)\exp(\pi J_{23} t_2) + \exp(-\pi J_{13} t_1)\exp(-\pi J_{23} t_2)\}. \quad [6.27]$$

Combining [6.26] (2QF) and [6.27] (3QF) according to [6.23] yields

$s(2QF)_{12} + 2s(3QF)_{12}$

$$= \frac{i}{8}\exp(\Omega_1 t_1)\sin(\pi J_{12} t_1)\exp(\Omega_2 t_2)\sin(\pi J_{12} t_2)$$

$$\times\{\exp(\pi J_{13} t_1)\exp(\pi J_{23} t_2) + \exp(-\pi J_{13} t_1)\exp(-\pi J_{23} t_2)\} \quad [6.28]$$

Thus, two components arising from the passive J_{13} and J_{23} couplings in [6.26] and [6.27] have been removed in [6.28].

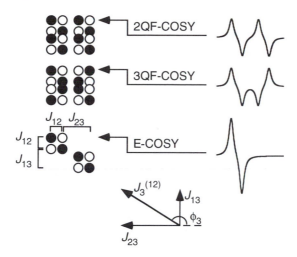

FIGURE 6.29 Schematic view of the I_1–I_2 cross-peak for a three-spin system during 2QF-COSY (top), 3QF-COSY (middle), and E-COSY (bottom) experiments. Filled and open circles indicate positive and negative components, respectively. Traces parallel to F_2 through each cross-peak are shown on the right, and were calculated for Lorentzian lines of 4-Hz half-height width, $J_{12} = 6$ Hz, $J_{13} = 10$ Hz, and $J_{23} = 15$ Hz. The vector construction at the bottom is described in the text.

The overall effect on the $I_1 \rightarrow I_2$ cross-peak is shown schematically in Fig. 6.29. The E-COSY cross-peak contains a superposition of two double-spin antiphase multiplets, with the displacement being equal to the size of the passive coupling in each dimension. The active coupling can be measured as the antiphase separation (as in Section 6.2.1.5). More importantly, the passive coupling can be measured from the displacements of the two multiplets because the peak-to-peak separation will not be subject to systematic errors associated with finite linewidth (Section 6.2.1.5). This feature allows use of the E-COSY method to determine scalar couplings much smaller than the linewidth, provided the appropriate cross-peaks can be generated.

A displacement vector, $J_3^{(12)}$, connecting peaks of like sign (constructed as shown at the bottom of Fig. 6.29) is associated with a particular passive spin (I_3 in this example). The angle ϕ_3 between the F_2 axis and the displacement vector reflects the sign of the product $J_{13} \times J_{23}$, and can distinguish the relative signs of these two coupling constants (14). Provided spectra are processed to have the diagonal running from top right to bottom left, and ϕ_3 is constrained to be between 0° and 180°, positive and negative products of $J_{13} \times J_{23}$ are indicated by $0° < \phi_3 < 90°$ and $90° < \phi_3 < 180°$, respectively.

Although this discussion has been limited to ¹H NMR, the E-COSY concept may be extended to include heteronuclear couplings. Small passive couplings (usually ¹H–¹H) may then be measured in F_2 between two components separated by a large heteronuclear coupling (one bond ¹H–X or X–X, where X is ¹⁵N or ¹³C). These applications of the E-COSY principle are discussed in more detail in Section 7.5.

A more thorough analysis extensible to spin systems of greater complexity involves consideration of single-transition basis operators within the energy-level diagram of the spin system of interest. Pairs of transitions give rise to each element of fine structure in a cross-peak and are described in the terms discussed in Section 2.6: regressive or progressive, connected or anticonnected. Depending on these qualities, the transition pairs display particular characteristics in the final spectrum, and the appearance of each multiple-quantum-filtered (MQF) spectrum, or combination of such spectra, can be deduced.

6.3.3.2 Experimental Protocol The majority of experimental details for acquisition of 2QF- and 3QF-COSY experiments have been covered in Sections 6.3.1.2 and 6.3.2.2; however, a few aspects are specific to the E-COSY experiment. The weights indicated by [6.23] indicate that the 3QF-COSY data should be scaled by a factor of 2 compared to the 2QF-COSY if both spectra have been acquired identically (i.e., spectral widths, number of transients, and number of t_1 increments). However, noise introduced by the addition (or subtraction) of the two data sets is minimized if no scaling is performed after acquisition; consequently, the 3QF-COSY spectrum is acquired with twice the number of transients as the 2QF-COSY and the two spectra simply are added. In order to complete the basic phase cycles of both experiments (16 and 24 transients) and acquire twice the number of transients in the 3QF-COSY, 48 and 96 transients must be collected for the 2QF- and 3QF-COSY, respectively. The spectra should be acquired in an interleaved fashion to avoid subtraction artifacts, and combined at the processing stage. In this mode of acquisition, the individual 2QF-COSY and 3QF-COSY experiments also can be analyzed. However, 144 transients must be collected for each t_1 increment in the E-COSY experiment, which requires considerable total acquisition time.

An alternative mode of acquisition uses phase cycling within a single experiment to perform the appropriate data combination between transients *(14)*. Many of the steps in the individual phase cycles of the 2QF- and 3QF-COSY experiments are identical, and need be performed only once. Table 6.1 indicates the number of transients that must be collected

TABLE 6.1

Phase Cycles for the E-COSY Experiment

Pulse phase	0°	60°	120°	180°	240°	300°
Receiver phase	0°	180°	0°	180°	0°	180°
Transients	4	3	1	0	1	3

The number of transients acquired for each phase increment in the combined E-COSY experiment. The phase shifts are applied to ϕ_1 and ϕ_2 in Fig. 6.20, while keeping the phase of ϕ_3 fixed (0°).

for each of the six phase increments required for a E-COSY experiment suitable for analysis of three-spin systems. The basic phase cycle contains 12 steps, and can be extended to 48 steps by performing CYCLOPS (Section 4.3.2.3). Although only 48 transients have to be acquired to complete the phase cycle in this method, separate analysis of the 3QF-COSY and 2QF-COSY is not possible.

Finally, the two antiphase multiplets must be resolved in both dimensions to permit accurate measurements of the coupling constants. Normally, the cross-peaks at the F_2 frequency of $^1H^\alpha$ are analyzed. These cross-peaks are separated in F_1 by the passive $^1H^{\beta'}-^1H^{\beta''}$ coupling, and given the large size of this interaction (15 Hz), sufficient t_1 increments to properly resolve the components are easily acquired. The spectra are acquired from D_2O solution to avoid interference from passive couplings to $^1H^N$, and may be folded in F_1 to improve the digital resolution (Section 6.2.1.2).

The sections of spectra shown in Fig. 6.30 were acquired as separate 2QF- and 3QF-COSY spectra in an interleaved fashion with 48 and 96 transients, respectively. The spectra were folded in F_1, and contained 512 increments of t_1 ($t_{1,max} = 82$ ms). The total acquisition time was 42 hs.

6.3.3.3 Processing Processing parameters for E-COSY data are identical for both methods of acquisition, and are generally outlined in Section 6.3.1.3. The $^1H^\alpha-^1H^\beta$ cross-peaks are of primary interest for measuring $^3J_{H^\alpha H^\beta}$, and are best resolved at the F_2 frequency of $^1H^\alpha$. In order to improve the accuracy of the measurement, the spectra should be extensively zero-filled in F_2 (to at least 0.5 Hz per point). The spectrum shown in Fig. 6.30 was processed with moderately shifted cosine bells and zero-filled from 2,048 to 16,384 points to give a final digital resolution of 0.38 Hz per point in F_2. Line-fitting procedures can be used to improve the accuracy of the coupling constant measurement, although coupling constant data are rarely interpreted in such detail that errors of 0.5 Hz are significant (Section 8.2.1).

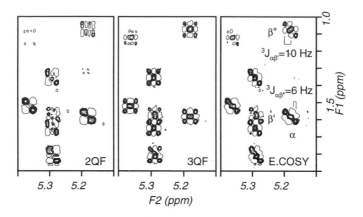

FIGURE 6.30 Section of the E-COSY spectrum of ubiquitin obtained by the coaddition of a 2QF-COSY (48 transients per t_1 increment) and a 3QF-COSY (96 transients per t_1 increment). Measurements of ${}^3J_{H^\alpha H^\beta}$ from the multiplet patterns are illustrated.

6.3.3.4 Information Content The principal information provided from the E-COSY is ${}^3J_{H^\alpha H^\beta}$. The analysis of the product operators has shown that co-addition of 2QF-COSY and 3QF-COSY is sufficient to provide an E-COSY pattern for the cross-peaks of a three-spin system. Longer spin systems have additional scalar coupling interactions with ^{1}H${}^\beta$ spins, and passive coupling components remain in an E-COSY spectrum constructed from 2QF-COSY and 3QF-COSY; higher orders of pQF-COSY must be included to remove these additional passive interactions. However, the passive components are present only in the $F_1({}^1H^\beta)$ dimension, and ${}^3J_{H^\alpha H^\beta}$ can still be measured from the displacements in the $F_2({}^1H^\alpha)$ dimension.

A section of the E-COSY of ubiquitin is shown in Fig. 6.30. Measurements are most accurate if taken from the displacement by the passive coupling since the artifacts described in Section 6.2.1.5 are not present. Thus, ${}^3J_{H^\alpha H^{\beta'}}$ is best measured from the ${}^1H^\alpha$–${}^1H^{\beta''}$ cross-peak and vice versa. In order to measure both ${}^1H^\alpha$–${}^1H^\beta$ coupling constants, the ${}^1H^\alpha$–${}^1H^{\beta'}$ and ${}^1H^\alpha$–${}^1H^{\beta''}$ cross-peaks must be resolved, which is not always the case in this crowded region of the spectrum.

6.3.3.5 Experimental Variants Linear combinations of COSY spectra acquired with mixing pulses less other 90° also produce E-COSY-type cross-peak fine structure. In the COSY-β spectrum of a I_1, I_2, I_3 spin system, the following operators give rise to a cross-peak representing coherence transfer from $I_1 \rightarrow I_2$ (the t_1 trigonometric factors have been ignored for simplicity):

$$-2I_{1z}I_{2y} \sin^2(\beta) - 4I_{1z}I_{2y}I_{3z} \sin^2(\beta) \cos(\beta). \qquad [6.29]$$

The second term of [6.29] would not be observed in a normal COSY experiment because the term at the end of t_1 from which it originates $(4I_{1y}I_{2z}I_{3z})$ would be converted entirely into unobservable magnetization $(4I_{1z}I_{2y}I_{3y})$ by the 90°_x mixing pulse (e.g., see [6.14]). The combination of single and double antiphase lineshapes, represented by the first and second terms in [6.29], respectively, is responsible for generating E-COSY fine structure (Section 6.3.3.1). Thus, combining COSY-β spectra with different rotation angles β suppresses the passive coupling components. Two main disadvantages of this approach are (1) the cross-peak intensity is low because of the dependence on $\sin^2(\beta)$ and (2) the diagonal is very intense and also contains dispersive components.

As is evident from [6.29], a single COSY-β experiment does have a degree of E-COSY character. The ratio of connected to unconnected transition intensities is maximal when $\beta \approx 35^\circ$. The undesired passive components are suppressed by more than 10-fold, which is adequate for many applications (4). Thus, COSY-35 provides a simple method for obtaining $^3J_{H^\alpha H^\beta}$. An example of a COSY-35 spectrum has been shown in Fig. 6.14 for the H_2O case of a glycine $^1H^N$–$^1H^\alpha$ peaks, and Fig. 6.31 compares a section of the $^1H^\alpha$–$^1H^\beta$ region in D_2O with the equivalent region of the E-COSY.

6.4 Multiple-Quantum Spectroscopy

The concept of multiple-quantum (MQ) coherence was introduced in Chapter 2, and MQ filtration of COSY spectra was described in Section 6.2.2. In this section, another useful application of MQ coherence is discussed. Although MQ states cannot induce directly an observable signal in the receiver coil, the pQ family of experiments indirectly observes MQ states with coherence level $p > 1$ during the t_1 period of a 2D experiment (33). The resulting spectra possess unique characteristics that complement the information obtained from COSY or pQF-COSY experiments.

The pulse sequence for a pQ experiment is shown in Fig. 6.32. Following the initial 90° pulse, antiphase coherence develops during a fixed spin-echo sequence (in contrast to COSY experiments in which antiphase

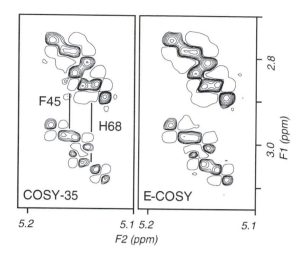

FIGURE 6.31 Comparison of the COSY-35 and E-COSY spectrum of ubiquitin in D$_2$O.

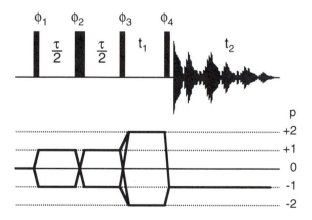

FIGURE 6.32 Pulse sequence and coherence level diagram for 2Q experiments. The basic 2Q phase cycle is as follows: $\phi_1 = \phi_2 = \phi_3 = x, y, -x, -y$; $\phi_4 = x$; receiver = $2(x, -x)$. The same pulse sequence is used for the 3Q experiment. The basic 3Q phase cycle is as follows: $\phi_1 = \phi_2 = 0°, 60°, 120°, 180°, 240°, 300°$; $\phi_3 = 90°, 150°, 210°, 270°, 330°, 30°$; $\phi_4 = 90°$; and receiver = $3(0°, 180°)$. Axial peak suppression and CYCLOPS phase cycling yields a 32-step phase cycle for 2Q and 48-step phase cycle for 3Q.

coherence develops during t_1). The 180° pulse serves to refocus chemical-shift evolution. Multiple-quantum coherence is generated by the second 90° pulse, and the precession of the desired pQ state is monitored during the t_1 period. The final 90° pulse creates observable single-quantum magnetization, which is recorded during t_2. The phases of the pulses in a pQ experiment are described by the following formula:

$$
\begin{aligned}
\phi_1 \text{ and } \phi_2: &\quad n\pi/p; \\
\phi_3: &\quad \phi_1 + \psi; \\
\phi_4: &\quad \psi; \\
\text{Receiver}: &\quad -n\pi + 2\psi,
\end{aligned}
\qquad [6.30]
$$

where ψ is 0 for even p and $\pi/2$ for odd p. In these expressions, n takes integer values from 0 to $2p$-1, thus the basic phase cycle consists of $2p$ steps. This can be expanded by performing CYCLOPS to yield a total of $8p$ steps (note the close similarity between this experiment and the pQF-COSY experiments (Section 6.3)).

   Although pQ experiments can be adjusted to observe a range of multiple quantum states, 2Q and 3Q experiments usually are adequate for studies of proteins.

6.4.1 2Q SPECTROSCOPY

6.4.1.1 Product Operator Analysis   Initially, evolution of a two-spin system during the 2Q pulse sequence will be treated. In this example, the simplest results are obtained if the fate of both I_1 and I_2 initial z-magnetization is calculated. At the end of the $90°_x$–$\tau/2$–$180°_x$–$\tau/2$–$90°_x$ sequence, the following operators are present at the start of the t_1 period:

$$
I_{1z} + I_{2z} \xrightarrow{\left(\frac{\pi}{2}\right)_x - \frac{\tau}{2} - \pi_x - \frac{\tau}{2} - \left(\frac{\pi}{2}\right)_x}
$$

$$
[I_{1z} + I_{2z}] \cos(\pi J_{12}\tau) + [2I_{1x}I_{2y} + 2I_{1y}I_{2x}] \sin(\pi J_{12}\tau). \quad [6.31]
$$

If the delay $\tau = 1/(2J_{12})$ the longitudinal terms, $I_{1z} + I_{2z}$, will vanish, leaving only the pure DQ$_y$ coherence, $2I_{1x}I_{2y} + 2I_{1y}I_{2x}$. However, in a real sample a range of values of J_{12} will be encountered, and not all magnetization can be converted into 2Q coherence; however, the phase cycling suppresses the residual longitudinal magnetization and retains only the 2Q coherence. In the rest of this analysis the longitudinal components will be ignored. During t_1, the two-quantum coherence evolves at the sum of the I_1 and I_2 chemical shifts, $\Omega_{12} = \Omega_1 + \Omega_2$, but does not evolve under the influence of J_{12} (Section 2.7). The final $90°_x$ pulse creates the product operators that evolve during t_2:

$$[2I_{1x}I_{2y} + 2I_{1y}I_{2x}] \sin(\pi J_{12}\tau) \xrightarrow{t_1} [2I_{1x}I_{2y} + 2I_{1y}I_{2x}] \cos(\Omega_{12}t_1) \sin(\pi J_{12}\tau)$$

$$- [2I_{1x}I_{2x} + 2I_{1y}I_{2y}] \sin(\Omega_{12}t_1) \sin(\pi J_{12}\tau)$$

$$\xrightarrow{\left(\frac{\pi}{2}\right)_x} [2I_{1x}I_{2z} + 2I_{1z}I_{2x}] \cos(\Omega_{12}t_1) \sin(\pi J_{12}\tau)$$

$$- [2I_{1x}I_{2x} + 2I_{1z}I_{2z}] \sin(\Omega_{12}t_1) \sin(\pi J_{12}\tau).$$

$$[6.32]$$

The last line of [6.32] contains a combination of longitudinal two-spin order, ZQ coherence, and 2Q coherence; none of these operators leads to observable magnetization during t_2. By analogy with the results obtained above for the COSY experiment, the $2I_{1x}I_{2z}$ and $2I_{1z}I_{2x}$ terms in [6.32] represent antiphase magnetization with peaks centered at $\Omega_1 \pm \pi J_{12}$ and $\Omega_2 \pm \pi J_{12}$ in F_2 (again with Ω in units of angular frequency). Both of the product operators are modulated by $\cos[(\Omega_1 + \Omega_2)t_1]$ and give rise to resonances at $\Omega_1 + \Omega_2$ in F_1. The lineshapes have no fine structure in this dimension. The appearance of a two-spin system in the 2Q spectrum is exemplified by the peaks from Tyr59 of ubiquitin Fig. 6.33.

More interesting results are obtained if an equivalent analysis is performed on a system containing the three spins I_1, I_2, and I_3 with coupling constants of J_{12}, J_{13} and J_{23}. During the initial spin-echo period, evolution of all three couplings will take place but chemical shifts will be refocused. Concentrating on the I_1 magnetization only

$$I_{1z} \xrightarrow{\left(\frac{\pi}{2}\right)_x - \frac{\tau}{2} - \pi_x - \frac{\tau}{2} - \left(\frac{\pi}{2}\right)_x}$$

$$I_{1z} \cos(\pi J_{12}\tau) \cos(\pi J_{13}\tau) + 2I_{1x}I_{2y} \sin(\pi J_{12}\tau) \cos(\pi J_{13}\tau)$$

$$+ 2I_{1x}I_{3y} \cos(\pi J_{12}\tau) \sin(\pi J_{13}\tau) - 4I_{1z}I_{2y}I_{3y} \sin(\pi J_{12}\tau) \sin(\pi J_{13}\tau). \quad [6.33]$$

The resulting longitudinal operator is suppressed by the phase cycle. The remaining three terms are mixtures of 2Q and ZQ coherence (Section 2.7.5) described by [6.15]–[6.17]. Once again, the phase cycling retains

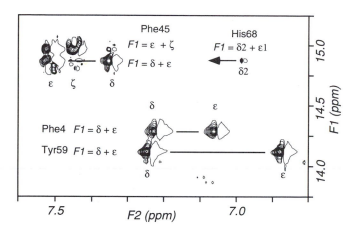

FIGURE 6.33 Example of the appearance of 2Q cross-peaks arising from two spin systems. The peak arising from aromatic-ring protons of Tyr59 is present at the bottom of the figure, and represents a two-spin system as $^4J_{H^{\delta 1}H^{\delta 2}}$ and $^4J_{H^{\epsilon 1}H^{\epsilon 2}}$ are not resolved. Although not formally two-spin systems, the cross-peaks from the aromatic-ring protons of Phe4 and Phe45 are also labeled. The small $^4J_{H^{\delta 2}H^{\epsilon 1}}$ of His68 also gives rise to a weak cross-peak. The spectrum was acquired from D$_2$O solution with an excitation delay, $\tau = 22$ ms.

only the 2Q operators. Allowing for evolution as described in Section 2.7.5, the following terms are present at the end of t_1:

$$[DQ_y^{12} + ZQ_y^{12}] \sin(\pi J_{12}\tau) \cos(\pi J_{13}\tau)$$

$$+ [DQ_y^{13} + ZQ_y^{13}] \cos(\pi J_{12}\tau) \sin(\pi J_{13}\tau)$$

$$+ 2I_{1z}[DQ_x^{23} - ZQ_x^{23}] \sin(\pi J_{12}\tau) \sin(\pi J_{13}\tau) \xrightarrow{t_1}$$

$$[DQ_y^{12} \cos(\pi K_{12}t_1) - 2DQ_x^{12}I_{3z} \sin(\pi K_{12}t_1)]$$

$$\times \cos(\Omega_{12}t_1) \sin(\pi J_{12}\tau) \cos(\pi J_{13}\tau)$$

$$+ [-DQ_x^{12} \cos(\pi K_{12}t_1) - 2DQ_y^{12}I_{3z} \sin(\pi K_{12}t_1)]$$

$$\times \sin(\Omega_{12}t_1) \sin(\pi J_{12}\tau) \cos(\pi J_{13}\tau)$$

$$+ [DQ_y^{13} \cos(\pi K_{13}t_1) - 2DQ_x^{13}I_{2x} \sin(\pi K_{13}t_1)]$$

$$\times \cos(\Omega_{13}t_1) \cos(\pi J_{12}\tau) \sin(\pi J_{13}\tau)$$

$$+ [-DQ_x^{13} \cos(\pi K_{13}t_1) - 2DQ_y^{13}I_{2z} \sin(\pi K_{13}t_1)]$$

$$\times \sin(\Omega_{13}t_1) \cos(\pi J_{12}\tau) \sin(\pi J_{13}\tau)$$

$$+ [2I_{1z}DQ_x^{23} \cos(\pi K_{23}t_1) - DQ_y^{23} \sin(\pi K_{23}t_1)]$$

$$\times \cos(\Omega_{23}t_1) \sin(\pi J_{12}\tau) \sin(\pi J_{13}\tau)$$

$$+ [2I_{1z}DQ_y^{23} \cos(\pi K_{23}t_1) - DQ_x^{23} \sin(\pi K_{23}t_1)]$$

$$\times \sin(\Omega_{23}t_1) \sin(\pi J_{12}\tau) \sin(\pi J_{13}\tau). \qquad [6.34]$$

In this expression, $K_{ij} = J_{ik} + J_{jk}$ is the DQ splitting and is equal to the sum of the coupling constants between the active spins (I_i and I_j) and the mutual passive spin (I_k).

Expansion of the right-hand side of [6.34] into single Cartesian operators indicates that, following the final 90_x° pulse, none of the terms 2, 4, and 6 lead to observable magnetization, and the terms 1, 3, and 5 yield

$$\frac{1}{2}[2I_{1x}I_{2z} + 2I_{1z}I_{2x}] \cos(\Omega_{12}t_1) \cos(\pi K_{12}t_1) \sin(\pi J_{12}\tau) \cos(\pi J_{13}\tau)$$

$$+ \frac{1}{2}[4I_{1x}I_{2x}I_{3y} - 4I_{1z}I_{2z}I_{3y}] \cos(\Omega_{12}t_1) \sin(\pi K_{12}t_1) \sin(\pi J_{12}\tau) \cos(\pi J_{13}\tau)$$

$$+ \frac{1}{2}[2I_{1x}I_{3z} + 2I_{1z}I_{3x}] \cos(\Omega_{13}t_1) \cos(\pi K_{13}t_1) \cos(\pi J_{12}\tau) \sin(\pi J_{13}\tau)$$

$$+ \frac{1}{2}[4I_{1x}I_{2y}I_{3x} - 4I_{1z}I_{2y}I_{3z}] \cos(\Omega_{13}t_1) \sin(\pi K_{13}t_1) \cos(\pi J_{12}\tau) \sin(\pi J_{13}\tau)$$

$$+ \frac{1}{2}[-4I_{1y}I_{2x}I_{3x} + 4I_{1y}I_{2z}I_{3z}] \cos(\Omega_{23}t_1) \cos(\pi K_{23}t_1) \sin(\pi J_{12}\tau) \sin(\pi J_{13}\tau)$$

$$+ \frac{1}{2}[2I_{2x}I_{3z} + 2I_{2z}I_{3x}] \cos(\Omega_{23}t_1) \sin(\pi K_{23}t_1) \sin(\pi J_{12}\tau) \sin(\pi J_{13}\tau). \qquad [6.35]$$

The terms proportional to $2I_{1x}I_{2z} + 2I_{1z}I_{2x}$ on line 1 of [6.35] are observable during t_2 and generate cross-peaks at frequencies Ω_1 and Ω_2 in F_2 with an antiphase splitting of J_{12}. Both of these peaks have an F_1 shift of $\Omega_1 + \Omega_2$ and are referred to as *direct peaks*. Because

$$\cos(\Omega_{12}t_1) \cos(\pi K_{12}t_1) = \frac{1}{2}[\cos(\Omega_{12}t_1 - \pi K_{12}t_1) + \cos(\Omega_{12}t_1 + \pi K_{12}t_1)],$$

$$[6.36]$$

the direct peaks have in-phase lineshapes in F_1. The direct peaks are phased to absorption in both dimensions, and the phases of other peaks in the spectrum are described relative to these direct peaks. Direct peaks between spins I_1 and I_3 are represented by line 3, and the direct peaks between spins I_2 and I_3 are on line 6 of [6.35]. In all cases the F_2 lineshapes are absorptive and antiphase with respect to the active coupling. The F_1 lineshape of the I_1–I_3 peaks is equivalent to that described above for the I_1–I_2 peaks, whereas the I_2–I_3 peaks are dispersive antiphase in this dimension.

The $2I_{1x}I_{2x}I_{3y}$ operator term on line 2 of [6.35] is a three-spin coherence that is not observable. The $2I_{1z}I_{2z}I_{3y}$ term describes an I_3 coherence that is antiphase with respect to I_1 and I_2 in F_2. Since both J_{12} and J_{13} are resolved, this is an observable term and will have a double antiphase dispersive lineshape in F_2; this lineshape and several others observed in 2Q spectra are displayed in Fig. 6.34. The cross-peak has an F_2 chemical shift of Ω_3 and is modulated at a frequency $\Omega_1 + \Omega_2$ during F_1. This resonance is described as a *remote peak*. Because of the $\cos(\Omega_{12}t_1)$ $\sin(\pi K_{12}t_1)$ modulation, this peak has a dispersive antiphase lineshape in F_1 (Fig. 6.34). The first terms on lines 4 and 5 of [6.35] are not observable, whereas the second terms describe the character of the I_1–I_3 and I_2–I_3 remote peaks, respectively. The F_2 lineshape of both peaks is double antiphase dispersive, while in F_1 the former is dispersive antiphase and the latter is absorptive in-phase. The positions of all the peaks described by [6.35] are depicted schematically in Fig. 6.35, and the lineshapes of these peaks are summarized in Table 6.2.

As a result of this analysis, several interesting points emerge for a spin system in which all three coupling constants are resolved: (1) the initial I_1 magnetization contributes to all nine cross-peaks in the spectrum; (2) peaks at a given F_1 and F_2 frequency can arise by three different pathways starting initially from I_1, I_2, or I_3 magnetization; (3) the three pathways contributing to a given peak all have the same F_2 lineshape (absorptive antiphase or dispersive double antiphase for direct or remote peaks, respectively); and (4) one of the pathways leading to a direct peak has dispersive antiphase character in F_1, whereas two of the pathways have this lineshape for the remote peaks. Consequently, all peaks will have dispersive tails in the F_1 dimension.

Generally, 2Q spectra are run with short mixing times ($\tau < 1/(2J_{jk})$) to avoid unnecessary loss of magnetization via relaxation, and $\sin(\pi J_{jk}\tau)$ and $\cos(\pi J_{jk}\tau)$ will both be positive. Because of the similarity of a negative dispersive antiphase peak and a positive absorptive in-phase peak (Fig. 6.34), the three contributions to a remote connectivity will add constructively in such experiments. Furthermore, the remote and direct peaks can be differentiated on the basis of the F_2 lineshape; the positive lobe of the absorptive antiphase direct peak will be downfield of the peak center whereas the major positive lobe of the dispersive double antiphase remote peak will be upfield of the peak center. Because of changes in sign of the trigonometric terms for long τ, these generalizations are not always valid,

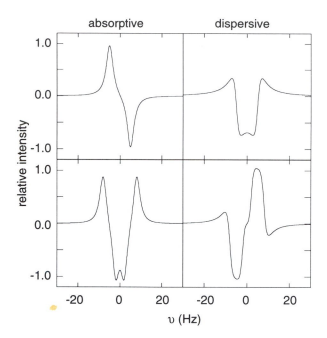

FIGURE 6.34 Lineshapes of single (top) and double (bottom) antiphase Lorentz-ian lines phased to absorption (left) and dispersion (right). These lineshapes are commonly found in cross-peaks in 2Q spectra. In all panels, the linewidths of the individual Lorentzian lines are 4.0 Hz, and a single line has an intensity of 1. The scalar coupling constants correspond to a 10-Hz coupling (top) or a 10- and a 6-Hz coupling (bottom). Note that in real spectra (poor digital resolution, linewidths comparable to the splittings and the presence of passive couplings), distinguishing between absorptive antiphase and dispersive double-antiphase lineshapes or be-tween dispersive antiphase and absorptive double antiphase lineshapes can be dif-ficult.

although the relative signs of the peaks do change in a predictable manner if the coupling constants are known (e.g., see Ref. *34*).

Finally, a linear three-spin system will be examined by considering the effect of setting one of the couplings to zero. I_1 may be considered a terminal spin in a linear system if $J_{13} = 0$. In this case, $\sin(\pi J_{13}\tau) = 0$ and terms such as $4I_{1z}I_{2z}I_{3y}$ are no longer observable since the antiphase state between I_1 and I_3 can never evolve into observable magnetization during t_2. Thus, of the nine terms described in Table 6.2, only lines 1 and 2 correspond to observable peaks; i.e., initial I_1 magnetization can contrib-ute to the direct peaks only between I_1 and I_2. Alternatively, I_1 may be

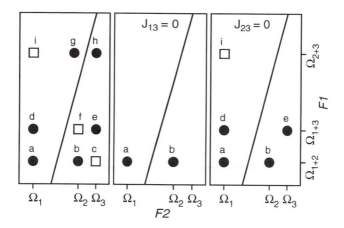

FIGURE 6.35 Contributions of initial I_1 magnetization to the peaks observed in the 2Q spectrum of a three-spin system. The filled circles and open squares represent direct and remote peaks, respectively, while the lines indicate the pseudodiagonal ($F_2 = 2F_1$). The left panel assumes that J_{12}, J_{13}, and J_{23} are resolved. If J_{13} (center) or J_{23} (right) is not resolved, then I_1 contributes to fewer peaks. The F_1 and F_2 lineshapes of the cross-peaks, labeled a–i, are described in Table 6.2.

considered the middle spin of a linear system if $J_{23} = 0$. In this case, initial I_1 magnetization contributes to the I_1–I_2 direct peaks (peaks a and b), I_1–I_3 direct peaks (peaks d and e), and the I_2–I_3 remote peak (peak i). Consideration of these results indicate that a remote peak is observed only when both of the spins contributing to the 2Q frequency have resolved couplings to the third passive spin whose frequency is measured in F_2. Results such as these form the basis of the selection rules that are discussed in more detail in Section 6.4.2.3.

6.4.1.2 Experimental Protocol As described above, the spin-echo sequence in Fig. 6.32 serves to generate antiphase magnetization. Thus, the choice of delay, τ, will depend on the magnitudes of the active scalar coupling constants. For 2Q spectra acquired for proteins in H_2O and D_2O solution, $\tau = 30$ ms compromises between coherent evolution and incoherent relaxation. In order to emphasize particular correlations, spectra with τ as short as 20 ms and as long as 80 ms have been reported in the literature (*22,34*).

The 180° pulse in the middle of the delay τ can be a source of several types of artifact in the final spectrum, many of which can be alleviated by the use of a composite 180° pulse of the form $90°_{\phi1}180°_{\phi1+\pi/2}90°_{\phi1}$ (Section

TABLE 6.2

Lineshapes in a 2Q Spectrum

Peak	Type	Chemical shift F_1	Chemical shift F_2	Lineshape F_1		Lineshape F_2	
a	Direct	$\Omega_1 + \Omega_2$	Ω_1	abs	in	abs	anti
b	Direct	$\Omega_1 + \Omega_2$	Ω_2	abs	in	abs	anti
c	Remote	$\Omega_1 + \Omega_2$	Ω_3	disp	anti	disp	anti $\times$ 2
d	Direct	$\Omega_1 + \Omega_3$	Ω_1	abs	in	abs	anti
e	Direct	$\Omega_1 + \Omega_3$	Ω_2	abs	in	abs	anti
f	Remote	$\Omega_1 + \Omega_3$	Ω_3	disp	anti	disp	anti $\times$ 2
g	Direct	$\Omega_2 + \Omega_3$	Ω_2	disp	anti	abs	anti
h	Direct	$\Omega_2 + \Omega_3$	Ω_3	disp	anti	abs	anti
i	Remote	$\Omega_2 + \Omega_3$	Ω_1	abs	in	disp	anti $\times$ 2

Shown are the lineshapes of the cross-peaks in a 2Q spectrum of a three-spin system that result from initial I_1 magnetization. The peak labels a–i refer to the schematic spectrum shown in Fig. 6.35. In the table, Ω_1, Ω_2, and Ω_3 are the resonance frequencies of spins I_1, I_2, and I_3, respectively; "abs" and "disp" indicate absorption and dispersion lineshapes, respectively; "in," "anti," and "anti $\times$ 2" refer to in-phase, antiphase, and doubly antiphase multiplet structures, respectively.

3.4.2) (*35*). In addition, spectra of high quality can usually be obtained without presaturation of the solvent resonance during the excitation period. This often leads to the observation of more intense correlations involving protons resonating close to the solvent signal (see the following paragraphs).

As discussed in Section 4.3.4.1, quadrature detection in the t_1 dimension of two-dimensional NMR spectra is achieved by shifting the phase of pulses prior to t_1 according to the TPPI, States, or TPPI–States protocols in order to shift the phase of the indirectly detected coherences by 90°. Multiple-quantum coherences are p-fold more sensitive to rf phase shifts than are single-quantum coherences. Thus, in a pQ experiment, the phases ϕ_1, ϕ_2, and ϕ_3 must be incremented by $\pi/(2|p|)$, or by $\pi/4$ for a 2Q experiment. The initial value of the t_1 sampling delay normally is set as short as possible to minimize phase differences between direct and remote peaks (*33*).

The choice of spectral width in F_1 of the 2Q experiment is not straightforward because resonance peaks appear at sums of the chemical shifts of coupled resonances, and some prior knowledge of the system is helpful. For work in H_2O solution, the spectrum must extend downfield of the largest sum of two coupled spins; most usually this will result from the most downfield $^1H^N$ or $^1H^\alpha$ resonance, or from aromatic-ring protons. This limit can be calculated if COSY or other correlation spectra have already been acquired, or else estimated from the sum of the most down-field $^1H^N$ and $^1H^\alpha$ resonances observed in a 1D spectrum. The upfield spectral limit can be calculated from the cross-peaks observed in the upfield region of the COSY (usually involving methyl resonances) or else estimated as twice the frequency of the most upfield resonance in the 1D spectrum. The carrier usually is positioned on-resonance with the water signal. The resulting large spectral width suggests that many t_1 points will have to be acquired to achieve the required resolution of resonances in the F_1 dimension. Alternatively, since the focus of ^1H NMR spectroscopy in H_2O is primarily the $^1H^N-^1H^\alpha$ correlations, the F_1 spectral width can be reduced and the carrier position shifted to span the spectral region expected for the $^1H^N-^1H^\alpha$ correlations only. In similar fashion, the digital resolution for 2Q spectra acquired from D_2O solution can be increased by setting the carrier position and spectral width to span resonances arising from the aliphatic resonances only.

The lineshape of all peaks in the F_1 dimension of the 2Q spectrum are either absorptive in-phase or dispersive anti-phase (Table 6.2 and Fig. 6.34); consequently, high digital resolution is not required in this dimension to prevent self-cancellation of the cross-peaks. Given the preceding discussion of product operators, the acquisition parameters chosen for F_2 evidently are very similar to those used for COSY spectra; thus, a high digital resolution is necessary to prevent self-cancellation of the absorptive antiphase components.

The 2Q spectra obtained from H_2O solution shown below were collected with 32 transients for each of 800 t_1 increments and a mixing period, τ, of 22 or 32 ms. The F_1 spectral width was 5600 Hz and $t_{1,max} = 71.4$ ms; this was sufficient to span all the $^1H^N-^1H^\alpha$ correlations, but resulted in folding of some of the upfield aliphatic side-chain correlations. An equivalent spectrum was recorded in D_2O solution with 600 t_1 increments, $\tau = 30$ ms, an F_1 spectral width of 5000 Hz ($t_{1,max} = 60$ ms), and with the transmitter in the center of the aliphatic region; aromatic resonances were folded.

6.4.1.3 Processing As for COSY spectra, unshifted sine-bell window functions are applied in F_2 of 2Q spectra acquired in H_2O solution to attenuate the residual solvent resonance; in D_2O solution, the window function can be phase shifted to increase sensitivity. Applying the sine bell over more points will decrease the S/N ratio but increase the resolution. The 2Q spectrum does not contain dispersive in-phase peaks in F_1; therefore, window functions need not be as strongly resolution-enhancing

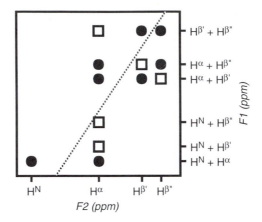

FIGURE 6.36 Schematic representation of the peaks expected in the 2Q spectrum of a $^1H^N$–$^1H^\alpha$–$^1H^{\beta_2}$ spin system ($^4J_{H^N H^\beta}$ is assumed to be zero). The filled circles and open squares represent direct and remote peaks, respectively, and the dotted line indicates the pseudodiagonal ($F_2 = 2F_1$)

as for COSY. A simple cosine bell to prevent truncation artifacts is usually sufficient. The center of the 2Q spectrum in F_1 is referenced to be double the carrier frequency. Phase parameters are derived in a similar fashion to those of the COSY experiment.

6.4.1.4 Information Content Like COSY spectra, 2Q spectra contain information about scalar coupling networks; however, 2Q spectra have unique features that circumvent some of the inherent problems of COSY experiments and provide additional information through both direct and remote correlations. Direct peaks occur at the sum of two chemical shifts of two coupled spins in F_1 and at the chemical shift of each spin in F_2. From the product operator analysis (Section 6.4.1.1), remote peaks also occur in systems of three or more coupled spins. Remote peaks occur at the sum of the frequencies of two actively coupled spins in F_1, and at the frequency of the third passively coupled spin in F_2. The peaks expected in the 2Q spectrum of an $^1H^{N}$–$^1H^{\alpha}$–$^1H_2^{\beta}$ spin system are presented schematically in Fig. 6.36. The presence of these peaks is formally described by selection rules (*33*) that are discussed more fully in Section 6.4.2.3.

The problems of COSY spectra that can be partially alleviated in 2Q spectroscopy fall into three categories: diagonal peaks, self-cancellation, and attenuation by solvent presaturation. The dispersive tails of diagonal peaks in COSY spectra can curtail observation of correlations between resonances with similar chemical shifts. The 2Q spectrum does not contain diagonal peaks, and observation of cross-peaks is facilitated. In this respect, 2Q spectra are comparable to 2QF-COSY spectra, provided that the latter have been acquired with sufficient t_1 points to allow resolution of the diagonal and cross-peaks at the same F_2 shift. Self-cancellation of COSY cross-peaks has been described in detail in Section 6.2.1.5. The attenuation that this causes is most severe in F_1, where the linewidth is determined by $t_{1,max}$. In 2Q spectroscopy, the peaks have absorptive in-phase and dispersive antiphase character in F_1, and are not subject to cancellation. Thus, for interactions with small scalar coupling constants or for proteins with large linewidths, 2Q spectra can display correlations that are unobserved in COSY spectra. Finally, presaturation of the solvent peak in the COSY experiment also attenuates correlations involving protons (usually $^1H^\alpha$) that resonate close to, or coincident with, the solvent resonance. In a 2Q experiment, even if presaturation is employed prior to the initial 90° pulse (Fig. 6.32), antiphase magnetization between $^1H^N$ and $^1H^\alpha$ still develops provided that the solvent resonance is not irradiated during the delay τ. Thus, 2Q coherence can still be generated for these resonances. In the COSY spectrum of ubiquitin, three $^1H^N$-$^1H^\alpha$ cross-peaks were absent because of $^1H^\alpha$ presaturation (Fig. 6.9). The equivalent region of the 2Q spectrum clearly reveals these correlations (Fig. 6.37).

Perhaps the biggest advantage of 2Q over COSY experiments is the presence of remote peaks, which occur in addition to direct peaks. For short values of τ, remote peaks are easily recognized by the 180° phase difference in F_2 relative to the direct peaks (see product operator analysis in Section 6.4.1.1). Because the remote F_1 frequency gives information about the sum of only two of the chemical shifts, at least one direct peak must also be observed at the same F_2 shift to determine all three chemical shifts of the spin system. However, the remote peaks are observed even if two of the coupled spins have degenerate chemical shifts. This feature makes 2Q spectroscopy one of the few reliable methods for identifying cases of chemical-shift degeneracy.

In addition to the important direct peaks with $^1H^\alpha$ (Fig. 6.37), amide protons are involved in two sets of remote connectivities involving glycine residues and all residues containing a $^1H^\beta$-proton (*36*). Glycine residues are unique in that their amide protons are directly coupled to two α-protons. The coupling within this group gives rise to three peaks at $F_2 = {}^1H^N$: the two direct peaks at $F^1 = {}^1H^N + {}^1H^{\alpha'}$ or $^1H^N + {}^1H^{\alpha''}$ and also a remote peak at $F_1 = {}^1H^{\alpha'} + {}^1H^{\alpha''}$. In COSY spectra, one of the two cross-peaks expected for glycine $^1H^N$–$^1H^\alpha$ residue is often missing as a result of either a small active coupling (leading to self-cancellation), overlap

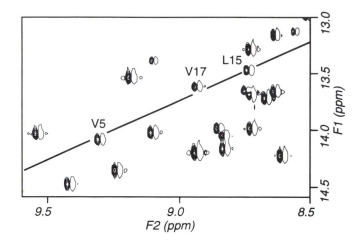

FIGURE 6.37 Section of the H_2O 2Q spectrum of ubiquitin (τ = 22 ms) showing part of the $^1H^N$–$^1H^\alpha$ region. The diagonal line represents the possible 2Q frequencies of $^1H^N$ coupled to $^1H^\alpha$ spins on resonance with the H_2O signal. The cross-peaks arising from Val5 and Leu15 were not observed in the COSY spectrum because of saturation of the $^1H^\alpha$ resonance (compare Fig. 6.9). The cross-peak arising from Arg54 is observed in the 2Q spectrum, but does not appear in the illustrated region. Because of partial saturation in the COSY, the cross-peak due to Val17 is much more intense in the 2Q spectrum.

with other $^1H^N$–$^1H^\alpha$ correlations, or degeneracy of the $^1H^{\alpha\prime}$ and $^1H^{\alpha\prime\prime}$ shifts; hence the observation of the remote peak in the 2Q spectrum provides an unequivocal assignment of these amino acid spin systems (Fig. 6.38). The lobes surrounding the central positive and negative components of the glycine 2Q remote peaks are not resolution-enhancement artifacts, but are actually the result of the natural dispersive double antiphase lineshape in F_2 and dispersive anti-phase lineshape in F_1 (see Fig. 6.34).

In all residues except glycine and proline, the $^1H^\alpha$ spin is coupled to both $^1H^N$ and at least one $^1H^\beta$; hence remote peaks at $F_1 = {}^1H^N + {}^1H^\beta$ are expected at $F_2 = {}^1H^\alpha$. Such peaks provide a useful means of correlating a $^1H^N$ spin directly with a $^1H^\beta$ spin in the same spin system. Unfortunately, the usefulness is compromised by the water stripe present at $F_2 = \delta(H_2O)$, which will obscure many $^1H^\alpha$ resonances; as a result, $^1H^N$–$^1H^\beta$ correlations are more readily established in TOCSY (Section 6.5) or relayed COSY (Section 6.2.2) spectra.

The $^1H^\alpha$–$^1H^\beta$ peaks form the main fingerprint region of interest in the upfield region of the 2Q spectrum and are most easily observed in D_2O solution at the F_2 shift of the $^1H^\alpha$ resonances. This region of the COSY

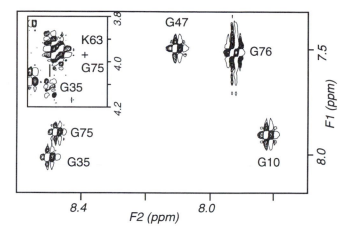

FIGURE 6.38 Section of the H_2O 2Q spectrum of ubiquitin (τ = 22 ms) showing the remote glycine peaks at F_1 = $^1H^{\alpha'}$ + $^1H^{\alpha''}$. The peaks from five of the six glycine residues are clearly present and labeled. The amide proton of the remaining glycine (Gly53) is unusually broad, and no cross-peaks involving it are observed in any of the correlation spectra. The inset shows the cross-peaks of Gly35 and Gly75 in the COSY spectrum. The degeneracy of $^1H^{\alpha'}$ and $^1H^{\alpha''}$ of Gly75 and the overlap with Lys63 $^1H^N$–$^1H^\alpha$ in the COSY are confirmed by the presence of the 2Q remote peak.

spectrum is normally crowded, rendering the observation of all $^1H^\alpha$–$^1H^\beta$ cross-peaks problematic. In addition, for β-methylene containing side chains, one of the $^1H^\alpha$–$^1H^\beta$ coupling constants frequently is small; thus, only one of the two expected peaks is observed in COSY spectra. Although the direct $^1H^\alpha$–$^1H^\beta$ peaks in the 2Q spectrum are equally crowded and possibly weak, the remote peaks are not; hence, observation of the remote peak and one direct peak allows assignment of all three resonance positions (Fig. 6.39).

Commonly, resonance positions within a side chain will be inferred from correlations relayed to the amide proton in TOCSY experiments acquired in H_2O. Although chemical-shift arguments allow assignment of the cross-peaks to a particular side-chain resonance, this process is fallible. Discriminating $^1H^\beta$ and $^1H^\gamma$ resonances within five-spin residues is one case where such a simple analysis commonly leads to incorrect side-chain assignments. Thus, the unambiguous determination of $^1H^\beta$ resonance positions in the 2Q spectrum provides a useful complement to other relay-based techniques (e.g., see Fig. 6.40). Indeed, during the analysis of spectra for this book, several $^1H^\beta$ resonances that had not been assigned previously were observed in MQ spectra (*37,38*).

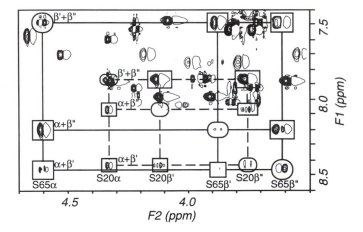

FIGURE 6.39 Example of remote and direct peaks for $C^\alpha H-C^\beta H_2$ fragments in the D_2O 2Q spectrum ($\tau = 30$ ms). Squares and ellipses indicate direct and remote correlations, respectively. The cross-peaks arising from Ser65 and Ser20 are connected by solid and dashed lines, respectively. The F_1 and F_2 frequencies are indicated adjacent to the lines connecting the cross-peaks. For this value of τ, the direct and remote peaks are distinguished by virtue of the opposite phase in F_2 (see text).

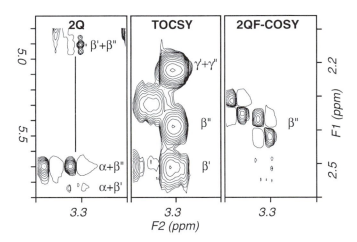

FIGURE 6.40 Cross-peaks between ¹Hα of Glu64 and its side-chain resonances. The ¹Hα–¹Hβ′ cross-peak is extremely weak in the 2QF-COSY because of the weak coupling between ¹Hα–¹Hβ′. All side-chain resonance positions are identified in the TOCSY ($\tau_m = 48$ ms); however, the ¹Hα–¹Hβ′ and ¹Hα–¹Hγ resonances cannot be unambiguously assigned. Observation of the remote peak in the 2Q spectrum ($\tau = 30$ ms) is required to identify the chemical shift of ¹Hβ′.

6.4.2 3Q Spectroscopy

As with 2QF- and 3QF-COSY, the 2Q experiment is just one of many experiments based on observation of multiple-quantum states. Multiple-quantum states higher than 3Q, although attainable theoretically, are rarely observed in protein systems. Thus, the 3Q experiment is the practical limit to pQ spectroscopy of proteins. The full-phase cycle for the 3Q experiment is shown in the caption to Fig. 6.32.

6.4.2.1 Product Operator Analysis   Because of the similarity of the initial parts of the 2Q and 3Q pulse sequences, evolution of the product operators is the same up to the end of the 90°_x–$\tau/2$–180°_x–$\tau/2$ sequence. At this point, a 90°_y pulse must be applied in order to generate odd orders of multiple quantum coherence. Thus, the following operators are obtained for a three spin system (I_1, I_2, and I_3):

$$I_{1z} \xrightarrow{\;(\pi/2)_x-\tau/2-\pi_x-\tau/2-(\pi/2)_y\;}$$

$$I_{1y}\cos(\pi J_{12}\tau)\cos(\pi J_{13}\tau) + 2I_{1z}I_{2x}\sin(\pi J_{12}\tau)\cos(\pi J_{13}\tau)$$

$$+2I_{1z}I_{3x}\cos(\pi J_{12}\tau)\sin(\pi J_{13}\tau) - 4I_{1y}I_{2x}I_{3x}\sin(\pi J_{12}\tau)\sin(\pi J_{13}\tau). \quad [6.37]$$

The 3Q phase cycle suppresses the first three terms of [6.37]. The fourth term is a mixture of $3Q_y$ and three-spin single-quantum coherence; a similar operator was encountered in the analysis of the 3QF-COSY experiment (Section 6.3.2.1). Once again, phase cycling retains only the pure 3Q term, and only its evolution needs to be considered during t_1:

$$3Q_y\sin(\pi J_{12}\tau)\sin(\pi J_{13}\tau)\xrightarrow{\;t_1\;}$$

$$3Q_y\cos(\Omega_{123}t_1)\sin(\pi J_{12}\tau)\sin(\pi J_{13}\tau)$$

$$-3Q_x\sin(\Omega_{123}t_1)\sin(\pi J_{12}\tau)\sin(\pi J_{13}\tau), \quad [6.38]$$

where $\Omega_{123} = \Omega_1 + \Omega_2 + \Omega_3$. Application of the final 90°_y pulse to the Cartesian single-operator components of [6.38] produces the following operators prior to t_2:

$$[-4I_{1z}I_{2y}I_{3z} - 4I_{1y}I_{2z}I_{3z} - 4I_{1z}I_{2z}I_{3y} + 4I_{1y}I_{2y}I_{3y}]$$

$$\times \cos(\Omega_{123}t_1)\sin(\pi J_{12}\tau)\sin(\pi J_{13}\tau)$$

$$+ [4I_{1z}I_{2z}I_{3z} - 4I_{1y}I_{2y}I_{3z} - 4I_{1z}I_{2y}I_{3y} - 4I_{1y}I_{2z}I_{3y}]$$

$$\times \sin(\Omega_{123}t_1)\sin(\pi J_{12}\tau)\sin(\pi J_{13}\tau). \quad [6.39]$$

The fourth term on line 1 and all four terms on line 3 of [6.39] are unobservable operators. The first three terms on line 1 give rise to observable peaks at the chemical shifts of each of the three spins in F_2 and at the sum of their chemical shifts in F_1. In F_2 each peak is doubly antiphase with respect to the other two spins, whereas in F_1 all three peaks are singlets without fine structure and can be phased to absorption.

The product operator analysis described above has been extended to include additional passive spins to give results applicable to other spin systems (39). Thus, the contribution of spin I_1 to the 3Q coherence ($I_1 + I_2 + I_3$) is described by

$$A_1 = \sin(\pi J_{12}\tau)\,\sin(\pi J_{13}\tau)\prod_{k=4}^{K}\cos(\pi J_{1k}\tau), \qquad [6.40]$$

in which I_1, I_2, and I_3 are active spins and all spins I_k, for $k > 3$, are passive spins. If all three active spins are passively coupled to a fourth spin, I_4, then I_4 also contributes to the 3Q coherence ($I_1 + I_2 + I_3$) in the following manner:

$$A_4 = \sin(\pi J_{14}\tau)\,\sin(\pi J_{24}\tau)\,\sin(\pi J_{34}\tau)\prod_{k=5}^{K}\cos(\pi J_{4k}\tau). \qquad [6.41]$$

These expressions have been used to generate excitation profiles as a function of τ for the spin systems and coupling constants commonly found in amino acids (39).

By considering evolution of the density matrix during a MQ experiment, Braunschweiler *et al.* (33) have described two selection rules for the peaks observed in such spectra: (1) multiple-quantum coherence involving a set of q spins can be transferred to the single-quantum transitions of a spin I only when the scalar couplings between I and *all* q spins are resolved (if I belongs to the set of q spins, only the couplings to the remaining q-1 spins need be resolved) and (2) multiple-quantum coherence involving two or more equivalent nuclei cannot be transferred to the single-quantum coherence of the equivalent nuclei because couplings between magnetically equivalent nuclei are ineffective in isotropic solution phase. As in the case of multiple-quantum-filtered COSY experiments, violations of the second selection rule by methyl groups in proteins are manifestations of multi exponential relaxation (27).

6.4.2.2 Experimental Protocol and Processing   Carrier positions and spectral widths are chosen in almost the same way as described above for the 2Q experiment. The principal difference is that now F_1 peak positions occur at the sum of three chemical shifts. Considering only the

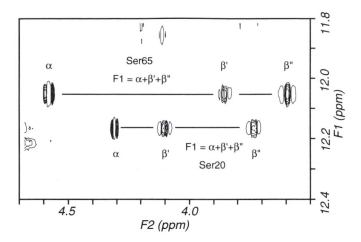

FIGURE 6.41 Demonstration of the spectral simplification achieved for a spin system containing three coupled spins in a 3Q spectrum; only three direct peaks are expected for such systems. Spins systems of Ser20 and Ser65 are shown (for comparison, the 2Q cross-peaks arising from these residues are depicted in Fig. 6.39).

aliphatic resonances (i.e., folding the aromatic signals in F_1), the downfield spectral limit will be at the sum of H^α, $H^{\beta'}$, and $H^{\beta''}$ chemical shifts. The largest sum usually arises from serine residues. As 3Q coherences are now evolving during t_1, the phases of all pulses prior to t_1 must be incremented by $\pi/(2|p|) = \pi/6$ in order to achieve quadrature detection (Section 4.3.4.1). Generally, $\tau = 30$ to 36 ms; additional details are given in Section 6.4.2.3. Other details of the experimental protocol are similar to those described above for the 2Q experiment (Section 6.4.1.2).

The sections of spectra shown in Fig. 6.41 were obtained in 16 hs on a 2-mM ubiquitin sample in D_2O solution using a preparation period of $\tau = 36$ ms; 48 transients were collected for each of 600 t_1 increments using a spectral width of 7000 Hz in F_1 ($t_{1,\mathrm{max}} = 43$ ms). The carrier was shifted from the water resonance to 2.0 ppm at the start of the pulse sequence, and the aromatic resonances were folded in F_1. The experiment was acquired with TPPI quadrature detection in F_1.

Window functions applied in F_2 are chosen in a similar fashion to the 2Q experiment (Section 6.4.1.3). Phase parameters are chosen to give a symmetrical three-lobed profile for direct peaks in this dimension (formally an absorptive double antiphase lineshape; Fig. 6.34). Given the predominantly absorptive lineshape in F_1, processing in this dimension should be sufficient to prevent truncation, and possibly to provide slight resolution enhancement (e.g., sine bells shifted by 60° to 90°). The center of the spectrum in F_1 is referenced to three times the carrier frequency.

6.4.2.3 Information Content The justifications for using 3Q experiments are 2-fold: (1) moving to a higher quantum state results in spectral simplification because some spin systems cannot attain the requisite level of coherence (much as the 3QF-COSY offers advantages over the 2QF-COSY) and (2) the remote peaks present in the 3Q spectrum offer unique assignment information. Even for spin systems that are present in *p*QF-COSY or 2Q spectra, the 3Q experiment may offer considerable simplification. For example, Fig. 6.41 depicts the correlations arising from the spin systems of Ser20 and Ser65; each spin system leads to three direct peaks only. Comparable regions of other spectra contain larger numbers of peaks; the COSY spectra contain six cross-peaks and three diagonal peaks, and the 2Q contains six direct and three remote peaks. The simplification makes observation of the spin systems of Ser20 and Ser65 very straightforward (cf. with Fig. 6.39).

The three main uses for the 3Q spectrum are (1) verification of ^{1}H$^\alpha$ and ^{1}H$^\beta$ resonance assignments made in other spectra, (2) identification of resonance positions for phenylalanine and tryptophan ring protons, and (3) identifying cases of resonance degeneracy at the end of longer spin systems (five-spin, lysine, arginine, and proline residues).

Amide protons of glycine residues contribute to a direct 3Q coherence ($F_1 = {}^1\mathrm{H}^N + {}^1\mathrm{H}^{\alpha\prime} + {}^1\mathrm{H}^{\alpha\prime\prime}$; $F_2 = {}^1\mathrm{H}^N$, ^{1}H$^{\alpha\prime}$, or ^{1}H$^{\alpha\prime\prime}$) that allows determination of all three chemical shifts; this information is obtained equally well from the remote peaks in a 2Q spectrum (Section 6.2.3.4). In the case of residues other than glycine or proline, an amide proton contributes to a direct ($F_1 = {}^1\mathrm{H}^N + {}^1\mathrm{H}^\alpha + {}^1\mathrm{H}^\beta$) peak, and if the amino acid contains a β-methylene group, a remote peak ($F_1 = {}^1\mathrm{H}^N + {}^1\mathrm{H}^{\beta\prime} + {}^1\mathrm{H}^{\beta\prime\prime}$). The former correlations have excitation maxima in the range $\tau = 30$–40 ms, whereas the latter are maximized for $\tau = 40$–50 ms (*39*). From the multiple-quantum selection rules, the 3Q coherence is observable only at $F_2 = {}^1\mathrm{H}^\alpha$ in both these cases. Not all of these correlations are observed in spectra acquired from H_2O solution because of interference from the water resonance.

The main peaks of interest in a 3Q spectrum acquired from D_2O solution are (1) coherences involving ^{1}H$^\alpha$ and ^{1}H$^\beta$ protons; (2) coherences arising from the aromatic resonances of tryptophan and phenylalanine residues; and (3) coherences involving the terminal protons of lysine (^{1}H$^\delta + {}^1$H$^{\varepsilon\prime} + {}^1H^{\varepsilon\prime\prime}$ and ^{1}H$^\delta + {}^1$H$^{\delta\prime\prime} + {}^1H^\varepsilon$), arginine, and proline residues (^{1}H$^\gamma + {}^1$H$^{\delta\prime} + {}^1$H$^{\delta\prime\prime}$ and ^{1}H$^{\gamma\prime} + {}^1$H$^{\gamma\prime\prime} + {}^1H^\delta$). Using the analytical expressions [6.40] and [6.41], values of τ to provide optimal intensity of these correlations can be deduced (*39*). Regardless of the size of $^3J_{\mathrm{H}^\alpha\mathrm{H}^\beta}$ and extensions of the spin system beyond $^1H^\beta$, the intensity of the $^1H^\alpha + {}^1H^{\beta'}$ $+ {}^1H^{\beta''}$ coherences are maximal at $\tau = 30$ ms except for serine residues, for which the smaller geminal $^1H^{\beta'}-{}^1H^{\beta''}$ coupling leads to maximal intensity at $\tau = 40$ ms. The linear and aromatic spin systems have a broad maximum centered at $\tau = 65$ ms. Thus, all types of correlations should be detectable for τ in the 30–40-ms range.

The efficiency of excitation for the terminal protons of longer side chains are more difficult to categorize because of the influence of contributions from passive spins. The various active and passive contributions have very different F_1 lineshapes that can lead to severe cancellation for some values of τ. This complex behavior also is very dependent on the exact linewidths of the resonances involved because the dispersive terms are also prone to self-cancellation. In summary, a value of $\tau = 36$ ms is optimal for most aliphatic side-chain correlations (*22*). Acquisition of experiments with longer or shorter values of τ may be necessary to observe all correlations involving the terminal groups of arginine, proline, and lysine. The work of Chazin should be consulted when optimizing τ for a particular correlation (*39*).

The aromatic region of 3Q spectra are usually quite simple as they contain peaks from phenylalanine or tryptophan residues only. Resonances from the aromatic ring of tyrosine are absent unless the $^4J_{H^{\delta 1}H^{\delta 2}}$ or $^4J_{H^{\varepsilon 1}H^{\varepsilon 2}}$ couplings are resolved. This region of the ubiquitin spectrum is further simplified by the degeneracy of H^ε and H^ζ of one of the two phenylalanine residues (Fig. 6.42). The 3Q correlations among resonances at the termini of longer side chains provides one of the few ways to positively identify resonance degeneracy (*39,40*). However, the number of possible 3Q frequencies can be large, and the ability to obtain useful assignment information will depend on the particular distributions of chemical shifts.

6.5 TOCSY

Soon after the introduction of the R-COSY and related experiments (Sections 6.2.2 and 6.2.3), a new method of obtaining relayed connectivities was introduced, namely, total correlation spectroscopy (TOCSY) (*41*). This technique is also known by the acronym HOHAHA (homonuclear Hartmann–Hahn) spectroscopy (*42*). Instead of relying on pulse-interrupted free-precession sequences to transfer coherence between antiphase states (Section 4.2.1.1), TOCSY utilizes isotropic mixing to transfer in-phase magnetization between spins via the strong scalar coupling Hamiltonian (Section 4.2.1.2), with the result that magnetization can be transferred through several couplings during the course of the mixing. In the absence

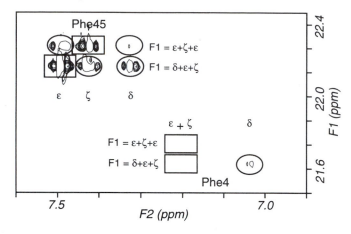

FIGURE 6.42 Section of the 3Q spectrum of ubiquitin in D₂O solution showing the aromatic side-chain resonances. Direct and remote peaks are indicated by rectangles and ellipses, respectively. Peaks from Phe45 are readily apparent, whereas the degeneracy of ¹Hᵉ and ¹Hᶜ for Phe4 prevents the generation of 3Q coherence hence the direct peaks are not observed. In the absence of resolved four-bond scalar couplings, no correlations are observed for Tyr59.

of relaxation, cross-peaks potentially are generated between all resonances within a spin system.

6.5.1 PRODUCT OPERATOR ANALYSIS

The TOCSY experiment can be performed in numerous ways. In the method described herein, after frequency labeling during t_1, magnetization is returned to the z axis for isotropic mixing using the DIPSI-2rc sequence, and then rotated to the transverse plane for detection (43). In addition to several advantages described below, this method also enables the addition of a Hahn echo at the end of the sequence to provide a flatter baseline (Section 3.6.4.2). The pulse sequence, coherence-level diagram, and phase cycle for such an experiment are shown in Fig. 6.43. The phase cycling used in this version of the TOCSY experiment is identical to that developed for the NOESY experiment, and is discussed in more detail in Section 6.6.

Following the $90_x^\circ - t_1 - 90_x^\circ$ sequence, the density operator is described by [6.1]. The I_{1x} and $2I_{1z}I_{2y}$ terms in [6.1] are eliminated by the TOCSY phase cycle. The multiple-quantum term $2I_{1x}I_{2y}$ is dephased by the rf inhomogeneity present during the isotropic mixing pulse sequence of duration τ_m (Section 3.5.3). During the isotropic mixing sequence, magnetization proportional to I_{1z} is transferred throughout the spin system via the

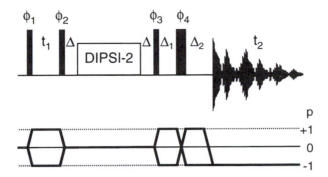

FIGURE 6.43 Pulse sequence and coherence-level diagram for the TOCSY experiment. A Hahn echo is included prior to detection. The basic eight steps of the phase cycle are $\phi_1 = 2(x, -x, -x, x)$; $\phi_2 = 2(x, x, -x, -x)$; $\phi_3 = (x)_8$; $\phi_4 = 4(y)$ $4(-y)$; and receiver $= 4(x, -x)$. The full 32-step phase cycle is completed by performing CYCLOPS on all pulses and the receiver. The delay Δ allows the transmitter power to be changed and is short enough ($\approx 20~\mu s$) to prevent development of NOE cross-peaks.

strong coupling Hamiltonian as described in Section 4.2.1.2:

$$-I_{1z} \cos(\Omega_I t_1) \cos(\pi J_{12} t_1) \xrightarrow{\tau_m} -\sum_{k=1}^{K} I_{kz} a_{1k}(\tau_m) \cos(\Omega_I t_1) \cos(\pi J_{12} t_1),$$

[6.42]

in which $a_{1k}(\tau_m)$ are mixing coefficients for transfer of magnetization through the spin system from spin I_1 to spin I_k, and zero-quantum terms (Section 4.2.1.2) have been ignored. Following the final 90°_x pulse and Hahn echo, the density operator prior to t_2 is given by

$$\sum_{k=1}^{K} I_{ky} a_{1k}(\tau_m) \cos(\Omega_I t_1) \cos(\pi J_{12} t_1).$$

[6.43]

The term proportional to I_{1y} represents the diagonal peak, and the terms proportional to I_{ky} for $k > 1$ represent cross-peaks. The observation of a cross-peak between two spins in the TOCSY experiment does not indicate that the spins are directly coupled; rather the cross-peak indicates that magnetization can be transferred between the two spins by a series of steps through two- and three-bond scalar couplings between mutually coupled spins.

The magnitude of a given cross-peak [governed by $a_{1k}(\tau_m)$] will depend on the topology of the spin system, the coupling constants between pairs of spins, the efficiency of the isotropic mixing sequence employed, and the rate of relaxation during the isotropic mixing pulse. Numerical calculations based on density matrix theory have been used to determine the transfer efficiencies for the coupling networks expected in naturally occurring amino acids using scalar coupling constants representative of a variety of conformations present in proteins (44). Such data are of much value in choosing an appropriate mixing time, as will be discussed in the next section.

Inspection of [4.16] indicates that the mixing process also leads to the transfer of zero-quantum coherence and gives rise to antiphase components in both the cross- and diagonal peaks. The 90° phase difference between the in-phase and antiphase components indicates that when the former are phased to absorption, the latter will produce dispersive tails spreading out from peaks in the TOCSY spectrum. Fortunately, the resonance linewidths present in most proteins are sufficiently broad that these antiphase components of the peak are reduced by self-cancellation. One advantage of performing the experiment in the manner indicated in Fig. 6.43 is that the size of the antiphase components may be further reduced by varying the delay following the isotropic mixing period over the course of the experiment. This process, known as z filtration (43,45), is analogous to techniques for suppression of zero-quantum peaks in NOESY spectra. A more complete discussion is given in Section 6.6.1.

The choice of appropriate isotropic mixing sequence has been the subject of much research. The transfer properties of WALTZ-16, MLEV-17, and DIPSI-2 are compared in Fig. 4.5. In the absence of relaxation effects, DIPSI-2 has coherence transfer properties superior to those of WALTZ-16 or MLEV-17 sequences (Section 4.2.1.2) (46). The most recent isotropic mixing schemes not only have desirable magnetization transfer properties but also minimize undesirable transfer via dipolar coupling during the mixing period [the rotating-frame Overhauser effect (ROE); Section 5.4.3]. ROE peaks are of sign opposite to those of the TOCSY peaks, and may attenuate or completely cancel TOCSY cross-peaks between proximal spins within the same spin system. The dipolar effects are particularly problematic during the long mixing times required to observe correlations in extended spin systems and are also more acute for large proteins because dipolar relaxation is more efficient. The "clean" isotropic mixing sequences take advantage of the difference in sign of the NOE (laboratory-frame Overhauser effect) and the ROE [5.120]. By placing suitable delays in the pulse train, NOEs develop that offset the ROE contributions. The two most widely used sequences in this respect are the DIPSI-2rc sequence (47) and the clean CITY sequence (48). Most recently, schemes have been designed that achieve compensation between

ROE and NOE effects without introducing delays during the mixing sequence *(49)*. The supercycle sequence of the "clean" DIPSI-2rc sequence is $R\bar{R}R\bar{R}$, in which

$$R = 180°-\Delta-140°-\overline{320°}-\Delta-\overline{90°}-270°-\Delta-20°-\overline{200°}-\Delta-\overline{85°}-30°-\overline{125°}-$$
$$\Delta-\overline{120°}-300°-\Delta-75°-\overline{255°}-\Delta-\overline{10°}-190°-\Delta-180°-\Delta,$$

and should be compared with the original DIPSI-2 sequence [4.24]. The delays Δ occur when (for ideal rotations) magnetization is aligned along the z axis to permit compensatory NOE cross-relaxation. The delay Δ has a nominal length equal to the length of a 143.87° pulse. None of these schemes are capable of suppressing transfer via chemical exchange (Section 5.6). The effect of exchange on the peaks observed in TOCSY spectra has been described by Feeney and coworkers *(50)*. Discrimination of chemical exchange peaks is discussed in more detail in Section 6.6.2.4.

6.5.2 EXPERIMENTAL PROTOCOL

In all the experiments described so far in this chapter, cross-peaks have had antiphase lineshapes in F_2. This fact, coupled with the severely resolution-enhancing window functions used during processing, means that little attention was paid to the quality of the baseline in the resulting spectra. If the spectrometer is reasonably well shimmed to minimize the residual water resonance, the baselines will be flat and the cross-peaks will be easily visible. In contrast, after the t_2 Fourier transformation of a 2D TOCSY spectrum, the peaks in each F_2 slice are in-phase and absorptive. Thus, TOCSY spectra have many similarities with 1D spectra, including the same problems of baseline distortion discussed in Section 3.3. After Fourier transformation in t_1, the baseline distortions, which tend to vary between t_1 increments, are manifested as alternating positive and negative ridges running parallel to the F_2 axis. In severe cases, the cross-peaks are obscured. In order to obtain the flattest baseline possible in absorptive homonuclear experiments such as TOCSY, the following points should be considered: (1) shim the sample with the specific aim of reducing the residual water signal remaining after presaturation (Section 3.6.2.2), (2) replace the final 90° pulse with a Hahn echo (Section 3.6.4.2), (3) adjust the preacquisition delay and receiver phase to remove the need for phase corrections in F_2 (Section 3.3.2.3); (4) digitally oversample in F_2 *(51)*, (5) set the audiofilter cutoff far enough beyond the spectral width to avoid baseline rolls (this slightly decreases sensitivity but often dramatically improves the baseline quality (Section 3.6.4.2), (6) set the sampling delay to avoid first-order phase corrections associated with precession during the finite length of the pulses bracketing t_1 (Section 3.4.1). More details on these topics can be found elsewhere in this text. Since the diagonal and cross-peaks in the TOCSY spectrum are predominantly in-phase and absorptive, the choice of $t_{1,max}$ is limited only by the desired resolution in F_1. Aside from the benefits of oversampling (52), sampling t_1 beyond $1.5/R_2$ to $2/R_2$ is unnecessary. Values of $t_{1,max}$ in the range of 40 to 60 ms are usually adequate.

Generally, isotropic mixing is used during a pulse sequence to achieve one of two goals: (1) efficient transfer of magnetization through a single three bond scalar coupling only or (2) maximum transfer of magnetization between resonances at extreme ends of a spin system. The former aim is required to identify all possible ^{1}H^N–^{1}H$^\alpha$ correlations, or if the isotropic mixing is part of a longer pulse sequence (Sections 6.5.5 and Sections 6.7). The latter objective intends to provide observable cross-peaks between all spins within a spin system and is critical for successful completion of the spin-system assignment stage of the sequential assignment process (Section 8.1.1). Both of these aims cannot be achieved simultaneously with a single mixing time because of the oscillatory nature of the magnetization transfer process (44); as the mixing time is increased, cross-peaks between directly coupled spins tend to increase in intensity quickly and then decrease before cross-peaks to more distant protons have even started to gain intensity, as illustrated for an isoleucine spin system in Fig. 6.44. During the assignment process, several TOCSY spectra are commonly acquired with different mixing times to avoid missing correlations that have a minimum in the transfer function at the mixing time used in any single experiment. During the mixing period, the supercycle comprising the mixing sequence must be executed an integral number of times. Given the rf field strengths commonly used in TOCSY experiments (10–12 kHz), the duration of a supercycle is 2–4 ms (depending on the sequence used). Therefore, the mixing time can be varied only in increments of this amount.

Simulations indicate that for residues with large $^3J_{H^NH^\alpha}$, transfer from ^{1}H^N to ^{1}H$^\alpha$ will be maximal at 30–50 ms, whereas transfer from ^{1}H^N to other side-chain protons increase approximately monotonically for mixing times up to 100 ms (44). In cases with smaller values of $^3J_{H^NH^\alpha}$, transfer from ^{1}H^N to all other protons is approximately monotonic up to 100 ms. Thus, maximum transfer from ^{1}H^N to ^{1}H$^\alpha$ will most likely be obtained with an isotropic mixing period 35–45 ms in duration. The simulations indicate that in order to maximize transfer to protons distant from ^{1}H^N, longer mixing times should be employed. However, relaxation during the

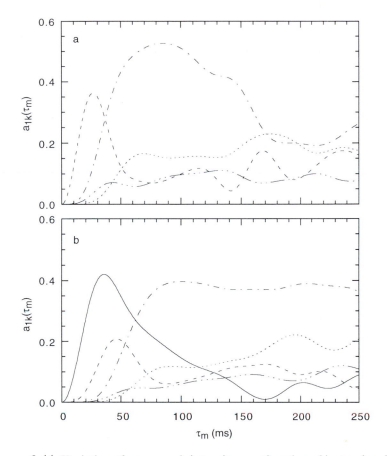

FIGURE 6.44 Variation of cross-peak intensity as a function of isotropic mixing time for an isoleucine spin system. (a) Cross-peak intensity for transfer from the $^1H^\alpha$ spin with the $^1H^N$ spin removed and (b) cross-peak intensity for magnetization transfer from the $^1H^N$ spin. The curves for the destination spins are (——) $^1H^\alpha$, (– – –) $^1H^\beta$, (– ··· –) $H^{\gamma 1}$, (– · –) $^1H^{\gamma 2}$, and (- - - -) $^1H^\delta$. The transfer functions were calculated using the following coupling constants: $^3J_{H^N H^\alpha} = 10.0$ Hz, and $^3J_{H^\alpha H^\beta} = 12$ Hz. Vicinal couplings to methyl groups were 6.7 Hz, all geminal couplings were -15 Hz, and all other vicinal couplings were 7 Hz. The effect of relaxation during the mixing was not considered.

mixing sequence, which was not incorporated into the simulations, reduces the intensity of all cross-peaks at longer mixing times. For a protein the size of ubiquitin, the maximal useful length of isotropic mixing is of the order of 100–120 ms; for larger proteins this limit is shorter. A usual course of action would be to acquire one TOCSY with a mixing time of

40–60 ms to obtain a high number of $^1H^N$–$^1H^\alpha$ correlations, and then to acquire a second TOCSY with as long a mixing time as possible (i.e., 75 ms or longer) without severe loss of signal intensity.

The peaks resulting from relay through several couplings are often of very low intensity in TOCSY spectra. Thus, total acquisition times for TOCSY experiments will generally be longer than for COSY, but shorter than for DR-COSY. One or two passes through the phase cycle described in Fig. 6.43 (32 or 64 scans, respectively) gives acceptable sensitivity for a protein in the 1–4-mM range. Assuming a $t_{1,\text{max}}$ of 50 ms and a reasonable F_1 spectral width, TOCSY experiments require total acquisition times of 10–20 hs.

The TOCSY spectra of ubiquitin were acquired using the DIPSI-2rc sequence with a rf field strength of ~12.5 kHz applied for 48, 83, or 102 ms (corresponding to 14, 24, or 28 complete cycles through the mixing sequence). In this study 32 transients were collected for each of 576 t_1 increments yielding a $t_{1,\text{max}}$ of 50 ms and total acquisition times of about 10.5 hs. Quadrature detection in F_1 was achieved with TPPI. The delays on either side of the isotropic mixing time facilitated a change only in transmitter power; z filtration (*43*) was not performed.

6.5.3 PROCESSING

Appropriate window functions for TOCSY spectra are far less resolution-enhancing than those used for COSY-type spectra. Sinebells shifted by 60–90° in F_1 and F_2 provide some degree of resolution enhancement while reducing truncation effects. The tradeoff between sensitivity and resolution can be fine-tuned by varying the width of the sine bell from 80 to 300 ms in t_2. Alternatively, matched exponential apodization or weak (i.e., not very resolution-enhancing) Lorentzian to Gaussian transformations can be performed. The Lorentzian to Gaussian transformations are beneficial when observing cross-peaks close to the diagonal; the intensity of the diagonal is such that the Lorentzian tails are very obtrusive. Commonly, spectra are processed with at least two different window functions to maximize sensitivity (for analyzing most of the spectrum) and enhance resolution (for analyzing heavily overlapped regions of the spectrum).

In addition to the precautions taken while setting up the experiment, other processing "tricks" (some of which are discussed in Chapter 3) can be beneficial in 2D spectra with absorptive lineshapes. In brief, these include (1) deconvolution to remove the residual water resonance in each FID (53); (2) linear prediction of the first (or first few) points of the FID; (3) removal of dc offset by adjusting the tail of the FID to have a zero average intensity (this is best performed after removal of the residual solvent signal); and (4) baseline correction with a functional form sufficient to remove dc offsets or correct tilting, bowing, or rolling of the baseline. In most cases if all other acquisition and processing precautions are taken, baseline corrections will not be needed.

The H_2O TOCSY spectra presented in this section were processed with weak Lorentzian-to-Gaussian transformations in t_1 and t_2. A cosine bell was also applied in t_1 to eliminate truncation artifacts. The solvent resonance was removed by deconvoluting the FID with a sine lineshape averaged over 32 points. Any dc offset was removed by subtracting the average intensity of the final 5% of each FID. The first point of each FID was linear predicted using the first 128 points (using the HSVD algorithm) and then scaled by 0.5 prior to Fourier transformation.

6.5.4 INFORMATION CONTENT

COSY-type experiments are very good at identifying sections of spin systems within the so-called fingerprint regions (Section 6.2.1.4). Connecting fragments from the different fingerprint regions is difficult in such spectra because of overlap and self-cancellation (due to complex fine structure) of the intervening methylene signals. TOCSY and other relayed experiments circumvent these problems by producing cross-peaks between the fingerprint resonances at opposite ends of extended spin systems. Because of the large chemical shift dispersion of the amide resonances, relayed connectivities involving these protons play a vital role in identification and assignment of spin-system type (Section 8.1.1). Thus, in cases where two spin systems have identical $^1H^\alpha$ chemical shifts, the complete spin systems can be assigned by observing correlations to the associated amide proton.

In TOCSY spectra acquired from H_2O solution, the most useful cross-peaks include correlations from $^1H^N$ to the methyl resonances of alanine, threonine, valine, leucine, or isoleucine; the $^1H^\delta$ or $^1H^\varepsilon$ resonances of arginine or lysine residues, respectively; and the $^1H^\beta$ of three-spin side chains, or $^1H^\beta$ and $^1H^\gamma$ of five-spin side chains. Because of the length of leucine, isoleucine, arginine, and lysine side chains, the interesting end-to-end cross-peaks are observed only for longer mixing times. Some examples of the correlations observed in these long side chains as the mixing time is increased are shown in Fig. 6.45. Commonly, not all of these correlations will be observed because small values of $^3J_{H^N H^\alpha}$ or small values for $^3J_{H^\alpha H^{\beta'}}$ and $^3J_{H^\alpha H^{\beta'}}$ will limit transfer from $^1H^N$ or $^1H^\alpha$. Provided that some $^1H^\alpha$ resonances are resolved, TOCSY spectra acquired from D_2O solution permit observation of cross-peaks between $^1H^\alpha$ and the spin-system termini.

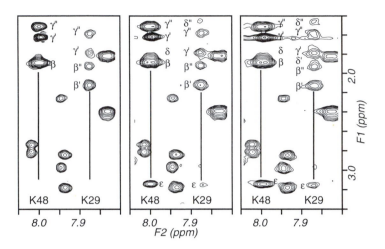

FIGURE 6.45 Sections of H₂O TOCSY spectra acquired with mixing times of 48 (left), 83 (center), and 102 ms (right). The cross-peaks observed to the amide protons of Lys29 and Lys48 are assigned at the different mixing times. Cross-peaks to the spin-system termini are observed only at the longer mixing times.

Analysis of TOCSY spectra is prone to two possible pitfalls. First, even if all side-chain resonances are observed, definitive assignment of a resonance to a particular position within the side chain may not be possible. This is particularly true for the methylene groups of five-spin side chains, arginine, and lysine. Estimates of the assignment from chemical-shift arguments, or by following TOCSY cross-peak intensity as a function of mixing time, are not infallible. Other correlation techniques that transfer coherence over a specific number of couplings are required to eliminate ambiguity (e.g., see Fig. 6.40). Second, the variation of cross-peak intensity with mixing time, including initial growth and then decay, may prevent observation of cross-peaks at certain mixing times.

In the 48-ms mixing time TOCSY experiment of ubiquitin, correlations between ¹H^N and the terminal protons of many shorter side chains are observed (i.e., transfer as far as ¹H$^\gamma$). However, very few correlations are observed from ¹H^N to ¹H$^\varepsilon$ of lysine or ¹H$^\delta$ of leucine (zero out of seven lysine residues and four out of nine leucine residues); many more such correlations are observed at 84 ms (five out of seven lysine residues and all nine leucine residues). Extending the mixing period to 102 ms leads to the observation of weak ¹H^N–¹H$^\varepsilon$ correlations for the two other lysine residues. For the seven isoleucine residues, cross-peaks from ¹H^N

to both terminal methyl groups are readily apparent after 48 or 84 ms of isotropic mixing. In two cases, the correlations from $^1H^N$ to the intervening $H^{\gamma 1}$ are observed only at 102 ms, and in a third case one of the $^1H^N$–$^1H^{\gamma 1}$ cross-peaks was not observed at all.

6.5.5 EXPERIMENTAL VARIANTS

The use of isotropic mixing has become a mainstay of ^1H NMR analysis of proteins. Four less common applications involving isotropic mixing are discussed in this section: pre-TOCSY experiments, TOCSY with very short acquisition times, sensitivity-enhanced TOCSY, and TOCSY with short mixing times.

The concept of using isotropic mixing prior to a pulse sequence was encountered earlier in the pre-TOCSY COSY experiment (Section 6.2.1.6). The idea behind this and all pre-TOCSY experiments is to use a short TOCSY mixing period to transfer magnetization to protons with resonances that have been attenuated by the solvent presaturation pulse (17). Thus, cross-peaks can be observed at the F_1 frequency of the water resonance. The process is potentially applicable to all homonuclear 2D experiments utilizing solvent presaturation, but is most commonly used in the COSY experiment. The introduction of the isotropic mixing period alters all cross-peak intensities in the final spectrum. This is particularly pertinent for NOESY spectra because cross-peak intensities are interpreted in a (semi)-quantitative fashion to generate distance restraints for protein structure calculations. In scalar correlation spectra, peak positions, and not intensities, are generally of interest, and perturbations from the pre-TOCSY sequence are unimportant.

Investigations of amide proton exchange with solvent play an important role in understanding protein structure and internal dynamics (54). Exchange rates can be obtained by rapidly transferring protein from H_2O to D_2O and repeatedly acquiring NMR spectra to follow the decay of amide ^1H signal as protons exchange with solvent deuterons. One-dimensional NMR spectra permit high temporal resolution (of the order of a few minutes); however, overlap of the $^1H^N$ resonances precludes detailed, site-specific analyses of the exchange rates. Two-dimensional spectra permit the cross-peaks from individual amide protons to be resolved; however, each spectrum must be acquired quickly (of the order of 30 min or less) and have a high sensitivity for amide protons. TOCSY spectra provide a useful compromise between these two factors (HSQC spectra (Section 7.1.1.1) of ^{15}N-labeled proteins are even more useful). A short $t_{1,max}$ (and short total acquisition time) can be used because the in-phase absorptive

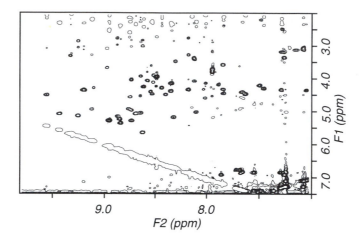

FIGURE 6.46 Section of a TOCSY experiment acquired in 15 min on a 2-mM ubiquitin sample in H_2O with a mixing time of 36 ms. Only a single contour is shown for the negative (folded) diagonal and cross-peaks. The pulse sequence of Fig. 6.43 was used without the Hahn echo sequence. The entire F_1 spectral width is shown. Four transients were collected for each of 160 t_1 experiments ($t_{1,max}$ = 29 ms). The phase cycle was ϕ_3 = receiver = $x, y, -x, -y$. Because of the short phase cycle, intense axial peaks (bottom edge of spectrum) and several other weak artifacts are present. In spite of the short acquisition time, virtually all of the expected ^{1}H^N–^{1}H$^\alpha$ correlations are observed. The negative peaks near the top of the spectrum result from correlations to side-chain protons that are folded in F_1.

lineshapes of the cross-peaks in TOCSY spectra are not subject to self-cancellation. The mixing time is kept short (no more than 40 ms) to maximize the intensity of ^{1}H^N–^{1}H$^\alpha$ correlations, and the F_1 resolution is maximized by reducing the spectral width in F_1 to cover the ^{1}H^N–^{1}H$^\alpha$ fingerprint region only (care should be taken to avoid folding or aliasing diagonal peaks into the region of interest). Finally, total acquisition times are kept short by performing only four or eight transients per increment. Bax and colleagues have discussed some of the experimental aspects of short 2D experiments (55). Figure 6.46 shows a section of a TOCSY spectrum acquired from a 2-mM H_2O sample of ubiquitin in 15 mins. The data consisted of four transients coadded for each of 160 t_1 experiments. The pulse sequence used was that of Fig. 6.43 without the final Hahn echo, since this would require additional phase cycling. The optimum four-step phase cycle (see Fig. 6.46 legend) was deduced empirically to determine which artifacts were dominant (and therefore required attenuation by the limited phase cycle) on the NMR spectrometer utilized. Although the acquisition time was short, most of the $^1H^N$–$^1H^\alpha$ correlations are visible.

In the version of the TOCSY experiment described above, after the initial 90°_x–t_1–90°_x period, a mixture of frequency-labeled I_{1z} and I_{1x} magnetization is created [6.1]. If an isotropic mixing sequence is applied orthogonally (i.e., all pulses in the sequence have phases y or $-y$), then magnetization transfer occurs independently by the pathways $I_{1z} \to I_{kz}$ [4.16] and $I_{1x} \to I_{kx}$ [4.22]. Thus, the TOCSY experiment is unusual in that equivalent coherence transfer is obtained simultaneously along two orthogonal axes. In the conventional experiment, phase cycling (43,45) or rf inhomogeneity effects (41,42) are used to remove one or other of the components and obtain amplitude-modulated phase-sensitive data. In Fig. 6.43, inversion of the phases ϕ_1 and ϕ_2 every second transient suppresses the signal from the $I_{1x} \to I_{kx}$ pathway. Cavanagh and Rance have shown that an improvement in S/N ratio of $\sqrt{2}$ is obtained by altering the phase cycle to retain both components entering the isotropic mixing period (56). Absorptive line-shapes are achieved by recording two independent data sets in which the phase of the 90° pulse prior to the mixing period differs by 180°. For example, in Fig. 6.43, the first two steps of the phase cycle ($\phi_1 = x$, $-x$; $\phi_2 = x$, x) are stored as one data set and the second two steps of the phase cycle ($\phi_1 = x$, $-x$; $\phi_2 = -x$, $-x$) are stored as a second data set. Addition of the two spectra retains the resonances arising from longitudinal magnetization at the start of the mixing period; (an effect identical to that obtained by the phase cycling for the conventional experiment). Subtraction of the two spectra retains the resonances arising from transverse magnetization at the beginning of the mixing period. After processing, the resonance peaks in the two spectra are identical, but the noise is statistically independent; therefore, addition produces a spectrum with a $\sqrt{2}$ gain in sensitivity over a conventional spectrum acquired with the same total acquisition time (56). Similar principles have been used in several heteronuclear pulse schemes to improve sensitivity (Section 7.1.3.2).

Because the rate of buildup of a TOCSY peak depends on the size of the scalar coupling constant, quantitative analysis of cross-peak intensities can be used to estimate coupling constants (57). The size of a TOCSY cross-peak depends on all the couplings within a spin system; therefore, extraction of single coupling constants must be performed with a degree of caution. Short mixing times (5–20 ms) must be used to ensure that magnetization transfers primarily through a single scalar coupling interaction; however, zero-quantum artifacts (see Section 6.6.1.1) can distort cross-peaks in such spectra. The rather crude estimates of scalar coupling constants obtained by this approach can be used to aid in the stereospecific assignments of β-methylene protons (58).

6.6 Cross-Relaxation NMR Experiments

6.6.1 NOESY

All the experiments described so far in this chapter have relied on magnetization or coherence transfer via scalar couplings and correlations are provided only between protons within the same amino acid residue. The sequential assignment process in an unlabeled protein sample is completed using the NOE to correlate protons that are close in space. Additionally, distance constraints for structure determination of proteins are derived primarily from NOE interactions. The pulse sequence for the nuclear Overhauser-effect spectroscopy (NOESY) experiment is shown in Fig. 6.47. Initially, a $90°-t_1-90°$ period frequency-labels the spins and returns the magnetization to the z axis. Magnetization transfer occurs via dipolar coupling for a period τ_m before observable transverse magnetization is created by the final $90°$ pulse. The final pulse can be replaced by a Hahn echo sequence with a concomitant improvement in the flatness of the baseline (Sections 3.6.4.2 and 6.5). A coherence-level diagram for this pulse sequence is presented in Fig. 6.47.

6.6.1.1 Product Operator Analysis The theory of magnetization transfer by dipolar coupling (the NOE) has been discussed in Sections 4.2.2 and Section 5.5. For a spin system containing two scalar-coupled spins, evolution through the $90°_x-t_1-90°_x$ sequence is described by [6.1]. Evolution of the I_{1z} term in [6.1] during τ_m is governed by the Solomon equations in which the initial condition is $-I_{1z}\cos(\Omega_1 t_1)\cos(\pi J_{12} t_1)$ and

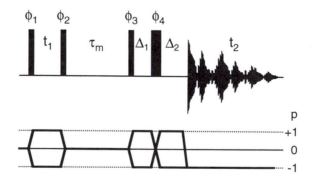

FIGURE 6.47 Pulse sequence and coherence level diagram for the NOESY experiment. A Hahn echo sequence is included prior to detection. Appropriate values for the delays are discussed in the text. The same 32-step phase cycle as for the TOCSY is employed (Fig. 6.43)

the equilibrium magnetization, I_{1z}^0, is rejected by phase cycling for axial peak suppression. If $K - 1$ spins (I_k for $k = 2, \ldots K$) are close in space to spin I_1 (this notation allows for the possibility that the scalar coupled spin, I_2, is dipolar-coupled to I_1 as well), the resulting evolution during τ_m is

$$-I_{1z} \cos(\Omega_I t_1) \cos(\pi J t_1) \xrightarrow{\tau_m}$$

$$-\sum_{k=1}^{K} I_{kz} a_{1k}(\tau_m) \cos(\Omega_1 t_1) \cos(\pi J_{12} t_1), \qquad [6.44]$$

in which $a_{1k}(\tau_m) = [\exp(-\mathbf{R}\tau_m)]_{1k}$ is the $(1, k)$th element of the matrix exponential and $\mathbf{R}$ is the matrix of rate constants ρ_i and σ_{ij} (Section 5.1.2). After the final 90° pulse and Hahn echo, the density operator terms that result from the longitudinal magnetization are given by

$$\sum_{k=1}^{K} I_{ky} a_{1k}(\tau_m) \cos(\Omega_1 t_1) \cos(\pi J_{12} t_1). \qquad [6.45]$$

The final spectrum contains I_1 diagonal peaks (the $k = 1$ term in [6.45]) and $I_1 \rightarrow I_k$ NOE cross-peaks, for $k > 1$. All the peaks are in-phase with respect to homonuclear scalar coupling in F_1 and F_2, and also can be phased to absorption in both dimensions.

The longitudinal magnetization that will give rise to the NOE cross-peaks has coherence level $p = 0$ during τ_m, and the phase cycle rejects other coherence levels during this period, including the single-quantum terms I_{1x} and $2I_{1x}I_{2z}$ in [6.1]. The second term of [6.1] is a mixture of ZQ_y^{12} ($p = 0$) and DQ_y^{12} ($p = 2$) coherences (Section 2.7.5). The double-quantum operator is suppressed by the phase cycling; however, the zero-quantum term survives. During τ_m the ZQ term will precess according to the difference in chemical shift of I_1 and I_2. The following terms will be generated by the final 90° pulse and Hahn echo:

$$-ZQ_y^{12} \cos(\Omega_1 t_1) \sin(\pi J_{12} t_1) \xrightarrow{\tau_m - (\pi/2)_x - \Delta_1 - \pi_y - \Delta_2}$$

$$+ \frac{1}{2} [2I_{1x}I_{2x} + 2I_{1z}I_{2z}] \cos(\Omega_1 t_1) \sin(\pi J_{12} t_1) \sin[(\Omega_1 - \Omega_2)\tau_m]$$

$$+ \frac{1}{2} [2I_{1z}I_{2x} - 2I_{1x}I_{2z}] \cos(\Omega_1 t_1) \sin(\pi J_{12} t_1) \cos[(\Omega_1 - \Omega_2)\tau_m]. \qquad [6.46]$$

The last line of [6.46] contains observable terms and therefore must be considered in an analysis of the NOESY spectrum. Such artifacts arise via a ZQ pathway and are referred to as zero-quantum peaks. These peaks are in antiphase in both dimensions, and are also in dispersion when the normal NOE peaks [6.45] are phased to absorption.

During analysis of a NOESY spectrum, the integrated intensity of a given cross-peak is interpreted in terms of the distance between the two protons giving rise to the peak (Sections 6.2.5.4 and 8.2.1). Clearly, from [6.46] the real NOE and ZQ peaks between two coupled spins appear at identical chemical shifts in F_1 and F_2. Although the net integrated intensity of the dispersive ZQ component is zero, accurately integrating the contributions from the dispersive tails of this component may not be possible, and errors in the measurement of the NOE cross-peak volume result; in addition, the antiphase dispersive tails can interfere with the integration of other cross-peaks in crowded regions of the spectrum. The magnitude of the ZQ component varies as $\cos[(\Omega_1 - \Omega_2)\tau_m]$, which depends on the chemical shifts of the spins involved and the mixing time. In addition, because the ZQ terms have transverse components during τ_m, relaxation is faster than for longitudinal magnetization, and the ZQ quantum component is reduced in intensity relative to the true NOE peak when a long mixing time is employed. The use of a z filter to suppress the zero-quantum terms is discussed in Section 6.6.1.2.

6.6.1.2 Experimental Protocol The peaks of interest in the NOESY spectrum are absorptive and in-phase in both dimensions [6.45]. Thus, all the precautions adopted in Section 6.5.2 to ensure flat baselines in TOCSY spectra should also be employed in the NOESY experiment. Obtaining a flat baseline is even more imperative in the NOESY spectrum. At the least, the intensities of the cross-peaks will be interpreted in a semiquantitative fashion; hence any offset in the baseline will lead to systematic errors in the NOE cross-peak volumes. The number of observable NOEs will have a direct bearing on the quality of structures produced from the data; hence obtaining a high S/N ratio is also important. Acquisition times of 24–48 hs are not uncommon, even with relatively concentrated samples.

Accurate quantitative analysis also requires minimization of any other effect that may systematically alter the intensity of peaks, especially when a detailed quantitative analysis is to be performed (e.g., see Sections 6.6.1.4 and 8.2). The recycle delay should be $3/R_1$, in which R_1 is the smallest longitudinal relaxation rate in the protein, to avoid steady-state effects that perturb the intensity of cross-peaks. Also, solvent presaturation should be avoided as a means of solvent suppression in these applications because the intensity of cross-peaks involving protons that resonate close to the water or labile protons that are exchanging with the solvent will be reduced. Methods that attempt to combat saturation effects are discussed in Section 6.6.1.5.

The theoretical time dependence of the NOE cross-peaks in the NOESY experiment (Section 5.1.2) suggests that the mixing time should be on the order of $1/R_1$ to maximize the intensities of NOE cross-peaks. A long mixing time also has the advantage that ZQ artifacts will be of low intensity. However, long mixing times will also allow multiple magnetization transfers, or spin diffusion, to contribute substantially to the cross-peak intensity. The origins and consequences of spin diffusion are illustrated for a three-spin system with the following relaxation rate matrix:

$$
\mathbf{R} = \begin{bmatrix} \rho_1 & \sigma_{12} & 0 \\ \sigma_{12} & \rho_2 & \sigma_{23} \\ 0 & \sigma_{23} & \rho_3 \end{bmatrix}.
\qquad [6.47]
$$

By construction, spins I_1 and I_3 are too far apart to have an appreciable dipolar coupling ($\sigma_{13} = 0$); thus direct magnetization transfer between I_1 and I_3 is not possible. The time dependence of the I_1 magnetization is given to third order in time by

$$
\langle I_{1z} \rangle (\tau_m) = \sum_{k=1}^{3} [\exp(-\mathbf{R}\tau_m)]_{1k} \langle I_{kz} \rangle (0)
$$

$$
\approx \sum_{k=1}^{3} \left[E_{1k} - R_{1k}\tau_m + \frac{1}{2} R_{1k}^2 \tau_m^2 - \frac{1}{6} R_{1k}^3 \tau_m^3 \right] \langle I_{kz} \rangle (0)
$$

$$
= \langle I_{1z} \rangle (0) \left\{ 1 - \rho_1 \tau_m + \frac{1}{2}(\rho_1^2 + \sigma_{12}^2)\tau_m^2 - \frac{1}{6}(\rho_1^3 + 2\rho_1\sigma_{12}^2 + \rho_2\sigma_{12}^2)\tau_m^3 \right\}
$$

$$
+ \langle I_{2z} \rangle (0) \left\{ -\sigma_{12}\tau_m + \frac{1}{2}(\rho_1 + \rho_2)\sigma_{12}\tau_m^2 \right.
$$

$$
\left. - \frac{1}{6}[(\rho_1^2 + \sigma_{12}^2)\sigma_{12} + (\rho_1 + \rho_2)\rho_2\sigma_{12} + \sigma_{12}\sigma_{23}^2]\tau_m^3 \right\}
$$

$$
+ \langle I_{3z} \rangle (0) \left\{ \frac{1}{2}\sigma_{12}\sigma_{23}\tau_m^2 - \frac{1}{6}(\rho_1 + \rho_2 + \rho_3)\sigma_{12}\sigma_{23}\tau_m^3 \right\}. \qquad [6.48]
$$

Each term in [6.48] can be assigned a physical interpretation; however, only three terms will be discussed in detail. The first-order term $-\sigma_{12}\tau_m$ $\langle I_{2z} \rangle (0)$ represents direct transfer of magnetization from spin I_2 to spin I_1 and gives rise to a cross-peak in the NOESY spectrum. In the initial rate regime, only this term contributes to the cross-peak intensity, and the cross-peak intensity is proportional to the cross-relaxation rate constant, σ_{12}. The second-order term $(1/2)\sigma_{12}\sigma_{23}\tau_m^2\langle I_{3z}\rangle(0)$ exemplifies spin diffusion. This term gives rise to a cross-peak between spins I_1 and I_3 by an indirect two-step transfer of $I_3 \rightarrow I_2 \rightarrow I_1$. In the quadratic time regime, the intensity of the spin-diffusion cross-peak depends on the product of the individual cross-relaxation rate constants. Finally, the third-order term $\rho_2\sigma_{12}^2\tau_m^3\langle I_{1z}\rangle(0)$ represents a back-transfer pathway $I_1 \rightarrow I_2 \rightarrow I_1$. The back-transfer has the effect of reducing the intensity of the cross-peak that would otherwise result from cross-relaxation between I_1 and I_2. Therefore, even for a two-spin system, outside the initial rate regime, NOE cross-peak intensities are not proportional to the cross-relaxation rate constants. The assumed linearity between the NOE cross-peak intensities and cross-relaxation rate constants some-times is called "the isolated two-spin approximation"; as the present discussion shows, this phrase is a misnomer.

As a consequence of spin diffusion, cross-peaks between pairs of protons that are far apart will gain intensity from magnetization that has been transferred via intervening spins, whereas cross-peak between pairs of protons that are close together will be decreased by the loss of magneti-zation to other nearby protons. Failure to adequately account for spin diffusion results in the derivation of inaccurate distance constraints be-tween pairs of protons; overly tight constraints derived from NOE cross-peaks dominated by spin diffusion lead to overly constrained and incorrect protein structures (Sections 6.6.1.4 and 8.2.1). Spin-diffusion effects may be minimized by using a short mixing time, but in these experiments all cross-peak intensities will be low, and ZQ artifacts will be emphasized. A compromise with mixing times of 50–150 ms provides reasonable cross-peak intensities that are not overly influenced by spin-diffusion or ZQ contributions. Dipolar relaxation is more efficient in systems with long rotational correlation times; hence a shorter mixing time is required to limit spin diffusion in large proteins.

The dispersive antiphase components observed in NOE cross-peaks between scalar-coupled protons can be suppressed by several methods. Perhaps the simplest procedure is simply to collect or process the data with limited resolution, since the line-broadening effect will cause self-cancellation of the antiphase components (Section 6.2.1.5). A common experimental method involves randomly varying the mixing time by a small amount from one t_1 increment to the next (59). By analogy to the multiple-quantum filters discussed in Section 6.3, this process is known as z-filtration (60). Because of the cosinusoidal dependence of the ZQ terms on τ_m, coaddition of these experiments leads to cancellation. The variation of τ_m should cover at least one period of the smallest difference of resonance frequencies of scalar-coupled peaks in the spectrum. Given the chemical shifts usually encountered in proteins, variation of τ_m by

20–30 ms will have the desired effect. Note that for very short mixing times, variation by this amount can have an undesirable effect on the intensity of the NOE peaks and may not be applicable. Alternatively, a variation in the ZQ precession phase is achieved by applying a 180° pulse at random positions (from one t_1 increment to the next) within a fixed mixing time (59).

The phase cycling used in the NOESY experiment must incorporate axial-peak-suppression phase cycling (of the first 90° pulse) and selection of $\Delta p = -1$ and $\Delta p = +1$ by phase cycling of the second 90° pulse. The basic phase cycle is four steps ($\phi_1 = x, -x, -x, x; \phi_2 = x, x, -x, -x$; and receiver $= x, -x, x, -x$). Alternatively, both pulses can be phase-cycled in synchrony to select $\Delta p = 0$ (Section 4.3.2.2). EXORCYCLE phase cycling is used for the Hahn echo, and CYCLOPS is applied to all pulses. Bodenhausen and coworkers have discussed phase cycles for NOESY experiments (61).

The NOESY spectra of ubiquitin were acquired from H_2O solution using the Hahn echo sequence of Fig. 6.47 with the sample delay set to avoid phase correction-associated baseline distortions in F_1 (Section 3.3.2.3). The solvent was saturated during the recycle delay and during the mixing time; 32 transients were collected for each of 576 increments of t_1 ($t_{1,max} = 50$ ms), a mixing time of 40 or 100 ms was used, and total acquisition time was 10.0 or 10.5 hs. No attempt was made to suppress ZQ contributions.

6.6.1.3 Processing The similarity of TOCSY and NOESY line-shapes dictates that processing of these two spectra is similar. All the details discussed in Section 6.2.4.3 are relevant to the present discussion. Accurate quantitative analysis of NOE cross-peak volumes places some restraints on the use of resolution-enhancing window functions as the relative peak intensity of peaks with different linewidths will not be preserved if the apodization functions have initial values different from unity (as in a sine bell). Errors of this kind should also be balanced against the problems associated with accurate measurement of the integrated intensity of a cross-peak with true Lorentzian tails. A variety of methods have been proposed to improve the quality of cross-peak volume extraction from NOESY spectra (62–65).

The NOESY spectra were processed in an identical fashion to the TOCSY spectra of Section 6.5.3.3 except that deconvolution of the residual water resonance was not necessary.

6.6.1.4 Information Content NOESY spectra provide a powerful means of elucidating conformational details of molecules in solution. The requirement that two protons be separated by less than 5 Å (or so) in

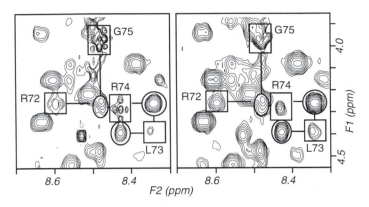

FIGURE 6.48 Comparison of sections of the 40 (left) and 100 ms (right) NOESY spectra of ubiquitin in H$_2$O solution. Several intraresidue and sequential ^{1}H^N–^{1}H$^\alpha$ cross-peaks are denoted by the rectangles and ellipses, respectively, and allow sequential assignment of several residues near the C-terminus. Because of the presence of zero-quantum artifacts, several of the intraresidue peaks contain dispersive antiphase components in the 40-ms mixing time experiment.

order to give rise to an NOE immediately allows a loose restraint to be placed on their separation. Furthermore, the size of the NOE depends inversely on the distance; hence the restraint can be shorter than 5 Å if the NOE is intense. In order to calculate the structure of a protein, many such restraints must be identified in an unambiguous fashion (Section 8.2.1). In most applications, NOE cross-peaks simply are placed into one of several size categories associated with an upper bound for the proton separation. More accurate calibration is difficult because of the complex relationship between NOE buildup, local correlation time, and the distribution of neighboring protons. Analysis of NOESY spectra with different mixing times (called a "buildup" or "τ_m series") allows the initial slope of the NOE buildup to be estimated and facilitates calibration. Methods for interpreting NOESY spectra are discussed in more detail in Section 8.2.1.

Figure 6.48 shows regions of the 40- and 100-ms NOESY spectra of ubiquitin in H$_2$O solution, and demonstrates that sequential ^{1}H$^\alpha$–^{1}H^N NOEs can be used to obtain sequence specific assignments. As expected for a protein of this size, the cross-peaks are all more intense at the longer mixing time. The residues at the C-terminus are conformationally mobile, have narrower linewidths, and therefore have very prominent dispersive contributions in the shorter-mixing-time spectrum (Section 6.6.1.1); self-cancellation reduces the impact of these components for other, broader resonances. Increasing the mixing time also leads to a decrease in the relative intensity of ZQ effects.

Besides cross-relaxation, chemical exchange can also lead to cross-peaks in NOESY spectra. In cases of slow exchange (on the chemical shift time scale) between two species (Section 5.6), a cross-peak is observed at the frequencies of a particular nucleus in the different sites if the exchange rate between the species is not slow compared to τ_m. For proteins, the chemical-exchange peaks will have the same sign as NOE cross-peaks (the same sign as the diagonal peaks, formally negative); hence discrimination of the two can be difficult. Very complicated spectra can result from combinations of exchange and cross-relaxation; in effect, these are spin-diffusion-type peaks involving two transfer steps. Identification of exchange effects is discussed in more detail in Section 6.6.2.4.

6.6.1.5 Experimental Variants   Suppressing the solvent resonance by presaturation unavoidably results in transfer of saturation to labile protons undergoing exchange with solvent, and attenuation of cross-peaks involving protons that resonate close to the solvent resonance (Section 3.5.1). Several alternatives to presaturation in 1D ^{1}H NMR have already been discussed (Section 3.5), and these methods are valuable for NOESY experiments. The final pulse of the NOESY experiment rotates longitudinal magnetization into the transverse plane much the same as in a 1D experiment; hence, this read pulse can be replaced with a sequence used for 1D-selective excitation.

The simplest method of acquiring NOESY spectra in H_2O solution without presaturation employs a jump-and-return observe sequence in place of the final 90° pulse (Section 3.5.2) (66). This experiment is usually referred to as a jump–return NOESY (JR-NOESY), and a pulse sequence is depicted in Fig. 6.49. Note that a Hahn echo cannot be incorporated into this experiment. A variety of more complicated schemes can also be incorporated into the NOESY experiment with some advantages over the jump–return version (67,68).

Experimental protocol for the JR-NOESY is similar to a NOESY experiment acquired with presaturation. The degree of solvent suppression can usually be improved by empirically varying the length and phase of the second pulse in the jump–return sequence by a few tenths of microseconds or a few degrees, respectively. The excitation maxima depend on Δ, and for most protein applications a value of 100 μs leads to optimal excitation of the amide resonances on a 500 MHz spectrometer.

F_1 phase discrimination in the experiment depicted in Fig. 6.49 will result in transverse water magnetization during τ_m for some increments of t_1 (this is valid for both the TPPI and hypercomplex methods; see Section 4.3.4). Unless τ_m is sufficiently long to allow radiation damping of the water resonance back to the z axis (relaxation is far too slow to be useful in this regard), solvent suppression will be very poor for these

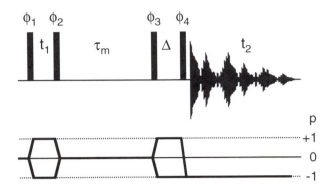

FIGURE 6.49 Pulse sequence and coherence-level diagram for the NOESY experiment with a jump–return observe pulse (JR-NOESY). The basic 8-step phase cycle is as follows: $\phi_1 = x, -x, y, -y, -x, x, -y, y$; $\phi_2 = x, x, y, y, -x, -x, -y, -y$; $\phi_3 = 8(x)$; $\phi_4 = 8(-x)$; and receiver $= x, -x, y, -y, -x, x, -y, y$. CYCLOPS is performed on all pulses and the receiver to yield a 32-step cycle. Shifting the rotation angle and phase of pulse ϕ_4 a few degrees can improve the degree of solvent suppression.

transients (69). Practically, τ_m must be 150 ms or longer to avoid these effects, and JR-NOESY spectra inevitably include contributions from spin diffusion. Modified sequences using pulsed field gradients to help suppress the water have been described recently (69).

The JR-NOESY spectrum described below was acquired from H_2O solution with 32 transients for each of 576 t_1 increments ($t_{1,max} = 50$ ms). A mixing time of 150 ms was used, and the total acquisition time was 11 hs. Sections of the JR-NOESY and NOESY with presaturation are depicted in Fig. 6.50. Many cross-peaks close to the water resonance are observed only in the JR-NOESY spectrum, and other peaks up to 0.2 ppm from the water line are noticeably more intense in the jump–return experiment (even allowing for the longer mixing time of this experiment). In Fig. 6.50 (and elsewhere for this spectrum) a large number of peaks occur at the exact F_1 frequency of the water resonance. These peaks can arise from a variety of mechanisms other than cross-relaxation with nonlabile protein protons coincident with the solvent resonance, including chemical exchange between water and labile ^{1}H^N or ^{1}H^O groups, NOEs from bound water to protein protons, or NOEs from protein hydroxyl protons on resonance with the water resonance.

In a second variant of the basic NOESY experiment, the mixing period includes both isotropic mixing and cross-relaxation periods (Fig. 6.51), and is known as a relayed NOESY. Experimental setup is identical to a

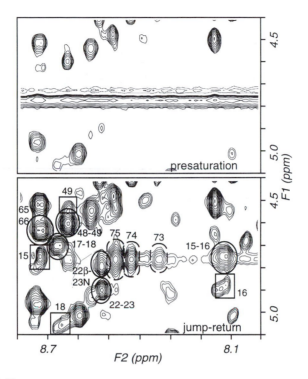

FIGURE 6.50 Comparison of NOESY spectra acquired from H_2O solution in which solvent was suppressed by presaturation (top) or selective excitation with a jump–return sequence (bottom). The spectra were collected under identical conditions except for the mixing times, which were 100 and 150 ms in presaturation and jump–return spectra, respectively. Intraresidue and sequential NOEs are denoted by rectangles and ellipses, respectively, with the peaks arising between $^1H^N$ and $^1H^\alpha$ unless otherwise noted. The three peaks surrounded by broken ellipses probably arise from exchange of amide protons with the solvent as residues 73–75 are close to the C-terminus and are flexible.

normal NOESY experiment. The aim is to use the isotropic mixing to transfer magnetization via a single scalar coupling, usually between $^1H^N$ and $^1H^\alpha$, prior to NOESY cross-relaxation. The isotropic mixing period normally is 20–30 ms in duration (Section 6.5.2). The choice of NOESY mixing time is based on a tradeoff between cross-peak intensity and contributions from spin diffusion (Section 6.6.1.2). The experiment discussed below was acquired with 32 transients for each of 576 t_1 increments (total acquisition time 10.5 hs); 27 ms of DIPSI-2rc isotropic mixing (*47*) was used prior to 100 ms of cross-relaxation.

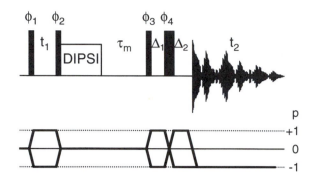

FIGURE 6.51 Pulse sequence and coherence-level diagram for the relayed NOESY experiment. The phase cycle is identical to that used in Figs. 6.47 and 6.43.

Analysis of a relayed NOESY experiment may help in the sequential assignment process in cases in which the ^{1}H$^\alpha$ protons of adjacent residues are degenerate. The sequential assignment process relies on the observation of NOEs from ^{1}H^N of residue i to ^{1}H^N, ^{1}H$^\alpha$, *and* ^{1}H$^\beta$ of residue $i-1$. For peptide sections in an extended conformation the sequential ^{1}H^N–^{1}H^N and ^{1}H^N–^{1}H$^\beta$ NOEs have low intensity; degeneracy of the ^{1}H$^\alpha$ resonances of adjacent residues will stymie the assignment procedure (the important sequential NOE will be overlapped with the intraresidue ^{1}H^N–^{1}H$^\alpha$ NOE of residue $i-1$). In the relayed NOESY experiment, ^{1}H^N and ^{1}H$^\beta$ resonance positions of residue $i-1$ are recorded in t_1, then this magnetization is passed on to ^{1}H$^\alpha$ of residue $i-1$ during the isotropic mixing and then to ^{1}H^N of residue i during the NOE mixing period. Intense sequential ^{1}H$^\beta$–^{1}H^N and ^{1}H^N–^{1}H^N NOE cross-peaks result from this process, allowing the sequential assignment process to continue. In the relayed NOESY spectrum, cross-peaks no longer have a direct dependence on the distance between the protons at the F_2 and F_1 frequencies of the cross-peak, and such data should be used for assignment purposes only. An example is shown in Fig. 6.52, where many sequential ^{1}H^N–^{1}H^N NOEs are readily observable in the relayed NOESY experiment, but are of negligible intensity in the normal NOESY experiment.

6.6.2 ROESY

Rotating frame Overhauser effect spectroscopy (ROESY) was first developed by Bothner-By and coworkers and was initially known by the acronym CAMELSPIN (cross-relaxation appropriate for minimolecules emulated by locked spins) (70). As both names suggest, the experiment

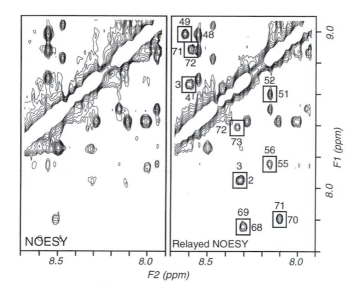

FIGURE 6.52 Comparison of sections of NOESY (left) and relayed NOESY (right) spectra. Both experiments were performed under identical conditions except the mixing period, which included 27 ms of DIPSI-2rc isotropic mixing in the relayed NOESY. Weak coherent irradiation was used to suppress the solvent before the experiment and during the 100 ms NOE mixing period. Relayed NOESY peaks surrounded by boxes indicate sequential $^1H^N$–$^1H^N$ NOEs between residues in the β-sheet of ubiquitin that are weak or not observed in the conventional NOESY experiment. The greater intensity allows sequential assignments to be made even if sequential $^1H^\alpha$ resonances are degenerate (as is the case for His68 and Leu69). The labels denote residue numbers of the amide protons contributing to each cross-peak. Relayed NOESY cross-peaks appear on only one side of the diagonal.

monitors cross-relaxation between spins that are spin-locked by the application of rf pulses (70,71). ROESY has the advantage that the rotating-frame Overhauser effect (ROE) cross-relaxation rate constant is positive for *all* rotational correlation times: the maximum size of the ROE varies from 0.38 for $\omega_0\tau_c \ll 1$ to 0.68 for $\omega_0\tau_c \gg 1$. Therefore, ROESY cross-peaks are observable even if $\omega_0\tau_c \approx 1$; in contrast, cross-peaks vanish in laboratory-frame NOESY experiments if $\omega_0\tau_c \approx 1$. ROESY is very useful in studies of peptides in which laboratory-frame NOEs are weak, but the experiment also has merits appropriate for the study of proteins. ROESY, NOESY (Section 6.6.1), and TOCSY (Section 6.5) are experimentally very similar; consequently, comparisons to NOESY and TOCSY will be made throughout this discussion. A more detailed discussion of relaxation in the rotating frame is given in Section 5.4.3.

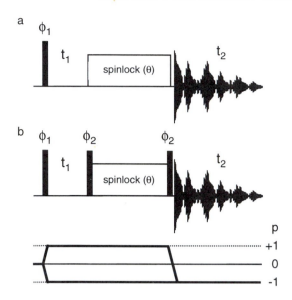

FIGURE 6.53 Pulse sequences and coherence-level diagram for the ROESY experiments. In (a) the basic phase cycle is $\phi_1 = x, -x$; and receiver $= x, -x$. The spin-lock phase $\theta = y$. In (b) the basic phase cycle is $\phi_1 = x, -x$; $\phi_2 = x$; and receiver $= x, -x$. The spin-lock phase $\theta = x$. The full-phase cycle is completed by performing CYCLOPS on all pulses and the receiver.

6.6.2.1 Product Operator Analysis The original version of the ROESY experiment simply consisted of a $90°-t_1-\tau_m-t_2$ sequence in which the spin-locking field during τ_m was provided by continuous low-power irradiation (2–4 kHz), as illustrated in Fig. 6.53a. For a scalar-coupled two-spin system, the evolution up to the mixing period is given by

$$I_{1z} \xrightarrow{(\pi/2)_x-t_1} -I_{1y}\cos(\Omega_1 t_1)\cos(\pi J_{12}t_1) + 2I_{1x}I_{2z}\cos(\Omega_1 t_1)\sin(\pi J_{12}t_1)$$
$$+ I_{1x}\sin(\Omega_1 t_1)\cos(\pi J_{12}t_1) + 2I_{1y}I_{2z}\sin(\Omega_1 t_1)\sin(\pi J_{12}t_1).$$

$$[6.49]$$

During the subsequent spin-locking period, any operators orthogonal to the rf field in a tilted rotating frame are dephased by rf inhomogeneity (Section 3.4.3). The x axes of the rotating and tilted reference frames are coincident; thus, all terms containing x operators are dephased. The transformation of z and y operators into the tilted frame is performed using [5.63]:

$$I_{1y} \Rightarrow I'_{1z} \sin \theta_1 + I'_{1y} \cos \theta_1$$

$$2I_{1y}I_{2z} \Rightarrow 2(I'_{1z} \sin \theta_1 + I'_{1y} \cos \theta_1)(I'_{2z} \cos \theta_2 - I'_{2y} \sin \theta_2)$$

$$= 2I'_{1z}I'_{2z} \sin \theta_1 \cos \theta_2 - 2I'_{1z}I'_{2y} \sin \theta_1 \sin \theta_2$$

$$+ 2I'_{1y}I'_{2z} \cos \theta_1 \cos \theta_2 - 2I'_{1y}I'_{2y} \cos \theta_1 \sin \theta_2,$$

[6.50]

in which θ_1 and θ_2 are the tilt angles for spins I_1 and I_2. The only terms that commute with the spin-lock Hamiltonian are proportional to I'_{1z} and $2I'_{1z}I'_{2z}$. If $K - 1$ spins (I_k for $k = 2, \ldots, K$) are close in space to spin I_1, the resulting evolution of the longitudinal magnetization is

$$-I'_{1z} \sin \theta_1 \cos(\Omega_1 t_1) \cos(\pi J_{12} t_1) \xrightarrow{\tau_m}$$

$$-\sum_{k=1}^{K} I'_{kz} a_{1k}(\tau_m) \sin \theta_1 \cos(\Omega_1 t_1) \cos(\pi J_{12} t_1),$$

[6.51]

in which $a_{1k}(\tau_m) = [\exp(-\mathbf{R}\tau_m)]_{1k}$ is the (1, k)th element of the matrix exponential and $\mathbf{R}$ is the matrix of rotating frame relaxation rate constants $R_{kk}(\theta_i)$ and $\sigma_{jk}(\theta_i, \theta_j)$ (Section 5.4.3). Transforming back from the tilted frame to the rotating frame yields the observable operators:

$$\sum_{k=1}^{K} I_{ky} a_{1k}(\tau_m) \sin \theta_1 \sin \theta_k \cos(\Omega_1 t_1) \cos(\pi J_{12} t_1).$$

[6.52]

The I_{1y} term represents a diagonal peak, and the remaining $K - 1$ terms represent cross-peaks. Diagonal peaks and cross-peaks have in-phase absorptive lineshapes in F_1 and F_2. In the usual methods of acquisition, the cross-peaks are of phase opposite to that of the diagonal because ρ_j and σ_{jk} are both positive (70).

The two-spin term $2I'_{1z}I'_{2z}$ does not cross-relax with other I_1 or I_2 spin operators during τ_m; however, the amplitude of the operator is reduced by relaxation (with relaxation rate constant designated R_{zz}). Transformation back into the rotating frame yields the observable operators:

$$(2I_{1y}I_{2z} \sin \theta_1 \cos \theta_2 + 2I_{1z}I_{2y} \cos \theta_1 \sin \theta_2)$$

$$\times \sin \theta_1 \cos \theta_2 \sin(\Omega_1 t_1) \sin(\pi J_{12} t_1) \exp(-R_{zz}\tau_m).$$

[6.53]

The limitations of the simple ROESY experiment are now evident: (1) amplitude of cross-peaks is reduced by a factor of $\sin\theta_1 \sin\theta_k$ and (2) two-spin order generates cross-peaks with antiphase lineshapes in both dimensions that distort the in-phase multiplet patterns expected for ROESY cross-peaks.

Griesinger and Ernst developed a simple and clever modification of the ROESY pulse sequence that overcomes these limitations (72). In this sequence (Fig. 6.53b) evolution through the $90°_x-t_1-90°_x$ block proceeds as described in [6.1]:

$$I_{1z} \xrightarrow{(\pi/2)_x-t_1-(\pi/2)_x} -I_{1z}\cos(\Omega_1 t_1)\cos(\pi J_{12}t_1) - 2I_{1x}I_{2y}\cos(\Omega_1 t_1)\sin(\pi J_{12}t_1)$$

$$+ I_{1x}\sin(\Omega_1 t_1)\cos(\pi J_{12}t_1) - 2I_{1z}I_{2y}\sin(\Omega_1 t_1)\sin(\pi J_{12}t_1).$$

[6.54]

The y operators are dephased by the x phase spin-lock rf field. Transformation of the z and x operators into the tilted frame yields

$$-I_{1z}\cos(\Omega_1 t_1)\cos(\pi J_{12}t_1) + I_{1x}\sin(\Omega_1 t_1)\cos(\pi J_{12}t_1) \Rightarrow$$

$$- (-I'_{1x}\sin\theta_1 + I'_{1z}\cos\theta_1)\cos(\Omega_1 t_1)\cos(\pi J_{12}t_1)$$

$$+ (I'_{1x}\cos\theta_1 + I'_{1z}\sin\theta_1)\sin(\Omega_1 t_1)\cos(\pi J_{12}t_1).$$

[6.55]

The only term that commutes with the spin-lock Hamiltonian is

$$-I'_{1z}(\cos\theta_1\cos(\Omega_1 t_1) - \sin\theta_1\sin(\Omega_1 t_1))\cos(\pi J_{12}t_1)$$

$$= -I'_{1z}\cos(\Omega_1 t_1 + \theta_1)\cos(\pi J_{12}t_1).$$

[6.56]

Cross-relaxation during τ_m yields

$$-I'_{1z}\cos(\Omega_1 t_1 + \theta_1)\cos(\pi J_{12}t_1) \xrightarrow{\tau_m}$$

$$- \sum_{k=1}^{K} I'_{kz}a_{1k}(\tau_m)\cos(\Omega_1 t_1 + \theta_1)\cos(\pi J_{12}t_1).$$

[6.57]

Transforming back from the tilted frame to the rotating frame and applying the last 90° pulse yields the observable operators:

$$\sum_{k=1}^{K} (I_{ky}\cos\theta_k - I_{kx}\sin\theta_k)a_{1k}(\tau_m)\cos(\Omega_1 t_1 + \theta_1)\cos(\pi J_{12}t_1).$$

[6.58]

The offset dependence of the ROESY cross-peaks appears in [6.58] as a phase error of θ_1 in t_1 and θ_k in t_2. Because θ_k is approximately linear for $0 \leq \Omega_k \leq \gamma B_1$ (Section 3.4.1), the resonance offset effects are compensated by phase correction during processing. No two spin operators that commute with the spin-lock Hamiltonian are created; therefore, the cross-peak multiplet structure is undistorted (minor contributions from evolution of ZQ coherences in the tilted frame have been ignored).

    Although the Griesinger–Ernst approach eliminates the offset dependence that arises from the projection of the spin operators between tilted and untilted frames, the magnitudes of cross-relaxation rate constants in a ROESY experiment also depend on resonance offset as shown by [5.111]. As a result, relaxation for off-resonance spins will contain a laboratory-frame component (i.e., an NOE) as well as a rotating frame component. Interestingly (and somewhat counterintuitively) for large biomolecules, the apparent offset-dependent cross-relaxation rate between two spins is actually most efficient for cross-peaks along the antidiagonal and least efficient for cross-peaks close to the diagonal away from the center of the spectrum (72). Any quantitative analysis of ROESY cross-peak intensities must consider the offset dependence of the rate constants.

A practical problem encountered in the ROESY experiment is that the spin-lock pulse is capable of inducing isotropic mixing (41). The TOCSY (or J cross-peaks) are of the same sign as the diagonal (Section 6.5); consequently, TOCSY transfer within a scalar-coupled system tends to cancel the cross-relaxation components and render quantitation of the ROE (and hence the interproton separation) difficult. More insidiously, cross-peaks that arise through consecutive TOCSY and ROE magnetization transfers have the same sign as the actual ROE peaks (73) and can be misinterpreted. Fortunately, a long, weak CW pulse is not efficient at achieving a Hartmann-Hahn match between two protons unless they are close in chemical shift or symmetrically disposed about the carrier position (Fig. 4.5a). Unambiguous ROE cross peaks can be identified by recording two ROESY spectra with very different rf carrier offsets (73). Development of pulse sequences that eliminate TOCSY transfer and generate pure ROE cross peaks is an area of active research (74).

6.6.2.2 Experimental Protocol and Processing   In order to achieve baselines flat enough to allow accurate quantitation of cross-peak intensities, the points discussed for the TOCSY and NOESY spectra (Sections 6.5.2 and 6.6.1.2) are also pertinent in the ROESY experiment. In addition, parameters must be chosen for the spin-lock mixing period. Sample heating and minimization of J transfer are jointly accommodated by using weak spin-lock field strengths (2–5 kHz), although the offset dependence of the ROE may be nontrivial in such cases. The ROE builds up at a rate twice that of the laboratory-frame NOE (Section 5.4.3); therefore, shorter mixing times are required to obtain ROE peaks comparable in size to their NOESY counterparts. One of the main uses for ROESY in protein spectroscopy is the avoidance of spin-diffusion effects, and mixing times are usually kept short (50–200 ms). The absorptive in-phase lineshape expected for diagonal and cross-peaks in the ROESY spectrum indicates that processing will be very similar to that already described for TOCSY and NOESY (Sections 6.5.3 and 6.6.1.3).

ROESY spectra were recorded with the pulse sequences of Fig.

6.53a,b. Thirty-two transients were collected for each of 512 t_1 increments ($t_{1,\mathrm{max}}$ = 44 ms). The spins were locked by a continuous low-power pulse (2.5-kHz field strength) of 40-ms duration. Sections of the spectra are shown in Fig. 6.54a,b. The increased cross-peak amplitude obtained with the pulse sequence of Fig. 6.53b is illustrated in Fig. 6.54c.

6.6.2.3 Information Content   Although resonance offset effects hinder quantitation of ROESY spectra, the ROESY experiment has several redeeming qualities for studies of proteins. Foremost, as discussed above, the ROE is always positive, and cross-peaks can be observed in ROESY spectra even if the peaks cannot be observed in NOESY spectra because $\omega_0 \tau_c \approx 1$.

A further advantage of ROESY over NOESY is that spin diffusion (or three-spin effects) produces contributions to cross-peaks that are of sign opposite that of the direct ROE peaks. Conceptually, the rotating-frame cross-relaxation rate constant is positive, and magnetization transfer between two spins occurs with inversion of sign. Thus, a diagonal peak and a cross-peak arising by a direct ROE between two spins have opposite signs. Transfer of the cross-peak magnetization to a third spin involves another change of sign. As a result, cross-peaks dominated by spin diffusion will be of the same sign as the diagonal. If a small three-spin interaction contributes to a ROESY cross-peak, the measured intensity is reduced and may be interpreted as a longer interproton separation. Consequently, the upper-bound restraint applied in structure calculations will not be overly restrictive.

The influence of spin diffusion in NOESY spectra is particularly pronounced for NOEs involving geminal methylene groups. Efficient spin diffusion between the $^1\mathrm{H}^{\beta''}$ and $^1\mathrm{H}^{\beta'}$ tends to equalize the intensity of NOEs to other protons even if the distances to $^1\mathrm{H}^{\beta''}$ and $^1\mathrm{H}^{\beta'}$ are not equal. Stereospecific assignment of β-methylene protons plays an important role in defining side-chain conformation, and depends heavily on estimating the relative sizes of intraresidue and sequential distances to $^1\mathrm{H}^{\beta''}$ and $^1\mathrm{H}^{\beta'}$ (75,76). The use of ROESY spectra for this process significantly reduces the chance of incorrectly making such assignments.

Another important facet of the ROESY experiment is that chemical-exchange peaks are of the same sign as the diagonal, i.e., opposite in sign to peaks arising from direct cross-relaxation. Thus, rotating frame experiments are invaluable in the study of dynamic processes involving slow exchange between two or more states. Protein–protein or peptide–protein interactions are one area where discrimination of cross-relaxation and chemical exchange is not possible from NOESY, but is apparent from ROESY data. As with chemical exchange in TOCSY spectra (50),

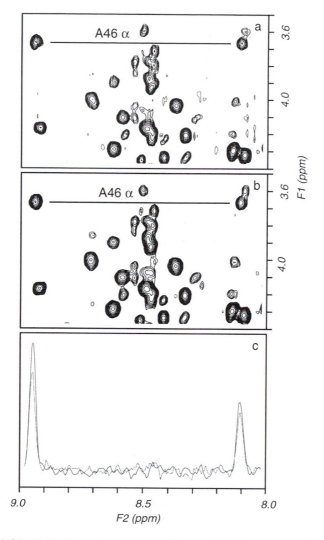

FIGURE 6.54 ROESY spectra of ubiquitin acquired from H_2O solution with a mixing time of 40 ms and a spin-lock field of 2.5 kHz. (a) Section of the spectrum acquired with the sequence of Fig. 6.53a. (b) Section of the spectrum acquired with the sequence of Fig. 6.53b. (c) Traces through the cross-peaks for the $^1H^\alpha$ of Ala46 for ($\cdots$) the spectrum shown in (a) and (——) the spectrum shown in (b).

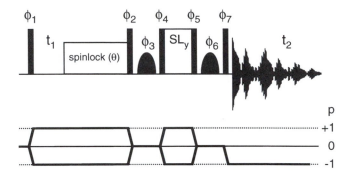

FIGURE 6.55 Pulse sequence and coherence-level diagram for a ROESY experiment acquired from H_2O solution without presaturation of the water resonance (68). The curved pulse shapes indicate selective 90° pulses at the water frequency; sinc-shaped pulse shapes perform well in this application. The phases are cycled as follows: $\phi_1 = 8(x)$; $\phi_2 = 2(x, x, -x, -x)$; $\phi_3 = 4(x, -x)$; $\phi_4 = 8(x)$; $\phi_5 = 8(-x)$; $\phi_6 = 4(x)\ 4(-x)$; $\phi_7 = 4(-x)\ 4(x)$; and receiver $= x, x, 4(-x), x, x$. This series of 8 steps is performed once with a mixing spin-lock pulse phase $\theta = -y$ and once with $\theta = y$. CYCLOPS is then performed on all pulses and the receiver simultaneously to produce a 64-step phase cycle. The selective pulses should be approximately 5 ms in duration, and a spin-lock pulse length of 1–2 ms serves to purge the residual solvent signal.

complex situations can arise where peaks result from both cross-relaxation and exchange.

6.6.2.4 Experimental Variants Just as in the JR-NOESY, ROESY experiments frequently are acquired without presaturation of the solvent resonance. A variety of suppression techniques have been devised that accomplish this (the jump–return sequence cannot be used effectively following the spin-lock pulse) (68), some of which were discussed earlier (Section 3.5.3). One possible pulse sequence is shown in Fig. 6.55.

6.7 ¹H 3D Experiments

Given the vast improvement in effective resolution between 1D and 2D NMR spectra, 3D NMR spectroscopy is a logical approach to increasing the effective resolution still further. The increase in dimensionality from 2D to 3D is achieved by inserting a second incrementable delay and mixing period immediately before the acquisition period of a normal 2D experiment as discussed in Section 4.5 and shown schematically in Fig. 4.19. The first example of a 3D NMR experiment useful in the study of proteins was reported in 1988 (77), and combined NOESY and TOCSY mixing with the measurement proton frequencies in all three dimensions.

The acquisition of 3D homonuclear NMR spectra introduces many technical challenges. The digital resolution in the indirect dimensions must be reasonably high because of the large number of protons and their relatively poor chemical-shift dispersion. Simultaneously, the total acquisition time must be minimized because two evolution delays must be incremented independently. As a consequence, such spectra are acquired with minimal digital resolution in t_1 and t_2 (usually less than 128 complex data points are collected) and with relatively few transients (never more than 16). The phase cycle must be chosen with great care so as to achieve the maximum degree of artifact suppression in the fewest steps. Even with optimization of conditions, homonuclear 3D experiments commonly require 4–8 days to acquire. Improvements in digital resolution (or a reduction in the total acquisition time) may be obtained by the use of selective pulses to limit the frequencies observed in one or both of the indirectly detected dimensions (78–80).

Aside from the technical issues, homonuclear 3D spectra are much more complicated than 2D spectra. For a protein the size of ubiquitin, 2000–3000 individual cross-peaks are observable in the 2D NOESY spectrum; in the corresponding homonuclear 3D NOESY-TOCSY spectrum, approximately five times as many cross-peaks might be observed because the magnetization transferred between spins by a particular NOE interaction (generating one cross-peak in the 2D spectrum) is transferred to all the spins in the same spin system by the isotropic mixing. In contrast, heteronuclear 3D and 4D spectra have numbers of cross-peaks similar to those of 2D experiments (Chapter 7). Thus, although homonuclear 3D experiments were developed first, heteronuclear multidimensional spectroscopy, because of its greater simplicity and superior resolving power, is preferable. Nevertheless, homonuclear 3D spectra can offer additional information in cases where isotopic labeling is not possible.

The main implementation of homonuclear 3D spectroscopy in the study of proteins combines NOESY and TOCSY mixing because both mechanisms transfer in-phase magnetization and produce lineshapes that do not suffer from self-cancellation under conditions of limiting digital resolution. The pulse sequence for a 3D NOESY-TOCSY experiment is shown in Fig. 6.56.

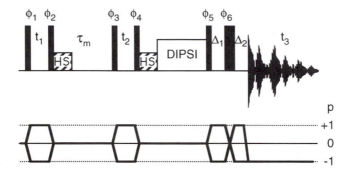

FIGURE 6.56 Pulse sequence and coherence-level diagram for a homonuclear 3D NOESY-TOCSY experiment. A Hahn echo sequence is included prior to detection. The DIPSI-2rc mixing pulses are of y phase, and the phases of the pulses are cycled as follows: $\phi_1 = 2(x, -x, y, -y)$; $\phi_2 = 2(-x, -x, -y, -y)$; $\phi_3 = -x, -x, -y, -y, x, x, y, y$; $\phi_4 = 2(x, x, y, y)$; $\phi_5 = 2(x, x, y, y)$; $\phi_6 = 2(x, x, y, y)$; and receiver $= x, -x, y, -y, -x, x, -y, y$. Typically, the homospoil pulses (cross-hatched boxes) are applied for 2–3 ms.

6.7.1 EXPERIMENTAL PROTOCOL

The precautions described above for obtaining flat baselines in 2D TOCSY (Section 6.5.2) and NOESY (Section 6.6.1.2) spectra are also relevant for 3D spectroscopy, with the additional limitation that the phase cycle must be kept to a bare minimum. The eight-step phase cycle described in Fig. 6.56 should be considered as a starting point for optimization of this experiment on a given spectrometer; the dominant artifacts will vary from one spectrometer to another, and some optimization may be necessary. Homospoil pulses are used to suppress artifacts arising from residual transverse magnetization that are not removed by the short phase cycle. The usefulness of the Hahn echo will depend on accurate measurement of the 180° pulse length because the pulse cannot be phase-cycled independently (a composite 180° pulse can be employed). Assuming that the artifact level is acceptable from 8 transients, then 256 increments can be collected for both t_1 and t_2 in just over a week of spectrometer time.

Acquisition of t_1–t_2 or t_1–t_3 slices provides a useful means for performing optimization prior to acquisition of the complete t_1–t_2–t_3 matrix. An example of the $t_3 = 0$ slice is shown in Fig. 6.57.

6.7.2 PROCESSING

Not only is the acquisition of a homonuclear 3D experiment time-consuming, but processing also provides challenges. In general, processing

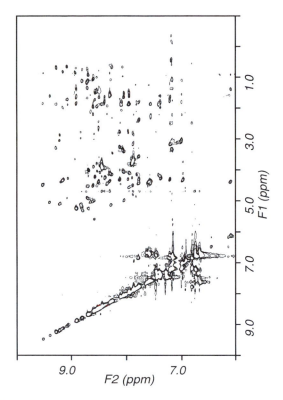

FIGURE 6.57 Section of the first t_1–t_2 slice ($t_3 = 0$) of a 3D NOESY-TOCSY experiment acquired from an H_2O solution of ubiquitin. The pulse sequence of Fig. 6.56 was used with a NOESY mixing time of 100 ms and 50 ms of DIPSI-2rc isotropic mixing. The solvent was suppressed by low-power coherent irradiation for 0.8 s prior to the pulse sequence and during the NOESY mixing period. Homospoil pulses of 1-ms duration followed by a 3-ms recovery delay were inserted at the beginning of each mixing period. TPPI was used for frequency discrimination in t_1, and 8 transients were collected for each of the 256 increments ($t_{1,max}$ = 22 ms). The acquisition time for this slice was 41 min, indicating that an entire 3D spectrum acquired with the same digital resolution in F_1 and F_2 would require 7.3 days. The spectrum was processed without linear prediction.

strategies discussed for TOCSY (Section 6.5.3) and NOESY (Section 6.6.1.3) experiments are applied to 3D spectra as well. However, even modest zero-filling can lead to excessively large data matrices, and using linear prediction or maximum entropy reconstruction to improve the resolution in F_1 and F_2 is a considerable computational burden. During the initial stages of resonance assignment, the peaks of most interest involve amide protons in F_3; therefore, processing can be simplified by saving only this region after the t_3 Fourier transformation.

6.7.3 INFORMATION CONTENT

Homonuclear 3D spectra are analyzed with the aim of obtaining resonance assignments or to obtain interproton distance restraints from the unambiguous assignment of NOE peaks. However, the spectra contain a vast number of redundant cross-peaks, and complete analysis (as might be performed for a 2D spectrum) is not an attractive proposition unless the process is assisted by computer automation. Alternatively, certain regions of the 3D spectrum may be analyzed in detail to resolve specific ambiguities arising from the 2D spectra.

The most intense peaks lie along the body diagonal at $F_1 = F_2 = F_3$ and arise from magnetization that has not been transferred during either mixing period. The next most intense peaks arise from transfer by either NOE *or* TOCSY but not both. The former occur in the $F_2 = F_3$, or NOE plane, whereas the latter occur in the $F_1 = F_2$, or TOCSY plane. The weakest cross-peaks arise from NOE *and* TOCSY transfer and are sometimes referred to as "real" or "true" 3D cross-peaks because they have no equivalent in 2D spectra. Further, the 3D cross-peaks in the $F_1 = F_3$ plane are referred to as "back-transfer peaks" as they arise from transfer from one proton to a second during the NOE mixing, and then *back* to the first proton during the isotropic mixing. Analysis of the 3D spectrum is most readily accomplished from 2D plots corresponding to F_1–F_2 or F_1–F_3 planes; the intersection of these planes with the NOE, TOCSY, and back-transfer planes gives rise to the NOE, TOCSY, and back-transfer lines, respectively, which play an important role in analysis of the spectrum (*81,82*).

Information additional to that present in 2D spectra is contained in the peaks with $F_1 \neq F_2 \neq F_3$. Note that the intensity of the peaks will depend on the efficiency of transfer via both dipolar relaxation *and* isotropic mixing. Since cross-peak intensity is not related directly to the NOE between the spins, calibration of interproton distances cannot be performed in a precise manner. The dependence of cross-peak intensity on the rate of TOCSY transfer can be especially problematic for the amide protons of helical residues; the small value of $^3J_{H^NH^\alpha}$ will reduce the size of all correlations, even if the interproton separation is relatively short.

Finally, although the 3D spectra may be able to resolve correlations that are overlapped in 2D spectra, correlations between two protons with identical chemical shifts still cannot be observed. Either four proton dimensions or a single proton and two heteronuclear dimensions (e.g., the HMQC-NOESY-HMQC; Section 7.2.4) are required to observe such cross-peaks.

6.7.4 EXPERIMENTAL VARIANTS

Several variants of homonuclear 3D experiments have been described that differ from the example described above in the choice of mixing schemes. Thus, the 3D TOCSY-NOESY has been proposed as a complementary method of performing the 3D NOESY-TOCSY experiment. Since both experiments have one TOCSY and one NOESY mixing period, the information content is identical, although cross-peaks containing equivalent information will be present in different regions of the spectra. Thus, if a particular cross-peak is obscured by some artifact in one type of spectrum, equivalent information may be present in an artifact-free region of the other experiment. In addition, a jump–return read pulse (Section 3.5.2) can be incorporated into the TOCSY-NOESY experiment, improving solvent suppression and allowing the observation of peaks close to the water resonance (*83,84*).

Homonuclear 3D spectra have also been used to study spin-diffusion pathways by utilizing NOESY transfer during both mixing periods in the 3D NOESY-NOESY experiment (*85*). In such a spectrum, the intensity of a true 3D cross-peak between spins I_1, I_2, and I_3 at $F_1 = \Omega_1$, $F_2 = \Omega_2$, and $F_3 = \Omega_3$ depends on the product of the cross-relaxation rates, σ_{12} and σ_{23}, and thus has intensity comparable to the spin-diffusion contribution to a peak between I_1 and I_3 in a 2D NOESY spectrum [6.48]. This is one of the few direct experimental methods for investigation of spin-diffusion pathways. 3D NOESY-NOESY spectra have been used for complete resonance assignments and identification of the secondary structure present in a protein (*86,87*). Finally, 3D TOCSY-TOCSY spectra have been used as an aid to automated resonance assignment routines (*88*). This experiment has the advantage that spectra of high quality can be obtained with very little phase cycling, and total acquisition times can be kept short.

References

1. D. S. Wishart and F. M. Richards, *FEBS Lett.* **293,** 72–80 (1991).
2. D. S. Wishart, B. D. Sykes, and F. M. Richards, *J. Mol. Biol.* **222,** 311–333 (1991).
3. J. Jeener. (Ampère Summer School, Basko Polje, Yugoslavia, 1971).
4. W. P. Aue, E. Bartholdi, and R. R. Ernst, *J. Chem. Phys.* **64,** 2229–2246 (1976).
5. D. Marion and K. Wüthrich, Biochem. *Biophys. Res. Commun.* **113,** 967–974 (1983).
6. D. Neuhaus, G. Wagner, M. Vasak, J. H. R. Kägi, and K. Wüthrich, *Eur. J. Biochem.* **151,** 257–273 (1985).
7. H. Widmer and K. Wüthrich, *J. Magn. Reson.* **74,** 316–336 (1987).
8. Y. Kim and J. H. Prestegard, *J. Magn. Reson.* **84,** 9–13 (1990).
9. S. Ludvigsen, K. V. Andersen, and F. M. Poulsen, *J. Mol. Biol.* **217,** 731–736 (1991).
10. L. J. Smith, M. J. Sutcliffe, C. Redfield, and C. M. Dobson, *Biochemistry* **30,** 986–996 (1991).

11. J. J. Titman and J. Keeler, *J. Magn. Reson.* **89,** 640-646 (1990).
12. L. McIntyre and R. Freeman, *J. Mag. Reson.* **96,** 425–431 (1992).
13. T. Szperski, P. Güntert, G. Otting, and K. Wüthrich, *J. Magn. Reson.* **99,** 552–560 (1992).
14. C. Griesinger, O. W. Sørensen, and R. R. Ernst, *J. Magn. Reson.* **75,** 747–492 (1987).
15. D. Marion and B. Bax, *J. Magn. Reson.* **80,** 528–533 (1988).
16. L. Mueller, *J. Magn. Reson.* **72,** 191–196 (1987).
17. G. Otting and K. Wüthrich, *J. Magn. Reson.* **75,** 546–549 (1987).
18. G. Eich, G. Bodenhausen, and R. R. Ernst, *J. Am. Chem. Soc.* **104,** 3731–3732 (1982).
19. A. Bax and G. Drobny, *J. Magn. Reson.* **61,** 306–320 (1985).
20. W. J. Chazin and K. Wüthrich, *J. Magn. Reson.* **72,** 358–363 (1987).
21. G. Wagner, *J. Magn. Reson.* **55,** 151–156 (1983).
22. W. J. Chazin, M. Rance, and P. E. Wright, *J. Mol. Biol.* **202,** 603–622 (1988).
23. U. Piantini, O. W. Sørensen, and R. R. Ernst, *J. Am. Chem. Soc.* **104,** 6800-6801 (1982).
24. M. Rance, O. W. Sørensen, G. Bodenhausen, G. Wagner, R. R. Ernst, and K. Wüthrich, *Biochem. Biophys. Res. Commun.* **117,** 479–485 (1983).
25. N. Müller, R. R. Ernst, and K. Wüthrich, *J. Am. Chem. Soc.* **108,** 6482–6496 (1986).
26. M. Rance, C. Dalvitt, and P. E. Wright, *Biochem. Biophys. Res. Commun.* **131,** 1094–1102 (1985).
27. N. Müller, G. Bodenhausen, K. Wüthrich, and R. R. Ernst, *J. Magn. Reson.* **65,** 531–534 (1985).
28. M. Rance and P. E. Wright, *Chem. Phys. Lett.* **124,** 572–575 (1986).
29. M. Rance, W. J. Chazin, C. Dalvit, and P. E. Wright, *Meth. Enzymol.* **176,** 114–134 (1989).
30. C. Griesinger, O. W. Sørensen, and R. R. Ernst, *J. Am. Chem. Soc.* **107,** 6394–6396 (1985).
31. Z. L. Mádi, C. Griesinger, and R. R. Ernst, *J. Am. Chem. Soc.* **112,** 2908–2914 (1990).
32. C. Griesinger, O. W. Sørensen, and R. R. Ernst, *J. Chem. Phys.* **85,** 6837–6852 (1986).
33. L. Braunschweiler, G. Bodenhausen, and R. R. Ernst, *Mol. Phys.* **48,** 535–560 (1983).
34. C. Dalvit, M. Rance, and P. E. Wright, *J. Magn. Reson.* **69,** 356–361 (1986).
35. M. Levitt and R. Freeman, *J. Magn. Reson.* **33,** 473–476 (1979).
36. G. Wagner and E. R. P. Zuiderweg, *Biochem. Biophys. Res. Commun.* **113,** 854–860 (1983).
37. D. L. Di Stefano and A. J. Wand, *Biochemistry* **26,** 7272–7281 (1987).
38. P. L. Weber, S. C. Brown, and L. Mueller, *Biochemistry* **26,** 7282–7290 (1987).
39. W. J. Chazin, *J. Magn. Reson.* **91,** 517–526 (1991).
40. W. J. Chazin, M. Rance, and P. E. Wright, *FEBS Lett.* **222,** 109–114 (1987).
41. L. Braunschweiler and R. R. Ernst, *J. Magn. Reson.* **53,** 521–528 (1983).
42. A. Bax and D. G. Davis, *J. Magn. Reson.* **65,** 355–360 (1985).
43. M. Rance, *J. Magn. Reson.* **74,** 557–564 (1987).
44. J. Cavanagh, W. J. Chazin, and M. Rance, *J. Magn. Reson.* **87,** 110-131 (1990).
45. R. Bazzo and I. D. Campbell, *J. Magn. Reson.* **76,** 358–361 (1988).
46. S. J. Glaser and G. P. Drobny, *Adv. Magn. Reson.* **14,** 35–58 (1990).
47. J. Cavanagh and M. Rance, *J. Magn. Reson.* **96,** 670-678 (1992).
48. J. Briand and R. R. Ernst, *Chem. Phys. Lett.* **185,** 276–285 (1991).
49. M. Kadkhodael, T.-L. Hwang, J. Tang, and A. J. Shaka, *J. Magn. Reson., Ser. A* **105,** 104–107 (1993).
50. J. Feeney, C. J. Bauer, T. A. Frenkiel, B. Birdsall, M. D. Carr, G. C. K. Roberts, and J. R. P. Arnold, *J. Magn. Reson.* **91,** 607–613 (1991).
51. M. A. Delsuc and J. Y. Lallemand, *J. Magn. Reson.* **69,** 504–507 (1986).

52. J.-M. Nuzillard and R. Freeman, *J. Magn. Reson., Ser. A* **110,** 252–256 (1994).
53. D. Marion, M. Ikura, and A. Bax, *J. Magn. Reson.* **84,** 425–430 (1989).
54. G. Wagner, *Quart. Rev. Biophys.* **16,** 1–57 (1983).
55. D. Marion, M. Ikura, R. Tschudin, and A. Bax, *J. Magn. Reson.* **85,** 393–399 (1989).
56. J. Cavanagh and M. Rance, *J. Magn. Reson.* **88,** 72–85 (1990).
57. F. Fogolari, G. Esposito, and P. Viglino, *J. Magn. Reson., Ser. A* **102,** 49–57 (1993).
58. G. M. Clore, A. Bax, and A. M. Gronenborn, *J. Biomol. NMR* **1,** 13–22 (1991).
59. S. Macura, Y. Huang, D. Suter, and R. R. Ernst, *J. Magn. Reson.* **43,** 259–281 (1981).
60. O. W. Sørensen, M. Rance, and R. R. Ernst, *J. Magn. Reson.* **56,** 527–534 (1984).
61. G. Bodenhausen, H. Kogler, and R. R. Ernst, *J. Magn. Reson.* **58,** 370-388 (1984).
62. W. Denk, R. Baumann, and G. Wagner, *J. Magn. Res.* **67,** 386–390 (1986).
63. V. Stoven, A. Mikou, D. Piveteau, E. Guittet, and J. Lallemand, *J. Magn. Reson.* **82,** 163–168 (1989).
64. J. Fejzo, Z. Zolnai, S. Macura, and J. L. Markley, *J. Magn. Reson.* **88,** 93–110 (1990).
65. G. H. Weiss, J. E. Kiefer, and J. A. Ferretti, *J. Magn. Reson.* **97,** 227–234 (1992).
66. P. Plateau and M. Guéron, *J. Am. Chem. Soc.* **104,** 7310–7311 (1982).
67. V. Sklenár and A. Bax, *J. Magn. Reson.* **75,** 378–383 (1987).
68. G. Otting and K. Wüthrich, *J. Am. Chem. Soc.* **111,** 1871–1875 (1989).
69. J. Stonehouse, G. L. Shaw, and J. Keeler, *J. Biomol. NMR* **4,** 799–805 (1994).
70. A. A. Bothner-By, R. L. Stephens, J.-M. Lee, C. D. Warren, and R. W. Jeanloz, *J. Am. Chem. Soc.* **106,** 811–813 (1984).
71. A. Bax and D. G. Davis, *J. Magn. Reson.* **63,** 207–213 (1985).
72. C. Griesinger and R. R. Ernst, *J. Magn. Reson.* **75,** 261–271 (1987).
73. D. Neuhaus and J. Keeler, *J. Magn. Reson.* **68,** 568–574 (1986).
74. T.-L. Hwang, M. Kadkhodaei, A. Mohebbi, and A. J. Shaka, *Magn. Reson. Chem.* **30,** S24–S34 (1992).
75. G. Wagner, W. Braun, T. F. Havel, T. Schaumann, N. Go, and K. Wüthrich, *J. Mol. Biol.* **196,** 611–639 (1987).
76. P. Güntert, W. Braun, M. Billeter, and K. Wüthrich, *J. Am. Chem. Soc.* **111,** 3997–4004 (1989).
77. H. Oschkinat, C. Grieinger, P. J. Kraulis, O. W. Sørensen, R. R. Ernst, A. M. Gronenborn, and G. M. Clore, *Nature* **332,** 374–376 (1988).
78. C. Griesinger, O. W. Sørensen, and R. R. Ernst, *J. Magn. Reson.* **73,** 574–579 (1987).
79. C. Griesinger, O. W. Sørensen, and R. R. Ernst, *J. Am. Chem. Soc.* **109,** 7227–7228 (1987).
80. H. Oschkinat, C. Ciesler, T. A. Holak, G. M. Clore, and A. M. Gronenborn, *J. Magn. Reson.* **83,** 450-472 (1989).
81. G. W. Vuister, R. Boelens, and R. Kaptein, *J. Magn. Reson.* **80,** 176–185 (1988).
82. G. W. Vuister, R. Boelens, A. Padilla, G. J. Kleywegt, and R. Kaptein, *Biochemistry* **29,** 1829–1839 (1990).
83. H. Oschkinat, C. Cieslar, A. M. Gronanborn, and G. M. Clore, *J. Magn. Reson.* **81,** 212–216 (1989).
84. J. P. Simorre and D. Marion, *J. Magn. Reson.* **94,** 426–432 (1991).
85. R. Boelens, G. W. Vuister, T. M. G. Koning, and R. Kaptein, *J. Am. Chem. Soc.* **111,** 8525–8526 (1989).
86. G. W. Vuister, R. Boelens, A. Padilla, and R. Kaptein, *J. Biomol. NMR* **1,** 421–438 (1991).
87. R. Bernstein, A. Ross, C. Cieslar, and T. A. Holak, *J. Magn. Reson., Ser. B* **101,** 185–188 (1993).
88. C. Cieslar, T. A. Holak, and H. Oschkinat, *J. Magn. Reson.* **89,** 184–190 (1990).

HETERONUCLEAR NMR EXPERIMENTS

The 2D and 3D ^{1}H NMR methods discussed in Chapter 6 are ineffective for proteins with molecular masses greater than approximately 10–12 kDa. The number of protons present in proteins scales approximately linearly with molecular mass. The rotational correlation times of globular proteins, and thus the linewidths of the NMR resonances, also increases linearly with molecular mass. The increased number and linewidth of the proton resonances in homonuclear ^{1}H NMR spectra result in extensive chemical-shift overlap and degeneracy. Conventional assignment procedures, based on observation of sequential NOE correlations (Chapter 8), become difficult or impossible. In addition, larger linewidths (larger spin–spin relaxation rate constants) result in decreased sensitivity for ^{1}H correlation experiments that rely on small (<10 Hz) homonuclear ^3J scalar couplings for coherence transfer (e.g., COSY, multiple-quantum, and TOCSY experiments).

Heteronuclear NMR spectroscopy (1–3) effectively circumvents these problems for proteins of molecular masses up to 25–30 kDa, provided that the proteins can be uniformly labeled with the NMR-active isotopes ^{13}C and ^{15}N (4); even larger proteins may be accessible by combining ^{15}N and ^{13}C labeling with fractional deuteration (5,6). Spectral resolution is improved by increasing the dimensionality of the NMR spectrum so that the highly overlapped ^{1}H resonances, present in ^{1}H 2D spectra, are separated in 3D and 4D spectra according to the better-resolved heteronuclear resonances. At the same time, the efficiency of coherence transfer is increased by utilizing relatively large one-bond and two-bond (^1J and ^2J) scalar coupling interactions between pairs of heteronuclei and between heteronuclei and their directly attached protons, rather than the relatively small ^{1}H homonuclear three-bond scalar coupling interactions.

In this chapter, the general principles of heteronuclear NMR spectroscopy are discussed. A selection of experiments are described that are necessary for resonance assignment and structure determination of isotopically labeled proteins. The summary of experiments presented is by no means complete; indeed, a given resonance correlation may be obtained using several different methods, and development of new experimental techniques continues apace.

7.1 Heteronuclear Correlation NMR Spectroscopy

All multidimensional heteronuclear NMR experiments correlate a heteronuclear resonance with a proton resonance by transfer of coherence (or polarization) between the heteronuclear (S) and proton (I) spins. Regardless of the specific protocol utilized to effect coherence transfer, the NMR experiment can start with excitation of either I or S spin polarization and must end with detection of either I or S spin magnetization. The overall sensitivity of heteronuclear correlation NMR experiments is proportional to (see [3.55] and [3.114])

$$S/N \propto \gamma_{ex}\gamma_{det}^{3/2}[1 - \exp(-R_{1,ex}T_c)], \qquad [7.1]$$

in which γ_{ex} and γ_{det} are the gyromagnetic ratios of each nucleus excited at the beginning of the sequence and detected at the end of the sequence, respectively; T_c is the recycle time of the experiment; and $R_{1,ex}$ is the spin–lattice relaxation rate constant of the excited nucleus (7). Therefore, *indirect* or *proton detection* is used whenever possible in order to maximize sensitivity. In these techniques, proton spin polarization initially is transferred to the heteronucleus, the desired heteronuclear spin manipulations are performed, and the heteronuclear coherence finally is transferred back to proton magnetization for detection. The gain in sensitivity compared to a correlation experiment in which proton magnetization is transferred to the S nucleus for detection is thus $n(\gamma_I/\gamma_S)^{3/2}$, in which γ_I and γ_S are the gyromagnetic ratios of the I and S nuclei, respectively, and n is the

a

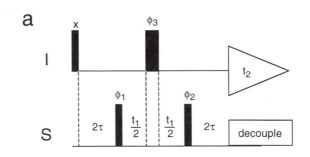

b

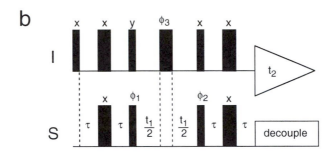

c

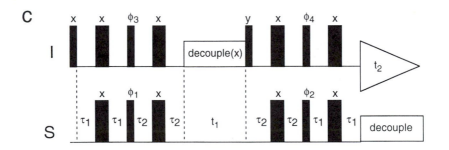

d

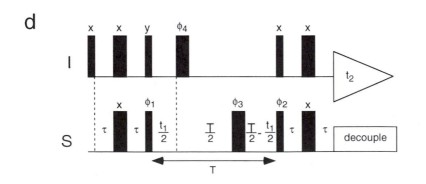

number of protons attached to the S nucleus. For $^1H-^{13}C$ correlations, the gain in sensitivity is approximately 24 for methyl protons, 16 for methylene protons, and 8 for methine protons, whereas for $^1H-^{15}N$ correlations of backbone amides the gain is about 30. The sensitivity gain relative to an experiment that starts with S-nucleus polarization and detects proton magnetization is simply the ratio of the gyromagnetic ratios, γ_I/γ_S; for $^1H-^{13}C$ correlations this ratio is about 4, and for $^1H-^{15}N$ correlations it is about 10. The larger spin–lattice relaxation rate constants of protons compared to heteronuclei ($R_{1I} > R_{1S}$) give an additional sensitivity advantage to the experiments starting with proton magnetization (i.e., the schemes involving $I \to S$ and $I \to S \to I$ transfers) because of the $[1 - \exp(-R_{1,ex}T_c)]$ factor in [7.1]. In practice, several factors can reduce the empirical gain in sensitivity, including relaxation during coherence transfer steps, incomplete coherence transfer in multispin systems (e.g., I_2S and I_3S systems), and resonance offset effects arising from the increased heteronuclear, relative to proton, chemical-shift range.

7.1.1 BASIC HETERONUCLEAR CORRELATION EXPERIMENTS

Two-dimensional proton-detected heteronuclear correlation experiments generally use HMQC or HSQC coherence transfer mechanisms.

FIGURE 7.1 Pulse sequences for the ^1H-detected heteronuclear correlation experiments. In all pulse sequence figures, thin bars represent 90° pulses and thick bars represent 180° pulses. The phase of each pulse is indicated above the bar. (a) The HMQC experiment, in which the phase cycling is $\phi_1 = x, -x$; $\phi_2 = 8(x), 8(-x)$; $\phi_3 = 2(x), 2(y), 2(-x), 2(-y)$; and receiver $= 2(x, -x, -x, x), 2(-x, x, x, -x)$. (b) The HSQC experiment, in which the phase cycling is $\phi_1 = x, -x$; $\phi_2 = 2(x)$, $2(-x)$; $\phi_3 = 4(y), 4(-y)$; and receiver $= 2(x, -x, -x, x)$. This phase cycle can be further extended by the inclusion of independent cycling of the 90°(^1H) pulses on either side of the t_1 period as in the decoupled-HSQC experiment. (c) The decoupled-HSQC experiment, in which the phase cycling is $\phi_1 = 2(x), 2(-x)$; $\phi_2 = 8(x)$, $8(-x)$; $\phi_3 = y, -y$; $\phi_4 = 4(x), 4(-x)$; and receiver $= x, -x, -x, x, 2(-x, x, x, -x), x, -x, -x, x$. (d) The constant-time HSQC experiment, in which the phase cycling is $\phi_1 = x, -x$; $\phi_2 = 8(x), 8(-x)$; $\phi_3 = 2(x), 2(y), 2(-x), 2(-y)$; $\phi_4 = 16(y)$, $16(-y)$; and receiver $= 2(x, -x, -x, x), 2(-x, x, x, -x)$. If desired, this 32-step phase cycle can be reduced to 8 steps by eliminating the cycling of ϕ_2 and using only the first 4 steps of the phase cycle of ϕ_3; an additional reduction by a factor of 2 can be obtained by eliminating cycling of ϕ_4. In each case the optimal value for 2τ is $1/(2J_{IS})$. Decoupling during t_2 can be achieved by using either GARP-1 or WALTZ-16 decoupling sequences. Decoupling during the t_1 evolution period of scheme c is best achieved using WALTZ-16 or DIPSI-2 sequences.

The two techniques are distinguished by whether the transferred coherence evolves during an indirect evolution period as heteronuclear multiple-quantum coherence (HMQC) or heteronuclear single-quantum coherence (HSQC) (8–11). Pulse sequences for these heteronuclear correlation experiments are illustrated in Fig. 7.1. As will be seen subsequently, the 2D HMQC and HSQC experiments are integral components of all heteronuclear 3D and 4D NMR experiments.

7.1.1.1 The HMQC Experiment A heteronuclear IS spin system ($I = {}^1$H, $S = {}^{15}$N, or ^{13}C), in which the I and S spins are directly covalently bonded, and the proton I is scalar-coupled to a remote proton, K, forms the basis for the following discussion. The homonuclear ^{1}H scalar coupling constant (J_{IK}) is assumed to be much smaller than J_{IS}. For this system, evolution during the HMQC scheme of Fig. 7.1a is described using the product operator formalism as follows:

$$I_z \xrightarrow{\frac{\pi}{2}(I_x+K_x)-2\tau-\frac{\pi}{2}S_x} -2I_xS_y$$

$$\xrightarrow{t_1/2-\pi(I_x+K_x)-t_1/2}$$
$$-2I_xS_y\cos(\Omega_St_1)\cos(\pi J_{IK}t_1) - 4I_yK_zS_y\cos(\Omega_St_1)\sin(\pi J_{IK}t_1)$$

$$\xrightarrow{\frac{\pi}{2}S_x-2\tau} -I_y\cos(\Omega_St_1)\cos(\pi J_{IK}t_1) + 2I_xK_z\cos(\Omega_St_1)\sin(\pi J_{IK}t_1), \qquad [7.2]$$

in which the delay 2τ is set to $1/(2J_{IS})$ (approximately 5.4 ms for one-bond ^{1}H–^{15}N $J_{NH} = 92$ Hz, and 3.6 ms for one-bond ^{1}H–^{13}C $J_{CH} = 140$ Hz). If desired, all delays of nominal length $1/(2J_{IS})$ can be shortened slightly to account for relaxation (Section 5.1.1). Only operators leading to observable terms have been propagated through the pulse sequence. Evolution of scalar coupling interactions other than J_{IS} during the periods 2τ has been ignored. Chemical-shift evolution of the I spin during the 2τ periods and during t_1 is refocused by the 180°(I) pulse. Heteronuclear multiple-quantum (MQ) coherence, represented by the $2I_xS_y$ operator, does not evolve under the influence of the active scalar coupling, J_{IS}, during the t_1 period (Section 2.7.5). Evolution under the homonuclear J_{IK} scalar coupling Hamiltonian is not refocused, because both proton spins I and K experience the effect of the nonselective 180°(I) pulse. The resulting correlation spectrum exhibits homonuclear J-coupling multiplet structure in the F_1 dimension. In addition, the F_2 lineshapes consist of the superposition of in-phase absorptive and antiphase dispersive components, represented by the I_y and $2I_xK_z$ operators in the last line of [7.2], respectively.

The antiphase dispersive component of the signal can be purged by inserting a 90°(I) pulse prior to the acquisition period. The purge pulse transforms the final operators in [7.2] to

$$-I_y \cos(\Omega_S t_1) \cos(\pi J_{IK} t_1) + 2I_x K_z \cos(\Omega_S t_1) \sin(\pi J_{IK} t_1) \xrightarrow{\left(\frac{\pi}{2}\right)(I_y + K_y)}$$

$$-I_y \cos(\Omega_S t_1) \cos(\pi J_{IK} t_1) - 2I_z K_x \cos(\Omega_S t_1) \sin(\pi J_{IK} t_1). \quad [7.3]$$

The antiphase I spin operator is transformed into an antiphase K spin operator. If the K spin resonance frequency occurs in an unimportant region of the spectrum (if it is not coupled to an S spin of interest), the IS correlation spectrum has pure in-phase absorptive lineshapes in both frequency dimensions.

Inclusion of homonuclear scalar coupling evolution during the periods 2τ is facilitated by using [2.117] to simplify the propagator for the pulse sequence prior to product operator analysis. The propagator is

$$\mathbf{U} = \exp[-i2\mathcal{H}\tau] \exp\left[-i\frac{\pi}{2}S_x\right] \exp[-i\mathcal{H}t_1/2] \exp[-i\pi(I_x + K_x)]$$

$$\times \exp[-i\mathcal{H}t_1/2] \exp\left[-i\frac{\pi}{2}S_x\right] \exp[-i2\mathcal{H}\tau]$$

$$\times \exp\left[-i\frac{\pi}{2}(I_x + K_x)\right], \quad [7.4]$$

in which the free-precession Hamiltonian is given by

$$\mathcal{H} = \Omega_I I_z + \Omega_S S_z + 2\pi J_{IS} I_z S_z + 2\pi J_{IK} I_z K_z. \quad [7.5]$$

Inserting $\mathbf{E} = \exp[i\pi(I_x + K_x)] \exp[-i\pi(I_x + K_x)]$ and applying [2.117] yields

$$\mathbf{U} = \exp[-i2\mathcal{H}\tau] \exp\left[-i\frac{\pi}{2}S_x\right] \exp[-i\mathcal{H}t_1/2] \exp[-i\pi(I_x + K_x)]$$

$$\times \exp[-\mathcal{H}t_1/2] \exp[i\pi(I_x + K_x)] \exp[-i\pi(I_x + K_x)]$$

$$\times \exp\left[-i\frac{\pi}{2}S_x\right] \exp[-i2\mathcal{H}\tau] \exp\left[-i\frac{\pi}{2}(I_x + K_x)\right]$$

$$= \exp[-i2\mathcal{H}\tau] \exp\left[-i\frac{\pi}{2}S_x\right] \exp[-i\mathcal{H}t_1/2]$$

$$\times \exp[-i(-\Omega_I I_z + \Omega_S S_z - 2\pi J_{IS} I_z S_z + 2\pi J_{IK} I_z K_z)t_1/2] \exp\left[-i\frac{\pi}{2}S_x\right]$$

$$\times \exp[-i(-\Omega_I I_z + \Omega_S S_z - 2\pi J_{IS} I_z S_z + 2\pi J_{IK} I_z K_z)2\tau]$$

$$\times \exp\left[i\frac{\pi}{2}(I_x + K_x)\right]$$

$$= \exp[-i2\pi J_{IK} I_z K_z(t_1 + 4\tau)] \exp[-i(\Omega_S S_z + 2\pi J_{IS} I_z S_z)2\tau]$$

$$\times \exp\left[-i\frac{\pi}{2}S_x\right] \exp[-i\Omega_S S_z t_1] \exp\left[-i\frac{\pi}{2}S_x\right]$$

$$\times \exp[-i(\Omega_S S_z - 2\pi J_{IS} I_z S_z)2\tau] \exp\left[i\frac{\pi}{2}(I_x + K_x)\right]. \qquad [7.6]$$

Evolution through the pulse sequence is represented as

$$I_z \xrightarrow{\left(-\frac{\pi}{2}\right)(I_x + K_x) - (\Omega_S S_z - 2\pi J_{IS} I_z S_z)2\tau - \frac{\pi}{2}S_x} -2I_x S_y$$

$$\xrightarrow{\Omega_S S_z t_1} -2I_x S_y \cos(\Omega_S t_1)$$

$$\xrightarrow{\frac{\pi}{2}S_x - (\Omega_S S_z + 2\pi J_{IS} I_z S_z)2\tau} -I_y \cos(\Omega_S t_1)$$

$$\xrightarrow{2\pi J_{IK} I_z K_z(t_1 + 4\tau)} -I_y \cos(\Omega_S t_1) \cos[\pi J_{IK}(t_1 + 4\tau)]$$

$$+ 2I_x K_z \cos(\Omega_S t_1) \sin[\pi J_{IK}(t_1 + 4\tau)]. \qquad [7.7]$$

As seen, evolution of the homonuclear scalar coupling during the 2τ periods introduces a phase error in the F_1 dimension of the 2D HMQC experiment. The magnitude of the phase error depends on J_{IK} and thus will vary nonlinearly throughout the spectrum. For a 10-Hz scalar coupling constant and $2\tau = 5.4$ ms, the phase error is 19.4°.

7.1.1.2 The HSQC Experiment In the HSQC experiment, illustrated in Fig. 7.1b, the INEPT (insensitive nuclei enhanced by polarization transfer) sequence introduced in Section 2.7.7.2 is used to transfer I spin polarization (I_z) into antiphase heteronuclear single-quantum (SQ) coherence ($2I_z S_y$). The antiphase heteronuclear SQ coherence evolves during the subsequent t_1 evolution period. A second INEPT sequence is used to transfer the frequency-labeled heteronuclear SQ coherence back to proton magnetization for detection. For a heteronuclear IS spin system, evolution through the pulse sequence is described as follows:

$$I_z \xrightarrow{\frac{\pi}{2}(I_x+K_x)-\tau-\pi(I_x+K_x),\pi S_x-\tau-\frac{\pi}{2}(I_y+K_y),\frac{\pi}{2}S_x} -2I_zS_y$$

$$\xrightarrow{t_1/2-\pi(I_x+K_x)-t_1/2} 2I_zS_y\cos(\Omega_S t_1) - 2I_zS_x\sin(\Omega_S t_1)$$

$$\xrightarrow{\frac{\pi}{2}(I_x+K_x),\frac{\pi}{2}S_x-\tau-\pi(I_x+K_x),\pi S_x-\tau} -I_x\cos(\Omega_s t_1) - 2I_yS_x\sin(\Omega_s t_1)$$

$$[7.8]$$

in which the delay $2\tau = 1/(2J_{IS})$ and evolution of the homonuclear ^{1}H scalar coupling interaction during the INEPT sequences has been ignored. The resultant term proportional to I_yS_x is unobservable multiple-quantum coherence. The 180°(I) pulse in the middle of the evolution period refocuses evolution of the proton–heteronuclear J_{IS} scalar coupling interaction. The $2I_zS_y$ operator present during t_1 commutes with the homonuclear ^{1}H scalar coupling Hamiltonian, and the F_1 lineshape does not contain contributions from ^{1}H scalar coupling interactions.

Evolution of the homonuclear scalar coupling interaction during the INEPT sequences is analyzed most easily by using [2.117] to simplify the propagator for the pulse sequence:

$$\mathbf{U} = \exp[-i\mathcal{H}\tau]\exp[-i\pi(I_x+K_x)]\exp[-i\pi S_x]\exp[-i\mathcal{H}\tau]$$

$$\times \exp\left[-i\frac{\pi}{2}(I_x+K_x)\right]\exp\left[-i\frac{\pi}{2}S_x\right]\exp[-i\mathcal{H}t_1/2]$$

$$\times \exp[-i\pi(I_x+K_x)]\exp[-i\mathcal{H}t_1/2]\exp\left[-i\frac{\pi}{2}S_x\right]$$

$$\times \exp\left[-i\frac{\pi}{2}(I_y+K_y)\right]\exp[-i\mathcal{H}\tau]\exp[i\pi(I_x+K_x)]\exp[-i\pi S_x]$$

$$\times \exp[-i\mathcal{H}\tau]\exp\left[-i\frac{\pi}{2}(I_x+K_x)\right]$$

$$= \exp[-i(2\pi J_{IS}I_zS_z + 2\pi J_{IK}I_zK_z)2\tau]\exp\left[i\frac{\pi}{2}(I_x+K_x)\right]$$

$$\times \exp\left[-i\frac{\pi}{2}S_x\right]\exp[-i(\Omega_S S_z + 2\pi J_{IK}I_zK_z)t_1]\exp\left[-i\frac{\pi}{2}S_x\right]$$

$$\times \exp\left[i\frac{\pi}{2}(I_y + K_y)\right]\exp[-i(2\pi J_{IS}I_zS_z + 2\pi J_{IK}I_zK_z)2\tau]$$

$$\times \exp\left[-i\frac{\pi}{2}(I_x + K_x)\right], \qquad\qquad [7.9]$$

prior to evaluating evolution through the pulse sequence as

$$I_z \xrightarrow{\frac{\pi}{2}(I_x+K_x)-(2\pi J_{IS}I_zS_z+2\pi J_{IK}I_zK_z)2\tau-\left(-\frac{\pi}{2}\right)(I_y+K_y),\frac{\pi}{2}S_x} -2I_zS_y\cos(2\pi J_{IK}\tau)$$

$$\xrightarrow{(-\Omega_S S_z+2\pi J_{IK}I_zK_z)t_1} -2I_zS_y\cos(\Omega_S t_1)\cos(2\pi J_{IK}\tau)$$

$$\xrightarrow{\frac{\pi}{2}(I_x+K_x),\frac{\pi}{2}S_x-(2\pi J_{IS}I_zS_z+2\pi J_{IK}I_zK_z)2\tau}$$

$$-I_x\cos(\Omega_S t_1)\cos^2(2\pi J_{IK}\tau) - I_yK_z\cos(\Omega_S t_1)\sin(4\pi J_{IK}\tau), \qquad [7.10]$$

in which only observable terms have been included. Evolution of the ^{1}H scalar coupling during the INEPT periods modulates the amplitude (but not the phase) of the in-phase absorptive resonance and introduces an antiphase dispersive contribution to the F_2 lineshape. As in the case of the HMQC experiment, the antiphase dispersive component can be purged by applying a $90^\circ_y(I)$ pulse prior to acquisition.

In Fig. 7.1b, the $90^\circ(I)$ pulse following the t_1 period is phase-shifted by 90° relative to the $90^\circ(I)$ pulse preceding the t_1 period. If the two pulses have the same y phase, evolution through the pulse sequence becomes

$$I_z \xrightarrow{\frac{\pi}{2}(I_x+K_x)-(2\pi J_{IS}I_zS_z+2\pi J_{IK}I_zK_z)2\tau-\left(-\frac{\pi}{2}\right)(I_y+K_y),-\frac{\pi}{2}S_x}$$

$$-2I_zS_y\cos(2\pi J_{IK}\tau) + 4I_yS_yK_x\sin(2\pi J_{IK}\tau)$$

$$\xrightarrow{(\Omega_S S_z+2\pi J_{IK}I_zK_z)t_1}$$

$$-2I_zS_y\cos(\Omega_S t_1)\cos(2\pi J_{IK}\tau) + 4I_yS_yK_x\cos(\Omega_S t_1)\sin(2\pi J_{IK}\tau)$$

$$\xrightarrow{\frac{\pi}{2}S_x-\frac{\pi}{2}(I_y+K_y)-(2\pi J_{IS}I_zS_z+2\pi J_{IK}I_zK_z)2\tau}$$

$$-I_y\cos(\Omega_S t_1)\cos(4\pi J_{IK}\tau) + 2I_xK_z\cos(\Omega_S t_1)\sin(4\pi J_{IK}\tau). \qquad [7.11]$$

Comparison of [7.10] and [7.11] demonstrates that the phase shift of the final $90^\circ(I)$ pulse reduces contributions from evolution of the homonuclear ^{1}H scalar coupling interaction during the initial INEPT period because $\cos^2(2\pi J_{IS}\tau) > \cos(4\pi J_{IS}\tau)$ and $\sin(4\pi J_{IS}\tau) < 2\sin(4\pi J_{IS}\tau)$.

7.1.1.3 The Decoupled HSQC Experiment HSQC experiments transfer coherence from I to S spins in the form of antiphase SQ coherence. In some experiments, net magnetization transfer from I to S spins in the form of in-phase SQ coherence is desirable, because of either the improved linewidth (discussed in Section 7.1.2.4) or the requirements of subsequent coherence transfer steps in an extended pulse sequence (see Section 7.4.3 for an example) (*10*). A decoupled HSQC experiment that achieves net magnetization transfer is illustrated in Fig.7.1c. A refocused INEPT sequence (Section 2.7.7.3) is used to transfer I spin polarization to in-phase S magnetization (*12*). The S magnetization evolves during the t_1 evolution period. Continuous proton decoupling is applied to suppress evolution under the IS scalar coupling Hamiltonian. A reverse refocused INEPT sequence transfers S spin magnetization back to I spin magnetization for detection.

Evolution through the decoupled HSQC pulse sequence for a scalar-coupled heteronuclear I_nS spin system can be described as follows (for simplicity, homonuclear scalar coupling to remote proton spins is ignored in the following derivation):

$$\sum_{k=1}^{n} I_{kz} \xrightarrow{\frac{\pi}{2}\Sigma I_{kx}-\tau_1-\pi\Sigma I_{kx},\pi S_x-\tau_1-\frac{\pi}{2}\Sigma I_{ky},\frac{\pi}{2}S_x} -\sum_{k=1}^{n} 2I_{kz}S_y$$

$$\xrightarrow{\tau_2-\pi\Sigma I_{kx},\pi S_x-\tau_2} nS_x \sin(2\pi J_{IS}\tau_2) \cos^{n-1}(2\pi J_{IS}\tau_2)$$

$$\xrightarrow{\;t_1\;} nS_x \cos(\Omega_S t_1) \sin(2\pi J_{IS}\tau_2) \cos^{n-1}(2\pi J_{IS}\tau_2)$$

$$\xrightarrow{\frac{\pi}{2}\Sigma I_{ky}-\tau_2-\pi\Sigma I_{kx},\pi S_x-\tau_2}$$

$$\sum_{k=1}^{n} n2I_{kz}S_y \cos(\Omega_S t_1) \sin^2(2\pi J_{IS}\tau_2) \cos^{2n-2}(2\pi J_{IS}\tau_2)$$

$$\xrightarrow{\frac{\pi}{2}\Sigma I_{kx},\frac{\pi}{2}S_x-\tau_1-\pi\Sigma I_{kx},\pi S_x-\tau_1} \sum_{k=1}^{n} nI_{kx} \cos(\Omega_S t_1)\Gamma_n^2(2\tau_2), \qquad [7.12]$$

in which $2\tau_1 = 1/(2J_{IS})$, and only observable terms arising from in-phase S_x magnetization have been shown. The coherence transfer function, $\Gamma_n(t)$, is

$$\Gamma_n(t) = \sin(\pi J_{IS}t) \cos^{n-1}(\pi J_{IS}t), \qquad [7.13]$$

and is graphed in Fig. 7.2. Complete refocusing cannot be obtained for I_nS spin systems with $n > 1$ because during the delays, $2\tau_2$, coherences

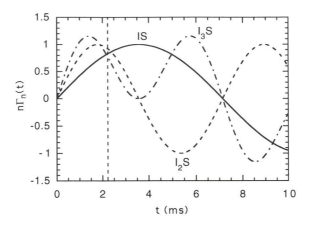

FIGURE 7.2 Plots of the refocused INEPT coherence transfer functions, $n\Gamma_n(t)$, for I_nS spin systems [7.13]. Results are shown for methine IS (—), methylene I_2S (–––), and methyl I_3S (–·–) groups with J_{CH} coupling constants of 140 Hz. The dashed vertical line at 2.2 ms indicates the optimal value of t to maximize $\Gamma_n(t)$ simultaneously for methine, methylene, and methyl carbons.

evolve under one active J_{IS} scalar coupling interaction and $n - 1$ passive J_{IS} scalar coupling interactions. The relative intensity of the I spin resonance in an IS spin system has a maximum value of 1.0 for $2\tau_2 = 1/(2J_{IS})$. The relative intensity of each I spin resonance in an I_2S spin system has a maximum value of 0.50 for $2\tau_2 = 1/(4J_{IS})$ and is nulled for $2\tau_2 = 1/(2J_{IS})$. The relative intensity for each I spin resonance in an I_3S spin system has a maximum value equal to approximately 0.45 for $2\tau_2 = 1/(5J_{IS})$ and is nulled for $2\tau_2 = 1/(2J_{IS})$. If the I spins are degenerate, as in a methyl moiety, the total signal intensity is obtained by summing the contributions from each I spin. The properties of the coherence transfer function can be used to edit a refocused HSQC experiment according to the I spin multiplicity. A compromise value of $2\tau_2 = 1/(3J_{IS})$ (2.4 ms for $^1J_{CH} = 140$ Hz) yields $n\Gamma_n(2\tau_2) = 0.87, 0.86,$ and 0.63 for $n = 1, 2, 3$.

Proton decoupling during t_1 is achieved by the use of a composite pulse decoupling scheme (e.g., WALTZ-16 or DIPSI-2; Section 3.4.3). The composite pulse decoupling must be applied *synchronously*; the decoupling sequence must begin at the same position within the decoupling supercycle (Section 3.4.3) for each recorded transient, because evolution of remote proton spins not scalar-coupled to the S spin must be identical during consecutive transients to obtain effective isotope filtration (Section 7.1.2.1).

Evolution under the passive scalar coupling interactions during the delays $2\tau_2$ generates $2^n - 1$ antiphase product operators in addition to the desired in-phase magnetization. For example, in an I_2S spin system, evolution of $-2I_{1z}S_y$ during the $2\tau_2$ delay of the forward refocused INEPT sequence gives

$$-2I_{1z}S_y \xrightarrow{\tau_2 - \pi\Sigma I_{kx}, \pi S_x - \tau_2} -2I_{1z}S_y \cos(2\pi J_{I_1S}\tau_2) \cos(2\pi J_{I_2S}\tau_2)$$

$$+ 4I_{1z}I_{2z}S_x \cos(2\pi J_{I_1S}\tau_2) \sin(2\pi J_{I_2S}\tau_2)$$

$$+ 2I_{2z}S_y \sin(2\pi J_{I_1S}\tau_2) \sin(2\pi J_{I_2S}\tau_2)$$

$$+ S_x \sin(2\pi J_{I_1S}\tau_2) \cos(2\pi J_{I_2S}\tau_2). \qquad [7.14]$$

The antiphase terms of the type $I_{1z}S_y$, $I_{2z}S_y$, and $I_{1z}I_{2z}S_x$, present in [7.14], are not efficiently dephased by a highly rf inhomogeneity-compensated composite pulse decoupling scheme (as discussed in Section 3.5.3, terms such as $4I_{1z}I_{2z}S_x$ are particularly insensitive to rf inhomogeneity). No net rotation of the I_{1z} and I_{2z} operators is expected for an integral number of cycles of an ideal composite pulse decoupling scheme applied to the I spins (ignoring homonuclear TOCSY transfer); however, the length of the evolution period t_1 generally will not be equivalent to an integral number of supercycles. If all the pulses of the composite decoupling sequence are applied along the $\pm x$ axis, the I spin operators experience a net rotation about the x axis by an angle α, given by the sum of the flip angles of the pulses in the applied fraction of the final supercycle. Transfer of antiphase S magnetization back to the I spin magnetization by the reverse INEPT sequence depends on α and therefore is a function of t_1. Subsequent Fourier transformation results in a pattern resembling t_1 noise for the I_2S and I_3S signals. The $90°_y(I)$ pulse immediately following the t_1 period in Fig. 7.1c suppresses this spurious magnetization transfer by converting the antiphase operators into multiple-quantum operators that are not refocused into observable proton magnetization by the reverse INEPT sequence.

7.1.1.4 The Constant-Time HSQC Experiment   Constant-time evolution periods were originally used in ^{1}H NMR spectroscopy to produce F_1-decoupled homonuclear NMR spectra (13–15), but subsequently have been employed in a number of heteronuclear 3D and 4D experiments. The constant-time HSQC (CT-HSQC) experiment (16,17), illustrated in Fig. 7.1d, differs from the generalized HSQC experiment (Fig. 7.1b), only in the way in which the heteronuclear SQ coherence evolves during the constant-time period, T, between the two INEPT sequences. The sequence fragment constituting the constant-time evolution period is

$$-t_1/2-180°(I, K)-T/2-180°(S)-(T - t_1)/2-$$ [7.15]

and has the propagator

$$
\begin{aligned}
\mathbf{U} &= \exp[-i\mathcal{H}(T - t_1)/2]\exp[-i\pi S_x]\exp[-i\mathcal{H}T/2] \\
&\quad \times \exp[-i\pi(I_x + K_x)]\exp[-i\mathcal{H}t_1/2] \\
&= \exp[-i2\pi J_{IK}I_zK_zT]\exp[-i\Omega_I I_z(T - t_1)] \\
&\quad \times \exp[i\Omega_S S_z t_1]\exp[-i\pi(I_x + K_x + S_x)].
\end{aligned}
$$ [7.16]

Evolution of the heteronuclear SQ coherence present following the initial INEPT sequence is given by

$$
\begin{aligned}
-\mathbf{U}2I_zS_y\mathbf{U}^{-1} &= -\exp(i\Omega_s t_1 S_z)2I_zS_y\exp(-i\Omega_s t_1 S_z) \\
&= -2I_zS_y\cos(\Omega_s t_1) + 2I_zS_x\sin(\Omega_s t_1).
\end{aligned}
$$ [7.17]

As shown by [7.16], the heteronuclear scalar coupling interaction, J_{IS}, is active for a total time period $(T/2 - t_1/2) - T/2 + t_1/2 = 0$; consequently, the S spin coherence remains antiphase with respect to the I spins. The homonuclear scalar coupling, J_{IK}, is active for the entire constant-time period, T, because both spins are affected equally by a nonselective 180° pulse. Consequently, evolution during the t_1 period is not modulated by homonuclear scalar coupling interactions, and the F_1 lineshape does not contain homonuclear multiplet structure. In the present example, this property is not particularly useful because the $2I_zS_y$ operator present during t_1 commutes with the ^{1}H homonuclear scalar coupling Hamiltonian. However, the same property exists for any homonuclear scalar coupling interaction and will be used to great advantage in ^{1}H–^{13}C HSQC spectra of fully ^{13}C-enriched proteins (Section 7.1.3.1).

Dephasing of the S spin coherence by magnetic field inhomogeneity during the constant-time period is refocused for a time $T - t_1$ by the 180°(S) pulse; consequently, relaxation of the heteronuclear single-quantum coherence is proportional to $\exp(-R_2T)\exp(-R_{\text{inhom}}t_1)$, in which R_2 is the homogeneous transverse relaxation rate of the SQ coherence (Section 7.1.2.4), and R_{inhom} is the inhomogeneous contribution to the total relaxation rate $R_2^* = R_2 + R_{\text{inhom}}$. In the constant-time HSQC experiment, $\exp(-R_2T)$ is a multiplicative factor reducing the intensity of the resonance signals, and $\exp(-R_{\text{inhom}}t_1)$ determines the F_1 linewidth (in practice, $R_{\text{inhom}}t_{1,\text{max}} \ll 1$, and the F_1 linewidth is determined principally by the apodization function employed). Clearly, $t_{1,\text{max}}$ cannot exceed T, and constant-time HSQC experiments invariably require compromise between resolution in F_1 (large values for T to maximize $t_{1,max}$) and sensitivity [small values of T to minimize $\exp(-R_2T)$].

Selected regions of the ^1H–^{15}N HMQC, HSQC, decoupled HSQC, and constant-time HSQC spectra of ubiquitin are compared in Fig. 7.3. The multiplet structure and dispersive contribution to the F_1 lineshapes associated with the homonuclear J coupling is clearly visible in the HMQC spectrum (Fig. 7.3a). The F_1 resolutions in the HSQC spectra (Fig. 7.3b–d) are clearly superior to the F_1 resolution of the HMQC spectrum (Fig. 7.3a). As predicted by [7.13], the NH_2 resonances are suppressed in the decoupled-HSQC pulse sequence for $2\tau_2 = 1/(2J_{IS})$ (Fig. 7.3c).

7.1.2 ADDITIONAL CONSIDERATIONS IN HMQC AND HSQC EXPERIMENTS

7.1.2.1 Phase Cycling and Artifact Suppression The minimum phase cycling required for HMQC and HSQC experiments comprises two steps necessary for *spectral editing* or *isotope filtration* and two steps required for frequency discrimination, or quadrature detection, in the indirectly detected dimension (Section 4.3.4).

Frequency discrimination is obtained by TPPI, States, or TPPI–States phase cycling (Section 4.3.4.1) of the $90°_{\phi_1}(S)$ pulse in the HMQC, HSQC, decoupled HSQC, and constant-time HSQC experiments (Fig. 7.1). In the HSQC experiments (Fig. 7.1b–d), the phase of the $180°(S)$ pulse preceding the $90°_{\phi_1}(S)$ pulse should be phase cycled in concert with the $90°_{\phi_1}(S)$ pulse for optimal results (*18*).

The initial t_1 sampling delay is adjusted to $1/(2SW)$ as described in Section 3.3.2.3 by adjusting $t_1(0)$ such that

$$1/(2SW) = 4\tau_{90(S)}/\pi + \tau_{180(I)} + t_1(0) \qquad [7.18]$$

for conventional evolution periods (e.g., Fig. 7.1a–c), and

$$1/(2SW) = t_1(0) \qquad [7.19]$$

for constant-time evolution periods (e.g., Fig. 7.1d).

The isotope filtration phase cycle is critically important for most heteronuclear experiments, because only signals from protons directly attached to the heteronucleus of interest (^{15}N or ^{13}C) are selected. Normally, changes in the coherence order of the S spin between $p_S = 0$ (corresponding to product operators containing a S_z component) and $p_S = \pm 1$ (corresponding to product operators containing a S_x or S_y component) are selected in a two-step phase cycle by simultaneously inverting the phase of a $90°(S)$ pulse and the receiver. Signals from I spins that are

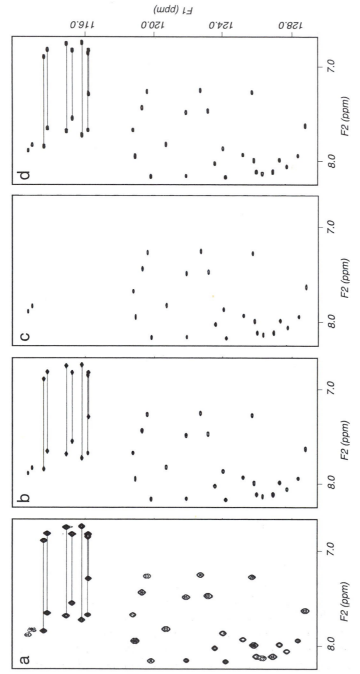

FIGURE 7.3 Comparison of selected regions from ^1H–^{15}N heteronuclear correlation spectra of ^{15}N-labeled ubiquitin, recorded using the schemes of Fig. 7.1a–c, respectively. Each spectrum was recorded in approximately the same total time with identical t_1 and t_2 acquisition times (348 and 164 ms, respectively). Each spectrum was processed similarly. No apodization function was applied in the F_1 dimension for the spectra shown in (a), (b) and (c); a cosine-bell apodization function was applied in the F_1 dimension of the spectrum shown in (d). In spectra (a), (b) and (d), NH_2 correlations are indicated by lines connecting the two nonequivalent proton resonances.

not scalar-coupled to S spins are unaffected by the phase shift of the $90°(S)$ pulse and are canceled by inversion of the receiver phase. Straightforward product operator analysis demonstrates that an isotope filter can be implemented in the HMQC, HSQC, decoupled HSQC, and constant-time HSQC experiments (Fig. 7.1) by use of a two-step $(x, -x)$ phase cycle on the $90°_{\phi_1}(S)$ pulse or the $90°_{\phi_2}(S)$ pulse together with a $(x, -x)$ receiver phase cycle. Time permitting, both pulses can be cycled independently to give a four step "double difference" isotope filter (*19*).

An isotope filter can be implemented alternatively in the constant-time HSQC experiment by selecting a coherence order change of $p_S = \pm 1$ to $p_S = \mp 1$ with a two-step (x, y) phase cycle applied to the $180°(S)$ pulse during T and a $(x, -x)$ receiver phase cycle. Although used infrequently (see Section 7.5.2 for an example), this technique has the advantage that the isotope filter can be implemented as part of EXORCYCLE phase cycling for artifact suppression (Section 4.3.2.3) without lengthening the overall phase cycle.

Additional phase cycling is utilized to eliminate artifacts resulting from pulse imperfections. The $180°(I)$ pulse in the middle of the t_1 evolution period of the HMQC experiment (Fig. 7.1a) and the $180°(S)$ pulse during the period T of the constant-time HSQC experiment (Fig. 7.1d) typically benefit from EXORCYCLE phase cycling (Section 4.3.2.3) to suppress artifacts from imperfect 180° pulses. If a 180° pulse is used simply to invert a longitudinal operator [e.g., the $180°(I)$ pulse applied to $I_z S_y$ in the t_1 period of HSQC experiments], artifacts caused by pulse imperfections are suppressed by inverting the phase of the 180° pulse in a two-step $(x, -x)$ phase cycle without changing the receiver phase. Time permitting, CYCLOPS phase cycling (Section 4.3.2.3) can be applied to the entire pulse sequence to eliminate quadrature artifacts.

The two-step $(x, -x)$ phase cycle applied to the $180°(I)$ pulse in the t_1 period of the HSQC experiment provides an important example of artifact suppression in heteronuclear correlation experiments. If the pulse has a nominal rotation angle of $\pi + \zeta$, rather than the ideal value of π, then evolution through the t_1 period is given by (for simplicity Ω_S is assumed to be zero):

$$-2I_zS_y \xrightarrow{t_1/2} -2I_zS_y \cos(\pi J_{IS}t_1/2) + S_x \sin(\pi J_{IS}t_1/2)$$

$$\xrightarrow{\pm(\pi+\zeta)I_x} -2I_zS_y \cos(\pi J_{IS}t_1/2) \cos(\pi + \zeta)$$

$$\pm 2I_yS_y \cos(\pi J_{IS}t_1/2) \sin(\pi + \zeta) + S_x \sin(\pi J_{IS}t_1/2)$$

$$\xrightarrow{t_1/2} 2I_zS_y\{-\cos^2(\pi J_{IS}t_1/2)\cos(\pi + \zeta) + \sin^2(\pi J_{IS}t_1/2)\}$$
$$+ S_x\{1 + \cos(\pi + \zeta)\}\cos(\pi J_{IS}t_1/2)\sin(\pi J_{IS}t_1/2)$$
$$\pm 2I_yS_y\cos(\Omega_I t_1/2)\cos(\pi J_{IS}t_1/2)\sin(\pi + \zeta)$$
$$\mp 2I_xS_y\sin(\Omega_I t_1/2)\cos(\pi J_{IS}t_1/2)\sin(\pi + \zeta). \qquad [7.20]$$

The resulting $2I_zS_y$ and $2I_xS_y$ operators in [7.20] are converted to observable proton magnetization by the reverse INEPT sequence:

$$2I_zS_y\{-\cos^2(\pi J_{IS}t_1/2)\cos(\pi + \zeta) + \sin^2(\pi J_{IS}t_1/2)\}$$
$$\mp 2I_xS_y\sin(\Omega_I t_1/2)\cos(\pi J_{IS}t_1/2)\sin(\pi + \zeta)$$
$$\xrightarrow{\text{INEPT}} - I_x\{-\cos^2(\pi J_{IS}t_1/2)\cos(\pi + \zeta) + \sin^2(\pi J_{IS}t_1/2)\}$$
$$\mp I_y\sin(\Omega_I t_1/2)\cos(\pi J_{IS}t_1/2)\sin(\pi + \zeta). \qquad [7.21]$$

This result can be simplified to

$$-I_x\{-\cos^2(\pi J_{IS}t_1 2)\cos(\pi + \zeta) + \sin^2(\pi J_{IS}t_1/2)\}$$
$$\mp I_y\sin(\Omega_I t_1/2)\cos(\pi J_{IS}t_1/2)\sin(\pi + \zeta)$$
$$= -I_x\{1 + \cos(\zeta)\}/2 + I_x\cos(\pi J_{IS}t_1)\{1 - \cos(\zeta)\}/2$$
$$\pm I_y\sin(\Omega_I t_1/2)\cos(\pi J_{IS}t_1/2)\sin(\zeta)$$
$$= -I_x\{1 - \zeta^2/4\} + I_x\cos(\pi J_{IS}t_1)\zeta^2/4$$
$$\pm I_y\sin(\Omega_I t_1/2)\cos(\pi J_{IS}t_1/2)\zeta, \qquad [7.22]$$

by applying standard trigonometric identities. The final result is obtained by assuming $\zeta \ll 1$. The first I_x term represents the desired heteronuclear correlation resonance peak, with slightly reduced intensity. The second I_x term represents an in-phase doublet split by the heteronuclear scalar coupling constant. The I_y term generates dispersive doublets at frequencies $+\Omega_I/2$ and $-\Omega_I/2$ (i.e., symmetrically positioned with respect to the main heteronuclear correlation resonance peak) with apparent splitting equal to $J_{IS}/2$ and amplitudes $\zeta/4$. For $\zeta = 20°$ (0.35 radians), the relative amplitudes of the resonance peaks are 0.97, 0.015, and 0.09. The I_y term in [7.22], which represents the largest artifact, is eliminated by the two-step phase cycle. The second I_x term cannot be eliminated by phase cycling (or field-gradient pulses); however, this term normally is small if the pulse lengths are measured carefully.

Artifacts associated with 180° pulses can also be eliminated by using field-gradient pulses (Section 4.3.3.2) and composite pulses (Section 3.4.2). Composite pulses are beneficial particularly for the 180°(S) pulses during the INEPT sequences of the HSQC experiment (Fig. 7.1b), because optimal inversion of the S_z operator component (over a typically wide spectral width) directly improves the sensitivity of the experiment.

7.1.2.2 ^{13}C Scalar Coupling and Multiplet Structure   The preceding analyses of the HMQC and HSQC experiments have indicated the multiplet structure arising as a result of heteronuclear *IS* and homonuclear *IK* scalar coupling interactions. For proteins enriched in ^{13}C, additional scalar coupling interactions must be considered.

In most of the other experiments designed for application to ^{13}C-labeled proteins, aliphatic ^{13}C$^{\alpha}$ and ^{13}CO spins are treated theoretically as different nuclear species because of the large difference in resonance frequencies ($\sim$100 ppm). The ^{13}C$^{\alpha}$ and ^{13}CO spins can be decoupled by applying selective 180° pulses or semiselective decoupling schemes to the ^{13}CO spins during t_1. However, rf fields applied at aliphatic ^{13}C frequencies can have significant effects at the carbonyl frequencies (and vice versa) that must be considered in practice. First, the carbonyl (aliphatic) pulses should minimally excite the aliphatic (carbonyl) spins, either by adjusting the power and duration of rectangular pulses as described in Section 3.4.1 or by using selective pulses (Section 3.4.4). Second, phase errors and frequency shifts caused by off-resonance effects of carbonyl (aliphatic) rf fields applied during aliphatic (carbonyl) ^{13}C evolution periods should be compensated in the pulse sequence whenever possible (Section 3.4.1).

In ^{1}H–^{15}N heteronuclear correlation NMR spectroscopy of ^{15}N/^{13}C double-labeled proteins, the ^{13}C–^{15}N scalar coupling interaction must be decoupled. In most instances, decoupling of ^{13}CO–^{15}N and aliphatic ^{13}C–^{15}N interactions is performed independently by using selective 180° pulses or semiselective composite pulse decoupling sequences. In ^{1}H–^{13}C heteronuclear correlation NMR spectroscopy of ^{15}N/^{13}C double-labeled proteins, an additional 180°(^{15}N) pulse is applied in the middle of the ^{13}C evolution period, or a broadband ^{15}N decoupling scheme is applied throughout t_1, in order to decouple the ^{15}N and ^{13}C spins.

These methods clearly will not work in the case of homonuclear ^{13}C–^{13}C couplings between aliphatic ^{13}C spins because all aliphatic ^{13}C spins will experience the effect of the inversion pulses. Therefore, the F_1 lineshapes in ^{1}H–^{13}C HSQC and HMQC spectra typically are composed of ^{13}C–^{13}C scalar-coupled multiplets. The multiplet structure can be collapsed by using a constant-time ^{13}C evolution period as described in Section 7.1.3.1.

7.1.2.3 Solvent Suppression ^{1}H–^{13}C HMQC and HSQC spectra of ^{13}C-labeled proteins normally are recorded using proteins in D_2O solution, and presaturation of the residual HDO solvent resonance usually is satisfactory. ^{1}H–^{13}C HMQC and HSQC spectra of ^{13}C/^{15}N-labeled proteins in H_2O solution can be recorded by using presaturation of the solvent resonance; however, pulsed field-gradient techniques for solvent suppression are preferable to avoid obscuring ^{1}H$^\alpha$–^{13}C$^\alpha$ correlations. For ^{1}H–^{15}N HMQC and HSQC spectra of ^{15}N- and ^{13}C/^{15}N-labeled proteins acquired in H_2O solution, presaturation of the solvent signal usually is avoided to minimize solvent saturation transfer to the amide protons that reduce the signal intensities (*20–22*).

Although jump–return versions of the HMQC experiment have been designed, the most effective solvent suppression schemes for ^{1}H–^{15}N heteronuclear NMR spectroscopy incorporate spin-lock purge pulses (*23*) or field-gradient pulses (*20,22,24*) (Section 3.5.3) into the HSQC experiment. Use of these techniques is facilitated because evolution under the ^{1}H–^{15}N scalar coupling Hamiltonian provides an efficient mechanism for independently manipulating the protein and solvent resonances. For example, evolution through the spin-echo portion of the initial INEPT sequence of the HSQC experiment yields

$$-I_y \xrightarrow{\tau - \pi I_x, \pi S_x - \tau} -2I_x S_z;$$

$$-W_y \xrightarrow{\tau - \pi W_x - \tau} W_y,$$

[7.23]

in which $I = {}^1H^N$ and $S = {}^{15}N$ are the solute spins, and W designates the solvent ^{1}H spins. The resulting solute and solvent ^{1}H operators are orthogonal. At this point, a purge pulse applied with x phase spin-locks the $2I_x S_z$ coherence and dephases the W_y coherence. Alternatively, a 90°_y pulse applied to the proton spins converts the $-2I_x S_z$ coherence into longitudinal two-spin order, $2I_z S_z$, without affecting the W_y operator. A subsequent z-axis field gradient or homospoil pulse selectively dephases the solvent magnetization without affecting the solute coherence. Finally, if a selective 90°_{-y} pulse is applied to the solvent spins prior to the first nonselective 90°_x pulse, then evolution through the initial INEPT sequence gives

$$I_z \xrightarrow{\text{INEPT}} -2I_z S_y;$$

$$W_z \xrightarrow{\frac{\pi}{2}W_{-y} - \text{INEPT}} W_z.$$

[7.24]

This pulse sequence fragment is the basis for "water flip-back" techniques, because the solvent magnetization is returned to the z axis by the INEPT sequence; similar concepts are used during the remainder of the pulse sequence to ensure that the solvent signal remained aligned along the z axis during the acquisition period.

7.1.2.4 Relaxation During HMQC and HSQC Experiments The effective F_1 linewidths of resonance signals in HMQC and HSQC spectra depend on the transverse relaxation rates of the coherences present during the chemical-shift evolution period plus the contributions from inhomogeneous broadening (*10,11*). Therefore, the linewidth in the F_1 dimension of a HMQC spectrum is determined by the relaxation rate of the heteronuclear MQ coherence, $2I_xS_y$; the F_1 linewidth of a HSQC spectrum is determined by the relaxation rate of the heteronuclear SQ coherence under free-precession conditions during t_1; and the F_1 linewidth of a decoupled HSQC spectrum is determined by the relaxation rate of in-phase SQ coherence in the absence of IS scalar coupling. The F_1 linewidth of a constant-time HSQC spectrum is determined only by inhomogenous broadening. In the following, the appropriate relaxation rate constants for the heteronuclear SQ and MQ operators are calculated using methods outlined in Chapter 5.

The average transverse relaxation rates of the S spin SQ coherence (during coherent decoupling), S spin SQ coherence (during free precession), heteronuclear MQ coherence (during free-precession), and I spin SQ coherence (during free precession) are given by

$$R_{2S} = R_2^{IS}(S) + R_2^{CSA}(S)$$

$$= \frac{d_{IS}}{24}\{4J(0) + J(\omega_I - \omega_S) + 3J(\omega_S) + 3J(\omega_I) + 6J(\omega_I + \omega_S)\}$$

$$+ \frac{d_{CSA}}{6}\{4J(0) + 3J(\omega_S)\}; \qquad\qquad [7.25]$$

$$\overline{R}_{2S} = \frac{1}{2}[R_2^{IS}(2I_zS^+) + R_2^{IS}(S) + R_1^{IK}(I)] + R_2^{CSA}(S)$$

$$= \frac{d_{IS}}{24}\{4J(0) + J(\omega_I - \omega_S) + 3J(\omega_S) + 3J(\omega_I) + 6J(\omega_I + \omega_S)\}$$

$$+ \frac{d_{CSA}}{6}\{4J(0) + 3J(\omega_S)\}$$

$$+ \frac{1}{24}\sum_k d_{ik}\{J(0) + 3J(\omega_I) + 6J(2\omega_I)\}; \qquad\qquad [7.26]$$

$$R_{2MQ} = \frac{1}{2}[R_2^{IS}(ZQ) + R_2^{IS}(DQ)] + R_2^{IK}(I) + R_2^{CSA}(S)$$

$$= \frac{d_{IS}}{24}\{J(\omega_I - \omega_S) + 3J(\omega_I) + 3J(\omega_S) + 6J(\omega_I + \omega_S)\}$$

$$+ \frac{d_{CSA}}{6}\{4J(0) + 3J(\omega_S)\}$$

$$+ \frac{1}{24}\sum_k d_{Ik}\{5J(0) + 9J(\omega_I) + 6J(2\omega_I)\}; \qquad [7.27]$$

$$\overline{R}_{2I} = \frac{1}{2}[R_2^{IS}(2I^+S_z) + R_2^{IS}(I)] + R_2^{IK}(I)$$

$$= \frac{d_{IS}}{24}\{4J(0) + J(\omega_I - \omega_S) + 3J(\omega_S) + 3J(\omega_I) + 6J(\omega_I + \omega_S)\}$$

$$+ \frac{1}{24}\sum_k d_{Ik}\{5J(0) + 9J(\omega_I) + 6J(2\omega_I)\}, \qquad [7.28]$$

respectively, in which the individual relaxation rate constants are obtained from Tables 5.5 and 5.7, $J(\omega)$ is given by [5.70], and the summations, Σ, include all the homonuclear $k \neq I$ spins. Equations [7.25]–[7.28] are subject to the following assumptions: (1) the S spin relaxes by dipole–dipole interactions with the directly attached I spin and through chemical-shift anisotropy (CSA), (2) the I spins relax through dipole–dipole interactions with the S spins and with k additional remote protons, (3) evolution of the scalar coupling interaction during free precession averages the relaxation rates of in-phase and antiphase operators as described in Section 5.4.2 (indicated by overbars), (4) coherent decoupling suppresses averaging by the scalar coupling interaction, and (5) $\omega_I \approx \omega_K$. The terms containing d_{IS} arise from heteronuclear dipolar coupling between the scalar-coupled I and S spins, terms containing d_{CSA} arise from CSA of the S spin, and the terms containing d_{Ik} reflect the homonuclear dipolar coupling between I spins (Section 5.4.1 and Section 5.4.4). In ^{13}C- or ^{13}C/^{15}N-labeled samples, the S spin (either ^{13}C or ^{15}N) has additional dipolar interactions with nearby, predominantly directly bonded, ^{13}C spins, designated as R spins. These interactions are smaller than the dipolar IS interaction by a factor of

$$\frac{d_{RS}}{d_{IS}} = \frac{\gamma_R^2 r_{IS}^6}{\gamma_I^2 r_{RS}^6}, \qquad [7.29]$$

and are neglected in the following discussion.

In the limit of slow overall tumbling, which typically applies for proteins, $\omega_I \tau_c \gg \omega_S \tau_c \gg 1$, and $J(0) \gg J(\omega_S) \gg J(\omega_I) \approx J(\omega_I \pm \omega_S)$. The relaxation rates are approximated by

$$R_{2S} = \frac{\tau_c}{15}[d_{IS} + 4d_{CSA}]; \qquad [7.30]$$

$$\overline{R}_{2S} = \frac{\tau_c}{15}\left[d_{IS} + 4d_{CSA} + \frac{1}{4}\sum_k d_{Ik}\right]; \qquad [7.31]$$

$$R_{2MQ} = \frac{\tau_c}{3}\left[\frac{4}{5}d_{CSA} + \frac{1}{4}\sum_k d_{Ik}\right]; \qquad [7.32]$$

$$\overline{R}_{2I} = \frac{\tau_c}{3}\left[\frac{d_{IS}}{5} + \frac{1}{4}\sum_k d_{Ik}\right]. \qquad [7.33]$$

For backbone amide moieties in proteins, $d_{CSA}/d_{IS} = 0.055$ (assuming $\Delta\sigma = -160$ ppm for ^{15}N and $B_0 = 11.74$ T) and $d_{IS}/\sum d_{Ik} \approx 0.3$. Therefore, $\overline{R}_{2I} > R_{2MQ} \gtrsim \overline{R}_{2S} \gtrsim R_{2S}$, and linewidths are narrower in the F_1 dimension of a ^{1}H–^{15}N HSQC spectrum than of a HMQC spectrum. Decoupling of the I and S spins during t_1 eliminates broadening due to longitudinal relaxation of I_z, and consequently results in even narrower linewidths in decoupled ^{1}H–^{15}N HSQC spectra. For ^{1}H–^{13}C methine moieties, $d_{CSA}/d_{IS} = 0.002$ (assuming $\Delta\sigma = 25$ ppm for ^{13}C and $B_0 = 11.74$ T) and $d_{IS}/\sum d_{Ik} \approx 1.4$. Therefore, $\overline{R}_{2I} > R_{2MQ} \approx \overline{R}_{2S} \approx R_{2S}$, and linewidth differences in ^{1}H–^{13}C HSQC and ^{1}H–^{13}C HMQC spectra are not as pronounced.

The different relaxation properties of the HMQC and HSQC experiments are emphasized in Fig. 7.4, which shows the t_1 interferograms through a selected amide proton resonance, and their corresponding Fourier transforms, for the spectra shown in Fig. 7.3. The observed linewidths in Fig. 7.4a–c are consistent with values of 4.9, 3.0, and 2.0 Hz respectively calculated from [7.32], [7.31], and [7.30] by using $\tau_c = 4.1$ ns. The dispersive contribution associated with the homonuclear J_{Ik} coupling is clearly visible in the HMQC spectrum (Fig. 7.4a). As expected, the interferogram for the constant-time HSQC experiment (Fig. 7.4d) exhibits very little decay, and the linewidth in the transformed spectrum is dominated by the apodization applied during processing.

7.1.2.5 Folding and Aliasing

For a fixed number of t_1 increments, the digital resolution of the F_1 dimension of a HSQC (or other) experiment generally can be increased by reduction of the spectral width to less

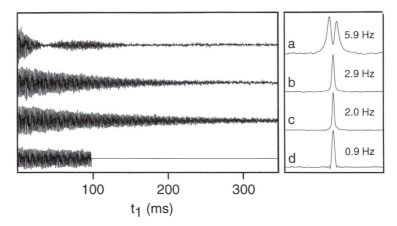

FIGURE 7.4 The t_1 interferograms, and their resulting Fourier transforms, taken through the amide proton resonance of Ile36 in the same ^1H–^{15}N heteronuclear correlation spectra as illustrated in Fig. 7.3a–d, respectively. The F_1 linewidth at half-height is indicated beside each peak. Linewidths were measured by curve-fitting the decay of the t_1 interferograms. For (d) the indicated linewidth represents inhomogenous broadening.

than the actual maximum frequency range of the resonances of interest. However, if the maximum number of t_1 increments is being utilized in a constant-time experiment ($t_{1,max} = T$), the digital resolution cannot be improved because reducing the spectral width requires that the number of t_1 increments be reduced to maintain $t_{1,max} \leq T$.

Whenever the spectral width is reduced, resonance signals from outside the acquired spectral window will be folded or aliased in the NMR spectrum (Sections 3.2.1 and 4.3.4.3). Data acquired with the hypercomplex methods (States or TPPI–States) are aliased, in which resonances upfield (downfield) of the acquired spectral window appear at the downfield (upfield) side of the spectrum. Folding, in which resonances upfield (downfield) of the acquired spectral window fold back into the spectrum at the upfield (downfield) edge, occurs for real acquisition (TPPI). If the initial sampling delay in the aliased dimension is set to $1/(2SW)$, peaks that have been folded or aliased an odd number of times have phase opposite those of peaks that have not been aliased or have been aliased an even number of times.

The effect of reducing the ^{13}C spectral width is shown schematically in Fig. 7.5 for hypercomplex and real acquisition in t_1. Clearly, the empirical relationship between ^1H and ^{13}C chemical shifts in proteins allows extensive aliasing via hypercomplex acquisition in the ^{13}C dimensions of 2D,

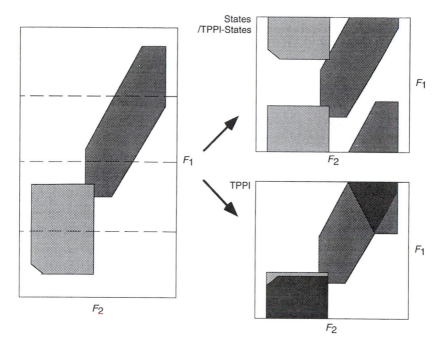

FIGURE 7.5 A schematic illustration of the effect of halving the F_1 spectral width in a ^{1}H–^{13}C heteronuclear correlation experiment, for both hypercomplex (States or TPPI–States) and real (TPPI) acquisition in t_1.

3D, and 4D NMR spectra. The ^{13}C spectral width can be readily set to as low as 20 ppm without adversely complicating the interpretation of the spectrum. Each apparent ^{13}C frequency in the final spectrum corresponds to several ^{13}C chemical shifts, separated by intervals equal to the acquired ^{13}C spectral width. Identification of the true ^{13}C chemical shift of a given cross-peak is determined from the associated aliphatic ^{1}H chemical shift together with knowledge of the sequential resonance assignments. Aliasing also can be used advantageously in ^{15}N correlation experiments (particularly to alias upfield Arg ^{15}N$^\varepsilon$ resonances).

The use of aliasing is illustrated in practice for the ^{1}H–^{13}C HSQC spectra of ubiquitin in Fig. 7.6. Both experiments were acquired using a simple HSQC pulse sequence (Fig. 7.1b) and the TPPI–States hypercomplex method of quadrature detection in t_1. The improvement in digital resolution can be seen clearly in the aliased spectrum acquired with the reduced t_1 spectral width as more of the homonuclear ^{13}C–^{13}C couplings are resolved in this spectrum.

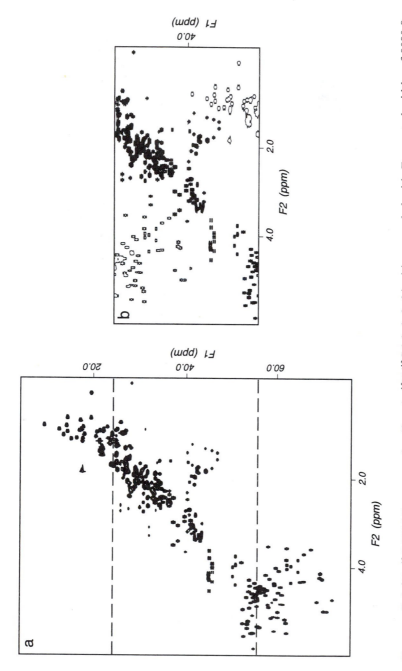

FIGURE 7.6 ^1H–^{13}C HSQC spectra of uniformly ^{15}N/^{13}C-labeled ubiquitin, recorded with F_1 spectral widths of 9090.9 Hz (72.3 ppm) (a) and 3971.4 Hz (31.6 ppm) (b). Negative cross-peaks in (b), which correspond to resonances that have been aliased in the F_1 dimension, are plotted with a single level only.

One of the main advantages of using a narrower ^{13}C spectral width is that a given digital resolution (set by $t_{1,max}$) can be achieved by using fewer t_1 increments. Consequently, for a fixed total acquisition time, more transients can be acquired per increment, more extensive phase cycling can be employed, and data storage requirements will be reduced. For given values of $t_{1,max}$ and the total acquisition time, reducing the spectral width and increasing the number of transients per increment does not affect sensitivity of the NMR experiment, because proportionally fewer increments are acquired. These features are particularly important when recording 3D and 4D spectra.

7.1.2.6 Processing HMQC and HSQC Experiments HMQC and HSQC experiments produce lineshapes that are predominately in-phase and absorptive in both the F_1 and F_2 dimensions. Accordingly, processing HMQC and HSQC spectra is relatively straightforward and does not require careful optimization of $t_{1,max}$ and window functions to avoid self-cancellation of antiphase lineshapes (Section 6.2.1.3). In F_2, the acquisition dimension, the FID will rarely be truncated, and exponential matched filtering or Lorentzian-to-Gaussian transformation are satisfactory. In the F_1 dimension, similar window functions are satisfactory if the interferograms are not truncated. More commonly, the interferograms will be truncated, and Kaiser, Hamming, or cosine-bell window functions provide satisfactory apodization (Section 3.3.2.2). Severely truncated data frequently are extended by linear prediction prior to Fourier transformation or are analyzed by maximum entropy reconstruction (Section 3.3.4). Interferograms in constant-time HSQC experiments are ideal for mirror-image linear prediction (Section 3.3.4.1).

7.1.3 VARIANT HSQC EXPERIMENTS

7.1.3.1 The Constant-Time ^1H–^{13}C HSQC Experiment As noted in Section 7.1.2.2, decoupling aliphatic ^{13}C–^{13}C scalar coupling interactions during the t_1 period of conventional ^1H–^{13}C HSQC or HMQC experiments is not feasible. Multiplet structure in the F_1 dimension due to these couplings is eliminated by using a constant-time ^{13}C evolution period (*16,17*). The constant-time ^1H–^{13}C HSQC (CT-HSQC) experiment illustrated in Fig. 7.7 differs from the pulse sequence of Fig. 7.1d by addition of ^{15}N coherent decoupling throughout the constant-time period T and addition of two carbonyl 180° pulses. Evolution under the ^{13}CO–^{13}C scalar coupling Hamiltonian is refocused during the initial t_1 fraction of the constant-time evolution period by the 180° pulse applied to the carbonyl spins, and during the remaining fraction of the constant-time period, $T - t_1$, by the

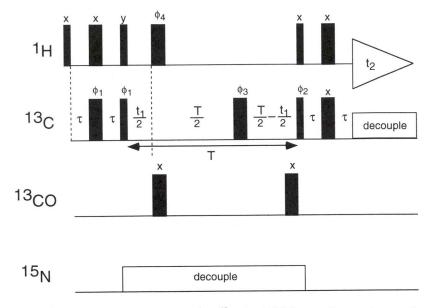

FIGURE 7.7 Pulse sequence for the ^{1}H–^{13}C CT-HSQC experiment. Appropriate phase cycling is $\phi_1 = x, -x$; $\phi_2 = 8(x), 8(-x)$; $\phi_3 = 2(x), 2(y), 2(-x), 2(-y)$; $\phi_4 = 16(y), 16(-y)$; and receiver $= 2(x, -x, -x, x), 2(-x, x, x, -x)$. If required, this 32-step phase cycle may be reduced by a factor of 2 by eliminating cycling of ϕ_4. The first and last 180°(^{13}C) pulses are best applied as composite pulses of the type 90°_x–180°_y–90°_x, in order to minimize resonance offset and rf inhomogeneity effects. ^{15}N decoupling during the constant-time evolution period can be accomplished using a WALTZ-16 or DIPSI-2 decoupling scheme, and ^{13}C decoupling during the t_2 ^{1}H data acquisition period can be accomplished with a GARP-1 decoupling scheme.

180° pulse applied to the aliphatic carbon spins. The ^{13}C–^{13}C scalar coupling interaction between the S spin and n other aliphatic ^{13}C spins, R, with coupling constant J_{CC}, is active during the entire period T. The net evolution of $2I_zS_y$ coherence during T is obtained by generalizing [7.17]

$$-2I_zS_y \xrightarrow{T} -2I_zS_y \cos(\Omega_S t_1) \cos^n(\pi J_{CC}T) + 2I_zS_x \sin(\Omega_S t_1) \cos^n(\pi J_{CC}T)$$

$$+ \text{antiphase terms.} \qquad [7.34]$$

The anti-phase terms (of the form $4I_zS_xR_{1z}$, $8I_zS_yR_{1z}R_{2z}$, etc.) are not converted back to observable proton magnetization by the reverse INEPT sequence following the constant-time evolution period, and can be ignored. The ^{13}C magnetization present at the end of the constant-time evolution period is modulated only by its chemical shift during t_1; therefore, cross-peaks in the 2D correlation spectrum appear as singlets in the F_1 dimension.

To achieve effective decoupling of the $^{13}C^\alpha$–^{13}CO scalar coupling interaction, the rectangular $180°(^{13}C^\alpha)$ pulse applied during T of Fig. 7.7 has a field strength given by $\gamma B_1 = \Omega/\sqrt{3}$, where Ω is the offset between the ^{13}CO and aliphatic ^{13}C carrier frequencies, and selective (sinc) $180°(^{13}CO)$ pulses are used (Section 7.1.2.2). The second $180°(^{13}CO)$ pulse has no effect on the product operator analysis of the pulse sequence and serves to refocus aliphatic ^{13}C evolution caused by the off-resonance effects of the first $180°(^{13}CO)$ pulse. If the second selective $180°(^{13}CO)$ pulse were omitted, the resulting spectrum would have a frequency-dependent phase error in the F_1 dimension (Section 3.4.1). Evolution of the $^{13}C^\alpha$–^{13}CO scalar coupling interaction is relatively unimportant during the short 2τ INEPT delays, so the aliphatic ^{13}C pulses may be applied at full power to maximize excitation of the aliphatic spins.

The uniformity of one-bond aliphatic ^{13}C–^{13}C couplings (J_{CC} ranges from 32 to 40 Hz) facilitates optimization of the length of T to maximize the $\cos^n(\pi J_{CC} T)$ factor in [7.34]. The experiment also can provide information on the number of aliphatic carbons attached to a given ^{13}C nucleus. As can be seen from [7.34], if $T = 1/J_{CC}$ the sign of the ^{13}C magnetization will be opposite for carbons coupled to an odd, relative to an even, number of other aliphatic carbons. If, on the other hand, $T = 2/J_{CC}$, all cross-peaks will have the same sign.

The resolution-enhancement and spectral-editing features of the CT-HSQC experiment are illustrated in Fig. 7.8, which compares ^1H–^{13}C HSQC spectra of uniformly $^{15}N/^{13}C$-labeled ubiquitin acquired using the conventional HSQC sequence of Fig. 7.1b with spectra acquired using the CT-HSQC scheme of Fig. 7.7 and constant-time evolution periods, T, of 27 and 54 ms. Expansions of the $^1H^\alpha$–$^{13}C^\alpha$ regions of the spectra are shown in Fig. 7.9. The conventional HSQC spectrum (Fig. 7.8a) and the CT-HSQC spectrum acquired with $T = 54$ ms (Fig. 7.8c) were recorded with identical digital resolution in the t_1 dimension (480 complex t_1 data points were acquired with $t_{1,max} = 52.8$ ms), whereas the CT-HSQC spectrum acquired using $T = 27$ ms (Fig. 7.8b) was recorded with one-half the digital resolution of the other two experiments (240 complex t_1 data points were acquired with $t_{1,max} = 26.4$ ms). Each spectrum was zero-filled to give a final F_1 digital resolution of 8.9 Hz.

The resolution obtained using $T = 2/J_{CC} \approx 54$ ms is clearly greater than that obtained using $T = 1/J_{CC} \approx 27$ ms. However, ^{13}C transverse relaxation during the constant-time period attenuates the observable signal by a factor of $\exp(-R_2 T)$. For proteins larger than ubiquitin, this attenua-

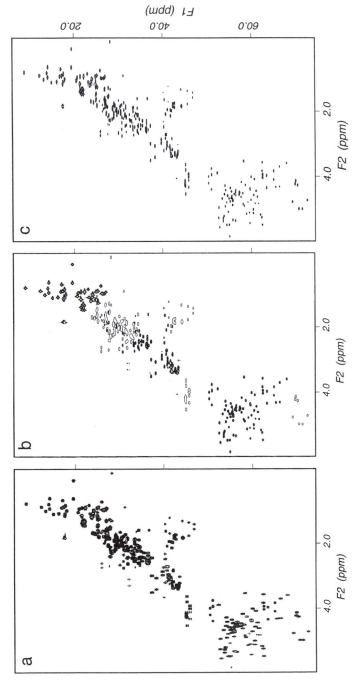

FIGURE 7.8 Comparison of the ^{1}H–^{13}C HSQC spectra of uniformly ^{15}N/^{13}C-labeled ubiquitin recorded using the conventional HSQC experiment (a), and the CT-HSQC experiment with $T = 27$ ms (b), and $T = 54$ ms (c). The CT-HSQC spectra were acquired with the initial sampling delay, $t_1(0)$, equal to zero, so that no phase correction was required in the F_1 dimension. Negative cross-peaks in (b), corresponding to ^{13}C nuclei that are coupled to zero or two other aliphatic carbons, are plotted with a single level only. Expansions of the ^{1}H$^\alpha$–^{13}C$^\alpha$ regions of these spectra are shown in Fig. 7.9.

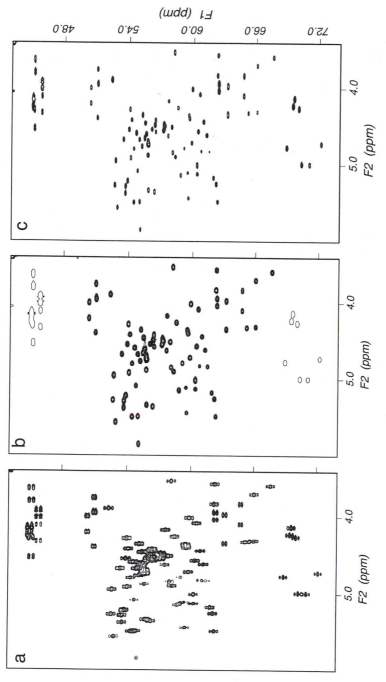

FIGURE 7.9 Expansions of the ^1H_$^{13}C^\alpha$ regions of the three spectra (a–c) shown in Fig. 7.8.

tion can be significant, even for $T = 27$ ms. To obtain the maximum ^{13}C resolution, the maximum number of t_1 increments must be acquired. In 3D or 4D experiments, the number of increments in the indirectly detected dimensions is limited by the time available for total acquisition of the spectrum and ^{13}C constant-time evolution periods are limited practically to 27 ms.

7.1.3.2 Sensitivity Improvement in Heteronuclear Correlation Experiments In the conventional HSQC experiment (Fig. 7.1b), heteronuclear SQ coherence evolves under the influence of the S spin chemical-shift Hamiltonian during the t_1 evolution period to yield *two* orthogonal terms proportional to $2I_zS_y$ and $2I_zS_x$ [7.8]. The second term is not refocused by the reverse INEPT sequence and does not contribute to the final observed magnetization [7.8] (this is essential if amplitude-modulated, pure phase spectra are to be recorded). Therefore, on average, one-half of the initial I spin polarization does *not* contribute to the detected signal.

Modification of the HSQC experiment to permit refocusing and detection of *both* orthogonal transverse magnetization components forms the basis of a class of heteronuclear correlation experiments, developed originally by Rance and coworkers, that can provide sensitivity improvements by factors of up to $\sqrt{2}$ relative to the conventional experiments (25–27). This technology has been denoted "preservation of equivalent pathways" (PEP) (28). The principle of the PEP technique for sensitivity improvement will be demonstrated for the PEP-HSQC experiment illustrated in Fig. 7.10a.

The evolution of the density operator for the enhanced PEP-HSQC sequence proceeds exactly as for the conventional HSQC experiment up to time point a in Fig. 7.10a. For an I_nS spin system, the evolution through the remainder of the pulse sequence is given by

$$2I_zS_y \cos(\Omega_S t_1) - 2I_zS_x \sin(\Omega_S t_1)$$

$$\xrightarrow{\frac{\pi}{2}I_x, \frac{\pi}{2}S_x} -2I_yS_z \cos(\Omega_S t_1) + 2I_yS_x \sin(\Omega_S t_1)$$

$$\xrightarrow{\tau - \pi I_y, \pi S_x - \tau} - I_x \cos(\Omega_S t_1) + \delta_{1,n} 2I_yS_x \sin(\Omega_S t_1)$$

$$\xrightarrow{\frac{\pi}{2}I_y, \frac{\pi}{2}S_y} I_z \cos(\Omega_S t_1) - \delta_{1,n} 2I_yS_z \sin(\Omega_S t_1)$$

$$\xrightarrow{\tau - \pi I_y, \pi S_y - \tau} - I_z \cos(\Omega_S t_1) + \delta_{1,n} I_x \sin(\Omega_S t_1)$$

$$\xrightarrow{-\frac{\pi}{2}I_x} - I_y \cos(\Omega_S t_1) - \delta_{1,n} I_x \sin(\Omega_S t_1), \qquad [7.35]$$

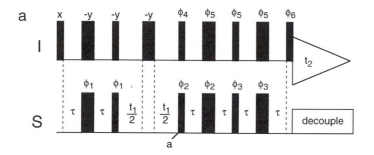

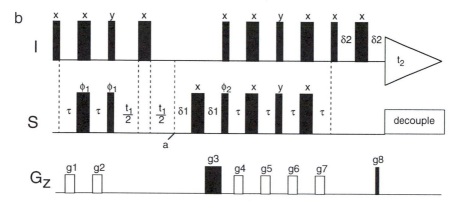

FIGURE 7.10 Pulse sequence for the (a) PEP-HSQC and (b) PFG-PEP-HSQC experiment. (a) In the original version, the phase cycling is $\phi_1 = x, -x$; $\phi_2 = 2(x)$, $2(-x)$; $\phi_3 = 2(y), 2(-y)$; $\phi_4 = x$; $\phi_5 = y$; $\phi_6 = -x$; and receiver $= x, -x, -x, x$. Quadrature detection in the F_1 dimension is performed using the TPPI–States method. In the modified version of the experiment the phase cycling is $\phi_1 = x$; $\phi_2 = x, y$; $\phi_3 = y, -x$; $\phi_4 = y, x$; $\phi_5 = x, -y$; $\phi_6 = -y, -x$; and receiver $= x$, $-x, -x, x$. For each t_1 increment, ϕ_1 and the receiver are inverted; no other quadrature detection scheme is required. (b) The phase cycling is $\phi_1 = x$; $\phi_2 = x$; and receiver $= x$. For each t_1 increment, ϕ_1 and the receiver are inverted; no other quadrature detection scheme is required. Additional phase cycling of the 180° pulses also can be added if required to suppress artifacts. The delays, 2τ, are set to $1/(2J_{IS})$. As discussed in the text, sensitivity improvement requires two experiments to be recorded, the first with the preceding phase cycles, and a second with inversion of the phase of the 90°(S) pulse immediately following the t_1 evolution period (ϕ_2 for the PEP-HSQC experiment and for the PFG-PEP-HSQC experiment). In the PFG-PEP-HSQC experiment, gradient g_8 is inverted at the same time as ϕ_2. The two data sets are stored separately and then combined as described in the text.

in which $\delta_{1,n} = \cos^{n-1}(2\pi J_{IS}\tau)$ for $2\tau = 1/(2J_{IS})$ results from evolution of the MQ coherence due to passive scalar coupling interactions. The sine-modulated component is stored as multiple-quantum coherence, whereas the cosine-modulated component is refocused to I spin magnetization; subsequently, the cosine-modulated I spin coherence is stored as longitudinal magnetization and the sine-modulated component is refocused to I spin coherence. Refocusing of both orthogonal signal components following t_1 is possible only for IS ($n = 1$) spin systems. For example, evolution through the pulse sequence for an I_2S spin system yields

$$2I_{1z}S_y \cos(\Omega_S t_1) - 2I_{1z}S_x \sin(\Omega_S t_1)$$

$$\xrightarrow{\frac{\pi}{2}I_x \cdot \frac{\pi}{2}S_x} -2I_{1y}S_z \cos(\Omega_S t_1) + 2I_{1y}S_x \sin(\Omega_S t_1)$$

$$\xrightarrow{\tau-\pi I_y,\pi S_x-\tau} -I_{1x} \cos(\Omega_S t_1) + 4I_{1y}I_{2z}S_y \sin(\Omega_S t_1)$$

$$\xrightarrow{\frac{\pi}{2}I_y,\frac{\pi}{2}S_y} I_{1z} \cos(\Omega_S t_1) + 4I_{1y}I_{2x}S_y \sin(\Omega_S t_1)$$

$$\xrightarrow{\tau-\pi I_y,\pi S_y-\tau} -I_{1z} \cos(\Omega_S t_1) - 4I_{1y}I_{2x}S_y \sin(\Omega_S t_1)$$

$$\xrightarrow{-\frac{\pi}{2}I_x} -I_y \cos(\Omega_S t_1) + 4I_{1z}I_{2x}S_y \sin(\Omega_S t_1). \qquad [7.36]$$

The first term on the last line of [7.36] is observable magnetization; in accordance with [7.35], the second term is unobservable multiple-quantum coherence.

The resultant $-I_y \cos(\Omega_S t_1)$ and $-\delta_{1,n}I_x \sin(\Omega_S t_1)$ terms in [7.35] describe orthogonal in-phase I spin magnetization components that have evolved at the frequency of the S spin during t_1. The final two terms in [7.35] represent superposed observable signals with a 90° phase difference in both dimensions of a 2D spectrum (resulting in phase-twisted lineshapes). In order to separate the two orthogonal terms and obtain purely absorptive lineshapes, a second experiment is performed in which the phase of the $90^\circ_{\phi_2}(S)$ pulse immediately following the t_1 evolution period is inverted. The relevant operator terms are (again beginning with the operators present at time a, given at the end of [7.8])

$$2I_zS_y \cos(\Omega_S t_1) - 2I_zS_x \sin(\Omega_S t_1)$$

$$\xrightarrow{\left(\frac{\pi}{2}\right)I_x \cdot \left(-\frac{\pi}{2}\right)S_x} 2I_yS_z \cos(\Omega_S t_1) + 2I_yS_x \sin(\Omega_S t_1)$$

$$\xrightarrow{\tau-\pi I_y,\pi S_x-\tau} I_x \cos(\Omega_S t_1) + \delta_{1,n}2I_yS_x \sin(\Omega_S t_1)$$

$$\xrightarrow{\left(\frac{\pi}{2}\right)I_y,\left(\frac{\pi}{2}\right)S_y} -I_z\cos(\Omega_S t_1) - \delta_{1,n}2I_y S_z \sin(\Omega_S t_1)$$

$$\xrightarrow{\tau-\pi I_y,\pi S_y-\tau} I_z\cos(\Omega_S t_1) - \delta_{1,n}I_x \sin(\Omega_S t_1)$$

$$\xrightarrow{\left(-\frac{\pi}{2}\right)I_x} I_y\cos(\Omega_S t_1) - \delta_{1,n}I_x \sin(\Omega_S t_1). \qquad [7.37]$$

Addition of the two data sets [7.35] and [7.37] gives the single observable term

$$-\delta_{1,n}2I_x \sin(\Omega_S t_1)\varepsilon_{MQ}, \qquad [7.38]$$

whereas subtraction of the two data sets yields the single observable term

$$-2I_y \cos(\Omega_S t_1)\varepsilon_I. \qquad [7.39]$$

The data set represented by [7.38] contains signals only from IS spin systems, because the extended reverse polarization transfer sequence implements a spin multiplicity filter. The coefficients of 2 in [7.38] and [7.39] arise because two acquisitions have been performed; in practice each experiment would be recorded with one-half of the total number of transients desired to maintain the same total acquisition time as the conventional experiment. The factors ε_{MQ} and ε_I have been introduced in [7.38] and [7.39] to account for the different relaxation rates of the sine- and cosine-modulated components. These factors are given by

$$\varepsilon_{MQ} = \exp(-2R_{2MQ}\tau) \qquad [7.40]$$

and

$$\varepsilon_I = \exp(-2R_{1I}\tau), \qquad [7.41]$$

in which R_{2MQ} is transverse relaxation rate constant for heteronuclear multiple-quantum coherence, and R_{1I} is the longitudinal relaxation rate constant of the I spin (26).

The data sets described by [7.38] and [7.39] differ in phase by 90° in both dimensions. The data sets can be transformed and phased into 2D heteronuclear correlation spectra with purely absorptive peak shapes in both dimensions (using either real or hypercomplex acquisition in t_1). The final sensitivity-enhanced spectrum is obtained by adding the two pure absorption spectra. The random noise in the two spectra described by [7.38] and [7.39] is independent and increases by $\sqrt{2}$ only in the sensitivity-enhanced spectrum. The achievable sensitivity improvement over a conventional HSQC experiment recorded in the same total time is

$$\varepsilon = \sqrt{2}(\varepsilon_I + \delta_{1,n}\varepsilon_{MQ})/2, \qquad [7.42]$$

which has a maximum value of $\sqrt{2}$ for an IS spin system.

The sensitivity improvement obtained using the PEP-HSQC pulse sequence of Fig. 7.10a, relative to the conventional HSQC pulse sequence of Fig. 7.1b, is illustrated for a ^{15}N-labeled ubiquitin sample in Fig. 7.11. The conventional HSQC spectrum was acquired using 16 transients per t_1 increment, whereas the two data sets required to produce the sensitivity-enhanced PEP-HSQC spectrum were acquired using 8 transients per t_1 increment. The total acquisition times for the conventional and PEP-HSQC spectra were therefore identical. Figure 7.11a shows F_2 cross sections through the cross-peak of Ile36 from both the conventional and enhanced HSQC spectra, together with the equivalent cross sections from the spectra resulting from addition (add) and subtraction (subtract) of the two data sets acquired using the PEP-HSQC pulse sequence. The final enhanced spectrum is simply the sum of the "add" and "subtract" spectra. Figure 7.11b shows similar cross sections through four cross-peaks corresponding to two NH_2 groups and thus illustrates the effect of the multiplicity filter of the "add" spectrum. As noted above, NH_2 resonances are suppressed in the "add" spectrum because only NH resonances are

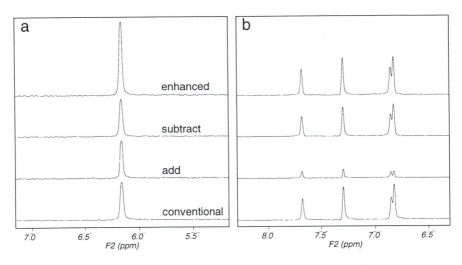

FIGURE 7.11 Cross sections through the NH resonances of Ile36 (a) and two NH_2 groups (b) from conventional (bottom) and sensitivity enhanced (top) HSQC spectra of ^{15}N-labeled ubiquitin. The middle traces were obtained from the spectra that result from addition (add) and subtraction (subtract) of the two data sets acquired with the PEP-HSQC pulse sequence; the enhanced spectrum was generated by linear combination of the "add" and "subtract" spectra. The slices are plotted on an absolute intensity scale. The sensitivity improvement of the "enhanced" spectrum over the "conventional" spectrum for the Ile36 peak is 1.39.

retained by the multiplicity filter implicit in the sensitivity enhanced sequence. In principle, these peaks should be completely absent from the "add" spectrum, but because of mismatching of the τ delays and the J_{NH} coupling, small residual components remain, as observed in Fig. 7.11b.

The original PEP method requires an increase in the length of the requisite phase cycle by a factor of 2 over the conventional pulse sequence, because a two-step phase cycle is necessary to deconvolute the orthogonal magnetization components. For example, the minimum phase cycle for the PEP-HSQC experiment discussed above is eight steps (two steps for isotope filtration, two steps for quadrature detection, and two steps for deconvolution of orthogonal magnetization components) compared with four steps for the conventional HSQC experiment (two steps for isotope filtration plus two steps for quadrature detection). The PEP method can be modified, by appropriate phase cycling (see the caption to Fig. 7.10) (29), to equalize the relaxation differences between the coherence transfer pathways used to refocus the orthogonal magnetization components. The sine- and cosine-modulated terms that result from addition and subtraction of the acquired data sets are given by

$$-I_x \sin(\Omega_S t_1)(\varepsilon_I + \delta_{1,n}\varepsilon_{MQ}); \qquad [7.43]$$

$$-I_y \cos(\Omega_S t_1)(\varepsilon_I + \delta_{1,n}\varepsilon_{MQ}), \qquad [7.44]$$

respectively. The two components have identical magnitudes and can be treated as a hypercomplex quadrature pair in the t_1 dimension (following a 90° zero-order phase correction in the acquisition dimension of the data described by [7.44]). If the data sets described by [7.38] and [7.39] were combined as a hypercomplex pair, the amplitude differences, ε_I and $\delta_{1,n}\varepsilon_{MQ}$, would result in quadrature artifacts in the F_1 dimension of the spectrum. The modified PEP-HSQC pulse sequence uses a minimum four-step phase cycle: two steps to balance the relaxation differences between the coherence-transfer pathways and obtain isotope filtration (which are performed simultaneously), and two steps for deconvolution of the orthogonal magnetization components. Separate phase cycling for quadrature detection is not necessary. Thus, the modified technique is readily applicable to 3D and 4D heteronuclear experiments in which the length of phase cycles must be minimized.

7.1.3.3 Gradient-Enhanced HSQC NMR Spectroscopy   Although pulsed field gradients can be used to generate frequency-discriminated pure absorption spectra by N/P selection, a signal-to-noise loss of $\sqrt{2}$ normally is incurred relative to techniques employing phase cycling (TPPI or hypercomplex methods) (Section 4.3.4.2). Kay and coworkers showed that pulsed field gradients can be used for coherence selection in a PEP-HSQC experiment without sacrificing sensitivity (*30*). A PFG-PEP-HSQC experiment is shown in Fig. 7.10b. Coherence selection is obtained using gradient pulses g_3 and g_8; the other gradient pulses are used to suppress artifacts associated with the 180° pulses (Section 4.3.3.2). The evolution of the density operator for the PFG-PEP-HSQC sequence proceeds exactly as for the conventional HSQC experiment up to time point a in Fig. 7.10b. Evolution through the first spin echo period yields

$$2I_z S_y \cos(\Omega_S t_1) - 2I_z S_x \sin(\Omega_S t_1) \xrightarrow{\delta_1 - \pi S_x - \delta_1}$$
$$- 2I_z S_y \cos[\Omega_S t_1 + \phi_S(z)] - 2I_z S_x \sin[\Omega_S t_1 + \phi_S(z)], \quad [7.45]$$

in which $\phi_S(z)$ is the spatially dependent phase acquired by the S spin coherence during the gradient pulse g_3. The PEP reverse INEPT sequence yields

$$-2I_z S_y \cos[\Omega_S t_1 + \phi_S(z)] - 2I_z S_x \sin[\Omega_S t_1 + \phi_S(z)] \xrightarrow{\text{PEP}}$$
$$I_y \cos[\Omega_S t_1 + \phi_S(z)]\varepsilon_I - \delta_{1,n} I_x \sin[\Omega_S t_1 + \phi_S(z)]\varepsilon_{\text{MQ}}. \quad [7.46]$$

Following the second spin-echo period and gradient pulse g_8, the I spin coherences acquire a spatially dependent phase $\phi_I(z)$. If the gradient pulses are adjusted such that $\phi_S(z) = \phi_I(z)$, the resulting observable magnetization is

$$\{I_y \cos(\Omega_S t_1) - I_x \sin(\Omega_S t_1)\}(\varepsilon_I + \delta_{1,n}\varepsilon_{\text{MQ}})/2. \quad [7.47]$$

Imbalance between the two coherence transfer pathways is suppressed by the N/P selection of the second field-gradient pulse. To deconvolute the orthogonal magnetization components, a second acquisition is performed in which the phase of the $90°_{\phi_2}(S)$ pulse immediately following the t_1 evolution period is inverted and the sign of gradient g_8 is reversed so that $\phi_S(z) = -\phi_I(z)$. The resulting observable magnetization is given by

$$\{-I_y \cos(\Omega_S t_1) - I_x \sin(\Omega_S t_1)\}(\varepsilon_I + \delta_{1,n}\varepsilon_{\text{MQ}})/2. \quad [7.48]$$

Addition and subtraction of these two data sets yields results equivalent to [7.43] and [7.44]. The resulting phase-sensitive PFG-PEP-HSQC spectrum has the same nominal sensitivity as the phase-cycled PEP-HSQC spectrum or *double* the sensitivity of a conventional PFG-HSQC spectrum. The $\sqrt{2}$ sensitivity loss associated with coherence selection by pulsed field gradients is recovered *and* the nominal $\sqrt{2}$ PEP sensitivity enhancement is obtained independently in the PFG-PEP-HSQC experiment. The gradient-enhanced pulse sequence is necessarily longer than the PEP-HSQC spectrum, because the gradient pulses required for coherence selection are placed within spin echoes in order to refocus chemical shift evolution. The sensitivity of the PFG-PEP-HSQC experiment is reduced relative to the PEP-HSQC experiment by a factor of $\exp[-2\delta_1 R_{2S} - 2\delta_2 R_{2I}]$, in which R_{2S} and R_{2I} are the relaxation rate constants for the S and I spin coherence present during the spin-echo periods (*31*).

7.2 Heteronuclear-Edited NMR Spectroscopy

Heteronuclear-edited NMR experiments represent the simplest use of heteronuclear spins to facilitate NMR spectroscopy of larger proteins (*1,2*). Three- and four-dimensional heteronuclear-edited NMR experiments resolve cross-peaks between ^{1}H spins according to the chemical shift of the heteronuclei bonded directly to the ^{1}H spins. A 3D heteronuclear-edited experiment consists of a homonuclear pulse sequence, usually a NOESY (Section 6.6.1) or TOCSY (Section 6.5) experiment, and a HSQC (or HMQC) pulse sequence (Section 7.1) catenated as discussed in Section 4.5 (*32*). A 4D heteronuclear-edited NOESY experiment consists of the catenation of a homonuclear NOESY pulse sequence and two HSQC (or HMQC) building blocks.

7.2.1 3D NOESY-HSQC SPECTROSCOPY

The basic IS ($I = {}^1$H, and $S = {}^{13}$C or ^{15}N) NOESY-HSQC experiment is illustrated in Fig. 7.12 (*33–35*). Up until time a, the sequence is a homonuclear NOESY experiment (Section 6.6.1) with S spin decoupling

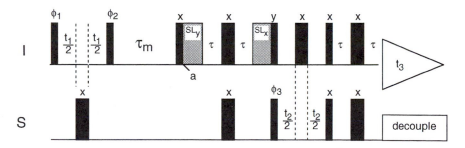

FIGURE 7.12 Pulse sequence for the 3D ^{1}H–^{15}N NOESY-HSQC experiment. The phase cycling is $\phi_1 = 2(x), 2(-x); \phi_2 = 4(x), 4(-x); \phi_3 = x, -x;$ and receiver $= x, -x, -x, x, -x, x, x, -x$. The spin-lock purge pulses (SL) are applied for 1–2 ms.

during the t_1 evolution period. Decoupling of the J_{IS} coupling interaction is achieved by application of a $180°(S)$ pulse at the midpoint of t_1 (illustrated), or by application of a composite decoupling pulse scheme throughout t_1. The $90°(I)$ pulse immediately preceding time a is equivalent to the first $90°(I)$ pulse in the HSQC experiment, and the remainder of the pulse sequence following time a is identical to the HSQC experiment (Fig. 7.1b). The product operator treatment of the NOESY-HSQC experiment is obtained by catenating the results for the NOESY and HSQC experiments. For example, for two protons spins I and K and a heteronuclear S spin that is scalar coupled only to the I spin, evolution through the pulse sequence is given by

$$I_z + K_z \xrightarrow{\frac{\pi}{2}(I_x+K_x)-t_1/2-\pi S_x-t_1/2-\frac{\pi}{2}(I_x+K_x)} I_z \cos(\Omega_I t_1) \cos(\pi J_{I\eta} t_1)$$

$$+ K_z \cos(\Omega_K t_1) \cos(\pi J_{K\lambda} t_1)$$

$$\xrightarrow{\tau_m} I_z\{a_{II}(\tau_m) \cos(\Omega_I t_1) \cos(\pi J_{I\eta} t_1)$$

$$+ a_{IK}(\tau_m) \cos(\Omega_K t_1) \cos(\pi J_{K\lambda} t_1)\}$$

$$\xrightarrow{\frac{\pi}{2}(I_x+K_x)-\tau-\pi(I_x+K_x),\pi S_x-\tau-\frac{\pi}{2}(I_y+K_y),\frac{\pi}{2}S_x}$$

$$-2I_z S_y\{a_{II}(\tau_m) \cos(\Omega_I t_1) \cos(\pi J_{I\eta} t_1)$$

$$+ a_{IK}(\tau_m) \cos(\Omega_K t_1) \cos(\pi J_{K\lambda} t_1)\}$$

$$\xrightarrow{t_2/2-\pi(I_x+K_x)-t_2/2} 2I_z S_y \cos(\Omega_S t_2)\{a_{II}(\tau_m) \cos(\Omega_I t_1) \cos(\pi J_{I\eta} t_1)$$

$$+ a_{IK}(\tau_m) \cos(\Omega_K t_1) \cos(\pi J_{K\lambda} t_1)\}$$

$$\xrightarrow{\frac{\pi}{2}(I_x+K_x),\frac{\pi}{2}S_x-\tau-\pi(I_x+K_x),\pi S_x-\tau}$$

$$-I_x \cos(\Omega_S t_2)\{a_{II}(\tau_m) \cos(\Omega_I t_1) \cos(\pi J_{I\eta} t_1)$$

$$+ a_{IK}(\tau_m) \cos(\Omega_K t_1) \cos(\pi J_{K\lambda} t_1)\} \qquad [7.49]$$

in which $2\tau = 1/(2J_{IS})$, $a_{II}(\tau_m)$, and $a_{IK}(\tau_m)$ are transfer functions for dipolar cross-relaxation (Sections 5.1.2 and 5.4.1), homonuclear scalar coupling interactions of the I and K spins with other protons are represented by terms containing $J_{I\eta}$ and $J_{K\lambda}$, and only terms leading to observable signals are derived. Magnetization originating on the K spin that is not transferred to the I spin during the NOESY mixing period is suppressed by the HSQC

isotope filter. If the K spins were coupled to S spins, analogous detectable K_x coherence terms would appear in [7.49]. The complete magnetization transfer pathway can be abbreviated as follows:

$$K_i \xrightarrow{\text{NOE}} I_j \xrightarrow{^1J_{IS}} S_j \xrightarrow{^1J_{IS}} I_j . \qquad [7.50]$$
$$(t_1) \qquad\qquad\quad (t_2) \qquad (t_3)$$

Coherence and magnetization transfer steps are indicated above the arrows, and the locations of the independent t_1, t_2, and t_3 evolution periods are indicated on the second line.

In order to obtain sufficient digital resolution in the 3D experiment, a large number of (t_1, t_2) experiments are required (typically 128×32 complex data points are acquired in t_1 and t_2, respectively); therefore, only a limited number of transients can be recorded for every (t_1, t_2) value if the entire experiment is to be completed in a reasonable time period. As a consequence, a short phase cycle must be utilized. The minimum phase cycling required for the 3D experiment involves four steps (two for the isotope filter and two for axial peak suppression in F_1); however, using the eight-step phase cycle illustrated in Fig. 7.12, a 3D spectrum can be acquired in 3–4 days. Quadrature detection in the t_1 and t_2 dimensions is achieved by shifting the phases of ϕ_1 and ϕ_3 independently in a TPPI or TPPI–States manner.

Processing NOESY-HSQC spectra follows the general methods outlined in Section 6.6.1.3 for processing NOESY spectra and Section 7.1.2.6 for processing HSQC spectra, apart from the more limited resolution of the F_1 and F_2 dimensions. In the final 3D spectrum, the $F_1(^1H)$–$F_3(^1H)$ projection corresponds to the $F_1(^1H)$–$F_2(^1H)$ region of a conventional 2D ^1H–^1H NOESY spectrum, and the $F_2(S)$–$F_3(^1H)$ projection corresponds to the $F_1(S)$–$F_2(^1H)$ dimensions of a 2D HSQC (or HMQC) correlation spectrum, as illustrated schematically in Fig. 7.13.

The NOESY-HSQC pulse sequence illustrated in Fig. 7.12 may be simplified by replacing the HSQC sequence with the HMQC sequence. From the previous discussion of the HSQC and HMQC experiments, such simplification may appear to be undesirable because of the superior resolution and relaxation properties of the HSQC experiment. However, the resolution in the heteronuclear dimension of a 3D experiment is typically limited by the digital resolution, so the detrimental effect of using the HMQC sequence is not as great in the 3D as in the 2D case. In addition, the simple HSQC sequence can be replaced with any of the variant HSQC pulse sequences (decoupled HSQC, constant-time HSQC, PEP-HSQC) as desired.

7.2.1.1 3D ^1H–^{15}N NOESY-HSQC   Presaturation of the solvent resonance normally is avoided in ^1H–^{15}N NOESY-HSQC NMR spectroscopy
of ^{15}N-labeled proteins. Instead water suppression is achieved by incorporating spin-lock purge pulses or field gradient pulses into the pulse sequence, as discussed in Section 7.1.2.3. The use of these techniques avoids
saturation of the $^1H^\alpha$ spins, and allows observation of important $^1H^N$–$^1H^\alpha$
cross peaks.

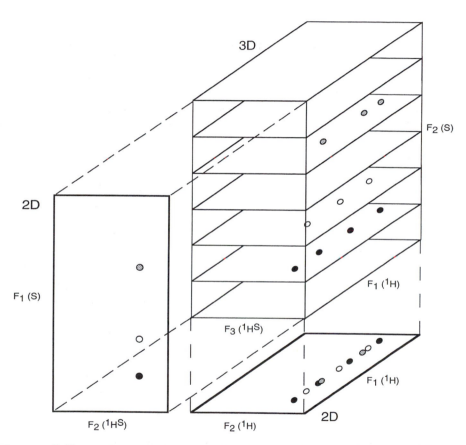

FIGURE 7.13 A schematic illustration showing the relationship between a 3D
heteronuclear-edited spectrum and 2D homonuclear and heteronuclear correlation
spectra. Cross-peaks representing three different spin systems, with degenerate
$^1H^S$ chemical shifts, but differing S spin chemical shifts, are indicated. The 3D
spectrum is represented as a series of $F_1(^1H)$–$F_3(^1H^S)$ slices edited by the chemical
shift of the directly attached S spins along the F_2 axis.

The advantage of not presaturating the water resonance in the ^1H–^{15}N NOESY-HSQC experiment is illustrated clearly by Fig. 7.14, which shows selected regions of $F_2(^{15}N)$ slices of a 3D NOESY-HSQC spectrum of ubiquitin compared with the equivalent regions from a 2D homonuclear NOESY spectrum. Several $^1H^\alpha$–$^1H^N$ cross-peaks can be seen close to the water resonance position in the NOESY-HSQC spectrum (Fig. 7.14a,b) that are not observable in the 2D homonuclear NOESY spectrum acquired using presaturation of the water resonance (Fig. 7.14c). These cross peaks would be difficult to observe if presaturation of the water resonance had been utilized in the 3D NOESY-HSQC experiment.

Although all protons are frequency-labeled during t_1, only magnetization from protons directly attached to ^{15}N is retained for detection during t_3, and therefore the only cross-peaks observed in the spectrum involve cross-relaxation to amide protons. Thus, the $F_1(^1H)$–$F_3(^1H)$ projection of the 3D spectrum corresponds to the $F_1(^1H)$–$F_2(^1H^N)$ region of a conventional 2D ^1H–^1H NOESY spectrum.

7.2.1.2 3D ^1H–^{13}C NOESY-HSQC   The ^1H–^{13}C NOESY-HSQC experiment, which typically is acquired in D_2O solution, provides NOE correlations between aliphatic protons (and between aliphatic and aromatic protons). In a ^{13}C-edited NOESY-HSQC spectrum the $F_1(^1H)$–$F_3(^1H)$ projection corresponds to the $F_1(^1H)$–$F_2(^1H^{aliphatic})$ region of a conventional 2D ^1H–^1H NOESY spectrum acquired in D_2O solution, and the $F_2(^{13}C)$–$F_3(^1H)$ projection corresponds to the 2D HSQC (or HMQC) correlation spectrum ($F_1(^{13}C)$–$F_2(^1H)$) (*36*).

For uniformly ^{13}C labeled proteins, the NOESY-HSQC experiment illustrated in Fig. 7.12 is modified to include ^{13}CO decoupling during the t_2 evolution period by using a selective composite pulse decoupling scheme such as SEDUCE-1 (*36*). Aliphatic ^{13}C decoupling during the ^1H t_1 evolution period can be accomplished by using a composite pulse decoupling scheme or more simply a composite 180°(^{13}C) pulse.

If the ^{13}C-edited NOESY-HSQC experiment utilizes a conventional HSQC sequence (Figs. 7.1b and 7.12), the maximum t_2 acquisition time must be kept shorter than $1/(2J_{CC})$ to avoid sensitivity losses due to resolution of the J_{CC} couplings in the F_2 dimension. This constraint usually is not limiting in a 3D experiment. For example, on a 500-MHz spectrometer, if 32 complex points are to be acquired in the t_2 dimension with an F_2 spectral width of 30–35 ppm, the value of $t_{2,max}$ will range from 8.5 to 7.3 ms compared with $1/(2J_{CC})$ = 14 ms. Of course, the CT-HSQC sequence (Fig. 7.7) also can be incorporated into the ^1H–^{13}C NOESY-HSQC experiment in order to completely eliminate line broadening due to J_{CC} scalar coupling (Section 7.1.3.1).

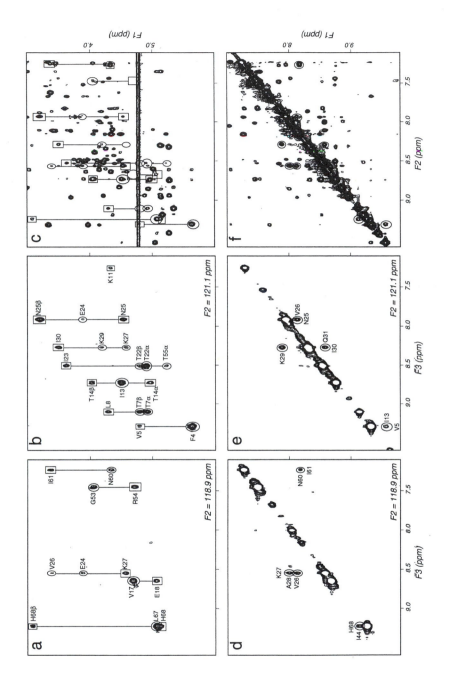

Assignment of NOE cross-peaks in a 3D ^{1}H–^{13}C NOESY-HSQC spectrum is greatly aided by the symmetry present in the spectrum. For two proximal protons, H_1 and H_2, NOE cross-peaks are expected at $[F_1(^1H_1), F_2(^{13}C_2), F_3(^1H_2)]$ and at $[F_1(^1H_2), F_2(^{13}C_1), F_3(^1H_1)]$. By searching the 3D spectrum for such symmetry-related peaks, the ^{13}C chemical shifts associated with both protons involved in the NOE interaction can therefore be identified. Knowledge of all four chemical shifts can potentially lead to an unambiguous NOE assignment. An example in which identification of the symmetry-related peak aids NOE assignment is included in Fig. 7.15, which shows selected regions from $F_1(^1H)$–$F_3(^1H)$ slices of a 3D ^{1}H–^{13}C NOESY-HSQC spectrum of ^{13}C-labeled ubiquitin, together with the equivalent region from a homonuclear ^{1}H–^{1}H NOESY spectrum of ubiquitin. The cross-peak between Lys6(^{1}H$^\alpha$) and Thr12(^{1}H$^\alpha$) is ambiguous in the homonuclear NOESY spectrum due to degeneracy of the ^{1}H$^\alpha$ chemical shifts of Lys6 and Thr66; the presence of this cross peak in Fig. 7.15a at the $F_3(^1H)$ chemical shift of Lys6, and its absence in Fig. 7.15b at the $F_3(^1H)$ chemical shift of Thr66, clearly support the assignment indicated.

7.2.2 3D TOCSY-HSQC Spectroscopy

In principle, the description of the ^{1}H–^{15}N NOESY-HSQC experiment given above also applies to the ^{1}H–^{15}N TOCSY-HSQC experiment, with the exception that the NOESY mixing period is replaced by a TOCSY isotropic mixing sequence (*1,3,34*). The information obtained from a 3D ^{1}H–^{15}N TOCSY-HSQC spectrum is the same as that obtained from the $F_1(^1H)$–$F_2(^1H^N)$ region of a conventional 2D ^{1}H–^{1}H TOCSY spectrum, but as illustrated schematically in Fig. 7.13, it is edited in a third dimension according to the ^{15}N chemical shift associated with the amide ^{1}H^N. In addition to providing intraresidue correlations that are important for the sequential assignment process, the ^{1}H–^{15}N TOCSY-HSQC experiment may be used to obtain qualitative estimates of $^3J_{H^\alpha H^\beta}$ coupling constants

---

Figure 7.14 Selected $F_1(^1H^\alpha)$–$F_3(^1H^N)$ regions from a 3D ^{1}H–^{15}N NOESY-HSQC spectrum of ^{15}N-labeled ubiquitin at $F_2(^{15}N)$ chemical shifts of 118.9 ppm (a) and 121.1 ppm (b), and the corresponding region from a 2D homonuclear NOESY spectrum (c) acquired using presaturation of the H_2O signal. Plots (d) and (e) correspond to $F_1(^1H^N)$–$F_3(^1H^N)$ regions at the same $F_2(^{15}N)$ chemical shifts as (a) and (b), respectively, and (f) corresponds to the equivalent region from the 2D homonuclear NOESY spectrum. Intraresidue NOEs are indicated by a box; inter-residue NOEs are indicated by ellipses.

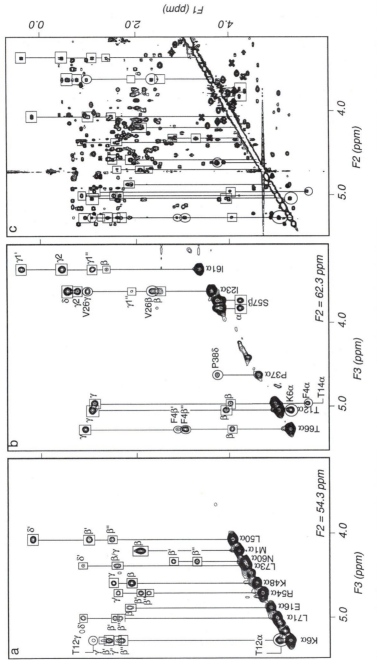

FIGURE 7.15 Selected $F_1(^1H)–F_3(^1H^\alpha)$ regions from a 3D $^1H–^{13}C$ NOESY-HSQC spectrum of ^{13}C-labeled ubiquitin in D_2O solution at $F_2(^{13}C)$ chemical shifts of 54.3 ppm (a) and 30.2/62.3 ppm (b), and the corresponding region from a 2D homonuclear NOESY spectrum (c) acquired using an unlabeled sample of ubiquitin in D_2O solution. Intraresidue NOEs are indicated by a box; interresidue NOEs are indicated by ellipses. The Lys6($^1H^\alpha$)–Thr12($^1H^\alpha$) cross-peak discussed in the text is located in the lower-left region of each spectrum.

from the relative intensities of well-resolved $^1H^N-^1H^\beta$ cross-peaks in spectra acquired with short mixing times (<35 ms) (*37*).

A pulse sequence for a $^1H-^{15}N$ TOCSY-HSQC experiment is shown in Fig. 7.16. Following the initial ^1H t_1 evolution period, a $90°(^1H)$ pulse returns the frequency-labeled magnetization to the $\pm z$ axis for the isotropic mixing period. The DIPSI-2rc isotropic mixing sequence (*38*) transfers ^1H magnetization from aliphatic spins to the corresponding intraresidue amide protons, while minimizing rotating-frame NOE effects. The $90°(^1H)$ pulse following the mixing sequence rotates the resulting z-magnetization back into the transverse plane, and is therefore analogous to the first pulse in an HSQC experiment, as discussed above for the $^1H-^{15}N$ NOESY-HSQC experiment. The remainder of the sequence is equivalent to a $^1H-^{15}N$ HSQC experiment. The complete magnetization transfer pathway is

$$I_i \xrightarrow{\text{TOCSY}} I_j \xrightarrow{^1J_{IS}} S_j \xrightarrow{^1J_{IS}} I_j. \qquad [7.51]$$
$$(t_1) \qquad\qquad (t_2) \qquad (t_3)$$

Sample $F_1(^1H)-F_3(^1H)$ planes from a TOCSY-HSQC spectrum of ubiquitin are compared with the equivalent region of a homonuclear TOCSY spectrum in Fig. 7.17.

An analogous $^1H-^{13}C$ TOCSY-HSQC experiment, for use with protein D_2O solutions, might also be designed. However, a significantly more sensitive experiment, which relies on coherence transfer via the large and uniform $^{13}C-^{13}C$ J couplings, rather than the smaller $^1H-^1H$ J couplings, gives the same information. These experiments, the HCCH-TOCSY, and the related HCCH-COSY, are discussed in Section 7.3.

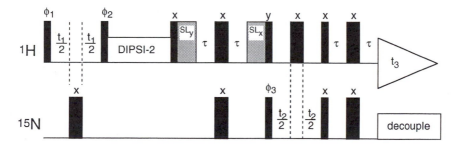

FIGURE 7.16 Pulse sequence for the 3D $^1H-^{15}N$ TOCSY-HSQC experiment. The phase cycling is $\phi_1 = 2(x), 2(-x)$; $\phi_2 = 4(x), 4(-x)$; $\phi_3 = x, -x$; receiver $= x$, $-x, -x, x, -x, x, x, -x$. The spin-lock purge pulses (SL) are applied for 1–2 ms.

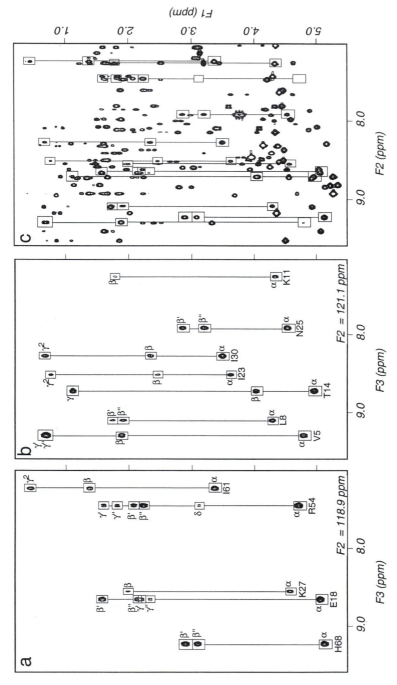

FIGURE 7.17 Selected $F_1(^1H)$–$F_3(H^N)$ regions from a 3D ^1H–^{15}N TOCSY-HSQC spectrum of ^{15}N-labeled ubiquitin at $F_2(^{15}N)$ chemical shifts of 118.9 ppm (a) and 121.1 ppm (b), and the corresponding region from a 2D homonuclear TOCSY spectrum (c) acquired using presaturation of the H_2O signal. Isotropic mixing times of 64 and 48 ms were used for the 3D TOCSY-HSQC and 2D TOCSY spectra, respectively.

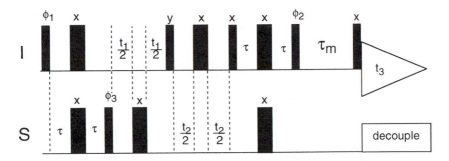

FIGURE 7.18 HSQC-NOESY pulse sequence. The phase cycling is $\phi_1 = x, -x$; $\phi_2 = x, x, -x, -x$; $\phi_3 = 4(x)\ 4(-x)$; and receiver $= x, -x, -x, x, -x, x, x, -x$.

7.2.3 3D HSQC-NOESY AND HSQC-TOCSY EXPERIMENTS

NOESY and HSQC (or HMQC) pulse sequences conceivably could be combined in the reverse order to yield a HSQC-NOESY (or HMQC-NOESY) experiment, illustrated in Fig. 7.18 (39). In this experiment, the F_1 and F_3 dimensions are exchanged relative to the NOESY-HSQC experiment; thus, the $F_1(S)-F_2(^1\text{H})$ projection corresponds to the 2D HSQC (or HMQC) correlation spectrum, and the $F_2(^1\text{H})-F_3(^1\text{H})$ projection corresponds to the $F_1(^1\text{H})-F_2(^1\text{H})$ region of a conventional 2D $^1\text{H}-^1\text{H}$ NOESY spectrum. Similarly, in a HSQC-TOCSY (or HMQC-TOCSY) experiment, the $F_1(S)-F_2(^1\text{H})$ projection corresponds to the 2D HSQC (or HMQC) correlation spectrum, and the $F_2(^1\text{H})-F_3(^1\text{H})$ projection corresponds to the $F_1(^1\text{H})-F_2(^1\text{H})$ region of a conventional 2D $^1\text{H}-^1\text{H}$ TOCSY spectrum. The coherence transfer pathway is given by

$$I_i \xrightarrow{\ ^1J_{IS}\ } S_i \xrightarrow{\ ^1J_{IS}\ } I_i \xrightarrow{\ \text{NOESY/TOCSY}\ } I_j. \qquad [7.52]$$
$$(t_1) \qquad\qquad (t_2) \qquad\qquad (t_3)$$

For ^{15}N-labeled proteins, these experiments have two advantages compared to the $^1\text{H}-^{15}\text{N}$ NOESY-HSQC and $^1\text{H}-^{15}\text{N}$ TOCSY-HSQC experiments (the differences are much less pronounced for $^1\text{H}-^{13}\text{C}$ heteronuclear-edited experiments). First, the $^1\text{H}-^1\text{H}$ planes in HSQC-NOESY and HSQC-TOCSY experiments have higher digital resolutions than the corresponding planes in NOESY-HSQC and TOCSY-HSQC experiments for a given experimental time, because the full ^1H spectral width is digitized during t_3 rather than t_1 and the narrower H^N spectral region is digitized in t_1 rather than t_3. Second, narrower F_1 linewidths result from evolution of heteronuclear multiple-quantum coherence during t_1 rather than ^{1}H single-quantum coherence (Section 7.1.2.4). However, the HSQC-NOESY and HSQC-TOCSY experiments have the disadvantage that a number of ^{1}H$^\alpha$ (F_3)–^{1}H^N (F_1) cross-peaks may be obscured by the intense residual water peak, unless very efficient water suppression can be obtained using field-gradient pulses (Section 3.5.3).

PEP sensitivity enhancement cannot be incorporated into the HSQC-NOESY experiment because the orthogonal magnetization components following t_2 cannot be converted simultaneously to longitudinal magnetization during the NOESY mixing period. In contrast, PEP sensitivity enhancement can be incorporated *independently* into the HSQC and TOCSY portions of the HSQC-TOCSY pulse sequence, which yields a theoretical sensitivity improvement of 2 compared to the conventional HSQC-TOCSY or TOCSY-HSQC experiment.

7.2.4 HMQC-NOESY-HMQC EXPERIMENTS

Although the 3D heteronuclear-edited NOESY spectra discussed above offer a vast improvement in resolution relative to 2D homonuclear NOESY spectra, the possibility of ambiguity still remains. One limitation of the 3D heteronuclear-edited NOESY experiments is that NOE correlations cannot be observed between protons with degenerate chemical shifts, because these cross peaks are coincident with the more intense *autocorrelation* or "diagonal" peak. Observation of such NOEs, which occur between both aromatic and aliphatic protons in 3D ^{1}H–^{13}C NOESY-HSQC spectra and between amide protons (particularly in proteins with a high helical content) in ^{1}H–^{15}N NOESY-HSQC spectra, is important for both resonance assignment and protein structure determination. Indeed, NOEs between aliphatic protons make up the majority of NOEs observed for proteins, and identification of as many of these NOEs as possible is essential. Additionally, in a ^{1}H–^{15}N NOESY-HSQC experiment, ambiguities related to ^{1}H^N chemical-shift degeneracy are removed provided that either the ^{1}H^N or ^{15}N chemical shifts of a given amide group can be resolved; however, ambiguities associated with overlap in the aliphatic ^{1}H region remain. Therefore, even if a given amide-aliphatic NOE cross-peak is fully resolved in the 3D ^{1}H–^{15}N NOESY-HSQC spectrum, unambiguous assignment on the basis of the aliphatic ^{1}H chemical shift alone may be impossible.

HMQC-NOESY-HMQC NMR spectroscopy provides a solution to these problems (the analogous HSQC-NOESY-HSQC experiments are seldom used because of the number of pulses involved) (*40*). These experi-

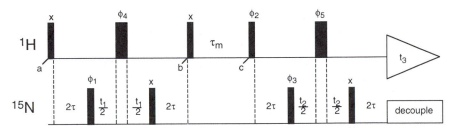

FIGURE 7.19 Pulse sequence for the 3D ^{1}H–^{15}N HMQC-NOESY-HMQC experiment. The phase cycling is $\phi_1 = x, -x$; $\phi_2 = 4(x), 4(-x)$; $\phi_3 = 2(x), 2(-x)$; $\phi_4 = 8(y), 8(-y)$; $\phi_5 = 16(y), 16(-y)$; and receiver $= x, -x, -x, x, -x, x, x, -x$.

ments are derived conceptually by overlapping HMQC-NOESY and NOESY-HMQC experiments or by catenating two HMQC experiments and a NOESY experiment. As the experiment names indicate, a ^{15}N/^{15}N HMQC-NOESY-HMQC combines ^{1}H–^{15}N HMQC, ^{1}H–^{1}H NOESY, and ^{1}H–^{15}N HMQC sequences; a ^{13}C/^{15}N HMQC-NOESY-HMQC combines ^{1}H–^{13}C HMQC, ^{1}H–^{1}H NOESY, and ^{1}H–^{15}N HMQC sequences; and a ^{13}C/^{13}C HMQC-NOESY-HMQC combines ^{1}H–^{13}C HMQC, ^{1}H–^{1}H NOESY and ^{1}H–^{13}C HMQC sequences. For the case of two protons I_1 and I_2 covalently bonded to two heteronuclei, S_1 and S_2, the first HMQC experiment correlates spins I_1 and S_1, magnetization is exchanged between I_1 and I_2 during the NOESY mixing period, and the second HMQC experiment correlates I_2 and S_2. Thus, each NOE cross-peak can be identified by up to four (two proton and two heteronuclear) chemical shifts in four independent frequency dimensions. In some instances, 3D versions of the HMQC-NOESY-HMQC experiments that correlate S_1, S_2 and I_2 in the three frequency dimensions are satisfactory, provided that the S_1 resonances are well resolved (see Section 7.2.5).

7.2.4.1 3D ^{15}N/^{15}N HMQC-NOESY-HMQC The 3D ^{1}H–^{15}N HMQC-NOESY-HMQC experiment illustrated in Fig. 7.19 is used to detect NOEs between amide protons with degenerate chemical shifts. In this experiment, the heteronuclear chemical shifts are labeled in the F_1 and F_2 dimensions, and the ^{1}H chemical shift is detected in the F_3 dimension *(41,42)*. The 4D version of this experiment is seldom used. The initial part of this experiment, between points *a* and *b*, is equivalent to a HMQC sequence (less the t_2 acquisition period) (Fig. 7.1a), and generates, at time

b, transverse $^1H^N$ magnetization that is modulated by the chemical shift of its attached ^{15}N nucleus as a function of t_1 [7.2]. The following $90°(^1H)$ pulse regenerates $^1H^N$ z-magnetization, which is transferred by cross-relaxation to proximal protons during the NOESY mixing time, τ_m. The second HMQC sequence converts any $^1H^N$ magnetization present following τ_m into heteronuclear multiple-quantum coherence for indirect detection of the associated ^{15}N chemical shifts during t_2 and direct detection of the $^1H^N$ frequencies during the acquisition period, t_3. The delays $2\tau = 1/(2^1J_{NH})$. The magnetization transfer pathway is

$$^1H_i^N \xrightarrow{\;^1J_{NH}\;} {}^{15}N_i \xrightarrow{\;^1J_{NH}\;} {}^1H_i^N \xrightarrow{\;NOE\;}$$
$$(t_1)$$

$$^1H_j^N \xrightarrow{\;^1J_{NH}\;} {}^{15}N_j \xrightarrow{\;^1J_{NH}\;} {}^1H_j^N.$$
$$(t_2) \qquad (t_3)$$

[7.53]

The 32-step phase cycle given in Fig. 7.19 is rather long by 3D standards, but because the digital resolution in the heteronuclear dimensions need not be as great as in proton dimensions, the total acquisition time can still be limited to ~4 days. Typically, the acquired 3D matrix consists of $64(t_1) \times 32(t_2) \times 512(t_3)$ complex data points. The first four steps of the phase cycle select signals that have arisen via $^1H^N$ multiple-quantum coherence, and the third step (the first and second set of four transients) eliminates artifacts that arise from single-quantum proton magnetization present during τ_m. The phase cycling of the two $180°(^1H)$ pulses simply serves to eliminate artifacts that result from imperfections in these pulses. If desired, phase cycling of these two pulses can be eliminated in order to reduce the overall experiment time, or to increase the digital resolution in the t_2 dimension by acquiring more increments with fewer transients per increment.

7.2.4.2 4D $^{13}C/^{15}N$ HMQC-NOESY-HMQC   The pulse sequence for the 4D $^{13}C/^{15}N$ HMQC-NOESY-HMQC experiment (*43*) is illustrated in Fig. 7.20. Conceptually, the experiment simply consists of a NOESY mixing period between two HMQC sequences; the first HMQC sequence is tuned to $^1H–^{13}C$ couplings, and the second is tuned to $^1H–^{15}N$ couplings. Magnetization is therefore transferred from ^{13}C-attached aliphatic protons

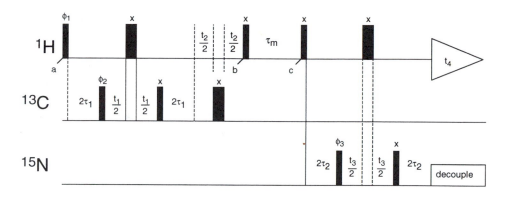

FIGURE 7.20 Pulse sequence for the 4D ^{13}C/^{15}N HMQC-NOESY-HMQC experiment. The phase cycling is $\phi_1 = x$; $\phi_2 = x, -x$; $\phi_3 = 2(x), 2(-x)$; receiver = x, $-x, -x, x$.

to ^{15}N-attached amide protons via the following pathway

$$^{1}H_i \xrightarrow{^{1}J_{CH}} \,^{13}C_i \xrightarrow{^{1}J_{CH}} \,^{1}H_i \xrightarrow{NOE}$$
$$\quad\;\; (t_1) \qquad\quad (t_2)$$

[7.54]

$$^{1}H_j^N \xrightarrow{^{1}J_{NH}} \,^{15}N_j \xrightarrow{^{1}J_{NH}} \,^{1}H_j^N.$$
$$\qquad (t_3) \qquad\quad (t_4)$$

Between times a and b in Fig. 7.20, the basic pulse sequence is similar to a 2D ^{1}H–^{13}C HMQC experiment (Section 7.1.1.1), except that the t_2 acquisition time of the 2D experiment has been substituted with an incremental t_2 evolution period for indirect detection of the aliphatic proton chemical shifts. The 180°(^{13}C) decoupling pulse applied in the middle of the t_2 evolution period should be applied as a composite pulse $(90°_x–180°_y–90°_x)$ in order to minimize resonance offset effects. The delays $2\tau_1 = 1/(2\,^{1}J_{CH})$. The 90°(^{1}H) pulse immediately following the initial ^{1}H–^{13}C HMQC sequence rotates the transverse ^{1}H magnetization to the z axis. During the subsequent NOESY mixing time, τ_m, magnetization can be transferred to proximal proton spins via dipolar couplings. The remainder of the pulse sequence, following time c, represents a $^1\mathrm{H}$–$^{15}\mathrm{N}$ HMQC sequence, with indirect detection of the $^{15}\mathrm{N}$ chemical shift during t_3, and finally detection of $^1\mathrm{H}^N$ during the acquisition time t_4. The delays $2\tau_2 = 1/(2\ ^1J_{\mathrm{NH}})$. Decoupling of the $^{13}\mathrm{CO}$ spins during t_1 and t_3 can be achieved using a suitable low-power composite pulse decoupling scheme, such as SEDUCE-1, as indicated in Fig. 7.20, or by application of selective $180°(^{13}\mathrm{CO})$ pulses at the midpoint of the evolution periods. Alternatively, $^{13}\mathrm{CO}$ decoupling may be omitted all together, because $t_{1,\mathrm{max}}$ and $t_{3,\mathrm{max}}$ are always less than $1/(2J_{\mathrm{C^\alpha CO}})$ and $1/(2J_{\mathrm{NCO}})$, respectively, and therefore evolution of these couplings do not greatly reduce the intensity.

In a 4D experiment every effort must be made to maximize the digital resolution, but at the same time keep the total acquisition time within reasonable bounds. The minimum phase cycling requires two steps for each heteronuclear-filter; therefore, the four-step phase cycle given in Fig. 7.20 is used. The double heteronuclear filtering in this experiment is very efficient at removing artifacts, including the intense "diagonal peaks" corresponding to magnetization that has not been transferred from a $^{13}\mathrm{C}$-attached proton to a $^{15}\mathrm{N}$-attached proton during the NOESY mixing time.

The initial t_1, t_2, and t_3 sampling delays are adjusted to $1/(2SW)$ as described in Sections 3.3.2.3 and 7.1.2.1. Alternatively, $t_2(0)$ can be set to zero if a compensatory delay equal in duration to the $180°(^{13}\mathrm{C})$ composite pulse is inserted prior to the first $90°(^{13}\mathrm{C})$ pulse (43). In this case, the $180°(^1\mathrm{H})$ pulse in the middle of t_1 refocuses the evolution during the delay and the $180°(^{13}\mathrm{C})$ composite pulse.

Quadrature detection in the t_1, t_2, and t_3 dimensions is achieved by shifting the phases of ϕ_1, ϕ_2, and ϕ_3 independently in a TPPI–States manner. A typical acquisition contains 8–16 (t_1) × 64 (t_2) × 8–16 (t_3) × 128–256 (t_4) complex data points. Processing of four-dimensional NMR experiments is discussed in Section 7.2.4.4.

Pulsed field gradients may be applied with particular advantage in the 4D $^{13}\mathrm{C}/^{15}\mathrm{N}$ HMQC-NOESY-HSQC experiment to suppress artifacts, eliminate the $\mathrm{H_2O}$ signal, and select for the coherence transfer pathway involving $^{15}\mathrm{N}$ magnetization (44). Gradient coherence selection is coupled with the PEP sensitivity improvement technology discussed in Section 7.1.3.3 to decrease the number of phase cycle steps by a factor of 2 relative to the nongradient experiment. The shorter phase cycle allows spectra to be recorded with increased resolution for a given total acquisition time.

7.2.4.3 4D $^{13}C/^{13}C$ HMQC-NOESY-HMQC   The pulse sequence for the 4D $^{13}\mathrm{C}/^{13}\mathrm{C}$ HMQC-NOESY-HMQC experiment (45) illustrated in Fig. 7.21 consists of a NOESY mixing period between two $^1\mathrm{H}$–$^{13}\mathrm{C}$ HMQC

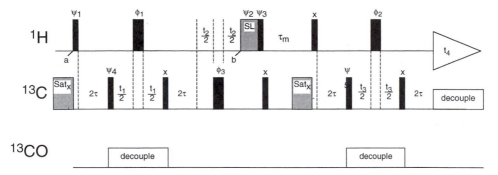

FIGURE 7.21 Pulse sequence for the 4D $^{13}\text{C}/^{13}\text{C}$ HMQC-NOESY-HMQC experiment. The phase cycling is $\psi_1 = x$; $\psi_2 = -y$; $\psi_3 = x$; $\psi_4 = x, -x$; $\psi_5 = 2(x), 2(-x)$; $\phi_1 = 4(x), 4(y)$; $\phi_2 = 4(x), 4(y)$; $\phi_3 = 2(x), 2(-x)$; and receiver $= x, -x, -x, x$. Quadrature detection in the t_1, t_2, and t_3 dimensions is achieved by incrementing independently the phases ψ_1, ψ_2, ψ_3, and ψ_4, and the receiver phase as discussed in the text and summarized in Table 7.1.

sequences. The delays $2\tau = 1/(2^1J_{\text{CH}})$. Magnetization is transferred via the following pathway:

$$^1\text{H}_i \xrightarrow{^1J_{\text{CH}}} {}^{13}\text{C}_i \xrightarrow{^1J_{\text{CH}}} {}^1\text{H}_i \xrightarrow{\text{NOE}}$$
$$(t_1) \qquad\qquad (t_2)$$

$$[7.55]$$

$$^1\text{H}_j \xrightarrow{^1J_{\text{CH}}} {}^{13}\text{C}_j \xrightarrow{^1J_{\text{CH}}} {}^1\text{H}_j.$$
$$(t_3) \qquad\qquad (t_4)$$

Comparison of Figs. 7.20 and 7.21 reveals several additional features of the $^{13}\text{C}/^{13}\text{C}$ HMQC-NOESY-HMQC experiment designed to eliminate spurious magnetization transfer pathways. These pathways were suppressed by the $^{13}\text{C}/^{15}\text{N}$ double heteronuclear filter in the $^{13}\text{C}/^{15}\text{N}$ HMQC-NOESY-HMQC experiment. In particular, artifactual peaks observed along the pseudo-diagonal with $F_2(^1\text{H}) = F_4(^1\text{H})$ and $F_1(^{13}\text{C}) \neq F_3(^{13}\text{C})$ would otherwise render identification of genuine NOEs between protons with degenerate chemical shifts extremely difficult (if not impossible).

The pulse sequence includes three purge pulses. The ~1-ms ^{13}C saturation pulse (Sat$_x$) applied immediately prior to the first 90°(^1H) pulse, time a, prevents the transfer of equilibrium magnetization originating on ^{13}C

spins to scalar-coupled ^1H spins. The ^1H spin-lock pulse ($SL_{\psi2}$), which is applied along the $-y$ axis for ~500 μs prior to the second $90°(^1H)$ pulse, dephases any ^1H magnetization that is not aligned with the y axis at time b. Residual ^{13}C magnetization is eliminated by the $\sim$1-ms ^{13}C saturation pulse applied at the end of τ_m.

Longitudinal two-spin order, $2H_zC_z$, can be present at the beginning of the NOESY mixing period, due to imperfections in the $180°(^{13}C)$ composite pulse applied in the middle of t_2, and consequent evolution of ^1H–^{13}C couplings during t_2. The $90°(^{13}C)$ pulse applied at the beginning of the NOESY mixing period does not affect the desired magnetization pathway [7.55]. This pulse converts $2H_zC_z$ to $-2H_zC_y$, which does not lead to observable magnetization.

The first four steps of the phase cycle correspond to independent cycling of the $90°(^{13}C)$ pulses at the beginning of the HMQC periods, in a fashion identical to that of the $^{13}C/^{15}N$ HMQC-NOESY-HMQC experiment discussed above; phase cycling of these pulses eliminates axial peaks in the F_1 and F_3 dimensions. To further prevent magnetization residing on ^1H spins at the beginning of the t_2 evolution period from being transferred to coupled ^{13}C spins, the $180°(^{13}C)$ composite pulse in the middle of t_2 is phase-cycled along the $\pm x$ axes together with the first $90°(^{13}C)$ pulse of the second HMQC sequence. The $180°(^1H)$ pulses in the middle of t_1 and t_3 are also cycled together in a two-step EXORCYCLE phase cycle in order to eliminate artifacts that give rise to false diagonal peaks that appear at $\pm\Omega_H$ in the ^{13}C dimensions. Simultaneous phase cycling of these two pulses is feasible because the artifacts being eliminated during the t_1 and t_3 evolution periods are strictly independent. The total phase cycle is eight steps, as indicated in Fig. 7.21. A typical acquisition is limited to 8 (t_1) × 64 (t_2) × 8 (t_3) × 128–256 (t_4) complex data points.

The TPPI–States method is used to obtain quadrature detection in the F_1, F_2, and F_3 dimensions. In t_1 and t_3, complex pairs are generated in the usual manner by incrementing ψ_4 and ψ_5 by 90° independently of each other. When t_1 is incremented, ψ_4 and the receiver phase are inverted. When t_3 is incremented, ψ_5 and the receiver phase are inverted. In order to eliminate quadrature artifacts in the F_2 dimension, the 90° phase shifts are applied to the second $90°(^1H)$ pulse (ψ_3) and its accompanying ^1H spin-lock pulse ($SL_{\psi2}$), while the 180° phase shifts are applied to the first $90°(^1H)$ pulse (ψ_1). Complex pairs are thus generated by decrementing ψ_2 and ψ_3 by 90°; each time t_2 is incremented, ψ_1 and the receiver phase are inverted. The schemes for obtaining quadrature in the indirectly detected dimensions of the 4D $^{13}C/^{13}C$ HMQC-NOESY-HMQC experiment are summarized in Table 7.1.

As with the $^{13}C/^{15}N$ HMQC-NOESY-HMQC experiment discussed

Table 7.1

Quadrature Detection in 4D $^{13}C/^{13}C$ HMQC-NOESY-NMQC Experiments

t_1 Dimension

ψ_4	t_1	Receiver
x	$t_1(0) + 2k_1\, \Delta t_1$	+
y	$t_1(0) + 2k_1 \Delta t_1$	+
$-x$	$t_1(0) + (2k_1 + 1)\, \Delta t_1$	−
$-y$	$t_1(0) + (2k_1 + 1)\, \Delta t_1$	−

t_2 Dimension

ψ_1	ψ_2	ψ_3	t_2	Receiver
x	$-y$	x	$t_2(0) + 2k_2\, \Delta t_2$	+
x	$-x$	$-y$	$t_2(0) + 2k_2 \Delta t_2$	+
$-x$	$-y$	x	$t_2(0) + (2k_2 + 1)\, \Delta t_2$	−
$-x$	$-x$	$-y$	$t_2(0) + (2k_2 + 1)\, \Delta t_2$	−

t_3 Dimension

ψ_5	t_3	Receiver
x	$t_3(0) + 2k_3\, \Delta t_3$	+
y	$t_3(0) + 2k_3\, \Delta t_3$	+
$-x$	$t_3(0) + (2k_3 + 1)\, \Delta t_3$	−
$-y$	$t_3(0) + (2k_3 + 1)\, \Delta t_3$	−

The index $k_i = 0, 1, \ldots, N_i/2 - 1$, in which N_i is the number of complex points acquired in the ith dimension ($i = 1, 2, 3$). The initial sampling delays are $t_i(0)$ and the sampling increments are Δt_i.

above, the initial t_1, t_2, and t_3 sampling delays are set to $1/(2SW)$ as described in Sections 3.3.2.3 and 7.1.2.1. Alternatively, $t_2(0)$ can be set to 0 if a delay equal in duration to the $180°(^{13}C)$ composite pulse is inserted prior to the first $90°(^{13}C)$ pulse.

The $^{13}C/^{13}C$ HMQC-NOESY-HMQC pulse sequence of Fig. 7.21 yields spectra that are relatively free of artifacts. Pulsed field gradients provide a more efficient method for suppressing undesired coherence transfer pathways (46,47). Unlike the 4D $^{13}C/^{15}N$ HMQC-NOESY-HSQC experiment (44), the PEP sensitivity improvement scheme (Section 7.1.3.2) does not compensate completely for the sensitivity loss associated with coherence selection by pulsed field gradients because, in addition to CH groups, CH_2 and CH_3 moieties must be detected (i.e., ^{13}C spins with $n > 1$ attached protons) (48). Nevertheless, incorporation of six PFG pulses into the 4D

$^{13}C/^{13}C$ HMQC-NOESY-HMQC experiment detailed above has enabled reduction of the phase cycle from eight to two steps (47). The phase cycling in the gradient experiment constitutes the two steps necessary for isotope filtration in the first HMQC transfer sequence. No isotope filtering is used after the NOESY mixing period; therefore, axial peaks occur at the edges of the spectrum in the F_3 dimension. If required, the phase cycle of the gradient experiment can be doubled to four steps to include isotope filtration during the second HMQC transfer sequence to eliminate the axial peaks. The shorter phase cycle (two-step or four-step) allows an increase in the resolution attainable in the indirectly detected dimensions for a given total acquisition time, or alternatively allows a reduction in the total acquisition time.

7.2.4.4 Processing 4D HMQC-NOESY-HMQC Spectra The acquired digital resolution in the indirectly detected dimensions of a 4D HMQC-NOESY-HMQC spectrum is necessarily low in order to keep the overall measuring time within reasonable limits (less than 7–8 days). In particular, the heteronuclear dimensions are limited to only 8–16 complex points (or slightly more if gradient-enhanced pulse sequences are used). Resolution enhancement of the severely truncated heteronuclear signals (t_1 and t_3) by either linear prediction or maximum entropy reconstruction (Section 3.3.4) is essential.

In maximum entropy reconstruction, the time-domain data for the ^1H dimensions, t_2 and t_4, are completely processed first, including apodization, zero-filling, Fourier transformation, and phasing. The imaginary parts of F_2 and F_4 are discarded following these initial steps. The heteronuclear (t_1, t_3) planes (for each (t_2, t_4) pair) are processed by using a two-dimensional maximum entropy algorithm to directly yield the final 4D spectrum (Section 3.3.4.2).

Analogously to maximum entropy processing, a two-dimensional linear prediction algorithm ideally would be used to increase the resolution in the (t_1, t_3) planes (49). However, as pointed out by Zhu and Bax, 2D linear prediction requires enormous amounts of computing time and is therefore (presently at least) impractical for 4D data sets (49). Instead, one-dimensional linear prediction routines are used to extend the time-domain data independently in both the t_1 and t_3 dimensions (45,50) by using the protocol presented in Table 7.2.

A data set containing 8 (t_1) × 64 (t_2) × 8 (t_3) × 128 (t_4) complex data points typically is processed by the maximum entropy reconstruction or linear prediction protocols to give a final spectrum comprising 32 × 128 × 32 × 256 real data points.

Table 7.2

Summary of Steps Used in Processing 4D ^{13}C/^{15}N and ^{13}C/^{13}C-Edited NOESY Data Sets

Step	Computation performed
1	Fourier transform in t_3(^{13}C or ^{15}N) dimension
2	Apodization, zero-filling, Fourier transformation, and phasing in t_2(^{1}H) and t_4(^{1}H) dimensions
3	Linear prediction of t_1(^{13}C) time-domain data
4	Apodization, zero-filling, Fourier transformation, and phasing in t_1(^{13}C) dimension
5	Inverse Fourier transform in t_3(^{13}C or ^{15}N) dimension
6	Linear prediction of t_3(^{13}C or ^{15}N) time-domain data
7	Apodization, zero-filling, Fourier transformation, and phasing in t_3(^{13}C or ^{15}N) dimension

Source: Adapted from Ref. 45.

7.2.5 RELATIVE MERITS OF 3D AND 4D HETERONUCLEAR-EDITED NOESY SPECTROSCOPY

The number of cross-peaks observable in these 3D and 4D heteronuclear-edited NOESY spectra is the same as is present in the 2D homonuclear NOESY spectra. Each ^{1}H–^{1}H NOE cross-peak in a 2D NOESY spectrum is separated into a third dimension by the chemical shift of the heteronucleus attached to one proton and, for 4D spectroscopy, into a fourth dimension by the chemical shift of the heteronucleus directly attached to the other proton. Therefore, the increased resolution associated with the extension to three or four dimensions is not accompanied by any increase in the complexity of the spectrum (unlike homonuclear 3D NMR spectroscopy; Section 6.7). In addition, the sensitivity of the 3D and 4D NOESY experiments is relatively high, even for larger proteins, because the through-bond coherence transfer steps are highly efficient (the heteronuclear couplings involved, $^1J_{CH}$ and $^1J_{NH}$, are significantly larger than the linewidths).

As a consequence, generally, equivalent information can be obtained from a set of complementary three-dimensional NMR experiments or a single four-dimensional experiment. For example, the information content of the two 3D ^{1}H–^{15}N NOESY-HSQC and 3D ^{13}C/^{15}N HMQC-NOESY-HMQC experiments theoretically is equivalent to that of a single 4D ^{13}C/^{15}N-edited HMQC-NOESY-HMQC experiment. However, no direct correlation can be made between the aliphatic ^{1}H and ^{13}C chemical shifts using the two 3D experiments, so the possibility of ambiguity remains, particularly as the two 3D experiments would be acquired at different times (with possible slight variations in conditions). On the other hand, the two 3D NMR experiments can be acquired with much greater resolutions in the indirect dimensions than can the 4D experiment, which facilitates more accurate determination of resonance frequencies.

Assuming that complete ^{1}H, ^{15}N, and ^{13}C assignments are available, the 4D ^{13}C/^{15}N-edited HMQC-NOESY-HMQC and 4D ^{13}C/^{13}C-edited HMQC-NOESY-HMQC experiments allow assignment of virtually all observable NOEs, because they eliminate most of the problems associated with resonance overlap. Because resonance overlap is eliminated, the 4D spectra are also amenable to automated assignment strategies.

7.3 ^{13}C–^{13}C Correlations: The HCCH-COSY and HCCH-TOCSY Experiments

The HCCH-COSY (*51–53*) (^{1}H–^{13}C–^{13}C–^{1}H correlation spectroscopy) and HCCH-TOCSY (*54,55*) (^{1}H–^{13}C–^{13}C–^{1}H total correlation spectroscopy) experiments are used in the assignment of aliphatic ^{1}H and ^{13}C resonances of ^{13}C-labeled proteins. These analogous experiments allow dispersion of the 2D ^{1}H–^{1}H COSY or TOCSY spectra, respectively, into a third (or fourth) ^{13}C frequency dimension by utilizing three magnetization transfer steps: first from a ^{1}H to its directly attached ^{13}C nucleus via the $^1J_{CH}$ coupling (~140 Hz), then from the ^{13}C to neighboring ^{13}C nuclei via the $^1J_{CC}$ couplings (32–40 Hz), and finally from ^{13}C back to the directly attached protons via the $^1J_{CH}$ coupling. For larger proteins the three-step magnetization transfer is significantly more efficient than transferring ^{1}H magnetization in a single step using the unresolved ^{1}H–^{1}H J couplings. In the HCCH-COSY experiment, ^{13}C magnetization transfer is achieved by using a 90° ^{13}C COSY mixing pulse (in fashion analogous to that for the 90° ^{1}H COSY mixing pulse in the ^{1}H–^{1}H COSY experiment discussed in Section 6.2.1.1), so that magnetization is transferred only from a ^{13}C nucleus to its directly bound neighbors; in the HCCH-TOCSY experiment, transfer is achieved by isotropic mixing of the ^{13}C spins, resulting in both direct and multiple-relayed magnetization transfers along the carbon side chain.

The amino acid spin system considered in the following sections consists of K (noncarbonyl) carbon spins, C^k for $k = 1, \ldots, K$, and a carbonyl spin, C'. The carbon spins are ordered so that C^1 and C^κ are the spins of interest. Carbon spin C^1 has M directly bonded protons, H_m^1 for $m = 1$ to M, and carbon spin C^κ has N_κ directly bonded protons, H_n^κ for $n = 1$ to N_κ. The ^{1}H and ^{13}C Larmor frequencies are Ω_{H^k} and Ω_{C^k}, respectively. The one-bond ^{1}H–^{13}C scalar coupling constants are designated as J_{CH}. The ^{13}C–^{13}C scalar coupling constants are designated as $J_{C^jC^k}$ and J_{C^kCO}; these interactions can be one-bond or multiple-bond depending on the context. Homonuclear ^{1}H–^{1}H scalar coupling interactions are unresolved in larger proteins and are not considered explicitly. For simplicity, the ^{15}N–^{13}C$^\alpha$ scalar coupling Hamiltonian is not considered. If desired, ^{15}N decoupling can be achieved by applying a composite pulse decoupling sequence during ^{13}C evolution periods. The free-precession Hamiltonian for the C^1 spin is given by

$$\mathcal{H}_1 = \Omega_{C^1}C_z^1 + \sum_{m=1}^{M} 2\pi J_{CH}H_{mz}^1 C_z^1 + \sum_{k=2}^{K} 2\pi J_{C^1C^k}C_z^1 C_z^k$$

$$+ 2\pi J_{C^1CO}C_z^1 C_z'. \quad [7.56]$$

A similar Hamiltonian can be written for C^κ. The net evolution through the HCCH-COSY and HCCH-TOCSY experiments is summarized as

$$\sum_{m=1}^{M} H_m^1 \xrightarrow{\;^1J_{CH}\;} C^1 \xrightarrow{\;^1J_{CC}\;} C^\kappa \xrightarrow{\;^1J_{CH}\;} \sum_{n=1}^{N_\kappa} H_n^\kappa. \quad [7.57]$$

$$(t_1) \qquad\qquad (t_2) \qquad\qquad (t_3)$$

Carbon nuclei C^1 and C^κ ($\kappa = 2$) are directly covalently bonded in the HCCH-COSY experiment (this restriction does not apply to the HCCH-TOCSY experiment). As a concrete example, if C^1 is the ^{13}C$^\alpha$ spin, and C^2 is the ^{13}C$^\beta$ spin of isoleucine, then $K = 5$, $M = 1$, $N_2 = 1$, C^3 is the ^{13}C$^{\gamma 1}$ spin, C^4 is the ^{13}C$^{\gamma 2}$ spin, C^5 is the ^{13}C$^\delta$ spin, $J_{C^1C^2}$ is a one-bond scalar coupling constant, $J_{C^1C^k}$ for $k > 2$ are (negligible) two- or three-bond scalar coupling constants, $J_{C^2C^3}$ and $J_{C^2C^4}$ are one-bond scalar coupling constants, $J_{C^2C^5}$ is a (negligible) two-bond scalar coupling constant, J_{C^1CO} is a one-bond scalar coupling constant, and J_{C^2CO} is a (negligible) two-bond scalar coupling constant.

7.3.1 HCCH-COSY

Figure 7.22a illustrates a simple HCCH-COSY pulse sequence (51–53). The basic principles behind this, and other, HCCH-type experiments will be described below using the product operator formalism.

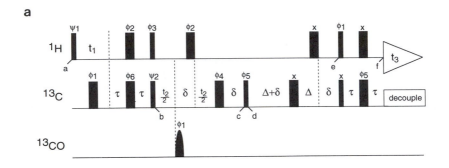

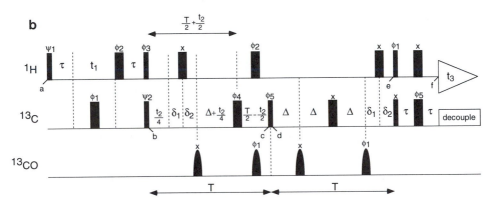

FIGURE 7.22 Pulse sequences for the 3D HCCH-COSY (a) and constant-time HCCH-COSY (b) experiments. The phase cycling for both experiments is given by $\psi_1 = x$; $\psi_2 = x$; $\phi_1 = 8(x), 8(-x)$; $\phi_2 = 4(x), 4(y), 4(-x), 4(-y)$; $\phi_3 = y, -y$; $\phi_4 = 2(x), 2(y), 2(-x), 2(-y)$; $\phi_5 = 4(x), 4(-x)$; $\phi_6 = 2(x), 2(-x)$; and receiver = $x, -x, -x, x, 2(-x, x, x, -x), x, -x, -x, x$. Quadrature detection in the t_1 and t_2 dimensions is achieved by incrementing independently the phases ψ_1 and ψ_2, respectively, and the receiver phase, in a TPPI–States manner.

Several shortcomings of this particular pulse sequence will be highlighted, that are alleviated in a constant-time version of the experiment (Fig. 7.22b).

The experiment begins at time a in Fig. 7.22a with longitudinal magnetization of the H^1_{mz} protons. At the end of the ^{1}H evolution period, t_1, this magnetization is transferred to the attached carbon via an INEPT sequence. The $180°(^{13}\text{C})$ decoupling pulse in the middle of t_1 ensures that the ^{1}H spins effectively are decoupled from ^{13}C spins during this evolution period. A composite pulse $(90°_x{-}180°_y{-}90°_x)$ is used to minimize resonance offset and rf inhomogeneity effects. For $2\tau = 1/(2J_{CH})$ the magnetization at time b is given by

$$\sigma_b = \sum_{m=1}^{M} 2H^1_{mz}C^1_y \cos(\Omega_{H^1}t_1). \qquad [7.58]$$

Between points b and c, only the Hamiltonian for the C^1 spin need be considered because σ_b commutes with the Hamiltonian for spin C^2. The propagator for the pulse sequence is given by

$$\mathbf{U} = \exp(-i\mathcal{H}_1\delta) \exp\left(-i\pi \sum_{k=1}^{K} C^k_x\right) \exp(-i\mathcal{H}_1 t_2/2) \exp\left(-i\pi \sum_{m=1}^{M} H^1_{mx}\right)$$

$$\times \exp(-i\mathcal{H}_1\delta) \exp(-i\pi C'_x) \exp(-i\mathcal{H}_1 t_2/2)$$

$$= \exp\left(-i \sum_{m=1}^{M} 4\pi J_{CH}\delta H^1_{mz}C^k_z\right) \exp\left(-i \sum_{k=2}^{K} 2\pi J_{C^1C^k}(2\delta + t_2)C^1_z C^k_z\right) \exp(i\Omega_{C^1}t_2)$$

$$\times \exp\left(-i\pi \sum_{k=1}^{K} C^k_x\right) \exp\left(-i\pi \sum_{m=1}^{M} H^1_{mx}\right) \exp(-i\pi C'_x), \qquad [7.59]$$

in which the last equality is obtained by applying [2.117]. The positioning of the ^{1}H and ^{13}C 180° pulses between points b and c results in proton decoupling during t_2. The $^{13}C^\alpha$ spins are decoupled from ^{13}CO spins during the entire $t_2 + 2\delta$ delay. The 180°(^{13}C) pulse refocuses chemical-shift evolution during the two δ delays. The magnetization at time c is described by

$$\sigma_c = \{-C^1_x \cos[\pi J_{C^1C^2}(t_2 + 2\delta)] - 2C^1_y C^2_z \sin[\pi J_{C^1C^2}(t_2 + 2\delta)]\}$$

$$\times \cos(\Omega_{H^1}t_1) \cos(\Omega_{C^1}t_2)\Pi_1(t_2 + 2\delta)M\Gamma_M(2\delta), \qquad [7.60]$$

in which $\Gamma_n(t)$ is given by [7.13] (for $J_{IS} = J_{CH}$), and

$$\Pi_j(t) = \prod_{k=3}^{K} \cos(\pi J_{C^jC^k}t) \qquad [7.61]$$

encapsulates the effect of passive ^{13}C–^{13}C couplings to spin C^j. Only terms that result in observable magnetization have been included in [7.61]. The 90°(^{13}C) pulse following time point c transfers the antiphase C^1 magnetization into antiphase C^2 magnetization in a COSY-like manner, giving, at time d:

$$\sigma_d = \{-C^1_x \cos[\pi J_{C^1C^2}(t_2 + 2\delta)] + 2C^1_z C^2_y \sin[\pi J_{C^1C^2}(t_2 + 2\delta)]\}$$

$$\times \cos(\Omega_{H^1}t_1) \cos(\Omega_{C^1}t_2)\Pi_1(t_2 + 2\delta)M\Gamma_M(2\delta). \qquad [7.62]$$

During the subsequent time interval, $2\Delta + 2\delta$, between points d and e, the propagator is

$$\mathbf{U} = \exp(-i[\mathcal{H}_1 + \mathcal{H}_2]\delta) \exp\left(-i\pi \sum_{m=1}^{M} H_{mx}^1\right) \exp\left(-i\pi \sum_{n=1}^{N_2} H_{nx}^2\right) \exp(-i[\mathcal{H}_1 + \mathcal{H}_2]\Delta)$$

$$\times \exp\left(-i\pi \sum_{k=1}^{K} C_x^k\right) \exp(-i[\mathcal{H}_1 + \mathcal{H}_2][\delta + \Delta])$$

$$= \exp\left(-i \sum_{m=1}^{M} 4\pi J_{CH}\delta H_{mz}^1 C_z^1\right) \exp\left(-i \sum_{n=1}^{N_2} 4\pi J_{CH}\delta H_{nz}^2 C_z^2\right)$$

$$\times \exp\left(-i \sum_{k=3}^{K} 4\pi J_{C^1C^k}(\delta + \Delta)C_z^1 C_z^k\right) \exp\left(-i \sum_{k=3}^{K} 4\pi J_{C^2C^k}(\delta + \Delta)C_z^2 C_z^k\right)$$

$$\times \exp(-i4\pi J_{C^1C^2}(\delta + \Delta)C_z^1 C_z^2)$$

$$\times \exp\left(-i\pi \sum_{k=1}^{K} C_x^k\right) \exp\left(-i\pi \sum_{m=1}^{M} H_{mx}^1\right) \exp\left(-i\pi \sum_{n=1}^{N_2} H_{nx}^2\right). \qquad [7.63]$$

The magnetization at time e is described by

$$\sigma_e = \left\{ -\sum_{m=1}^{M} 2H_{mz}^1 C_y^1 \cos[\pi J_{C^1C^2}(t_2 + 2\delta)] \cos[2\pi J_{C^1C^2}(\delta + \Delta)]\Pi_1(2\delta + 2\Delta)\Gamma_M(2\delta) \right.$$

$$\left. -\sum_{n=1}^{N_2} 2H_{nz}^2 C_y^2 \sin[\pi J_{C^1C^2}(t_2 + 2\delta)] \sin[2\pi J_{C^1C^2}(\delta + \Delta)]\Pi_2(2\delta + 2\Delta)\Gamma_{N_2}(2\delta) \right\}$$

$$\times \cos(\Omega_{H^1}t_1) \cos(\Omega_{C^1}t_2)\Pi_1(t_2 + 2\delta)M\Gamma_M(2\delta). \qquad [7.64]$$

The remainder of the experiment is a reverse INEPT sequence. At the start of the detection period, f, the magnetization is described by

$$\sigma_f = \left\{ \sum_{m=1}^{M} 2H_{mx}^1 \cos[\pi J_{C^1C^2}(t_2 + 2\delta)] \cos[2\pi J_{C^1C^2}(\delta + \Delta)]\Pi_1(2\delta + 2\Delta)\Gamma_M(2\delta) \right.$$

$$\left. + \sum_{n=1}^{N_2} 2H_{nx}^2 \sin[\pi J_{C^1C^2}(t_2 + 2\delta)] \sin[2\pi J_{C^1C^2}(\delta + \Delta)]\Pi_2(2\delta + 2\Delta)\Gamma_{N_2}(2\delta) \right\}$$

$$\times \cos(\Omega_{H^1}t_1) \cos(\Omega_{C^1}t_2)\Pi_1(t_2 + 2\delta)M\Gamma_M(2\delta). \qquad [7.65]$$

For carbons with at least one passive coupling partner, $^{13}C-^{13}C$ coherence transfer is optimized by setting $2\Delta + 2\delta = 1/(4^1J_{CC})$. To maximize both $\Gamma_M(2\delta)$ and $\Gamma_{N_2}(2\delta)$ simultaneously for methine, methylene, and methyl carbons, $2\delta \approx 2.2$ ms (Fig. 7.2). The first term in [7.65] represents the autocorrelation or "diagonal peak," and the second term represents the cross-peak resulting from coherence transfer from the H_m^1 protons to the H_n^2 protons by the pathway [7.57].

The principal disadvantage to this pulse sequence is that the efficiency of ^{13}C–^{13}C magnetization transfer between J-coupled carbons depends on t_2 via $\sin[\pi J_{C^1C^2}(t_2 + 2\delta)]\Pi_1(t_2 + 2\delta)$. This has three consequences: (1) magnetization transfer is not optimal, (2) the lineshape in the ^{13}C dimension (F_2) is not purely absorptive because the term $2\pi J_{C^1C^2}\delta \approx 0.24$ radians (14°) represents a phase shift, and (3) the lineshape in F_2 is a multiplet with the active $J_{C^1C^2}$ coupling antiphase and passive $J_{C^1C^k}$ couplings in-phase. The antiphase, partially dispersive, character of the lineshape reduces sensitivity and resolution of the spectrum.

7.3.2 CONSTANT-TIME HCCH-COSY

The ^{13}C–^{13}C magnetization transfer can be optimized independently of t_2, and the multiplet structure in the F_2 dimension can be collapsed by using the constant-time HCCH–COSY experiment shown in Fig. 7.22b (53). The same spin system is considered.

The modified ^{1}H evolution period and INEPT sequence between time points a and b in the constant-time HCCH-COSY pulse sequence reduces the number of 180°(^{13}C) pulses from two to one and generates heteronuclear multiple-quantum coherence, $2H_{mx}^1 C_y^1$, during the t_1 evolution period. Reducing the number of 180°(^{13}C) pulses reduces artifacts from pulse imperfections (56). The heteronuclear multiple-quantum coherence also relaxes more slowly than the single-quantum proton coherence (Section 7.1.2.4). Ignoring artifacts and relaxation, the magnetization at time b is identical to the magnetization present at time b of the original HCCH-COSY experiment [7.58]. These modifications also could be incorporated into the original pulse sequence (Fig. 7.22a).

Between points b and c, the propagator for the pulse sequence is given by

$$\mathbf{U} = \exp(-i\mathcal{H}_1(T - t_2)/4) \exp\left(-i\pi \sum_{m=1}^{M} H_{mx}^1\right) \exp(-i\pi C_x') \exp(-i\mathcal{H}_1(T - t_2)/4)$$

$$\times \exp\left(-i\pi \sum_{k=1}^{K} C_x^k\right) \exp(-i\mathcal{H}_1(t_2/4 + \Delta)) \exp(-i\pi C_x') \exp(-i\mathcal{H}_1\delta_2)$$

$$\times \exp\left(-i\pi \sum_{m=1}^{M} H_{mx}^1\right) \exp(-i\mathcal{H}_1(t_2/4 + \delta_1))$$

$$= \exp\left(-i \sum_{m=1}^{M} 4\pi J_{CH} \delta_2 H_{mz}^1 C_z^1\right) \exp\left(-i \sum_{k=2}^{K} 2\pi J_{C^1 C^k} T C_z^1 C_z^k\right)$$

$$\times \exp(i\Omega_{C^1} t_2) \exp\left(-i\pi \sum_{k=1}^{K} C_x^k\right) \qquad [7.66]$$

for $\Delta = T/4 = \delta_1 + \delta_2$ and $2\delta_2 = 2.2$ ms. Evolution due to $^{13}C-^{13}C$ scalar coupling interactions occurs during the entire constant-time evolution period, whereas C^1 chemical-shift evolution proceeds during t_2 only. The selective 180°(CO) pulses remove the effects of one-bond $^{13}C-^{13}CO$ scalar coupling interaction (selective pulses also can be applied simultaneously to the aromatic ^{13}C spins to remove the effects of scalar coupling between the $^{13}C^\beta$ and $^{13}C^\gamma$ spins of aromatic residues). The duration of the selective pulses must be minimized in order to maximize the attainable resolution, for a given value of T. The magnetization present at time c is described by

$$\sigma_c = \{C_x^1 \cos(\pi J_{C^1 C^2} T) + 2C_y^1 C_z^2 \sin(\pi J_{C^1 C^2} T)$$

$$\times \cos(\Omega_{H^1} t_1) \cos(\Omega_{C^1} t_2) \Pi_1(T) M \Gamma_M(2\delta_2). \qquad [7.67]$$

Magnetization transfer through the remainder of the sequence is essentially the same as for the non-constant-time version of the experiment. At the start of the final acquisition period, f, the magnetization is described by

$$\sigma_f = \left\{ -\sum_{m=1}^{M} H_{mx}^1 \cos^2(\pi J_{C^1 C^2} T) \Pi_1(T) \Gamma_M(2\delta_2) \right.$$

$$\left. -\sum_{n=1}^{N_2} H_{nx}^2 \sin^2(\pi J_{C^1 C^2} T) \Pi_2(T) \Gamma_N(2\delta_2) \right\}$$

$$\times \cos(\Omega_{H^1} t_1) \cos(\Omega_{C^1} t_2) \Pi_1(T) M \Gamma_M(2\delta_2). \qquad [7.68]$$

Magnetization transfer between scalar coupled carbons is independent of t_2 and can be optimized by setting the duration of T. The maximum t_2 acquisition time, and therefore digital resolution, is limited to be less than T, whereas values of T significantly longer than $1/(4J_{CC})$ reduce sensitivity because of the passive carbon couplings. In practice, a value of $T \sim 7.8$ ms gives close to optimal transfer and sufficient digital resolution in the ^{13}C dimension (if the spectral width is chosen to give appropriate aliasing of resonances in this dimension; Section 7.1.2.5). At the same time, purely absorptive, singlet Lorenztian lineshapes are obtained in the F_2 dimension.

Fig. 7.23 shows an example $F_1(^1H)-F_3(^1H)$ slice from a constant-time HCCH-COSY spectrum of ^{13}C-labeled ubiquitin. The F_2 spectral width is only ~ 32 ppm; therefore, aliasing has occurred and the displayed slice

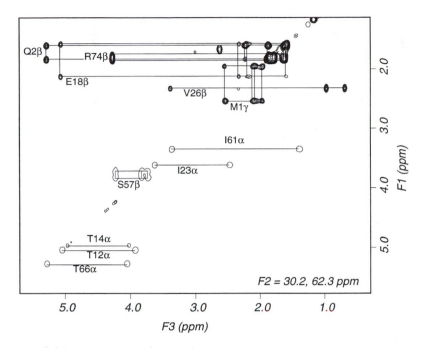

FIGURE 7.23 A selected $F_1(^1H)$–$F_3(^1H)$ slice from a constant-time HCCH-COSY spectrum of ^{13}C-labeled ubiquitin in D_2O solution, acquired using the pulse sequence illustrated in Fig. 7.22b. Negative cross-peaks, which correspond to resonances that have been aliased in the $F_2(^{13}C)$ dimension, are plotted with a single level only; these peaks have $F_2(^{13}C)$ chemicals shifts of 62.3 ppm. The labels indicate the assignment in the $F_1(^1H)$ and $F_2(^{13}C)$ dimensions.

corresponds to two ^{13}C chemical shifts (30.2 and 62.3 ppm); those resonances that have been aliased in the $F_2(^{13}C)$ dimension have phases to opposite those resonances that have not been aliased.

7.3.3 HCCH-TOCSY

The HCCH-TOCSY experiment (54,55) is similar to the HCCH-COSY experiment, except that the 90°(^{13}C) mixing pulse is replaced by an isotropic mixing scheme that results in both direct and relayed magnetization transfer along the carbon side chain. The HCCH-TOCSY pulse sequence illustrated in Fig. 7.24 combines features from the two HCCH-COSY pulse sequences (Fig. 7.22a,b).

Up to time b, the HCCH-TOCSY sequence is equivalent to the sequence shown in Fig. 7.22b, and the magnetization present at time b is

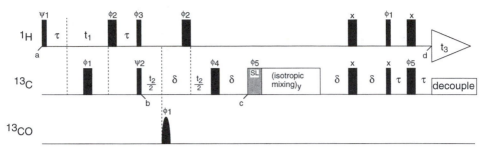

FIGURE 7.24 Pulse sequence for the HCCH-TOCSY experiment. The phase cycling for this experiment is given by $\psi_1 = x$; $\psi_2 = x$; $\phi_1 = 8(x)$, $8(-x)$; $\phi_2 = 4(x)$, $4(y)$, $4(-x)$, $4(-y)$; $\phi_3 = y$, $-y$; $\phi_4 = 2(x)$, $2(y)$, $2(-x)$, $2(-y)$; $\phi_5 = 4(x)$, $4(-x)$; and receiver $= x$, $-x$, $-x$, x, $2(-x, x, x, -x)$, x, $-x$, $-x$, x. Quadrature detection in the t_1 and t_2 dimensions is achieved by incrementing independently the phases ψ_1 and ψ_2, respectively, and the receiver phase, in a TPPI–States manner.

described by [7.58]. The following sequence, up to time c, is equivalent to the sequence shown in Fig. 7.22a, and the relevant magnetization present at time c is given by [7.60]. The short ($\sim$2-ms) spin-lock "trim pulse" (SL) applied along the x axis following time c defocuses the $2C_y^l C_z^k$ antiphase coherence that is not parallel to the effective field. The subsequent isotropic mixing sequence (Section 4.2.1.2) transfers the in-phase C_x^l magnetization to its neighbors within the carbon spin system, via the ^{13}C–^{13}C scalar coupling interactions. The remainder of the experiment comprises a refocused reverse INEPT sequence that transfers the in-phase ^{13}C magnetization back to the attached protons for detection. The final magnetization prior to acquisition is given by

$$\sigma_d = \left\{ \sum_{m=1}^{M} H_{mx}^l \Pi_1(2\delta)\Gamma_M(2\delta)a_{11}(\tau_m) + \sum_{k=2}^{K} \sum_{n=1}^{N} H_{nx}^k \Pi_2(2\delta)\Gamma_N(2\delta)a_{1k}(\tau_m) \right\}$$
$$\times \cos(\Omega_H^l t_1) \cos(\Omega_{C^l} t_2) \cos[\pi J_{C^l C^2}(t_2 + 2\delta)] \cos(2\pi J_{C^l C^2}\delta)$$
$$\times \Pi_1(t_2 + 2\delta)M\Gamma_M(2\delta), \qquad [7.69]$$

in which $a_{11}(\tau_m)$ and $a_{1k}(\tau_m)$ are the isotropic mixing coefficients and τ_m is the mixing time.

An efficient broadband isotropic mixing scheme, such as DIPSI or FLOPSY sequences (57–60), must be used for the HCCH-TOCSY experiment, because the ^{13}C chemical shifts are dispersed over a wide frequency range. The rate of coherence transfer from one carbon to its neighbor depends on the magnitude of the effective scalar coupling during the mixing time. For coupled ^{13}C spins with significantly different chemical

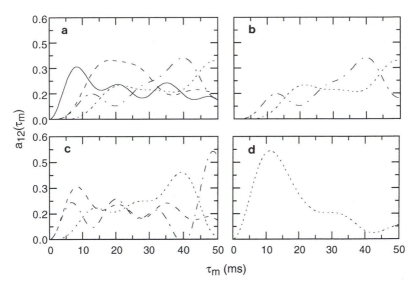

FIGURE 7.25 Isotropic mixing for ¹³C spins in isoleucine. Transfer functions for ideal isotropic mixing under the strong coupling Hamiltonian are shown for magnetization transfer for magnetization originating on the (a) ¹³Cᵅ, (b) ¹³Cᵝ, and (c) ¹³Cᵞ², and (d) ¹³Cᵞ¹ spins. The curves for the destination spins are (—) Cᵝ, (– – –) Cᵞ², (– · –) Cᵞ¹, and (- - - -) Cᵟ. The one-bond scalar coupling constants were assumed to be 35 Hz. Relaxation and resonance offset effects have been neglected.

shifts (e.g., Thr ¹³Cᵝ and ¹³C$_\gamma$, and Ala ¹³C$_\alpha$ and ¹³C$_\beta$), the magnitude of the effective coupling can be reduced substantially and the rate of ¹³C magnetization transfer during the isotropic mixing period is reduced correspondingly.

To determine the optimum ¹³C isotropic mixing time to be used in an HCCH-TOCSY experiment, rates of the carbon–carbon magnetization transfer must be known for the amino acid spin systems of interest. Calculations for the DIPSI-3 sequence (54,61) indicate that an isotropic mixing time of 20–35 ms is optimal for the observation of ¹³C relayed connectivities. Fig. 7.25 shows the calculated net intraresidue ¹³C–¹³C magnetization transfer for an isoleucine amino acid spin system as a function of the mixing time during an ideal isotropic sequence.

Selected regions from a HCCH-TOCSY spectrum of ¹³C-labeled ubiquitin, illustrating the assignment of the side-chain resonances of Ile23, are shown in Fig. 7.26. Cross-peaks corresponding to ¹³Cᵅ, ¹³Cᵞ², and ¹³Cᵟ in the F_2 dimension have been aliased, and appear with phase opposite to those of the cross-peaks corresponding to ¹³Cᵝ, and ¹³Cᵞ¹. The low

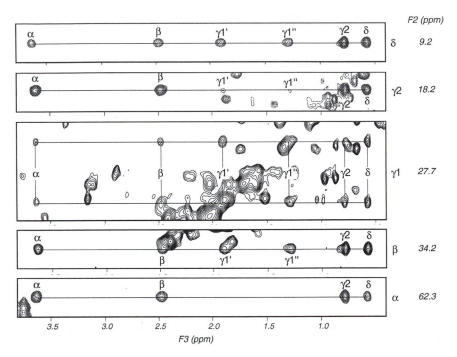

FIGURE 7.26 Selected regions from $F_2(^{13}C)$ slices of a 22.5-ms mixing-time 3D HCCH-TOCSY spectrum of ^{13}C-labeled ubiquitin, showing correlations originating from the $^1H^\alpha$, $^1H^\beta$, $^1H^{\gamma 1}$, $^1H^{\gamma 2}$, and $^1H^\delta$ of Ile23. Isotropic mixing was achieved using a DIPSI-2 sequence with a 7.7-kHz rf field strength.

intensity of the $^1H^\alpha$–$^1H^{\gamma 1}$ and $^1H^{\gamma 2}$–$^1H^{\gamma 1}$ correlations in this 22.5-ms-mixing-time HCCH-TOCSY spectrum is to be expected from Fig. 7.25.

7.4 3D Triple-Resonance Experiments

Three- and four-dimensional heteronuclear triple-resonance experiments correlate backbone $^1H^N$, ^{15}N, $^1H^\alpha$, $^{13}C^\alpha$, and ^{13}CO (and side-chain $^1H^\beta$ and $^{13}C^\beta$) spins using one-bond and two-bond scalar coupling interactions. These experiments constitute an alternative to the classical sequential resonance assignment strategy based on observation of characteristic short-range NOEs (Chapter 8). A large number of triple-resonance pulse schemes have been published since the original description of this approach in application to calmodulin (16.7 kDa), by Bax and coworkers in 1990 (62,63). As of 1994, the largest monomeric proteins to be assigned using these methods are in the 27–28-kDa range (*64,65*); assignment of a 31.4-kDa homodimeric protein has also been achieved (*66*).

The nomenclature established for triple-resonance experiments is more-or-less systematic. The spins that are frequency labeled during the indirect evolution periods or the acquisition period are listed using HN, N, HA, CA, CO, HB, and CB to represent the $^1H^N$, ^{15}N, $^1H^\alpha$, $^{13}C^\alpha$, ^{13}CO, $^1H^\beta$, and $^{13}C^\beta$ spins, respectively. Spins through which coherence is transferred, but not frequency-labeled, are given in parentheses. For example, a triple-resonance experiment utilizing the following coherence transfers:

$$^1H^N \rightarrow \underset{(t_1)}{^{15}N} \rightarrow ^{13}CO \rightarrow \underset{(t_2)}{^{13}C^\alpha} \rightarrow ^{13}CO \rightarrow \underset{(t_3)}{^{15}N} \rightarrow ^1H^N \qquad [7.70]$$

might be called a (HN)N(CO)CA(CO)(N)NH experiment. However, this unwieldy moniker can be shortened by using the following conventions. First, the experiment is a so-called "out and back" pulse sequence in which the initially excited proton spin and the detected proton spin are identical. Omitting the back-transfer steps from the name yields the shorter form, HNN(CO)CA, without introducing ambiguity because the $^{13}C^\alpha$ would never be the detected spin, and the presence of a back-transfer pathway to the $^1H^N$ spin thereby is implied. Second, the designation of the H^N spin is redundant, because the transfer from $^1H^N \leftrightarrow {}^{15}N$ is the only available step. Thus, HN can be abbreviated as H without complication to yield the final name, HN(CO)CA, for this experiment. This abbreviated name describes equally well an experiment that rearranges the labeling periods as

$$^1H^N \rightarrow \underset{(t_1)}{^{15}N} \rightarrow ^{13}CO \rightarrow \underset{(t_2)}{^{13}C^\alpha} \rightarrow \underset{(t_3)}{^{13}CO} \rightarrow ^{15}N \rightarrow ^1H^N. \qquad [7.71]$$

The order in which the frequency labeling is performed is easily determined from the pulse sequence. A four-dimensional version of this experiment that includes a CO evolution period would be designated HNCOCA.

Table 7.3 summarizes the correlations that are observed, and the scalar couplings utilized for coherence transfer, in several useful triple-resonance experiments. These experiments offer alternative ways of establishing sequential backbone connectivities, and at least two and often more independent pathways can be found to support a given sequential assignment, without any knowledge of the spin-system type. Knowledge of some of the spin systems involved is required to "align" the assignments with the protein amino acid sequence; this information may be obtained from the HCCH-COSY and HCCH-TOCSY experiments (Section 7.3),

TABLE 7.3

Triple-Resonance Experiments Used for Sequential Resonance Assignment[a]

Experiment	Correlations observed	Magnetization transfer	J Couplings[b]	Ref.
HNCA	$^1H_i^N-^{15}N_i-^{13}C_i^\alpha$ $^1H_i^N-^{15}N_i-^{13}C_{i-1}^\alpha$		$^1J_{NH}$ $^1J_{NC^\alpha}$ $^2J_{NC^\alpha}$	63,69,67
HN(CO)CA	$^1H_i^N-^{15}N_i-^{13}C_{i-1}^\alpha$		$^1J_{NH}$ $^1J_{NCO}$ $^1J_{C^\alpha CO}$	70,67
H(CA)NH	$^1H_i^\alpha-^{15}N_i-^1H_i^N$ $^1H_i^\alpha-^{15}N_{i+1}-^1H_{i+1}^N$		$^1J_{C^\alpha H^\alpha}$ $^1J_{NC^\alpha}$ $^2J_{NC^\alpha}$ $^1J_{NH}$	56
HNCO	$^1H_i^N-^{15}N_i-^{13}CO_{i-1}$		$^1J_{NH}$ $^1J_{NCO}$	63,67,71
HN(CA)CO	$^1H_i^N-^{15}N_i-^{13}CO_i$ $^1H_i^N-^{15}N_i-^{13}CO_{i-1}$		$^1J_{NH}$ $^1J_{NC^\alpha}$ $^2J_{NC^\alpha}$ $^1J_{C^\alpha CO}$	72

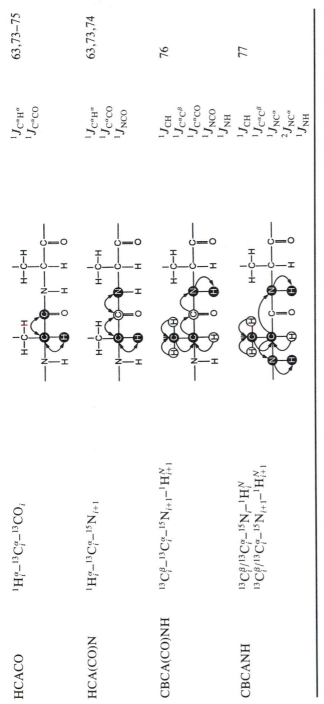

Experiment	Diagram	Pathway	Coupling constants	Ref.
HCACO		$^1H_i^\alpha - ^{13}C_i^\alpha - ^{13}CO_i$	$^1J_{C^\alpha H^\alpha}$ $^1J_{C^\alpha CO}$	63,73–75
HCA(CO)N		$^1H_i^\alpha - ^{13}C_i^\alpha - ^{15}N_{i+1}$	$^1J_{C^\alpha H^\alpha}$ $^1J_{C^\alpha CO}$ $^1J_{NCO}$	63,73,74
CBCA(CO)NH		$^{13}C_i^\beta / ^{13}C_i^\alpha - ^{15}N_i - ^1H_i^N$ $^{13}C_i^\beta / ^{13}C_i^\alpha - ^{15}N_{i+1} - ^1H_{i+1}^N$	$^1J_{CH}$ $^1J_{C^\alpha C^\beta}$ $^1J_{C^\alpha CO}$ $^1J_{NCO}$ $^1J_{NH}$	76
CBCANH		$^{13}C_i^\beta / ^{13}C_i^\alpha - ^{15}N_i - ^1H_i^N$ $^{13}C_i^\beta / ^{13}C_i^\alpha - ^{15}N_{i+1} - ^1H_{i+1}^N$	$^1J_{CH}$ $^1J_{C^\alpha C^\beta}$ $^1J_{NC^\alpha}$ $^2J_{NC^\alpha}$ $^1J_{NH}$	77

[a] Only the experiments analyzed in the text are listed. A more extensive tabulation of triple-resonance (and other) NMR experiments is presented elsewhere (120).

[b] $^1J_{NH} \sim 91$ Hz, $^1J_{NC^\alpha} \sim 7\text{-}11$ Hz, $^2J_{NC^\alpha} \sim 4\text{-}9$ Hz, $^1J_{NCO} \sim 15$ Hz, $^1J_{C^\alpha CO} \sim 55$ Hz, $^1J_{CH}(^1J_{C^\alpha H^\alpha}, ^1J_{C^\beta H^\beta}) \sim 140$ Hz, $^1J_{C^\alpha C^\beta} \sim 35$ Hz.

from a ^{1}H–^{15}N TOCSY-HSQC experiment (Section 7.2.2), or from knowledge of ^{13}C$^\alpha$ and ^{13}C$^\beta$ chemical shifts. Assignment strategies using triple-resonance experiments are summarized in Chapter 8.

The experiments listed in Table 7.3 are discussed in more detail below to demonstrate the basic principles of triple-resonance NMR spectroscopy. The set of experiments presented is certainly not complete, and new triple-resonance pulse sequences, based on the principles described herein, continue to be developed. In the following discussion, the nuclear spins of the ith amino acid residue are designated H_i^N for the amide proton spin, N_i for the amide ^{15}N spin, H_i^κ and C_i^κ for aliphatic proton and ^{13}C spins ($\kappa = \alpha, \beta$, etc.), and C_i' for the carbonyl ^{13}C spin. One-bond scalar coupling constants are designated J_{CH}, J_{NH}, $J_{C^\kappa C^\eta}$ ($\kappa, \eta = \alpha, \beta, \gamma$, etc.), $J_{C^\alpha CO}$, and J_{NCO} (interresidue ^{15}N–^{13}CO scalar coupling). Intraresidue and interresidue ^{13}C$^\alpha$–^{15}N scalar coupling constants are designated $^1J_{C^\alpha N}$ and $^2J_{C^\alpha N}$, respectively. Aspects of data acquisition and processing common to all triple-resonance NMR experiments are discussed in Section 7.4.6.

7.4.1 A PROTOTYPE TRIPLE-RESONANCE EXPERIMENT: HNCA

The HNCA experiment correlates the amide ^{1}H and ^{15}N chemical shifts with the intraresidue ^{13}C$^\alpha$ shift, by making use of the relatively small one-bond ^{15}N–^{13}C$^\alpha$ J coupling (7–11 Hz) to establish J correlations between the ^{15}N and ^{13}C$^\alpha$ spins (63,67). In addition, this experiment also provides sequential connectivities by transferring coherence from the ^{15}N spins to the ^{13}C$^\alpha$ of the preceding residue via the interresidue two-bond ^{15}N–^{13}C$^\alpha$ J coupling, which can be as large as 9 Hz. The HNCA experiment utilizes out-and-back coherence transfer, in which the ^{1}H^N magnetization excited initially in the pulse sequence also is detected during t_3. A simple HNCA pulse sequence (63) is illustrated in Fig. 7.27a. Analysis of this pulse sequence using the product operator formalism will reveal the basic principles of triple-resonance NMR spectroscopy. Figure 7.27 also includes more sophisticated versions of the same experiment that overcome several shortcomings of the original. The more advanced pulse sequences utilize many of the concepts introduced elsewhere in this text and serve to illustrate the iterative process by which NMR experiments are refined.

7.4.1.1 A Simple HNCA Experiment   The design of the HNCA pulse sequence shown in Fig. 7.27a is particularly straightforward, and evolution through the sequence can be evaluated by inspection. Magnetization originating on amide protons is transferred to the directly attached ^{15}N spins via an INEPT sequence. For $2\tau = 1/(2J_{NH})$ the resulting antiphase ^{15}N magnetization at time a in Fig. 7.27a is represented by

$$\sigma_a = -2H_{iz}^N N_{iy} \qquad [7.72]$$

Chemical-shift evolution of ^{15}N proceeds during the subsequent t_1 evolution period. Evolution due to scalar coupling interactions between the ^{15}N spin and ^{1}H^N, ^{13}C$^\alpha$, and ^{13}CO spins is eliminated by 180° refocusing pulses applied in the middle of the t_1 period. The magnetization present at time b is described by

$$\sigma_b = -2H_{iz}^N N_{iy} \cos(\Omega_{N_i} t_1). \qquad [7.73]$$

On some spectrometers, simultaneous application of the 180°(^{13}C$^\alpha$) and 180°(^{13}CO) pulses may not be feasible. Omitting the 180°(^{13}CO) pulse from the experiment introduces an additional $\cos(\pi J_{NCO} t_1)$ factor in [7.73]; fortunately, this does not greatly attenuate the signal because the maximum acquisition time in the ^{15}N dimension ($t_{1,max}$) is typically chosen to be less than $1/(2J_{NCO})$. Indeed, the original description of this experiment did not include ^{13}CO decoupling in t_1 (63).

Following t_1 evolution, the ^{15}N magnetization becomes antiphase with respect to the coupled ^{13}C$^\alpha$ spins during the delay δ. Evolution due to ^{1}H–^{15}N couplings also proceeds during this delay; therefore, the duration of δ is restricted to be an integral multiple of $1/J_{NH}$, so that the ^{15}N magnetization is antiphase with respect to its coupled proton at the end of the delay. Scalar coupling evolution involving the ^{13}CO spins is refocused by the 180°(^{13}CO) pulse in the middle of the t_2 evolution period. Ignoring evolution of the ^{13}CO scalar coupling interactions during δ, the relevant components of the density operator present at time c are given by

$$\sigma_c = \{4H_{iz}^N N_{ix} C_{iz}^\alpha \Gamma_1(\delta) + 4H_{iz}^N N_{ix} C_{(i-1)z}^\alpha \Gamma_2(\delta)\} \cos(\Omega_{N_i} t_1) \cos(\pi J_{NH} \delta),$$
$$[7.74]$$

in which $\Gamma_1(t)$ and $\Gamma_2(t)$ are coherence transfer functions for intra- and interresidue scalar connectivities between the ^{15}N and ^{13}C$^\alpha$ spins, respectively:

$$\Gamma_1(t) = \sin(\pi \,^1J_{C^\alpha N} t) \cos(\pi \,^2J_{C^\alpha N} t); \qquad [7.75]$$
$$\Gamma_2(t) = \cos(\pi \,^1J_{C^\alpha N} t) \sin(\pi \,^2J_{C^\alpha N} t).$$

As shown in Fig. 7.28, these functions have two relatively broad maxima centered around 28 and 35 ms for α-helical and β-sheet structures, respectively (68). As shown in Fig. 7.29, the total amplitudes of the operators in [7.74], $\Gamma_1(\delta) \cos(\pi J_{NH} \delta)$ and $\Gamma_2(\delta) \cos(\pi J_{NH} \delta)$, have maxima for δ equal

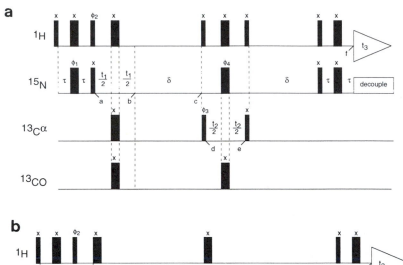

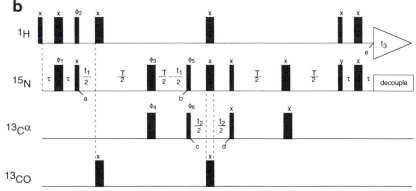

FIGURE 7.27 Pulse sequences for the HNCA experiment. (a) A non-constant-time HNCA experiment, in which the phase cycling is $\phi_1 = x, -x$; $\phi_2 = y, -y$; $\phi_3 = 2(x), 2(-x)$; $\phi_4 = 4(x), 4(y), 4(-x), 4(-y)$; and receiver $= x, -x, -x, x, -x, x, x, -x$. (b) A constant-time HNCA experiment, in which the phase cycling is $\phi_1 = x, -x$; $\phi_2 = y, -y$; $\phi_3 = 4(x), 4(y), 4(-x), 4(-y)$; $\phi_4 = 16(x), 16(-x)$, $\phi_5 = 16(y), 16(-y)$; $\phi_6 = 2(x), 2(-x)$; and receiver $= 2(x, -x, -x, x, -x, x, x, -x)$, $2(-x, x, x, -x, x, -x, -x, x)$. (c) A ^{1}H-decoupled constant-time HNCA experiment with phase cycling $\phi_1 = x, -x$; $\phi_2 = y, -y$; $\phi_3 = 4(x), 4(y), 4(-x), 4(-y)$; $\phi_4 = \phi_5 = 16(x), 16(-x)$; $\phi_6 = 2(x), 2(-x)$; receiver $= 2(x, -x, -x, x, -x, x, x, -x)$, $2(-x, x, x, -x, x, -x, -x, x)$. (d) A PFG-PEP-HNCA experiment with phase cycling $\phi_1 = y, -y$; $\phi_2 = 2(x), 2(-x)$; $\phi_3 = 4(x), 4(y), 4(-x), 4(-y)$; and receiver $= 2(x, -x, -x, x, -x, x, x, -x)$. The data are acquired and processed as described for the PFG-PEP-HSQC experiment (Section 7.1.3.3).

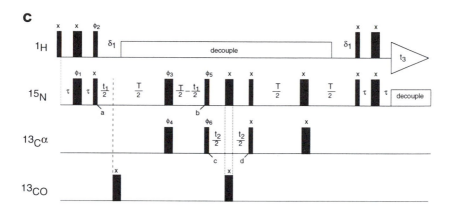

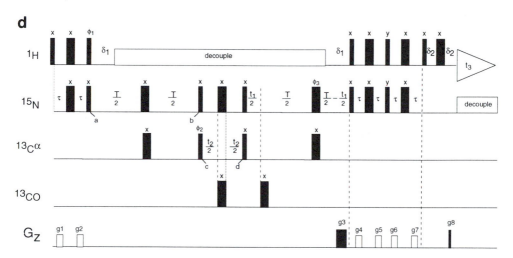

FIGURE 7.27—*Continued*

to ~22 ms and ~33 ms. The positions of these maxima correspond to $2/J_{NH}$ and $3/J_{NH}$, respectively, and are not significantly affected by the small variations of the $^{13}C^{\alpha}$–^{15}N coupling constants with local secondary structure. In order to minimize relaxation losses, the value of δ should be set to ~22 ms, for which $\cos(\pi J_{NH}\delta) \approx 1$.

The 90° pulses applied to both ^{1}H and $^{13}C^{\alpha}$ spins immediately following time c creates multiple-quantum three-spin $^1H^N$–^{15}N–$^{13}C^{\alpha}$ coherence represented by

$$\sigma_d = \{4H_{iy}^N N_{ix} C_{iy}^{\alpha}\Gamma_1(\delta) + 4H_{iy}^N N_{ix} C_{(i-1)y}^{\alpha}\Gamma_2(\delta)\}\cos(\Omega_{N_i} t_1)\cos(\pi J_{NH}\delta).$$

$$[7.76]$$

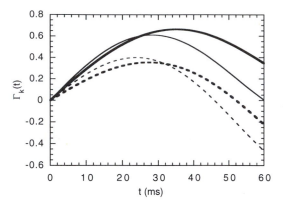

FIGURE 7.28 Plots of the HNCA coherence transfer functions, $\Gamma_k(T)$, for (——) $k = 1$ or (– – –) $k = 2$ [7.75]. Thin and thick lines show the results for α-helical and β-sheet conformations, respectively. Nominal values of $^1J_{C^\alpha N}$ and $^2J_{C^\alpha N}$ for an α-helical conformation are assumed to be 9.6 and 6.4 Hz, respectively, and in a β-sheet conformation are assumed to be 10.9 and 8.3 Hz, respectively (68).

During the subsequent t_2 evolution period, both ^{1}H and ^{15}N chemical shifts are refocused by the 180° pulses applied to these spins; effectively, the three-spin coherence evolution depends only on the ^{13}C$^\alpha$ chemical shift. The three-spin coherence does not evolve under the influence of the active scalar couplings between the spins (Section 2.7.5). Application of the 180°(^{13}CO) pulse at the midpoint of the t_2 evolution period, however,

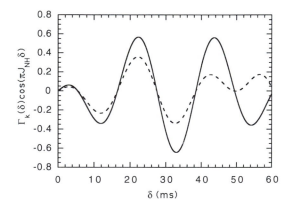

FIGURE 7.29 Plots of the HNCA coherence transfer functions, $\Gamma_k(\delta)$ cos(π $J_{NH}\delta$) for (——) $k = 1$ or (– – –) $k = 2$. The plots were calculated for $^1J_{C^\alpha N}$ and $^2J_{C^\alpha N}$, coupling constants of 10 and 7 Hz, respectively, and a J_{NH} coupling constant of 91 Hz.

ensures that evolution due to ^{13}CO scalar coupling is refocused; the small F_2 phase errors resulting from the off-resonance effects of the $180°(^{13}CO)$ pulse are readily corrected when processing the data (Section 3.4.1). The scalar coupling interactions between $^{13}C^\alpha$ and $^{13}C^\beta$ spins do, however, remain active during the t_2 evolution period. At the end of the t_2 period, e, the magnetization is described by

$$\sigma_e = \{4H_{iy}^N N_{ix} C_{iy}^\alpha \cos(\Omega_{C_i^\alpha} t_2)\Gamma_1(\delta) + 4H_{iy}^N N_{ix} C_{(i-1)y}^\alpha \cos(\Omega_{C_{i-1}^\alpha} t_2)\Gamma_2(\delta)\}$$

$$\times \cos(\Omega_{N_i} t_1) \cos(\pi J_{C^\alpha C^\beta} t_2) \cos(\pi J_{NH}\delta). \quad\quad [7.77]$$

Alternatively, the antiphase ^{15}N magnetization present at time c can be transferred into antiphase $^{13}C^\alpha$ magnetization by application of 90° pulses to the ^{15}N and $^{13}C^\alpha$ spins, rather than to the ^1H and $^{13}C^\alpha$ spins as described. This alternative method, which is described in Section 7.4.1.2 for a constant-time version of the HNCA experiment, might be thought to offer a distinct advantage because the signal decay in the t_2 dimension is determined by the transverse relaxation rate of single-quantum $^{13}C^\alpha$ coherence rather than the three-spin $^1H^N - ^{15}N - ^{13}C^\alpha$ multiple-quantum coherence. However, as discussed by Grzesiek and Bax (67), the $^{13}C^\alpha$ spins remain coupled to the $^{13}C^\beta$ spins during the t_2 evolution period [7.77], and the acquisition time in the t_2 dimension is therefore kept shorter than $1/(2J_{C^\alpha C^\beta})$, about 8–10 ms in practice. Over this time period, the difference in the intrinsic relaxation rates of transverse $^{13}C^\alpha$ magnetization and three-spin $^1H^N - ^{15}N - ^{13}C^\alpha$ coherence is of little consequence, and the two methods of transferring coherence from ^{15}N to $^{13}C^\alpha$ are equivalent.

The magnetization present following the t_2 evolution period [7.77] is transferred back to observable $^1H^N$ magnetization by a pathway that is the reverse of that described above, with the exception that the t_1 evolution period is omitted. The 90° pulses immediately following time e converts the three-spin $^1H^N - ^{15}N - ^{13}C^\alpha$ coherence back into antiphase ^{15}N magnetization, which rephases with respect to its coupled $^{13}C^\alpha$ spin during the delay δ. The ^{15}N remains antiphase with respect to its coupled amide ^1H because, as discussed above, the duration of δ is set to be an integral multiple of $1/J_{NH}$. The final series of pulses represent a reverse INEPT sequence and result in observable $^1H^N$ magnetization described by

$$\sigma_f = H_{ix}^N\{\cos(\Omega_{C_i^\alpha} t_2)\Gamma_1^2(\delta) + \cos(\Omega_{C_{i-1}^\alpha} t_2)\Gamma_2^2(\delta)\}$$

$$\times \cos(\Omega_N t_1) \cos(\pi J_{C^\alpha C^\beta} t_2) \cos^2(\pi J_{NH}\delta). \quad\quad [7.78]$$

A representative $F_2(^{13}C^\alpha) - F_3(^1H^N)$ slice from a HNCA spectrum of $^{15}N/^{13}C$-labeled ubiquitin is shown in Fig. 7.30. The observed $^1H^N - ^{13}C^\alpha$ correlations are labeled according to the $^{13}C^\alpha$ resonance; in this example, the stronger correlations correspond to the intraresidue connectivities whereas the weaker correlations correspond to sequential interresidue

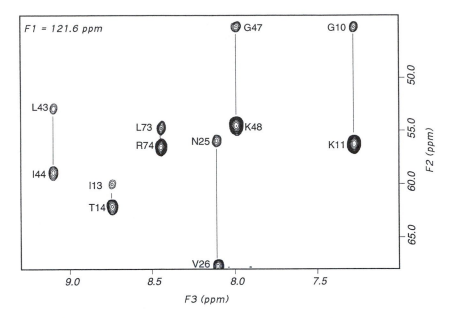

FIGURE 7.30 A selected $F_2(^{13}C^\alpha)$–$F_3(^1H^N)$ slice, at an $F_1(^{15}N)$ chemical shift of 121.6 ppm, from an HNCA spectrum of $^{15}N/^{13}C$-labeled ubiquitin.

connectivities. Empirically, average values of $^1J_{C^\alpha N} > {}^2J_{C^\alpha N}$; however, the ranges of intra- and interresidue scalar coupling constants observed in proteins overlap, and intensities of the resonance peaks can be affected by differences in relaxation rates. Consequently, identification of the intra- and interresidue connectivities on the basis of resonance intensities in the HNCA experiment is not infallible.

7.4.1.2 The CT-HNCA Experiment The modified version of the HNCA experiment illustrated in Fig. 7.27b includes a "constant-time" period during which evolution of the ^{15}N chemical shift and evolution of the ^{15}N–$^{13}C^\alpha$ scalar coupling interaction occur simultaneously. In the original HNCA experiment (Fig. 7.27a), the ^{15}N chemical shift and ^{15}N–$^{13}C^\alpha$ scalar coupling evolution periods are independent. The main advantage of the constant-time evolution scheme, as discussed in more detail below, is a reduction in relaxation losses and a concomitant improvement in sensitivity.

During the first part of the CT-HNCA experiment, magnetization originating on amide protons is transferred to their directly attached ^{15}N spins via an INEPT sequence, in a fashion identical to that for the original experiment discussed above. The magnetization present at time a in Fig. 7.27b is therefore represented by [7.72].

Between points a and b, the pulse sequence is similar to the constant-time periods used in the constant-time HSQC (Section 7.1.1.4) and constant-time ^{1}H–^{13}C HSQC (Section 7.1.3.1) experiments. By similar reasoning, evolution due to the ^{15}N–^{13}C$^\alpha$ J coupling interaction ($^1J_{C^\alpha N}$, $^2J_{C^\alpha N}$) occurs during the entire constant-time evolution period, T, whereas ^{15}N chemical-shift evolution proceeds only during t_1. The ^{15}N spins are effectively decoupled from both the ^{1}H and ^{13}CO spins by the application of 180° pulses to these nuclei; consequently, the ^{15}N magnetization remains antiphase with respect to its coupled proton. The magnetization at time b is described by

$$\sigma_b = \{4H_{iz}^N N_{ix} C_{iz}^\alpha \Gamma_1(T) + 4H_{iz}^N N_{ix} C_{(i-1)z}^\alpha \Gamma_2(T)\} \cos(\Omega_{N_i} t_1), \quad [7.79]$$

in which $\Gamma_1(T)$ and $\Gamma_2(T)$ are the coherence transfer functions [7.75]. If relaxation during T is ignored, intraresidue coherence transfer is optimized by selecting T to maximize $\Gamma_1(T)$ (Fig. 7.28). The amplitude of the signal is further reduced, however, by a factor of exp(-RT), in which R is the average relaxation rate constant for the $H_{iz}^N N_{ix} C_{iz}^\alpha$ operator. In practice T is typically chosen to be 22–28 ms.

The magnetization present at the end of the constant-time delay, T, is transferred by the simultaneous ^{15}N and ^{13}C$^\alpha$ 90° pulses into antiphase ^{13}C$^\alpha$ magnetization at time c:

$$\sigma_c = \{4H_{iz}^N N_{iz} C_{iy}^\alpha \Gamma_1(T) + 4H_{iz}^N N_{iz} C_{(i-1)y}^\alpha \Gamma_2(T)\} \cos(\Omega_{N_i} t_1). \quad [7.80]$$

The three 180° pulses applied at the midpoint of the t_2 evolution period serve to refocus ^{1}H, ^{15}N, and ^{13}CO scalar coupling evolution. The ^{13}C$^\alpha$ and ^{13}C$^\beta$ spins are scalar-coupled during t_2; thus, the acquisition time in the t_2 dimension should be shorter than $1/(2J_{C^\alpha C^\beta})$, or about 8–10 ms in practice. At the end of the t_2 evolution period, d, the relevant operators are

$$\sigma_d = \{4H_{iz}^N N_{iz} C_{iy}^\alpha \cos(\Omega_{C_i^\alpha} t_2)\Gamma_1(T) + 4H_{iz}^N N_{iz} C_{(i-1)y}^\alpha \cos(\Omega_{C_{i-1}^\alpha} t_2)\Gamma_2(T)\}$$

$$\times \cos(\Omega_{N_i} t_1) \cos(\pi J_{C^\alpha C^\beta} t_2). \quad [7.81]$$

This magnetization is then transferred back into observable amide ^{1}H magnetization by a pathway the reverse of that described above. At the beginning of the acquisition period, the observable magnetization is described by

$$\sigma_e = H_{ix}^N \{\cos(\Omega_{C_i^\alpha} t_2)\Gamma_1^2(T) + \cos(\Omega_{C_{i-1}^\alpha} t_2)\Gamma_2^2(T)\} \cos(\Omega_N t_1) \cos(\pi J_{C^\alpha C^\beta} t_2). \quad [7.82]$$

For $T = \delta = 2/J_{NH}$, [7.82] is identical to that derived in [7.78] for the original non-constant-time version of the HNCA experiment. As mentioned above, however, the advantage of the constant-time evolution scheme is found when relaxation effects are considered. In the original HNCA experiment (Fig. 7.27a) signal decays as t_1 increases; in the CT-HNCA experiment (Fig. 7.27b) signal does not decay as t_1 increases. For $T = \delta$, the gain in sensitivity that can be obtained from using the constant-time evolution scheme is approximately

$$G = \frac{t_{1,max}}{\int_0^{t_{1,max}} \exp(-t_1 \overline{R}_{2N})\, dt_1} = \frac{t_{1,max}\overline{R}_{2N}}{1 - \exp(-t_{1,max}\overline{R}_{2N})}, \qquad [7.83]$$

in which $\overline{R}_{2N}$ is the transverse relaxation rate of the antiphase ^{15}N magnetization ($2H_{iz}^N N_{iy}$) present during the t_1 evolution period of the non-constant-time experiment [7.31], and $t_{1,max}$ is the maximum duration of the t_1 evolution period. As with any constant-time evolution scheme, $t_{1,max}$ is limited to be less than (or equal to) the constant-time period, T. For ubiquitin, $\overline{R}_{2N} \approx 9.9$ s^{-1} (calculated as described in Section 7.1.2.4 for $\tau_c = 4.1$ ns), and for $t_{1,max} = 18.6$ ms, [7.83] predicts a sensitivity gain of 10% in the constant-time experiment. Larger gains would be expected for larger proteins with larger relaxation rate constants.

7.4.1.3 The Decoupled CT-HNCA Experiment   The sensitivity of the preceding CT-HNCA experiment (and related pulse sequences) can be further increased by the introduction of synchronous broadband ^1H decoupling (67,69), as illustrated in Fig. 7.27c. At the end of the first delay, $\delta_1 = 1/(2J_{NH})$, the ^{15}N magnetization is refocused to be in-phase with respect to its attached proton. Application of synchronous broadband ^1H decoupling (e.g., WALTZ-16 or DIPSI-2) prevents the creation of any ^{15}N or ^{13}C quantum states that are antiphase with respect to their attached proton. Contributions to the ^{15}N and ^{13}C linewidths from ^1H longitudinal relaxation are therefore eliminated, since both the ^{15}N and ^{13}C transverse magnetization remains in-phase (Section 7.1.1.3). The resulting decrease in the apparent R_{2N} and R_{2C} relaxation rate constants reduces the signal loss caused by ^{15}N and ^{13}C transverse relaxation during delays between the two $^1H-^{15}N$ INEPT sequences. The magnetization at time c is therefore described by

$$\sigma_c = \{-2N_{iz}C_{iy}^\alpha \Gamma_1(T) - 2N_{iz}C_{(i-1)y}^\alpha \Gamma_2(T)\}\cos(\Omega_{N_i} t_1). \qquad [7.84]$$

This expression can be compared directly with [7.76] and [7.80] from the HNCA and CT-HNCA experiments, respectively, which contain multiple-quantum ($4H_{iy}^N N_{ix}C_{iy}^\alpha$) or anti-phase ($4H_{iz}^N N_{iz}C_{iy}^\alpha$) states with respect to ^1H spins.

7.4.1.4 The Gradient-Enhanced HNCA Experiment The HNCA, and many other triple-resonance experiments, include an indirect evolution period for amide ^{15}N spins and detect amide proton spins during the acquisition period. These experiments are particularly easy to modify for coherence selection using pulsed field gradients and PEP sensitivity enhancement (Section 7.1.3.3). A pulse sequence for a decoupled PFG-PEP-HNCA experiment is shown in Fig. 7.27d (a nongradient PEP-HNCA could be designed as well). This experiment is very similar to that shown in Fig. 7.27c, with the following exceptions: (1) ^{15}N frequency labeling is performed during the second period, T, rather than the first; (2) the reverse INEPT sequence of Fig. 7.27c is replaced by a PEP reverse INEPT sequence; (3) field-gradient pulses are used for coherence selection and quadrature detection in the ^{15}N evolution period; and (4) pulsed field-gradient (PFG) pulses are used to suppress artifacts associated with 180° pulses. Acquisition and processing of the PFG-PEP-HNCA experiment is performed as described for the PFG-PEP-HSQC experiment (Section 7.1.3.3).

7.4.2 A COMPLEMENTARY APPROACH: THE HN(CO)CA EXPERIMENT

The HN(CO)CA experiment provides sequential correlations between the amide ^{1}H and ^{15}N chemical shifts of one amino acid residue and the ^{13}C$^\alpha$ chemical shift of the preceding residue by transferring coherence via the intervening ^{13}CO spin (*67,70*). The same sequential information is obtained from the HNCA experiment (Section 7.4.1); however, the HNCA experiment does not always distinguish intraresidue and interresidue connectivities because the $^1J_{C^\alpha N}$ and $^2J_{C^\alpha N}$ coupling constants can be of comparable magnitude, or the intraresidue and interresidue ^{13}C$^\alpha$ chemical shifts may be coincidentally degenerate. The HN(CO)CA experiment circumvents these problems by providing sequential correlations exclusively. In addition, the sensitivity of the HN(CO)CA experiment is greater than that of the HNCA for larger proteins, because the relay of magnetization via the one bond J_{NCO} and $J_{C^\alpha CO}$ scalar coupling interactions is more efficient than transfer via the relatively small two-bond $^2J_{C^\alpha N}$ scalar coupling interaction.

The salient features of the CT-HN(CO)CA experiment illustrated in Fig. 7.31 are discussed below. Following the initial INEPT-type transfer of magnetization from the amide protons to their directly attached ^{15}N spins, the magnetization present at time a in Fig. 7.31 is represented by

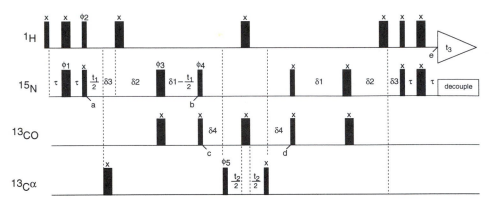

FIGURE 7.31 Pulse sequence for a constant-time HN(CO)CA experiment. The phase cycle is $\phi_1 = x, -x$; $\phi_2 = y, -y$; $\phi_3 = 4(x), 4(y), 4(-x), 4(-y)$; $\phi_4 = x, -x$; $\phi_5 = 2(x), 2(-x)$; and receiver $= 2(x), 4(-x), 2(x)$.

$$\sigma_a = -2H_{iz}N_{iy} \quad \text{for} \quad 2\tau = 1/(2J_{NH}). \quad [7.85]$$

During the subsequent constant-time evolution period, $2\delta_1 = 2(\delta_2 + \delta_3)$, the propagator is

$$\mathbf{U} = \exp\left(-i\frac{\pi}{2}C'_{(i-1)x}\right)\exp\left(-i\frac{\pi}{2}N_{ix}\right)\exp[-i\mathcal{H}(\delta_1 - t_1/2)]$$

$$\times \exp(-i\pi C'_{(i-1)x})\exp(-i\pi N_{ix})\exp[-i\mathcal{H}\delta_2]\exp\left(-i\pi\sum H_{kx}^{\kappa}\right)$$

$$\times \exp[-i\mathcal{H}\delta_3]\exp\left(-i\pi\sum C_{kx}^{\kappa}\right)\exp[-i\mathcal{H}t_1/2]$$

$$= \exp\left(-i\frac{\pi}{2}C'_{(i-1)x}\right)\exp\left(-i\frac{\pi}{2}N_{ix}\right)\exp(-i4\pi J_{NCO}\delta_1 N_{iz}C'_{(i-1)z})$$

$$\times \exp(-i4\pi J_{NH}\delta_3 H_{iz}^N N_{iz})\exp(-i\Omega_{N_i}t_2 N_{iz})$$

$$\times \exp\left(-i\pi\sum H_{kx}^{\kappa}\right)\exp(-i\pi\sum C_{kx}^{\kappa})\exp(-\pi C'_{ix})\exp(-\pi N_{ix}),$$

$$[7.86]$$

in which $\exp(-i\pi\sum H_{kx}^{\kappa})$ represents a nonselective proton 180° pulse and $\exp(-i\pi\sum C_{kx}^{\kappa})$ represents a nonselective carbon 180° pulse (with summations over k and κ implied); only chemical-shift and scalar coupling interactions affecting the ^{15}N spin have been included in the Hamiltonian, and the last line is obtained by using [2.117]. Evolution due to $^1H–^{15}N$ coupling occurs only during $2\delta_3$, whereas evolution due to $^{15}N–^{13}CO$ coupling occurs during the entire period, $2\delta_1$. For $2\delta_3 = 1/(2J_{NH})$, the antiphase ^{15}N operator is refocused to become in-phase with respect to the amide proton at the end of the constant-time period. Chemical-shift evolution of the ^{15}N coherence occurs only during t_1. The duration of the constant-time evolution period, $2\delta_1$, can be adjusted from $1/(3J_{NCO})$ to $1/(2J_{NCO})$, independently of the $^1H–^{15}N$ coupling evolution period, $2\delta_3$, as required to minimize relaxation losses. Typically, $2\delta_1 \approx 22$ ms $[1/(3J_{NCO})]$. The magnetization present at time b is described by

$$\sigma_b = 2N_{iy}C'_{(i-1)z} \cos(\Omega_{N_i} t_1) \sin(2\pi J_{NCO}\delta_1) \sin(2\pi J_{NH}\delta_3). \quad [7.87]$$

The pair of simultaneous 90° pulses transfers [7.87] into antiphase carbonyl magnetization to give, at time c,

$$\sigma_c = 2N_{iz}C'_{(i-1)y} \cos(\Omega_{N_i} t_1) \sin(2\pi J_{NCO}\delta_1) \sin(2\pi J_{NH}\delta_3). \quad [7.88]$$

The product operator in [7.88] is in-phase with respect to the amide proton spin. If this ^{13}CO magnetization were antiphase, its transverse relaxation rate (and the relaxation rates of the product operators during t_2) would be increased as a result of contributions from proton longitudinal relaxation, and the signal observed would be correspondingly decreased. This would be the case if the amide protons were decoupled during the constant-time evolution period, as in the CT-HNCA experiment [7.79].

During the subsequent $\delta_4–t_2–\delta_4$ period the ^{13}CO spin is correlated with its attached $^{13}C^\alpha$ spin in a HMQC-like manner. Ignoring evolution due to relatively small scalar couplings between $^{13}C^\alpha$ and ^{15}N and between ^{13}CO and protons, which are not refocused by the 180° pulses at the midpoint of t_2, the coherence present at time d is

$$\sigma_d = -2N_{iz}C'_{(i-1)y} \cos(\Omega_{N_i}t_1) \cos(\Omega_{C^\alpha_{i-1}}t_2) \cos(\pi J_{C^\alpha C_\beta}t_2)$$
$$\times \sin^2(\pi J_{C^\alpha CO}\delta_4) \sin(2\pi J_{NCO}\delta_1) \sin(2\pi J_{NH}\delta_3). \quad [7.89]$$

The delay δ_4 is set to $1/(3J_{C^\alpha CO})$ to $1/(2J_{C^\alpha CO})$. The $^{13}C^\alpha$ and $^{13}C^\beta$ spins remain coupled during t_2; therefore, in order to avoid sensitivity losses due to resolved $J_{C^\alpha C^\beta}$ scalar couplings in the F_2 dimension, the t_2 acquisition time must be shorter than $1/(2J_{C^\alpha C^\beta})$, or about 8–10 ms in practice. The off-resonance effect of the 180°(^{13}CO) pulse in the middle of the t_2 evolution period results in a small frequency-dependent phase shift in this dimension that can be corrected approximately during data processing (Section 3.4.1). Magnetization is transferred back along the same pathway to yield

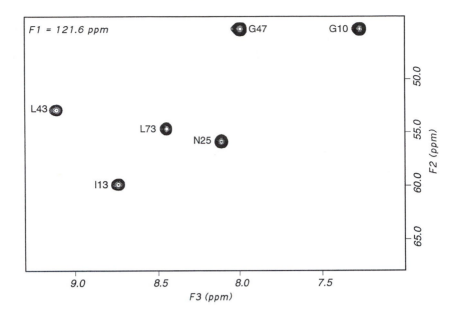

121.6 ppm, from an HN(CO)CA spectrum of ^{15}N/^{13}C-labeled ubiquitin. This slice may be compared directly with the equivalent region from the HNCA spectrum of ubiquitin illustrated in Fig. 7.30.

the observable magnetization at time e:

$$\sigma_e = H_{ix}^N \cos(\Omega_{N_i} t_1) \cos(\Omega_{C_{i-1}^\alpha} t_2) \cos(\pi J_{C^\alpha C^\beta} t_2)$$

$$\times \sin^2(\pi J_{C^\alpha CO} \delta_4) \sin^2(2\pi J_{NCO} \delta_1) \sin^2(2\pi J_{NH} \delta_3). \qquad [7.90]$$

As in the HNCA experiment, the sensitivity of the HN(CO)CA experiment can be improved by the use of synchronous broadband proton decoupling after the antiphase term at time a, [7.85], has been allowed to refocus and by incorporating PEP sensitivity enhancement.

A representative $F_2(^{13}C^\alpha)$–$F_3(^1H^N)$ slice from a HN(CO)CA spectrum of ^{15}N/^{13}C-labeled ubiquitin is shown in Fig. 7.32. This slice was taken at the same $F_1(^{15}N)$ chemical shift as the $F_2(^{13}C^\alpha)$–$F_3(^1H^N)$ slice from the HNCA spectrum of ubiquitin shown in Fig. 7.30. By comparison of these spectra, the interresidue correlations in the HNCA spectrum can be unambiguously distinguished from the intraresidue correlations, because only the interresidue connectivities are observed in the HN(CO)CA experiment.

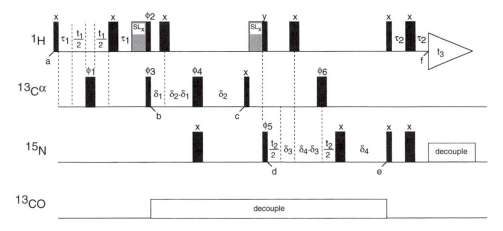

FIGURE 7.33 Pulse sequence for a constant-time H(CA)NH experiment. The phase cycle is $\phi_1 = x, -x$; $\phi_2 = y, -y$; $\phi_3 = 2(x), 2(-x)$; $\phi_4 = 4(x), 4(y), 4(-x)$, $4(-y)$; $\phi_5 = 16(x), 16(-x)$; $\phi_6 = 8(x), 8(-x)$; and receiver $= 2(x, -x, -x, x, -x, x, x, -x), 2(-x, x, x, -x, x, -x, -x, x)$.

7.4.3 A STRAIGHT-THROUGH TRIPLE-RESONANCE EXPERIMENT: H(CA)NH

The H(CA)NH experiment (56) correlates $^1H^\alpha$ chemical shifts with intra-residue amide ^1H and ^{15}N chemical shifts, by making use of the one-bond $^{15}N-^{13}C^\alpha$ J coupling (7–11 Hz). This experiment also provides sequential connectivities by transferring coherence from the $^{13}C^\alpha$ spins to the ^{15}N of the following residue via the interresidue two-bond $^{15}N-^{13}C^\alpha$ J coupling (4–9 Hz). The H(CA)NH experiment differs fundamentally from the HNCA experiment because one-way "straight-through" or "out-and-stay" transfer of magnetization from $^1H^\alpha$ to ^{15}N spins is obtained via successive through-bond transfer between the directly coupled $^1H^\alpha-^{13}C^\alpha$, $^{13}C^\alpha-^{15}N$, and $^{15}N-^1H^N$ pairs. At each transfer step, net magnetization transfer (rather than coherence transfer to antiphase spin states) must be obtained using refocused INEPT sequences. The product operator of the H(CA)NH experiment illustrated in Fig. 7.33 is described in the following paragraphs.

Beginning with longitudinal $^1H^\alpha$ magnetization at time a ($\sigma_a = H_{iz}^\alpha$), transverse $^1H^\alpha$ magnetization is frequency labeled during t_1 and transferred in an INEPT-like manner to its directly attached $^{13}C^\alpha$, to yield at time b:

$$\sigma_b = 2H_{iz}^\alpha C_{iy}^\alpha \cos(\Omega_{H_i^\alpha} t_1), \qquad [7.91]$$

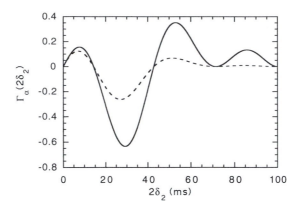

FIGURE 7.34 Plots of the H(CA)NH coherence transfer function, $\Gamma_\alpha(2\delta_2)$ [7.93], for $^1J_{C^\alpha N}$, $^2J_{C^{\alpha 2}N}$, and $^1J_{C^\alpha C^\beta}$ coupling constants of 10, 7, and 35 Hz, respectively. The solid line corresponds to the transfer function in the absence of relaxation; the dashed line includes the relaxation term $\exp(-2R_{2,C^\alpha}\delta_2)$ for $^{13}C^\alpha$ linewidths of $1/(\pi T_{2,C^\alpha}) = 10$ Hz.

for $2\tau_1 = 1/(2J_{CH})$ (~3.5 ms). A spin-lock purge pulse (SL_x) is applied immediately prior to the two simultaneous 90° pulses to suppress the water signal (Section 3.5.3). With the exception of the spin-lock purge pulse, the identical concatenated t_1 evolution period and INEPT magnetization transfer scheme was used in the CT-HCCH-COSY experiment (Fig. 7.22b).

Following the simultaneous 90° ^{1}H and $^{13}C^\alpha$ pulses, the antiphase $^{13}C^\alpha$ magnetization, [7.91], is refocused with respect to the $^1H^\alpha$ spins during the period $2\delta_1$, while dephasing due to $^{15}N-^{13}C^\alpha$ one-bond and two-bond J couplings and $^{13}C^\alpha-^{13}C^\beta$ one-bond J coupling occurs during $2\delta_2$. For $^{13}C^\alpha$ directly coupled to a single proton (i.e., all amino acids except glycine), the relevant components of the density operator at time c are

$$\sigma_c = \{-2C_{iy}^\alpha N_{iz}\Gamma_1(2\delta_2) - 2C_{iy}^\alpha N_{(i-1)z}\Gamma_2(2\delta_2)\}$$
$$\times \cos(\Omega_{H_i^\alpha}t_1)\sin(2\pi J_{CH}\delta_1)\cos(2\pi J_{C^\alpha C^\beta}\delta_2), \qquad [7.92]$$

in which the coherence transfer functions, $\Gamma_1(t)$ and $\Gamma_2(t)$, are given by [7.75]. The optimal value of $2\delta_1 = 1/(2J_{CH})$. Intraresidue coherence transfer is optimized by selecting δ_2 to maximize

$$\Gamma_\alpha(2\delta_2) = \Gamma_1(2\delta_2)\cos(2\pi J_{C^\alpha C^\beta}\delta_2)\exp(-2R_{2C_i^\alpha}\delta_2), \qquad [7.93]$$

in which $R_{2C_i^\alpha}$ is the $^{13}C^\alpha$ transverse relaxation rate. This function, which is plotted in Fig. 7.34, has a maximum value for $\delta_2 = 12$ to 13 ms.

The $90°(^{13}C^{\alpha})$ pulse immediately following c creates $^{13}C^{\alpha}-^{15}N$ two-spin order. The subsequent ^{1}H spin-lock and $90°_y$ pulses further suppress the residual water resonance. The two-spin order subsequently is transferred to antiphase ^{15}N coherence by the $90°(^{15}N)$ pulse to give, at time d,

$$\sigma_d = \{2C_{iz}^{\alpha}N_{iy}\Gamma_1(2\delta_2) + 2C_{iz}^{\alpha}N_{(i+1)y}\Gamma_2(2\delta_2)\}\cos(\Omega_{H_i^{\alpha}}t_1)$$

$$\times \sin(2\pi J_{CH}\delta_1)\cos(2\pi J_{C^{\alpha}C^{\beta}}\delta_2). \qquad [7.94]$$

During the subsequent evolution period between points d and e, the propagator is

$$\mathbf{U} = \exp[-i\mathcal{H}\delta_4]\exp\left(-i\pi\sum N_{kx}\right)\exp[-i\mathcal{H}t_2/2]\exp\left(-i\pi\sum C_{kx}^{\alpha}\right)$$

$$\times \exp[-i\mathcal{H}(\delta_4 - \delta_3)]\exp\left(-i\pi\sum H_{kx}^{\kappa}\right)\exp[-i\mathcal{H}(\delta_3 + t_2/2)]$$

$$= \sum \{\exp(-i4\pi J_{NH}\delta_3 H_{kz}^{N}N_{kz})\exp(-i4\pi^1 J_{C^{\alpha}N}\delta_4 C_{kz}^{\alpha}N_{kz})$$

$$\times \exp(-i4\pi^2 J_{C^{\alpha}N}\delta_4 C_{(k-1)z}^{\alpha}N_{kz})\exp(-i\Omega_{N_k}t_2 N_{kz})\}$$

$$\times \exp\left(-i\pi\sum N_{kx}\right)\exp\left(-i\pi\sum C_{kx}^{\alpha}\right)\exp\left(-i\pi\sum H_{kx}^{\kappa}\right), \qquad [7.95]$$

in which only chemical-shift and scalar coupling interactions affecting the ^{15}N spins have been included in the Hamiltonian and the last line is obtained by using [2.117]. The antiphase ^{15}N coherences refocus with respect to the active $^{15}N-^{13}C$ scalar coupling and dephase as a result of passive $^{15}N-^{13}C$ scalar coupling during $2\delta_4$. Thus, $2C_{iz}^{\alpha}N_{iy}$ refocuses under the intraresidue $^1J_{C^{\alpha}N}$ scalar coupling interaction and dephases under the $^2J_{C^{\alpha}N}$ scalar coupling to the $^{13}C^{\alpha}$ spin of the preceding residue, whereas $2C_{iz}^{\alpha}N_{(i+1)y}$ refocuses under the interresidue scalar coupling interaction and dephases due to the intraresidue coupling. The ^{15}N magnetization also defocuses with respect to the amide proton during the time period $2\delta_3$. The relevant coherences present at time e are described by

$$\sigma_e = \{-2H_{iz}^{N}N_{iy}\cos(\Omega_{N_i}t_2)\Gamma_1(2\delta_2)\Gamma_1(2\delta_4)$$

$$-2H_{(i+1)z}^{N}N_{(i+1)y}\cos(\Omega_{N_{i+1}}t_2)\Gamma_2(2\delta_2)\Gamma_2(2\delta_4)\}$$

$$\times \cos(\Omega_{H_i^{\alpha}}t_1)\sin(2\pi J_{CH}\delta_1)\cos(2\pi J_{C^{\alpha}C^{\beta}}\delta_2)\sin(2\pi J_{NH}\delta_3). \quad [7.96]$$

Intraresidue coherence transfer through this segment of the experiment is maximized by setting $2\delta_3 = 1/(2J_{NH})$ ($\sim$5.4 ms) and adjusting $2\delta_4$ to

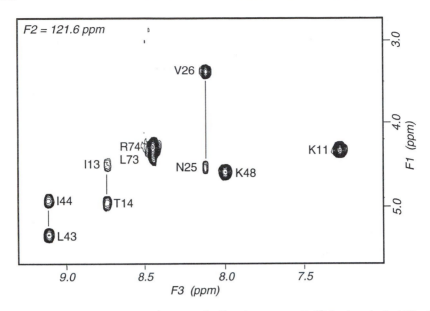

FIGURE 7.35 A selected $F_1(^1H^\alpha)$–$F_3(^1H^N)$ slice, at an $F_2(^{15}N)$ chemical shift of 121.6 ppm, from a H(CA)NH spectrum of $^{15}N/^{13}C$-labeled ubiquitin. Resonances for Gly10 and Gly47 are suppressed for $2\tau_1 = 1/(2J_{CH})$.

maximize $\Gamma_1(2\delta_4)$. This transfer function is identical to that plotted in Fig. 7.28 for the HNCA experiment [7.75]. When ^{15}N transverse relaxation, R_{2N}, is considered, a value of $2\delta_4$ of ~23 ms is close to optimal.

The final INEPT step in the pulse sequence transfers the antiphase ^{15}N magnetization to in-phase $^1H^N$ magnetization to yield, at time f,

$$\sigma_f = \{H_{ix}^N \cos(\Omega_{N_i} t_2)\Gamma_1(2\delta_2)\,\Gamma_1(2\delta_4) + H_{(i+1)x}^N \cos(\Omega_{N_{i+1}} t_2)\Gamma_2(2\delta_2)\Gamma_2(2\delta_4)\}$$

$$\times \cos(\Omega_{H_i^\alpha} t_1)\sin(2\pi J_{CH}\delta_1)\cos(2\pi J_{C^\alpha C^\beta}\delta_2)\sin(2\pi J_{NH}\delta_3) \qquad [7.97]$$

for $2\tau_2 = 1/(2J_{NH})$. Thus, magnetization that originated as $^1H^\alpha$ magnetization has been transferred to and is detected as amide $^1H^N$ magnetization.

A representative $F_1(^1H^\alpha)$–$F_3(^1H^N)$ slice from a H(CA)NH spectrum of $^{15}N/^{13}C$-labeled ubiquitin is shown in Fig. 7.35. The observed $^1H^N$–$^1H^\alpha$ correlations are labeled according to the $^1H^\alpha$ resonance. In this example, the stronger correlation corresponds to the intraresidue connectivities and the weaker correlation corresponds to sequential interresidue connectivities. Identification of the intra- and interresidue connectivities on the basis of resonance intensities in the H(CA)NH experiment has the same caveats as the HNCA experiment (Section 7.4.1).

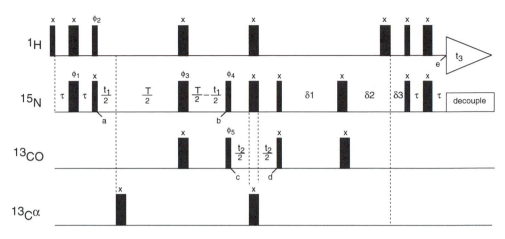

FIGURE 7.36 Pulse sequence for a constant-time HNCO experiment. The phase cycle is $\phi_1 = x, -x$; $\phi_2 = y, -y$; $\phi_3 = 4(x), 4(y), 4(-x), 4(-y)$; $\phi_4 = x, -x$; $\phi_5 = 2(x), 2(-x)$; and receiver $= 2(x), 4(-x), 2(x)$.

7.4.4 OTHER TRIPLE-RESONANCE EXPERIMENTS FOR BACKBONE CORRELATIONS

The HNCA, HN(CO)CA, and H(CA)NH experiments illustrate many of the principles utilized in triple resonance NMR spectroscopy. These experiments alone seldom are sufficient to establish complete backbone sequential connectivities, and do not indicate the enormous variety of triple-resonance experiments that have been developed. For completeness, a number of additional experiments are presented briefly in this section.

7.4.4.1 HNCO The HNCO experiment correlates the amide ^{1}H and ^{15}N chemical shifts of one amino acid with the ^{13}CO chemical shift of the preceding residue, by using the one-bond ^{15}N–^{13}CO J coupling ($\sim$15 Hz) to establish the sequential correlation *(63,67)*. The sequential connectivities provided by this experiment are particularly useful in conjunction with interresidue connectivities provided by the HN(CA)CO (Section 7.4.4.2) and HCACO (Section 7.4.4.3) experiments. A detailed description of alternative pulse sequences for the HNCO experiment has been given by Muhandiram and Kay *(71)*.

The CT-HNCO pulse sequence illustrated in Fig. 7.36 is closely analogous to the CT-HNCA and CT-HN(CO)CA experiments. The CT-HNCO substitutes an evolution period for the carbonyl spins instead of the

HMQC-type magnetization transfer from the ^{13}CO to the $^{13}C^\alpha$ spins used in the HN(CO)CA experiment. The important coherences present at times a–e of the pulse sequence are

$$\sigma_a = -2H_{iz}^N N_{iy};$$ [7.98]

$$\sigma_b = 2N_{iy} C'_{(i-1)z} \cos(\Omega_{N_i} t_1) \sin(\pi J_{NCO} T) \sin(\pi J_{NH} T);$$ [7.99]

$$\sigma_c = -2N_{iz} C'_{(i-1)y} \cos(\Omega_{N_i} t_1) \sin(\pi J_{NCO} T) \sin(\pi J_{NH} T);$$ [7.100]

$$\sigma_d = -2N_{iz} C'_{(i-1)y} \cos(\Omega_{N_i} t_1) \cos(\Omega_{CO_{i-1}} t_2) \sin(\pi J_{NCO} T) \sin(\pi J_{NH} T);$$

[7.101]

$$\sigma_e = H_{ix}^N \cos(\Omega_{N_i} t_1) \cos(\Omega_{CO_{i-1}} t_2) \sin(\pi J_{NCO} T) \sin(\pi J_{NH} T)$$

$$\times \sin(2\pi J_{NCO} \delta_1) \sin(2\pi J_{NH} \delta_3),$$ [7.102]

in which $2\tau = 1/(2J_{NH})$, $2\delta_1 = 2(\delta_2 + \delta_3)$, and $2\delta_3 = 1/(2J_{NH})$. Typically $2\delta_1 \approx 1/(3J_{NCO})$ (22 ms) instead of $1/(2J_{NCO})$ to reduce relaxation losses. The time T must be an odd multiple of $1/(2J_{NH})$ [e.g., $5/(2J_{NH}) = 27$ ms] to ensure that the ^{15}N magnetization is in-phase with respect to the amide proton spins at the end of the constant-time evolution period.

In contrast to the CT-HNCA experiment (Fig. 7.27b), antiphase ^{15}N magnetization present at a rephases with respect to the proton spins under the influence of the ^{15}N–^1H coupling during the constant-time evolution period. Therefore, evolution of the $2N_{iz} C'_{(i-1)y}$ coherence during t_2 does not contain contributions from relaxation of the amide proton spin. The constant-time evolution scheme described here also differs from that described in Section 7.4.2 for the HN(CO)CA experiment, in which the duration of the constant-time period was not limited to be an odd multiple of $1/(2J_{NH})$. Either scheme may be used in each experiment, because if relaxation effects are ignored, the final results are very similar. The sensitivity of this scheme may be further improved by the use of synchronous broadband proton decoupling in an analogous way as for the decoupled CT-HNCA experiment.

7.4.4.2 HN(CA)CO   The HN(CA)CO experiment provides intraresidue correlations between the amide ^1H, ^{15}N, and ^{13}CO chemical shifts by using the one-bond ^{15}N–$^{13}C^\alpha$ and $^{13}C^\alpha$–^{13}CO J couplings to transfer coherence (72). In addition, this experiment can also provide sequential connectivities from the ^{15}N spins to the ^{13}CO of the preceding residue via the interresidue two-bond ^{15}N–$^{13}C^\alpha$ J coupling. When used in conjunction with the HNCO experiment, which gives the sequential correlations only

(Section 7.4.4.1), the HN(CA)CO experiment provides a method for sequentially assigning the amide ^{1}H, ^{15}N, and ^{13}CO resonances.

A constant-time version of a HN(CA)CO pulse sequence is illustrated in Fig. 7.37. The important product operator terms at times a–f of the pulse sequence are given by

$$\sigma_a = -2H_{iz}^N N_{iy}; \tag{7.103}$$

$$\sigma_b = \{2N_{iy}C_{iz}^\alpha \Gamma_1(T) + 2N_{iy}C_{(i-1)z}^\alpha \Gamma_2(T)\}\cos(\Omega_{N_i}t_1); \tag{7.104}$$

$$\sigma_c = \{2N_{iz}C_{iy}^\alpha \Gamma_1(T) - 2N_{iz}C_{(i-1)y}^\alpha \Gamma_2(T)\}\cos(\Omega_{N_i}t_1); \tag{7.105}$$

$$\sigma_d = \{4N_{iz}C_{iz}^\alpha C_{iy}' \Gamma_1(T) + 4N_{iz}C_{(i-1)z}^\alpha C_{(i-1)y}' \Gamma_2(T)\}\cos(\Omega_{N_i}t_1)\Gamma_3(2\delta_2); \tag{7.106}$$

$$\sigma_e = \{4N_{iz}C_{iz}^\alpha C_{iy}' \cos(\Omega_{CO_i}t_2)\Gamma_1(T)$$
$$+ 4N_{iz}C_{(i-1)z}^\alpha C_{(i-1)y}' \cos(\Omega_{CO_{i-1}}t_2)\Gamma_2(T)\}\cos(\Omega_{N_i}t_1)\Gamma_3(2\delta_2); \tag{7.107}$$

$$\sigma_f = H_{ix}^N\{\cos(\Omega_{CO_i}t_2)\Gamma_1^2(T) + \cos(\Omega_{CO_{i-1}}t_2)\Gamma_2^2(T)\}\cos(\Omega_{N_i}t_1)\Gamma_3^2(2\delta_2), \tag{7.108}$$

in which $\delta = 1/(2J_{NH})$, and the coherence transfer functions, $\Gamma_1(t)$ and $\Gamma_2(t)$, are given by [7.75] and

$$\Gamma_3(t) = \sin(\pi J_{C^\alpha CO}t)\cos(\pi J_{C^\alpha C^\beta}t). \tag{7.109}$$

As with the CT-HNCA experiment, the length of the constant-time evolution period, T, is typically chosen to be 22–28 ms (Section 7.4.1.2). A value of 3.0–3.5 ms is optimal for δ_2.

The CT-HN(CA)CO pulse sequence illustrated in Fig. 7.37 differs slightly from the originally reported sequence (72), in which a refocused INEPT sequence (Section 2.7.7.3) is used to generate in-phase ^{15}N magnetization prior to the constant-time evolution period, and to convert in-phase ^{15}N magnetization back into detectable ^{1}H magnetization immediately prior to acquisition. The experiment discussed here achieves the same magnetization state at times b and e (ignoring relaxation effects), but does so in less total evolution time and with four less 180° pulses; for these reasons the sequence presented here is likely to be more sensitive than the originally proposed sequence.

7.4.4.3 HCACO

The HCACO experiment correlates intraresidue ^{1}H$^\alpha$, ^{13}C$^\alpha$, and ^{13}CO chemical shifts (63,73). Given that this experiment

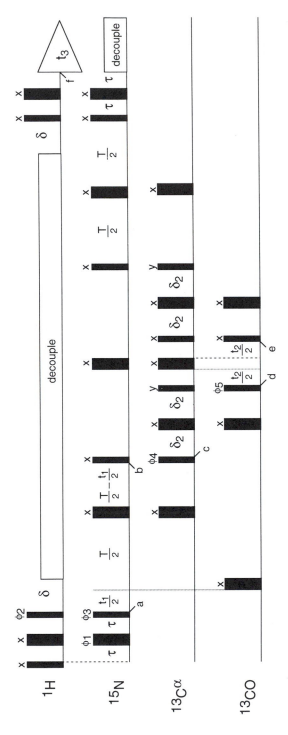

FIGURE 7.37 Pulse sequence for a constant-time HN(CA)CO experiment. The phase cycle is $\phi_1 = x,\ -x$; $\phi_2 = y,\ -y$; $\phi_3 = 8(x)$, $8(-x)$; $\phi_4 = 2(x)$, $2(-x)$; $\phi_5 = 4(x)$, $4(-x)$; and receiver $= x,\ -x,\ -x,\ x,\ 2(-x, x, x, -x),\ x,\ -x,\ -x,\ x$.

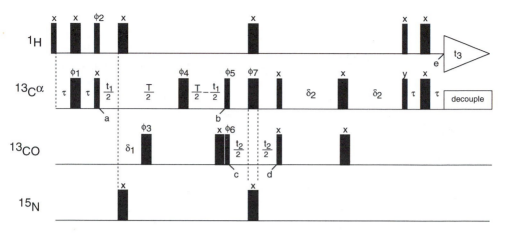

FIGURE 7.38 Pulse sequence for a constant-time HCACO experiment. The phase cycle is $\phi_1 = x, -x$; $\phi_2 = y, -y$; $\phi_3 = 8(x), 8(-x)$; $\phi_4 = 8(x), (8y), 8(-x), 8(-y)$; $\phi_5 = 4(y), 4(-y)$; $\phi_6 = 2(x), 2(-x)$; $\phi_7 = 4(x), 4(-x)$; receiver = $x, -x, -x, x$, $2(-x, x, x, -x), x, -x, -x, x$. For experiments acquired with $T \approx 27$ ms, the phase cycle may be reduced to eight steps by elimination of the EXORCYCLE phase cycling of ϕ_4, thus allowing more t_1 increments to be acquired for a given total acquisition time.

does not detect amide ^{1}H chemical shifts, and that some ^{1}H$^\alpha$ chemical shifts are close to the water resonance, the HCACO experiment usually is performed using protein samples in D$_2$O solution.

A constant-time version of the HCACO experiment (73–75) is illustrated in Fig. 7.38. The important product operator terms at times a–e of the pulse sequence are given by

$$\sigma_a = -2H^\alpha_{iz}C^\alpha_{iy};\qquad\qquad [7.110]$$

$$\sigma_b = 4H^\alpha_{iz}C^\alpha_{ix}C'_{iz}\cos(\Omega_{C^\alpha_i}t_1)\Gamma_1(2\delta_1, T);\qquad\qquad [7.111]$$

$$\sigma_c = 4H^\alpha_{iz}C^\alpha_{iz}C'_{iy}\cos(\Omega_{C^\alpha_i}t_1)\Gamma_1(2\delta_1, T);\qquad\qquad [7.112]$$

$$\sigma_d = 4H^\alpha_{iz}C^\alpha_{iz}C'_{iy}\cos(\Omega_{C^\alpha_i}t_1)\cos(\Omega_{CO_i}t_2)\Gamma_1(2\delta_1, T);\qquad\qquad [7.113]$$

$$\sigma_e = H^\alpha_{ix}\cos(\Omega_{C^\alpha_i}t_1)\cos(\Omega_{CO_i}t_2)\sin(2\pi J_{C^\alpha CO}\delta_2)\Gamma_1(2\delta_1, T),\qquad [7.114]$$

in which $2\tau = 1/(2J_{CH})$ (~3.4 ms) and, $2\delta_2 \approx 1/(3J_{C^\alpha CO})$ (~6 ms) in order to minimize loss of magnetization due to relaxation and dephasing caused by passive couplings to the ^{13}C$^\beta$ spins. The coherence transfer function is

$$\Gamma_1(t, T) = \sin(\pi J_{C^\alpha CO} t) \cos(\pi J_{C^\alpha C^\beta} T). \qquad [7.115]$$

The magnitude of [7.115] is maximized by setting $2\delta_1 = 1/(2J_{C^\alpha CO})$ (~9 ms) and $T = 1/J_{CC}$ (~27 ms). For larger proteins, in which relaxation losses are significant, the sensitivity of this experiment may be increased by setting $T = 2\delta_1 \approx 7$ ms. However, reducing the duration of the constant-time period, T, also reduces the maximum resolution obtainable in the F_1 dimension, because the maximum value of t_1 cannot exceed T.

The second $180°(^{13}CO)$ pulse, applied immediately before time b, simply serves to eliminate the phase error caused by the off-resonance effects of the first $180°(^{13}CO)$ pulse on the transverse $^{13}C^\alpha$ magnetization, and thus ensures pure cosinusoidal modulation in the t_1 dimension (Section 3.4.1). The power of the $180°(^{13}C^\alpha)$ pulse applied in the middle of the $2\delta_2$ period is reduced to a rf field strength of ~2 kHz, in order to further reduce dephasing caused by passive $^{13}C^\alpha - ^{13}C^\beta$ couplings during this interval. Such a semiselective pulse gives good inversion of the $^{13}C^\alpha$ spins, but poor inversion for $^{13}C^\beta$ spins that resonate outside the $^{13}C^\alpha$ region, and hence results in (partial) decoupling of these spins. Amino acids that have $^{13}C^\beta$ resonances close to the $^{13}C^\alpha$ resonances (e.g., Ser and Thr) remain coupled, and [7.114] contains an additional factor equal to $\cos(2\pi J_{C^\alpha C^\beta} \delta_2)$.

If required, ^{15}N can be decoupled from the $^{13}C^\alpha$ spins by application of $180°(^{15}N)$ pulses as shown in Fig. 7.38 or by the application of broadband ^{15}N decoupling throughout the constant-time period, T, and the following t_2 evolution period.

In the CT-HCACO experiment illustrated in Fig. 7.38 the linewidth in the ^{13}CO (F_2) dimension is determined in part by the relaxation rate of $H_{iz}^\alpha C_{iz}^\alpha C_{iy}'$. Relaxation of this term is clearly influenced by $^1H^\alpha$ longitudinal relaxation, which can significantly increase the decay rate of the antiphase ^{13}CO magnetization. As with the previously discussed triple-resonance experiments, the sensitivity of the HCACO experiment can be improved by allowing the antiphase term at the beginning of the constant-time evolution period [7.110] to rephase prior to the application of synchronous broadband proton decoupling (75), in a manner analogous to that for the HNCA and HN(CA)CO pulse sequences illustrated in Figs. 7.27c and 7.37.

7.4.4.4 HCA(CO)N   The HCA(CO)N experiment establishes backbone sequential connectivities by correlating the $^1H^\alpha$ and $^{13}C^\alpha$ chemical shifts of one residue with the ^{15}N chemical shift of the following residue (63,73,74). A simple extension of the CT-HCACO experiment (Fig. 7.38), in which the ^{13}CO chemical-shift evolution period (t_2) is replaced by a $^{13}CO-^{15}N$ HMQC-type sequence, yields the CT-HCA(CO)N pulse sea

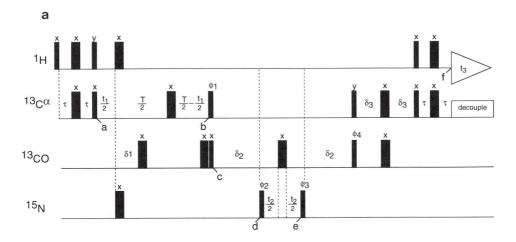

b

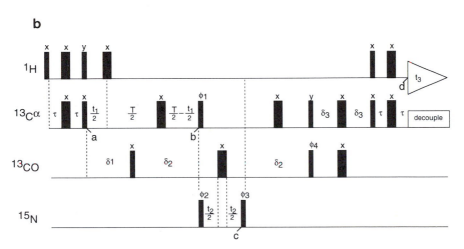

FIGURE 7.39 Pulse sequences for the (a) original and (b) modified constant-time HCA(CO)N experiments, as discussed in the text. The phase cycle for both experiments is $\phi_1 = 4(y), 4(-y)$; $\phi_2 = x, -x$; $\phi_3 = 8(x), 8(-x)$; $\phi_4 = 2(x), 2(-x)$; and receiver $= x, -x, -x, x, 2(-x, x, x, -x), x, -x, -x, x$.

quence illustrated in Fig. 7.39a (*73,74*). This pulse sequence is similar to the CT-HN(CO)CA experiment discussed in Section 7.4.2.

The evolution in the CT-HCA(CO)N experiment up to time c (Fig. 7.39a) is identical to that found above for the HCACO experiment (Fig. 7.38). The magnetization present at times a, b, and c in Fig. 7.39a is therefore described by operators [7.110], [7.111], and [7.112], respectively. The relevant operators at times d–f are

$$\sigma_d = -8H_{iz}^{\alpha}C_{iz}^{\alpha}C_{ix}'N_{(i+1)z}\cos(\Omega_{C_i^{\alpha}}t_1)\sin(\pi J_{NCO}\delta_2)\Gamma_1(2\delta_1, T); \qquad [7.116]$$

$$\sigma_e = -8H_{iz}^{\alpha}C_{iz}^{\alpha}C_{ix}'N_{(i+1)y}\cos(\Omega_{C_i^{\alpha}}t_1)\cos(\Omega_{N_{i+1}}t_2)\sin(\pi J_{NCO}\delta_2)\Gamma_1(2\delta_1, T);$$

$$[7.117]$$

$$\sigma_f = H_{ix}^{\alpha}\cos(\Omega_{C_i^{\alpha}}t_1)\cos(\Omega_{N_{i+1}}t_2)\sin(\pi J_{NCO}\delta_2)\sin(\pi J_{C^{\alpha}CO}\delta_3)\Gamma_1(2\delta_1, T),$$

$$[7.118]$$

in which $\Gamma_1(2\delta_1, T)$ is the constant-time coherence transfer function, given by [7.115]; the value of δ_2 is typically set to $1/(3J_{NCO})$, or even slightly less (~18 ms), in order to minimize relaxation losses; and the delay δ_3 is typically set to ~$1/(3J_{C^{\alpha}CO})$ (~6 ms) in order to minimize magnetization losses due to relaxation, and dephasing caused by coupling to $^{13}C^{\beta}$. Evolution due to $^{13}C^{\alpha}$–^{13}CO coupling and ^{13}CO chemical-shift evolution during δ_2 have been ignored in deriving [7.116] since these effects are refocused by the 180°(^{13}CO) pulse present at the midpoint of t_2. The 180°($^{13}C^{\alpha}$) pulse at the midpoint of the $2\delta_3$ interval is a semiselective pulse, with an rf field strength of ~2 kHz, to reduce the effects of passive $^{13}C^{\alpha}$–$^{13}C^{\beta}$ couplings during this interval (see Section 7.4.4.3). Amino acids that have $^{13}C^{\beta}$ resonances close to the $^{13}C^{\alpha}$ resonances (e.g., Ser and Thr) remain coupled, and [7.118] will contain an additional factor equal to $\cos(2\pi J_{C^{\alpha}C^{\beta}}\delta_3)$.

The overall length of the CT-HCA(CO)N pulse sequence is significantly greater than the CT-HCACO pulse sequence, from which it was derived, due to the presence of the ^{13}CO–^{15}N HMQC magnetization transfer steps ($2\delta_2$). In order to minimize relaxation losses, the duration of the constant-time evolution period is reduced to $T = 2\delta_1 \approx 7$ ms. As mentioned above for the HCACO experiment, reduction of the constant-time evolution period also reduces the maximum resolution obtainable in the F_1 dimension.

If increased *experimental* resolution in the F_1 dimension is required, a pulse sequence has been developed that shortens the overall duration of the CT-HCA(CO)N experiment by overlapping the first δ_2 delay and the constant-time period, T (Fig. 7.39b) (74). Overlapping the two evolution periods reduces relaxation losses so that a longer value of T can be used. The important product operator terms at times a–d in the pulse sequence are given by

$$\sigma_a = -2H_{iz}^{\alpha}C_{iy}^{\alpha}; \qquad [7.119]$$

$$\sigma_b = 8H_{iz}^{\alpha}C_{ix}^{\alpha}C_{ix}'N_{(i+1)z}\cos(\Omega_{C_i^{\alpha}}t_1)\sin(\pi J_{NCO}\delta_2)\Gamma_1(2\delta_1, T); \qquad [7.120]$$

$$\sigma_c = 8H^{\alpha}_{iz}C^{\alpha}_{iz}C'_{ix}N_{(i+1)z}\cos(\Omega_{C^{\alpha}_i}t_1)\cos(\Omega_{N_{i+1}}t_2)\sin(\pi J_{NCO}\delta_2)\Gamma_1(2\delta_1, T);$$

[7.121]

$$\sigma_d = H^{\alpha}_{ix}\cos(\Omega_{C^{\alpha}_i}t_1)\cos(\Omega_{N_{i+1}}t_2)\sin^2(\pi J_{NCO}\delta_2)\sin(\pi J_{C^{\alpha}CO}\delta_3)\Gamma_1(2\delta_1, T),$$

[7.122]

in which $\Gamma_1(2\delta_1, T)$ is the constant-time coherence transfer function, given by [7.115]. Suitable values of δ_1, δ_2, and T, to maximize [7.122], are 9 ms ($\sim 1/(2J_{C^{\alpha}CO})$), 18 ms [slightly less than $\sim 1/(3J_{NCO})$] and 27 ms ($\sim 1/J_{C^{\alpha}C^{\beta}}$), respectively.

During the constant-time period in the CT-HCA(CO)N pulse sequence illustrated in Fig. 7.39b, evolution due to the $^{13}C^{\alpha}$–^{13}CO coupling ($J_{C^{\alpha}CO}$) occurs only during δ_1. Following the 90°(^{13}CO) pulse, which creates two-spin $^{13}C^{\alpha}$–^{13}CO multiple-quantum (MQ) coherence, the ^{15}N–^{13}CO coupling is free to evolve during δ_2. Note, however, that the $^{13}C^{\alpha}$–^{13}CO coupling does not evolve during this period because the two spins are in a MQ state (Section 2.7.5). Chemical-shift evolution of the C'_{ix} operator during δ_2 is refocused by the 180°(^{13}CO) pulse in the middle of the t_2 period and has not been included in [7.120]. Unlike the scheme shown in Fig. 7.39a, ^{15}N decoupling cannot be obtained easily during t_1; the effect of $J_{NC^{\alpha}}$ coupling evolution during t_1 has been ignored in deriving [7.120]. The 180°($^{13}C^{\alpha}$) pulse at the midpoint of the second δ_2 delay is necessary to refocus the evolution due to $^{13}C^{\alpha}$-^{13}CO coupling, because, unlike Fig. 7.39a, the two δ_2 delays are not symmetrically disposed about the 180°(^{13}CO) pulse in the middle of t_2.

The sensitivity of the HCA(CO)N scheme illustrated in Fig. 7.39b, for δ_1, δ_2, and T, equal to 9, 18, and 27 ms, respectively, is inevitably somewhat lower than that obtained using the pulse sequence of Fig. 7.39a, with $T = 2\delta_1 \approx 7$ ms. However, the F_1 digital resolution attainable using the pulse sequence in Fig. 7.39 is significantly higher because $t_{1,max}$ is limited to be less than T. Spectra with satisfactory signal-to-noise ratios may be obtained with either scheme, in as little as 30–40 hs, for proteins up to at least 20 kDa.

7.4.5 CORRELATIONS WITH THE C^{β}/H^{β} SPINS

The triple-resonance experiments discussed thus far provide sequential connectivities along the peptide backbone. The HCCH experiments (Section 7.3) provide connectivities of the aliphatic side-chains of individual amino acid residues. Complete assignments can be obtained if the backbone assignments and the side-chain assignments can be connected using either the $^1H^\alpha$ or $^{13}C^\alpha$ spins. Congestion in the $^1H^\alpha$–$^{13}C^\alpha$ region of the spectrum often renders this approach difficult. One solution to the problem combines HCCH-type magnetization transfer from the side chain to the $^{13}C^\alpha$ spin (using either COSY or TOCSY transfers) with a straight-through magnetization transfer from the $^{13}C^\alpha$ spin to the amide moiety of the following residue (76–85). The side-chain assignments are connected thereby to the better resolved amide $^1H^N$ and ^{15}N spins. Two such experiments, CBCA(CO)NH and CBCANH, are discussed in the following sections.

The topology of the amino acid spin system affects coherence transfer in the CBCA(CO)NH and CBCANH more strongly than triple-resonance experiments that only correlate backbone spins. In the following, b is the number of $^{13}C^\beta$ nuclei in the spin system, g is number of resonant aliphatic $^{13}C^\gamma$ nuclei in the spin system, m is the number of $^1H^\alpha$ spins, and n is the number of $^1H^\beta$ spins in the spin system. The value of b is 0 for glycine and 1 for all other amino acids; the value of g is 0 for alanine, aspartic acid, asparagine, cysteine, and serine, 2 for valine and isoleucine, and 1 for all other residues (aromatic $^{13}C^\gamma$ spins may be perturbed by the aliphatic ^{13}C pulses); m is 2 for glycine and 1 for all other amino acids; and n is 1 for valine, isoleucine and threonine and 2 for all other amino acids.

7.4.5.1 CBCA(CO)NH

The CBCA(CO)NH experiment correlates both the $^{13}C^\alpha$ and $^{13}C^\beta$ resonances of an amino acid residue with the amide ^1H and ^{15}N resonances of the following residue (76). These correlations are extremely useful if significant $^{13}C^\alpha$–$^1H^\alpha$ chemical-shift degeneracy exists. In addition, the $^{13}C^\alpha$ and $^{13}C^\beta$ chemical shifts provide information on the type of amino acid preceding each amide (Chapter 8). As with the HCA(CO)N and HN(CO)CA experiments, interresidue correlations are established by transferring coherence via the intervening ^{13}CO spin. The pulse sequence for the CBCA(CO)NH experiment, which incorporates two constant-time evolution periods, is illustrated in Fig. 7.40. The relevant components of the density operator at the indicated time points in the pulse sequence are given by

$$\sigma_a = -2H_{iz}^\alpha C_{iy}^\alpha - 2H_{iz}^\beta C_{iy}^\beta; \tag{7.123}$$

$$\sigma_b = C_{ix}^\alpha \cos(\Omega_{C_i^\alpha} t_1) \cos^b(\pi J_{C^\alpha C^\beta} T_{AB}) \, m\Gamma_m(\delta_1)$$

$$+ 2C_{iy}^\beta C_{iz}^\alpha \cos(\Omega_{C_i^\beta} t_1) \sin(\pi J_{C^\alpha C^\beta} T_{AB})$$

$$\cos^g(\pi J_{C^\beta C^\gamma} T_{AB}) \, n\Gamma_n(\delta_1); \tag{7.124}$$

$$\sigma_c = C_{ix}^\alpha \cos(\Omega_{C_i^\alpha} t_1) \cos^b(\pi J_{C^\alpha C^\beta} T_{AB}) m\Gamma_m(\delta_1)$$

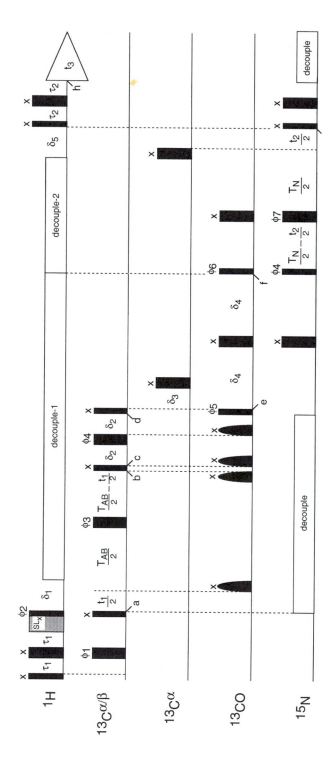

FIGURE 7.40 Pulse sequence for the CBCA(CO)NH experiment. The phase cycle is $\phi_1 = x, -x$; $\phi_2 = y, -y$; $\phi_3 = 8(x), 8(y), 8(-x)$, $8(-y)$; $\phi_4 = 4(x), 4(-x)$; $\phi_5 = 2(x), 2(-x)$; $\phi_6 = x$ plus the off-resonance phase error calculated using [3.72] (see text); $\phi_7 = 8(x)$, $8(-x)$; and receiver $= x, -x, -x, x, 2(-x, x, x, -x), x, -x, -x, x$.

$$-2C_{iz}^{\beta}C_{iy}^{\alpha}\cos(\Omega_{C_i^{\beta}}t_1)\sin(\pi J_{C^{\alpha}C^{\beta}}T_{AB})$$

$$\times\cos^g(\pi J_{C^{\beta}C^{\gamma}}T_{AB})n\Gamma_n(\delta_1);\qquad\qquad [7.125]$$

$$\sigma_d = 2C_{iy}^{\alpha}C_{iz}'\cos(\Omega_{C_i^{\alpha}}t_1)\cos^b(\pi J_{C^{\alpha}C^{\beta}}T_{AB})m\Gamma_m(\delta_1)\Gamma_{1A}(\delta_2)$$

$$+2C_{iy}^{\alpha}C_{iz}'\cos(\Omega_{C_i^{\beta}}t_1)\sin(\pi J_{C^{\alpha}C^{\beta}}T_{AB})$$

$$\times\cos^g(\pi J_{C^{\beta}C^{\gamma}}T_{AB})n\Gamma_n(\delta_1)\Gamma_{1B}(\delta_2);\qquad\qquad [7.126]$$

$$\sigma_e = -2C_{iz}^{\alpha}C_{iy}'\cos(\Omega_{C_i^{\alpha}}t_1)\cos^b(\pi J_{C^{\alpha}C^{\beta}}T_{AB})m\Gamma_m(\delta_1)\Gamma_{1A}(\delta_2)$$

$$-2C_{iz}^{\alpha}C_{iy}'\cos(\Omega_{C_i^{\beta}}t_1)\sin(\pi J_{C^{\alpha}C^{\beta}}T_{AB})$$

$$\times\cos^g(\pi J_{C^{\beta}C^{\gamma}}T_{AB})n\Gamma_n(\delta_1)\Gamma_{1B}(\delta_2);\qquad\qquad [7.127]$$

$$\sigma_f = 2C_{iy}'N_{(i+1)z}\{\cos(\Omega_{C_i^{\alpha}}t_1)\cos^b(\pi J_{C^{\alpha}C^{\beta}}T_{AB})m\Gamma_m(\delta_1)\Gamma_{1A}(\delta_2)$$

$$+\cos(\Omega_{C_i^{\beta}}t_1)\sin(\pi J_{C^{\alpha}C^{\beta}}T_{AB})$$

$$\times\cos^g(\pi J_{C^{\beta}C^{\gamma}}T_{AB})n\Gamma_n(\delta_1)\Gamma_{1B}(\delta_2)\}\Gamma_2(\delta_3,\delta_4);\qquad [7.128]$$

$$\sigma_g = 2H_{(i+1)z}^N N_{(i+1)y}\{\cos(\Omega_{C_i^{\alpha}}t_1)\cos^b(\pi J_{C^{\alpha}C^{\beta}}T_{AB})m\Gamma_m(\delta_1)\Gamma_{1A}(\delta_2)$$

$$+\cos(\Omega_{C_i^{\beta}}t_1)\sin(\pi J_{C^{\alpha}C^{\beta}}T_{AB})\cos^g(\pi J_{C^{\beta}C^{\gamma}}T_{AB})n\Gamma_n(\delta_1)\Gamma_{1B}(\delta_2)\}$$

$$\times\cos(\Omega_{N_{i+1}}t_2)\sin(\pi J_{NCO}T_N)\Gamma_2(\delta_3,\delta_4);\qquad\qquad [7.129]$$

$$\sigma_h = H_{(i+1)x}^N\{\cos(\Omega_{C_i^{\alpha}}t_1)\cos^b(\pi J_{C^{\alpha}C^{\beta}}T_{AB})m\Gamma_m(\delta_1)\Gamma_{1A}(\delta_2)$$

$$+\cos(\Omega_{C_i^{\beta}}t_1)\sin(\pi J_{C^{\alpha}C^{\beta}}T_{AB})\cos^g(\pi J_{C^{\beta}C^{\gamma}}T_{AB})n\Gamma_n(\delta_1)\Gamma_{1B}(\delta_2)\}$$

$$\times\cos(\Omega_{N_{i+1}}t_2)\sin(\pi J_{NCO}T_N)\Gamma_2(\delta_3,\delta_4),\qquad\qquad [7.130]$$

in which $2\tau_1 = 1/(2J_{CH})$, $T_{AB} \approx 6.6$ ms to maximize coherence transfer for both $^{13}C^{\alpha}$ and $^{13}C^{\beta}$ spins, $T_N = 22$ ms, and $\delta_5 = 2\tau_2 = 1/(2J_{NH})$. The magnitudes of the coherence transfer functions, $\Gamma_n(\delta_1)$ and $\Gamma_m(\delta_1)$, are optimized for $\delta_1 = 2.2$ ms ([7.13] and Fig. 7.2). The magnitudes of the coherence transfer functions

$$\Gamma_{1A}(\delta_2) = \cos^b(2\pi J_{C^{\alpha}C^{\beta}}\delta_2)\sin(2\pi J_{C^{\alpha}CO}\delta_2);$$

$$[7.131]$$

$$\Gamma_{1B}(\delta_2) = \sin(2\pi J_{C^{\alpha}C^{\beta}}\delta_2)\sin(2\pi J_{C^{\alpha}CO}\delta_2)$$

are optimized simultaneously by setting $\delta_2 \approx 3.7$ ms. The coherence transfer function $\Gamma_2(\delta_3, \delta_4)$ is

$$\Gamma_2(\delta_3, \delta_4) = \sin(2\pi J_{C^\alpha CO}\delta_3)\sin(2\pi J_{NCO}\delta_4), \qquad [7.132]$$

in which $\delta_4 \approx 1/(6J_{NCO})$ to $1/(4J_{NCO})$, as required to minimize relaxation losses, and $\delta_3 = 1/(4J_{C^\alpha CO})$ (~ 4.5 ms). In the original description of this experiment, a value of $\delta_4 = 11.4$ ms was demonstrated to be suitable for a 31-kDa protein (76).

All $^{13}C^{\alpha/\beta}$ pulses in the CBCA(CO)NH experiment are applied near the center of the $^{13}C^\alpha$ and $^{13}C^\beta$ chemical-shift range in order to maximize excitation of the $^{13}C^\alpha$ and $^{13}C^\beta$ spins, and the power of these pulses is adjusted in order to minimize their effect on the ^{13}CO spins (i.e., for 90° pulses $\gamma B_1 = \Omega/\sqrt{15}$, and for 180° $\gamma B_1 = \Omega/\sqrt{3}$; Section 3.4.1). The two 180°($^{13}C^\alpha$) pulses in this experiment are applied in the center of the $^{13}C^\alpha$ chemical-shift region to maximize their effect on these nuclei, and the power of these pulses is adjusted such that they do not perturb the ^{13}CO spins (i.e., $\gamma B_1 = \Omega\sqrt{3}$; Section 3.4.1). At time f the position of the ^1H carrier, and hence the frequency of the ^1H broadband decoupling, is shifted from the water resonance to the center of the amide region.

The second shaped selective 180°(^{13}CO) pulse, applied immediately before time b, serves to eliminate phase error induced by the off-resonance effects of the first such pulse on the transverse ^{13}C magnetization (Section 3.4.1), and thus ensures pure cosinusoidal modulation in the t_1 dimension. The 180°($^{13}C^\alpha$) pulse applied between time points e and f results in a change in the ^{13}CO phase due to the off-resonant effect of this pulse (Section 3.4.1). This phase error, which can be approximated using [3.72], can be compensated for by an adjustment of the phase of the 90°(^{13}CO) pulse immediately following time f (ϕ_6); the phase of this pulse should be set to x plus the off-resonance phase error, in order to fully transfer ^{13}CO magnetization to the ^{15}N spins.

Selected $F_2(^{15}N)$ slices from a CBCA(CO)NH spectrum of ubiquitin are shown in Fig. 7.41. Two resonances are observed at the $F_1(^{13}C)$ frequencies of the $^{13}C^\alpha$ and $^{13}C^\beta$ spins of residue i and $F_3(^1H^N)$ frequency of residue $i + 1$ for all amino acid residues except glycine (which has only a single resonance corresponding the $^{13}C^\alpha$ spin).

With relatively minor modifications, the CBCA(CO)NH experiment can be converted to an experiment that correlates the $^1H^N$ and ^{15}N resonances of one residue with the $^1H^\alpha$ and $^1H^\beta$ resonances of the preceding residue; such a pulse scheme has been called the HBHA(CBCACO)NH experiment (81). However, this experiment alone does not provide unambiguous assignment of the $^1H^\alpha$ and $^1H^\beta$ resonances, because such assignments generally cannot be made on the basis of chemical shifts alone.

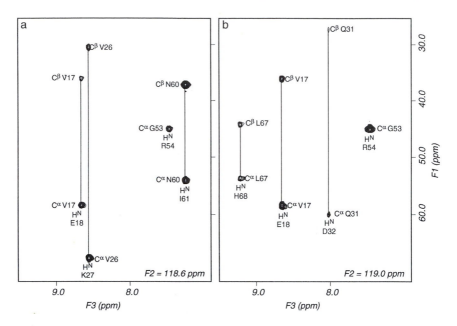

FIGURE 7.41 CBCA(CO)NH spectrum of ubiquitin. Selected (adjacent) $F_1(^{13}C)$–$F_3(^1H^N)$ slices from a 600-MHz 3D CBCA(CO)NH spectrum of $^{15}N/^{13}C$-labeled ubiquitin at $F_2(^{15}N)$ chemical shifts of 118.6 ppm (a) and 119.0 ppm (b). The interresidue cross-peaks are labeled.

7.4.5.2 CBCANH The CBCANH experiment correlates the $^{13}C^\alpha$ and $^{13}C^\beta$ resonances with the amide ^1H and ^{15}N resonances of the same residue and the amide ^1H and ^{15}N resonances of the succeeding residue via the $^1J_{C^\alpha N}$ and $^2J_{C^\alpha N}$ couplings, respectively (77). For proteins up to ~20 kDa, this experiment alone can provide virtually complete sequential assignment of the $^1H^N$, ^{15}N, $^{13}C^\alpha$, and $^{13}C^\beta$ resonances, because in addition to the sequential connectivities, the $^{13}C^\alpha$ and $^{13}C^\beta$ chemical shifts provide information on the amino acid type. The pulse sequence for the CBCANH experiment is illustrated in Fig. 7.42. As with the closely related CBCA(CO)NH experiment discussed above, the CBCANH experiment utilizes two constant-time evolution periods.

The evolution in the CBCANH experiment up to time c (Fig. 7.42) is identical to that found for the CBCA(CO)NH experiment (Fig. 7.40). The magnetization present at times a, b, and c in Fig. 7.42 is therefore described by operators [7.123], [7.124], and [7.125], respectively. The relevant components of the density operator at the time points d–g in the pulse sequence are given by

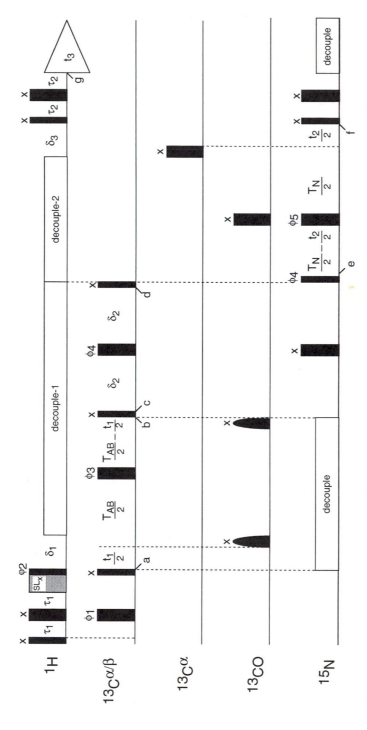

FIGURE 7.42 Pulse sequence for the CBCANH experiment. The phase cycle is $\phi_1 = x, -x$; $\phi_2 = y, -y$; $\phi_3 = 8(x), 8(y), 8(-x)$, $8(-y)$; $\phi_4 = 2(x), 2(-x)$; $\phi_5 = 4(x), 4(-x)$; receiver $= 2(x, -x, -x, x), 2(-x, x, x, -x)$.

$$\sigma_d = 2C_{iy}^{\alpha}N_{iz}\{\cos(\Omega_{C_i^{\alpha}}t_1)\cos^b(\pi J_{C^{\alpha}C^{\beta}}T_{AB})m\Gamma_m(\delta_1)\Gamma_{1A}(\delta_2)$$
$$+ \cos(\Omega_{C_i^{\beta}}t_1)\sin(\pi J_{C^{\alpha}C^{\beta}}T_{AB})\cos^g(\pi J_{C^{\beta}C^{\gamma}}T_{AB})n\Gamma_n(\delta_1)\Gamma_{1B}(\delta_2)\}$$
$$+ 2C_{iy}^{\alpha}N_{(i+1)z}\{\cos(\Omega_{C_i^{\alpha}}t_1)\cos^b(\pi J_{C^{\alpha}C^{\beta}}T_{AB})m\Gamma_m(\delta_1)\Gamma_{2A}(\delta_2)$$
$$+ \cos(\Omega_{C_i^{\beta}}t_1)\sin(\pi J_{C^{\alpha}C^{\beta}}T_{AB})\cos^g(\pi J_{C^{\beta}C^{\gamma}}T_{AB})n\Gamma_n(\delta_1)\Gamma_{2B}(\delta_2)\}; \quad [7.133]$$

$$\sigma_e = -2C_{iz}^{\alpha}N_{iy}\{\cos(\Omega_{C_i^{\alpha}}t_1)\cos^b(\pi J_{C^{\alpha}C^{\beta}}T_{AB})m\Gamma_m(\delta_1)\Gamma_{1A}(\delta_2)$$
$$+ \cos(\Omega_{C_i^{\beta}}t_1)\sin(\pi J_{C^{\alpha}C^{\beta}}T_{AB})\cos^g(\pi J_{C^{\beta}C^{\gamma}}T_{AB})n\Gamma_n(\delta_1)\Gamma_{1B}(\delta_2)\}$$
$$- 2C_{iz}^{\alpha}N_{(i+1)y}\{\cos(\Omega_{C_i^{\alpha}}t_1)\cos^b(\pi J_{C^{\alpha}C^{\beta}}T_{AB})m\Gamma_m(\delta_1)\Gamma_{2A}(\delta_2)$$
$$+ \cos(\Omega_{C_i^{\beta}}t_1)\sin(\pi J_{C^{\alpha}C^{\beta}}T_{AB})\cos^g(\pi J_{C^{\beta}C^{\gamma}}T_{AB})n\Gamma_n(\delta_1)\Gamma_{2B}(\delta_2)\}; \quad [7.134]$$

$$\sigma_f = 2H_{iz}^{N}N_{iy}\{\cos(\Omega_{C_i^{\alpha}}t_1)\cos^b(\pi J_{C^{\alpha}C^{\beta}}T_{AB})m\Gamma_m(\delta_1)\Gamma_{1A}(\delta_2)$$
$$+ \cos(\Omega_{C_i^{\beta}}t_1)\sin(\pi J_{C^{\alpha}C^{\beta}}T_{AB})\cos^g(\pi J_{C^{\beta}C^{\gamma}}T_{AB})n\Gamma_n(\delta_1)\Gamma_{1B}(\delta_2)\}$$
$$\times \cos(\Omega_{N_i}t_2)\Gamma_1(T_N)$$
$$+ 2H_{(i+1)z}^{N}N_{(i+1)y}\{\cos(\Omega_{C_i^{\alpha}}t_1)\cos^b(\pi J_{C^{\alpha}C^{\beta}}T_{AB})m\Gamma_m(\delta_1)\Gamma_{2A}(\delta_2)$$
$$+ \cos(\Omega_{C_i^{\beta}}t_1)\sin(\pi J_{C^{\alpha}C^{\beta}}T_{AB})\cos^g(\pi J_{C^{\beta}C^{\gamma}}T_{AB})n\Gamma_n(\delta_1)\Gamma_{2B}(\delta_2)\}$$
$$\times \cos(\Omega_{N_{i+1}}t_2)\Gamma_2(T_N); \quad [7.135]$$

$$\sigma_g = H_{ix}^{N}\{\cos(\Omega_{C_i^{\alpha}}t_1)\cos^b(\pi J_{C^{\alpha}C^{\beta}}T_{AB})m\Gamma_m(\delta_1)\Gamma_{1A}(\delta_2)$$
$$+ \cos(\Omega_{C_i^{\beta}}t_1)\sin(\pi J_{C^{\alpha}C^{\beta}}T_{AB})\cos^g(\pi J_{C^{\beta}C^{\gamma}}T_{AB})n\Gamma_n(\delta_1)\Gamma_{1B}(\delta_2)\}$$
$$\times \cos(\Omega_{N_i}t_2)\Gamma_1(T_N)$$
$$+ H_{(i+1)x}^{N}\{\cos(\Omega_{C_i^{\alpha}}t_1)\cos^b(\pi J_{C^{\alpha}C^{\beta}}T_{AB})m\Gamma_m(\delta_1)\Gamma_{2A}(\delta_2)$$
$$+ \cos(\Omega_{C_i^{\beta}}t_1)\sin(\pi J_{C^{\alpha}C^{\beta}}T_{AB})\cos^g(\pi J_{C^{\beta}C^{\gamma}}T_{AB})n\Gamma_n(\delta_1)\Gamma_{2B}(\delta_2)\}$$
$$\times \cos(\Omega_{N_{i+1}}t_2)\Gamma_2(T_N) \quad [7.136]$$

for $2\tau_1 = 1/(2J_{CH})$ and $2\tau_2 = 1/(2J_{NH})$. The coherence transfer functions $\Gamma_{1A}(\delta_2)$, $\Gamma_{2A}(\delta_2)$, $\Gamma_{1B}(\delta_2)$, and $\Gamma_{2B}(\delta_2)$ are given by

$$\Gamma_{1A}(\delta_2) = \sin(2\pi\,^1J_{C^{\alpha}N}\delta_2)\cos(2\pi\,^2J_{C^{\alpha}N}\delta_2)\cos^b(2\pi J_{C^{\alpha}C^{\beta}}\delta_2);$$

$$\Gamma_{2A}(\delta_2) = \cos(2\pi\,^1J_{C^{\alpha}N}\delta_2)\sin(2\pi\,^2J_{C^{\alpha}N}\delta_2)\cos^b(2\pi J_{C^{\alpha}C^{\beta}}\delta_2);$$

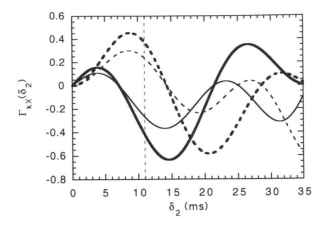

FIGURE 7.43 Plots of the CBCANH coherence transfer functions, $\Gamma_{1A}(\delta_2)$ (–, thick), $\Gamma_{2A}(\delta_2)$ (–, thin), $\Gamma_{1B}(\delta_2)$ (---, thick), and $\Gamma_{2B}(\delta_2)$ (---, thin) [7.137], for $^1J_{C^\alpha N^1}$, $^2J_{C^\alpha N^2}$, and $^1J_{C^\alpha C^\beta}$ coupling constants of 10, 7, and 35 Hz, respectively. The dashed vertical line at 11 ms indicates the optimal value of δ_2 to maximize all four transfer functions simultaneously.

$$\Gamma_{1B}(\delta_2) = \sin(2\pi\,^1J_{C^\alpha N}\delta_2)\cos(2\pi\,^2J_{C^\alpha N}\delta_2)\sin(2\pi J_{C^\alpha C^\beta}\delta_2);$$

$$\Gamma_{2B}(\delta_2) = \cos(2\pi\,^1J_{C^\alpha N}\delta_2)\sin(2\pi\,^2J_{C^\alpha N}\delta_2)\sin(2\pi J_{C^\alpha C^\beta}\delta_2). \qquad [7.137]$$

The magnitudes of these transfer functions are optimized simultaneously by setting δ_2 to 11 ms, as indicated in Fig. 7.43. The coherence transfer functions $\Gamma_1(T_N)$ and $\Gamma_2(T_N)$ are given by [7.75]. As indicated by Fig. 7.28, $\Gamma_1(T_N)$ is maximized by setting T_N to 22–28 ms. Following time d, the position of the ^{1}H carrier and hence the frequency of the ^{1}H broadband decoupling is shifted from the water resonance to the center of the amide region. As usual, the $180°(^{13}C^{\alpha/\beta})$ pulse power is adjusted such that it does not perturb the ^{13}CO spins.

Each given amide proton will have correlations to its intraresidue $^{13}C^\alpha$ and $^{13}C^\beta$ nuclei and to the $^{13}C^\alpha$ and $^{13}C^\beta$ of the preceding residue. With the exception of glycine residues, for which the exponent b is 0, the $\Gamma_{1A}(\delta_2)$ and $\Gamma_{2A}(\delta_2)$ functions have signs opposite to those of $\Gamma_{1B}(\delta_2)$ and $\Gamma_{2B}(\delta_2)$ (Fig. 7.43). This feature is helpful in the final spectrum since the $^{13}C^\alpha$ resonances of glycines and all $^{13}C^\beta$ resonances have phases opposite to those of the other $^{13}C^\alpha$ resonances. This feature of the CBCANH is particularly useful for discriminating between the $^{13}C^\alpha$ and $^{13}C^\beta$ resonances of serine residues, which resonate in the same spectral region. A CBCANH spectrum of ubiquitin is shown in Fig. 7.44. Identification of the intra- and interresidue connectivities on the basis of resonance intensities in the

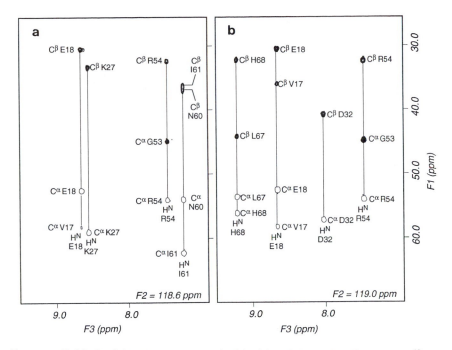

FIGURE 7.44 CBCANH spectrum of ubiquitin. Selected (adjacent) $F_1(^{13}C)$–$F_3(^1H^N)$ slices from a 600-MHz 3D CBCANH spectrum of $^{15}N/^{13}C$-labeled ubiquitin at $F_2(^{15}N)$ chemical shifts of 118.6 ppm (a) and 119.0 ppm (b). Correlations to $^{13}C^\beta$ nuclei and the $^{13}C^\alpha$ of glycines are plotted with multiple contours, whereas those to the remaining $^{13}C^\alpha$ nuclei, which have opposite phase, are plotted with a single contour. Cross-peak assignments are indicated.

CBCANH experiment is subject to the same uncertainties discussed for the HNCA experiment (Section 7.4.1).

As with the CBCA(CO)NH experiment discussed above, relatively minor modifications can be made to the CBCANH experiment, to produce an experiment that provides correlations to the $^1H^\alpha$ and $^1H^\beta$ resonances rather than the $^{13}C^\alpha$ and $^{13}C^\beta$ resonances; this experiment has been called the HBHA(CBCA)NH experiment (85). In contrast to the HBHA(CBCA-CO)NH experiment (81) mentioned above, the $^1H^\alpha$ and $^1H^\beta$ resonances in an HBHA(CBCA)NH spectrum are easily distinguished as they have opposite phases (the $^1H^\alpha$ resonances of glycine residues also have phases opposite to those of all other $^1H^\alpha$ resonances).

In combination, the HBHA(CBCA)NH, HBHA(CBCACO)NH, CBCANH, and CBCA(CO)NH experiments can provide complete sequential assignments of the $^1H^N$, ^{15}N, $^1H^\alpha$, $^{13}C^\alpha$, $^1H^\beta$, and $^{13}C^\beta$ resonances for

proteins up to about 20 kDa. A major limitation of the CBCANH and HBHA(CBCA)NH experiments, however, is that they are relatively insensitive. The sensitivity of these two experiments is limited by the transverse relaxation rate of $^{13}C^\alpha$ magnetization. For larger proteins (> 20 kDa), observation of a complete set of interresidue correlations is therefore unlikely.

7.4.6 ADDITIONAL CONSIDERATIONS FOR TRIPLE-RESONANCE EXPERIMENTS

Unless otherwise noted above, $^{13}C^\alpha$ (or ^{13}CO) pulses are applied with the transmitter frequency in the middle of the $^{13}C^\alpha$ (or ^{13}CO) region of the spectrum. The field strengths of the $^{13}C^\alpha$ 90° and 180° pulses are adjusted to minimally excite the ^{13}CO spins (i.e., $\gamma B_1 = \Omega/\sqrt{15}$ or $\Omega/\sqrt{3}$ for 90° and 180° pulses, respectively, in which Ω is the frequency difference between the $^{13}C^\alpha$ and ^{13}CO spectral regions; Section 3.4.1). Similarly, the ^{13}CO pulses are applied as selective shaped pulses or as weak rectangular pulses, with field strengths adjusted such that the $^{13}C^\alpha$ spins remain unperturbed. Nonresonant effects of the ^{13}CO pulses are compensated using the techniques presented in Section 3.4.1 and discussed in the descriptions of individual triple resonance experiments.

Frequency-discrimination in indirectly detected dimensions usually is achieved using the TPPI–States protocol (Section 4.3.4.1). The phase shifts are applied to *all* the pulses preceding the evolution period that are applied to the spin whose coherence is being frequency-labeled. Normally, the initial value of the sampling delay is adjusted to exactly $1/(2SW)$ as described in Section 3.3.2.3.

Most triple-resonance experiments (particularly constant-time versions) yield in-phase absorptive lineshapes. Consequently, the discussion of processing HSQC spectra (Section 7.12.6) is applicable. The data are truncated in the t_1 and t_2 dimensions of triple-resonance experiments, and linear prediction or maximum entropy reconstruction usually will be utilized. A secondary problem is that most experiments using a non-constant-time frequency labeling of the $^{13}C^\alpha$ spins contain a contribution from passive coupling to the $^{13}C^\beta$ spins. In these cases, the total acquisition time is limited to less than $1/(2J_{CC})$ to avoid resolution of the scalar coupling and loss of sensitivity. The effect of the coupling can be eliminated by multiplication of the interferogram by the function $\cos^{-1}(\pi J_{CC}t_1)$ or $\cos^{-1}(\pi J_{CC}t_2)$ as relevant prior to linear prediction or maximum entropy reconstruction.

The preceding analyses of triple resonance NMR experiments have focused on 3D versions of the experiments; however, most of the 3D experiments can be converted into 4D experiments by inserting an additional evolution period. Generally, equivalent information can be obtained from a pair of complementary 3D NMR experiments or a single 4D experiment. However, as the triple-resonance spectra of ubiquitin demonstrate, 3D triple-resonance experiments typically yield very well-resolved spectra. A fourth frequency dimension is unnecessary for distinguishing overlapping resonances (in contrast, significant overlap can persist in 3D heteronuclear-edited NOESY spectra; see Section 7.2). The greater resolution in the indirect dimensions of 3D experiments, compared with 4D experiments, facilitates more accurate determination of resonance frequencies during the assignment process (Chapter 8). The 4D triple-resonance experiments do have the advantage that assignments are obtained from a single spectrum, whereas the two 3D experiments would be acquired at different times (with possible slight variations in conditions).

7.5 Measurement of Scalar Coupling Constants

Homo- and heteronuclear three-bond J coupling constants are important for establishing local backbone and side-chain conformations in proteins; in particular, information on the backbone dihedral angle ϕ and the side-chain torsion angle χ_1 can be obtained from measurement of such coupling constants. Measurement of $^3J_{H^NH^\alpha}$ and $^3J_{H^\alpha H^\beta}$ from homonuclear spectra by either direct methods (i.e., direct measurement or iterative fitting of in-phase or antiphase splittings) (Section 6.2.1.5) or from homonuclear E-COSY spectra (Section 6.3.3), was discussed in Chapter 6. For larger proteins, accurate measurement of homonuclear J couplings by these methods is difficult because the couplings involved are smaller than the proton resonance linewidths. In recent years a large number of alternative experiments to measure three-bond J coupling constants from isotopically labeled protein samples that circumvent these problems have been published. The new methods may be classified into three distinct categories: (1) direct measurement of resolved J couplings from heteronuclear-edited spectra (*86,87*); (2) E-COSY-like methods for measurement of unresolved J couplings, in which a well-resolved ^1J coupling in one dimension of an *n*-dimensional experiment allows resolution of two components of the cross-peaks that are separated by the ^3J coupling of interest (*88–110*); and (3) quantitative J correlation, in which the coupling constant is determined from the intensity ratio of two cross-peaks (*111–119*).

There is a great deal of similarity between different published experiments within these three experimental categories. The most common

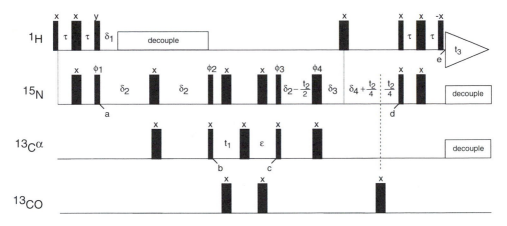

FIGURE 7.45 Pulse sequence for the HNCA-J experiment. The phase cycle is $\phi_1 = x, -x$; $\phi_2 = 2(x), 2(-x)$; $\phi_3 = 8(x), 8(-x)$; $\phi_4 = 4(x), 4(y), 4(-x), 4(-y)$; and receiver $= x, -x, -x, x, 2(-x, x, x, -x), x, -x, -x, x$.

schemes for measurement of homo- and heteronuclear three-bond J coupling constants to date are based on the E-COSY principle; an example of this class of experiment, designed to measure $^3J_{H^NH^\alpha}$ coupling constants, is discussed below. An alternative method for measurement of $^3J_{H^NH^\alpha}$ coupling constants using the quantitative J correlation class of experiments is discussed later.

7.5.1 HNCA-J EXPERIMENT

The first example of a heteronuclear triple-resonance correlation experiment to measure $^3J_{H^NH^\alpha}$ coupling constants using the heteronuclear E-COSY principle was an HNCA-J experiment (*88,91*). Indeed, this was also the first HNCA-type experiment to be published. Further modification of the original sequence (*92,95,96,103,106*) has lead to the HNCA-J experiment illustrated in Fig. 7.45 (*109,110*). The basic principle behind this E-COSY-like experiment is identical to that described previously for the homonuclear E-COSY experiment (Section 6.3.3): the $^{13}C^\alpha$ and $^1H^N$ spins are correlated without disturbing the spin state of the $^1H^\alpha$ nucleus. As a result, the large one-bond $J_{C^\alpha H^\alpha}$ coupling resolved in one dimension allows the $^3J_{H^NH^\alpha}$ coupling to be measured in another dimension. The (optional) ^{15}N dimension simply improves the spectral resolution by separating the $^{13}C^\alpha$-$^1H^N$ correlations into a third dimension. A more detailed product operator description of the HNCA-J experiment is given in the following paragraphs.

The HNCA-J experiment illustrated in Fig. 7.45 begins by transferring magnetization originating on amide protons to their directly bonded ^{15}N nuclei via an INEPT sequence. For $2\tau = 1/(2J_{NH})$, the magnetization present at time a is represented by

$$\sigma_a = -2H_z^N N_y. \qquad [7.138]$$

A second INEPT sequence transfers the ^{15}N magnetization to its coupled ^{13}C$^\alpha$ spins. In addition, the ^{15}N magnetization refocuses to become in-phase with respect to its attached proton during the delay $\delta_1 = 1/(2J_{NH})$. Application of synchronous broadband ^{1}H decoupling (e.g., WALTZ-16 or DIPSI-2) after δ_1 ensures that the ^{15}N spin remains decoupled from its attached proton. As with the previously discussed triple-resonance experiments (Section 7.4.6), the power and duration of the ^{13}C$^\alpha$ pulses are adjusted so that they do not perturb the ^{13}CO spins (Section 3.4.1). Focusing on the intraresidue correlations, and ignoring the interresidue correlations, the relevant component of the density operator at time b is given by

$$\sigma_b = -2N_z C_y^\alpha \Gamma_1(2\delta_2), \qquad [7.139]$$

in which the coherence transfer function $\Gamma_1(2\delta_2)$ is given by [7.75]. For the purpose of measuring the $^3J_{H^N H^\alpha}$ coupling, $2\delta_2$ is adjusted to optimize $\Gamma_1(2\delta_2)$ for intraresidue coherence transfer only; after allowing for the additional relaxation term, $\exp(-2\delta_2 R_{2N})$, $2\delta_2$ is typically set to 20–30 ms.

At this stage, [7.139] is rewritten using a mixture of Cartesian operators for the ^{15}N spin and single element operators for the ^{13}C$^\alpha$ and ^{1}H$^\alpha$ spins (Section 2.7.2) to give

$$\sigma_b = -N_z[C_y^\alpha H^\alpha(\alpha) - C_y^\alpha H^\alpha(\beta)]\Gamma_1(2\delta_2), \qquad [7.140]$$

in which $H^\alpha(\alpha)$ and $H^\alpha(\beta)$ correspond to the $|\alpha\rangle$ and $|\beta\rangle$ spin states of the ^{1}H nucleus, respectively. Thus, $C_y^\alpha H^\alpha(\alpha)$ and $C_y^\alpha H^\alpha(\beta)$ represent the ^{13}C$^\alpha$ spins whose attached ^{1}H$^\alpha$ spins are in the $|\alpha\rangle$ and $|\beta\rangle$ states. Using [2.208], $C_y^\alpha H^\alpha(\alpha)$ and $C_y^\alpha H^\alpha(\beta)$ can be expanded to give the equivalent Cartesian representations:

$$C_y^\alpha H^\alpha(\alpha) = C_y^\alpha \left(H_z^\alpha + \frac{1}{2}\mathbf{E} \right) = C_y^\alpha H_z^\alpha + \frac{1}{2}C_y^\alpha \qquad [7.141]$$

and

$$C_y^\alpha H^\alpha(\beta) = C_y^\alpha \left(H_z^\alpha - \frac{1}{2}\mathbf{E} \right) = C_y^\alpha H_z^\alpha - \frac{1}{2}C_y^\alpha. \qquad [7.142]$$

Evolution of the single-element operators can be followed using either the product operator rules, or more simply by direct matrix manipulation using the matrix representations given in Table 2.2, together with the relationships given in [7.141] and [7.142].

Following time b the $^{13}C^{\alpha}$ chemical shift evolves without proton decoupling during t_1. The application of $180°(^{15}N)$ and $180°(^{13}CO)$ pulses at the midpoint of the t_1 evolution period ensures decoupling of these spins; however, the passive scalar coupling interactions between $^{13}C^{\alpha}$ and $^{13}C^{\beta}$ spins evolve during this period. The $180°(^{13}C^{\alpha})$ and $180°(^{13}CO)$ pulses applied following t_1 and prior to time c provide a method to compensate for the off-resonance effect of the first $180°(^{13}CO)$ pulse on the transverse $^{13}C^{\alpha}$ magnetization (Section 3.4.1). The $180°(^{15}N)$ pulse applied in the middle of the ε delay decouples the $^{13}C^{\alpha}$ and ^{15}N spins. Setting the delay ε equivalent to the initial value of t_1, $t_1(0)$, ensures that no phase correction is necessary in the F_1 dimension; $t_1(0)$ need be long enough only to accommodate the $180°(^{13}CO)$ pulse. The relevant components of the density operator present at time c are described by

$$\sigma_c = -N_z\{C_y^{\alpha}H^{\alpha}(\alpha) \cos[(\Omega_{C^{\alpha}} + \pi J_{C^{\alpha}H^{\alpha}})t_1$$
$$- C_y^{\alpha}H^{\alpha}(\beta) \cos[(\Omega_{C^{\alpha}} - \pi J_{C^{\alpha}H^{\alpha}})t_1]\} \cos(\pi J_{C^{\alpha}C^{\beta}}t_1)\Gamma_1(\delta_2). \quad [7.143]$$

The frequencies of the two multiplet components are now separated in the F_1 dimension by the large one-bond $J_{C^{\alpha}H^{\alpha}}$ scalar coupling constant. During the remainder of the experiment the spin states of the $^1H^{\alpha}$ nuclei effectively are unchanged, and the two correlations observed in the F_1 dimension are treated separately.

Following time c, the antiphase $^{13}C^{\alpha}$ magnetization is transferred back to ^{15}N by the simultaneous $90°$ $^{13}C^{\alpha}$ and ^{15}N pulses. During the following constant-time evolution period of length $2\delta_2$, in which $\delta_2 = \delta_3 + \delta_4$, and $2\delta_4 = 1/(2J_{NH})$, the anti-phase ^{15}N magnetization refocuses with respect to $^{13}C^{\alpha}$ (during $2\delta_2$) and dephases with respect to its attached $^1H^N$ (during $2\delta_4$). The ^{15}N chemical-shift evolution proceeds during t_2 only. The magnetization present at time d is thus represented by

$$\sigma_d = N_y\{H_z^N H^{\alpha}(\alpha) \cos[(\Omega_{C^{\alpha}} + \pi J_{C^{\alpha}H^{\alpha}})t_1]$$
$$- H_z^N H^{\alpha}(\beta) \cos[(\Omega_{C^{\alpha}} - \pi J_{C^{\alpha}H^{\alpha}})t_1]\}$$
$$\times \cos(\Omega_N t_2) \cos(\pi J_{C^{\alpha}C^{\beta}}t_1)\Gamma_1^2(2\delta_2). \quad [7.144]$$

The final set of pulses represent a reverse INEPT sequence that transfers the antiphase magnetization back into observable $^1H^N$ magnetization. Note that the inclusion of the final $90°(^1H)$ pulse results in zero net rotation of the $^1H^{\alpha}$ spins (and all other protons not directly bound to ^{15}N) between times b and e. In addition, this pulse also returns any $^1H^N$ magnetization that is not aligned along the x axis to the z axis, and thus effectively purges phase errors resulting from the $^3J_{H^N H^\alpha}$ coupling that is active during the τ delays (Section 7.1.1.2). For $2\tau = 1/(2J_{NH})$, the magnetization present at time e is described by

$$\sigma_e = \frac{1}{2}\{H_x^N H^\alpha(\alpha) \cos[(\Omega_{C^\alpha} + \pi J_{C^\alpha H^\alpha})t_1]$$

$$- H_x^N H^\alpha(\beta) \cos[(\Omega_{C^\alpha} - \pi J_{C^\alpha H^\alpha})t_1]\}$$

$$\times \cos(\Omega_N t_2) \cos(\pi J_{C^\alpha C^\beta} t_1)\Gamma_i^2(2\delta_2). \qquad [7.145]$$

During acquisition ^{15}N spins are decoupled from the $^1H^N$ spins. Decoupling of the ^{13}C spins ($^{13}C^\alpha$ and ^{13}CO) also is desirable in order to remove the effects of the small $^2J_{H^N C^\alpha}$, $^2J_{H^N CO}$, and $^3J_{H^N C^\alpha}$ couplings. The coherences represented by σ_e evolve under the chemical-shift Hamiltonian during t_3 to yield an observable signal, after forming the trace with the observation operator, proportional to

$$\frac{1}{4}\{\cos[(\Omega_{C^\alpha} + \pi J_{C^\alpha H^\alpha})t_1] \exp[i(\Omega_{H^N} - \pi^3 J_{H^N H^\alpha})t_3]$$

$$+ \cos[(\Omega_{C^\alpha} - \pi J_{C^\alpha H^\alpha})t_1] \exp[i(\Omega_{H^N} + \pi^3 J_{H^N H^\alpha})t_3]\}$$

$$\times \cos(\Omega_N t_2) \cos(\pi J_{C^\alpha C^\beta} t_1)\Gamma_i^2(2\delta_2). \qquad [7.146]$$

From [7.146], two components corresponding to the $|\alpha\rangle$ and $|\beta\rangle$ spin states of $^1H^\alpha$ are observed for each intraresidue $F_1(^{13}C^\alpha)$–$F_3(^1H^N)$ cross peak. The F_1 displacement of the two components corresponds to $^1J_{CH}$, and the F_3 displacement corresponds to $^3J_{H^N H^\alpha}$.

The E-COSY principle requires that the $^1H^\alpha$ nuclear spin does not change spin states between the t_1 and t_3 time periods. In practice, nuclear spin relaxation exchanges the $^1H^\alpha$ spin between states $|a\rangle$ and $|b\rangle$ with two consequences: (1) relaxation during the t_3 period results in self-decoupling of the spins (Section 5.4.2), and (2) relaxation during the fixed time period between b and d results in the superposition of the E-COSY and natural multiplet structures in the final spectrum. Both effects mediated by spin relaxation reduce the size of the apparent $^3J_{H^N H^\alpha}$ scalar coupling constant.

Measurement of $^3J_{H^N H^\alpha}$ by the heteronuclear E-COSY technique is illustrated for a cross-peak from a selected $F_2(^{15}N)$ slice in the HNCA-J spectrum of ubiquitin in Fig. 7.46. The asymmetry in the peak shape is a consequence of relaxation between points b and d in the pulse sequence, as discussed above.

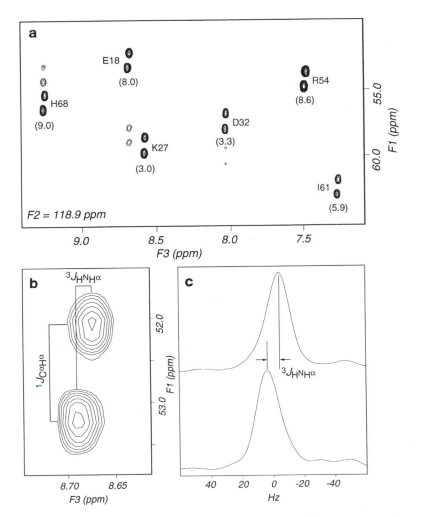

FIGURE 7.46 A selected $F_1(^{13}C^{\alpha})$–$F_3(^1H^N)$ slice, at an $F_1(^{15}N)$ chemical shift of 118.9 ppm, from a HNCA-J spectrum of ^{15}N/^{13}C-labeled ubiquitin. The intraresidue cross-peaks are labeled with the residue name and the measured $^3J_{H^NH^{\alpha}}$ coupling constants in parentheses. The weaker unlabeled cross-peaks are due to interresidue correlations. The E-COSY-like cross-peak pattern is highlighted in (b) for the cross-peak of Glu18. The $^3J_{H^NH^{\alpha}}$ coupling constants were determined as illustrated in (c), which shows rows taken through the maxima of the two cross-peak components of Glu18. The peak displacements were measured following inverse Fourier transformation, zero-filling to 16384 points, and retransformation.

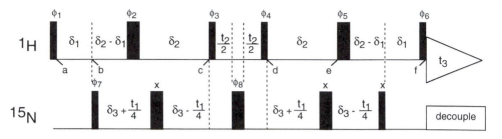

FIGURE 7.47 Pulse sequence for the HNHA experiment. The phase cycle is $\phi_1 = x$; $\phi_2 = 4(x), 4(-x), 4(y), 4(-y)$; $\phi_3 = x, -x$; $\phi_4 = 2(x), 2(-x)$; $\phi_5 = 4(x)$, $4(y), 4(-x), 4(-y)$; $\phi_6 = y, -y$; $\phi_7 = x$; $\phi_8 = 16(x), 16(y)$; and receiver $= 4(x)$, $8(-x), 4(x), 4(-x), 8(x), 4(-x)$. The delay δ_1 is set to 4.5 ms, and the delay δ_2 is set to 13.05 ms.

7.5.2 HNHA EXPERIMENT

An alternative method to the HNCA-J experiment for measurement of $^3J_{H^N H^\alpha}$ coupling constants relies on a quantitative analysis of the diagonal to cross-peak intensity ratio in a ^{15}N-edited $^1H^N$–$^1H^\alpha$ correlation experiment. This technique is referred to as the HNHA experiment because intraresidue $^1H^N$, ^{15}N, and $^1H^\alpha$ resonances are correlated by using the pulse sequence illustrated in Fig. 7.47 (*116*). This experiment using the product operator formalism is described in the paragraphs that follow.

The initial part of the pulse sequence transfers $^1H^N$ magnetization to heteronuclear multiple-quantum coherence in an HMQC-type manner, giving, at time b

$$\sigma_b = 2H_x^N N_y \sin(\pi J_{NH}\delta_1). \qquad [7.147]$$

In addition, the transverse $^1H^N$ magnetization present at time a dephases as a result of homonuclear $^1H^N$–$^1H^\alpha$ scalar coupling that is active for a total duration $2\delta_2$, between times a and c. Chemical-shift evolution of the ^{15}N spins occurs in a constant-time period between points b and c ($2\delta_3$) for a duration of $t_1/2$. The multiple-quantum coherence does not evolve under the influence of the active $^1H^N$–^{15}N coupling during the constant-time period (Section 2.7.5). The magnetization present at time c can therefore be represented by

$$\sigma_c = \{-2H_x^N N_y \cos(2\pi\,^3J_{H^N H^\alpha}\delta_2) - 4H_y^N H_z^\alpha N_y \sin(2\pi\,^3J_{H^N H^\alpha}\delta_2)\}$$
$$\times \cos(\Omega_N t_1/2) \sin(\pi J_{NH}\delta_1). \qquad [7.148]$$

The subsequent 90°(^{1}H) pulse converts the antiphase $^1H^N$ magnetization to antiphase $^1H^\alpha$ magnetization. Following a short t_2 evolution period, the antiphase $^1H^\alpha$ terms are converted back to antiphase $^1H^N$ magnetization to give, at time d

$$\sigma_d = \{-2H_x^N N_y \cos(\Omega_{H^N} t_2) \cos(2\pi^3 J_{H^N H^\alpha} \delta_2)$$

$$- 4H_y^N H_z^\alpha N_y \cos(\Omega_{H^\alpha} t_2) \sin(2\pi^3 J_{H^N H^\alpha} \delta_2)\}$$

$$\times \cos(\Omega_N t_1/2) \sin(\pi J_{NH} \delta_1). \qquad [7.149]$$

During the following rephasing period, between time points d and f, the ^{15}N chemical-shift evolution is continued for an additional period $t_1/2$, the $^1H^N-^1H^\alpha$ scalar coupling is again active and the $^{15}N-^1H^N$ MQ coherence is converted back to observable $^1H^N$ magnetization. The observable magnetization present at time f is represented by

$$\sigma_f = -H_y^N\{-\cos(\Omega_{H^N} t_2) \cos^2(2\pi^3 J_{H^N H^\alpha} \delta_2) + \cos(\Omega_{H^\alpha} t_2) \sin^2(2\pi^3 J_{H^N H^\alpha} \delta_2)\}$$

$$\times \cos(\Omega_N t_1) \sin^2(\pi J_{NH} \delta_1). \qquad [7.150]$$

The 90°(^1H) purge pulse applied immediately before acquisition ensures that only in-phase $^1H^N$ magnetization contributes to the observed signal. The analysis given above has neglected the effects of heteronuclear J_{NH^α} two- and three-bond couplings that are active for limited parts of the pulse sequence. In the first part of the sequence (i.e., prior to t_2 between times b and c), these couplings are active for a period corresponding to $2\delta_3 - t_1/2 - \delta_1$. In the latter part of the experiment, the J_{NH^α} couplings are active for a time $2\delta_3 + t_1/2 - \delta_1$ if $t_1/2 < \delta_1$, and for $2\delta_3 - t_1/2 + \delta_1$ if $t_1/2 > \delta_1$. The effects of these couplings are identical for both terms in [7.150] and can be accounted for by the inclusion of additional cosine terms; for instance, for $t_1/2 < \delta_1$, the $^2J_{NH^\alpha}$ coupling contributes the following terms to the observed magnetization:

$$\cos[\pi^2 J_{NH^\alpha}(2\delta_3 - t_1/2 - \delta_1)] \cos[\pi^2 J_{NH^\alpha}(2\delta_3 + t_1/2 - \delta_1)]. \qquad [7.151]$$

Analogous expressions can be written for $t_1/2 > \delta_1$, and for the $^3J_{NH^\alpha}$ couplings. In practice, however, these coupling constants are small, such that $2\delta_3 \ll 1/J_{NH^\alpha}$, and therefore have little effect on the observed intensity.

Equation [7.150] indicates that two peaks, an autocorrelation or "diagonal" peak at $F_1(^{15}N)-F_2(^1H^N)-F_3(^1H^N)$ and a cross-peak at $F_1(^{15}N)-F_2(^1H^\alpha)-F_3(^1H^N)$, with opposite phases will be observed for each amino acid spin system (except glycine). The lineshapes of these peaks in the F_1 and F_3 dimensions are determined by identical factors. The intrinsic linewidths of the "diagonal" and cross-peaks in the F_2 dimension, however, are determined by the relaxation rates of the transverse $^1H^N$ and $^1H^\alpha$ magnetization present during t_2. If the lineshapes of the diagonal and

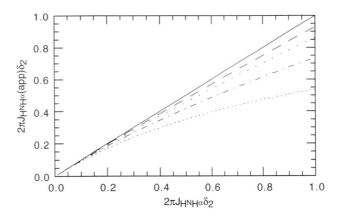

FIGURE 7.48 Effects of relaxation on the HNHA experiment. The apparent coupling constant calculated using [7.152] is plotted (in dimensionless units) against the actual coupling constant calculated using [7.155] (also in dimensionless units). Results are shown for $\chi = R_{1,H^\alpha}/(2\pi J_{H^N H^\alpha})$ equal to (——) 0, (− −) 0.1, ($\cdot\cdot\cdot$) 0.2, ($\cdot-\cdot$) 0.4 and ($\cdots$) 0.8.

cross-peaks in the F_2 dimension are assumed to be identical (but see below for a discussion of relaxation rates), the intensity ratio of these peaks then provides a measure of the magnitude of the $^3J_{H^N H^\alpha}$ coupling constant:

$$\frac{S_{\text{cross}}}{S_{\text{diagonal}}} = -\frac{\sin^2(2\pi \, ^3J_{H^N H^\alpha}\delta_2)}{\cos^2(2\pi \, ^3J_{H^N H^\alpha}\delta_2)} = -\tan^2(2\pi \, ^3J_{H^N H^\alpha}\delta_2). \qquad [7.152]$$

In practice, the intensity of the cross-peak resonance is reduced relative to the "diagonal" resonance because the antiphase component of the $4H_y^N H_z^\alpha N_y$ operator leads to faster relaxation during the $2\delta_2$ periods compared with the pure multiple-quantum $2H_x^N N_y$ operator. Ignoring cross correlation, to a first approximation, the relaxation rates of the two operators are given by

$$R_2(2H_x^N N_y) = R_{2H^N} + R_{2MQ}; \qquad [7.153]$$

$$R_2(4H_y^N H_z^\alpha N_y) = R_{2H^N} + R_{2MQ} + R_{1H^\alpha}. \qquad [7.154]$$

The time dependence of the in-phase and antiphase operators during the $2\delta_2$ delays, including relaxation effects, poses a problem identical to that presented in Section 5.4.2 for relaxation in a scalar-coupled spin system. Using the results presented in Section 5.4.2, [7.152] is modified to

$$\frac{S_{cross}}{S_{diagonal}} = -\left\{\frac{\tan[(1-\chi^2)^{1/2}\zeta]}{(1-\chi^2)^{1/2} + \chi\tan[(1-\chi^2)^{1/2}\zeta]}\right\}^2, \qquad [7.155]$$

in which the dimensionless parameters, χ and ζ, are given by

$$\chi = R_{1H^\alpha}/(2\pi\,^3J_{H^NH^\alpha}); \qquad \zeta = 2\pi\,^3J_{H^NH^\alpha}\delta_2. \qquad [7.156]$$

Equation [7.155] reduces to [7.152] as expected if χ approaches zero. If R_{1H^α} is known from experimental measurements or from calculations, [7.155] can be solved numerically to determine values of $^3J_{H^NH^\alpha}$ that are

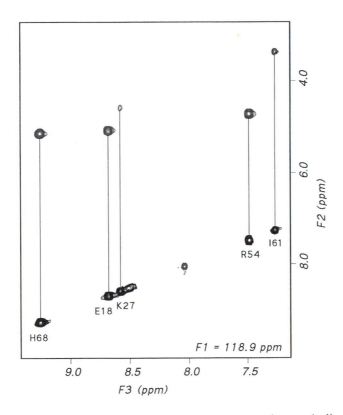

FIGURE 7.49 HNHA spectrum of ubiquitin. A selected $F_2(^1H)$–$F_3(^1H^N)$ slice, at an $F_1(^{15}N)$ chemical shift of 118.9 ppm, from a HNHA spectrum of ^{15}N-labeled ubiquitin acquired with $\delta_1 = 4.5$ ms and $2\delta_2 = 26.1$ ms. The intraresidue peaks are labeled with their residue names. The cross-peaks have phase opposite that of the "diagonal" peaks. From the cross-peak to "diagonal" peak intensity ratios the following uncorrected $^3J_{H^NH^\alpha}$ coupling constant values were obtained: His68 8.9 Hz; Glu18 8.2 Hz; Lys27 3.1 Hz; Arg54 8.5 Hz; Ile61 6.4 Hz.

TABLE 7.4

$^3J_{H^NH^\alpha}$ Scalar Coupling Constants

			HNHA	
Residue	COSY	HNCA-J	uncorrected	corrected
Glu18	9.3	8.0	8.2	8.8
Lys27	4.8	3.0	3.1	3.4
Asp32	4.3	3.3	3.4	3.7
Arg54	9.7	8.6	8.5	9.2
Ile61	6.9	5.9	6.7	7.3
His68	9.8	9.0	8.9	9.6

All values are given in hertz. Scalar coupling constants were determined from COSY spectra by line fitting (Section 6.2.1.5). HNCA-J results were measured from the spectrum as illustrated in Fig. 7.46. Uncorrected HNHA results were calculated from [7.152]. Corrected HNHA results were calculated from [7.155] assuming $R_1(H^\alpha) = 6.4 \text{ s}^{-1}$.

more accurate than those determined from [7.152]. The differences between the actual value of $^3J_{H^NH^\alpha}$ and the apparent values calculated using [7.152] are shown in Fig. 7.48. As shown, the effect of relaxation is always to reduce the apparent scalar coupling constant measured using [7.152].

A selected $F_1(^{15}N)$ slice of an HNHA spectrum of ubiquitin is shown in Fig. 7.49. Results obtained for $^3J_{H^NH^\alpha}$ from COSY, HNCA-J and HNHA experiments are summarized in Table 7.4. As shown, the values of $^3J_{H^NH^\alpha}$ obtained from COSY spectra are consistently larger than values obtained from HNCA-J or HNHA spectra. Values obtained from the HNHA experiment using [7.155] are approximately 7.5% larger than values obtained using [7.152].

References

1. S. W. Fesik and E. R. P. Zuiderweg, *Quart. Rev. Biophys.* **23**, 97–131 (1990).
2. G. M. Clore and A. M. Gronenborn, *Prog. NMR Spectrosc.* **23**, 43–92 (1991).
3. A. Bax and S. Grzesiek, *Acc. Chem. Res.* **26**, 131–138 (1993).
4. L. P. McIntosh and F. W. Dahlquist, *Quart. Rev. Biophys.* **23**, 1–38 (1990).
5. D. M. LeMaster, *Quart. Rev. Biophys.* **23**, 113–174 (1990).
6. T. Yamazaki, W. Lee, C. H. Arrowsmith, D. R. Muhandiram, and L. E. Kay, *J. Am. Chem. Soc.* **116**, 11655–11666 (1994).
7. R. R. Ernst, G. Bodenhausen, and A. Wokaun, "Principles of Nuclear Magnetic Resonance in One and Two Dimensions," pp. 1–610. Clarendon Press, Oxford, 1987.
8. L. Müller, *J. Am. Chem. Soc.* **101**, 4481–4484 (1979).
9. G. Bodenhausen and D. J. Ruben, *Chem. Phys. Lett.* **69**, 185–189 (1980).
10. A. Bax, M. Ikura, L. E. Kay, D. A. Torchia, and R. Tschudin, *J. Magn. Reson.* **86**, 304–318 (1990).

11. T. J. Norwood, J. Boyd, J. E. Heritage, N. Soffe, and I. D. Campbell, *J. Magn. Reson.* **87,** 488–501 (1990).
12. D. P. Burum and R. R. Ernst, *J. Magn. Reson.* **39,** 163–168 (1980).
13. A. Bax, A. F. Mehlkopf, and J. Smidt, *J. Magn. Reson.* **35,** 373–377 (1979).
14. A. Bax and R. Freeman, *J. Magn. Reson.* **44,** 542–561 (1981).
15. M. Rance, G. Wagner, O. W. Sørensen, K. Wüthrich, and R. R. Ernst, *J. Magn. Reson.* **59,** 250-261 (1984).
16. J. Santoro and G. C. King, *J. Magn. Reson.* **97,** 202–207 (1992).
17. G. W. Vuister and A. Bax, *J. Magn. Reson.* **98,** 428–435 (1992).
18. A. Hammarström and G. Otting, *J. Magn. Reson., Ser. A* **109,** 246–249 (1994).
19. J. Cavanagh and J. Keeler, *J. Magn. Reson.* **77,** 356–362 (1988).
20. S. Grzesiek and A. Bax, *J. Am. Chem. Soc.* **115,** 12593–12594 (1993).
21. Y.-C. Li and G. T. Montelione, *J. Magn. Reson., Ser. B* **101,** 315–319 (1993).
22. J. Stonehouse, G. L. Shaw, J. Keeler, and E. D. Laue, *J. Magn. Reson., Ser. A* **107,** 178–184 (1994).
23. B. A. Messerle, G. Wider, G. Otting, C. Weber, and K. Wüthrich, *J. Magn. Reson.* **85,** 608–613 (1989).
24. G. W. Vuister, R. Boelens, R. Kaptein, R. E. Hurd, B. John, and P. C. M. van Zijl, *J. Am. Chem. Soc.* **113,** 9688–9690 (1991).
25. J. Cavanagh, A. G. Palmer, P. E. Wright, and M. Rance, *J. Magn. Reson.* **91,** 429–436 (1991).
26. A. G. Palmer, J. Cavanagh, P. E. Wright, and M. Rance, *J. Magn. Reson.* **93,** 151–170 (1991).
27. A. G. Palmer, J. Cavanagh, R. A. Byrd, and M. Rance, *J. Magn. Reson.* **96,** 416–424 (1992).
28. J. Cavanagh and M. Rance, *Annu. Rep. NMR Spectrosc.* **27,** 1–58 (1993).
29. M. Akke, P. A. Carr, and A. G. Palmer, *J. Magn. Reson., Ser. B* **104,** 298–302 (1994).
30. L. E. Kay, P. Keifer, and T. Saarinen, *J. Am. Chem. Soc.* **114,** 10663–10665 (1992).
31. G. Kontaxis, J. Stonehouse, E. D. Laue, and J. Keeler, *J. Magn. Reson., Ser. A* **111,** 70–76 (1994).
32. C. Griesinger, O. W. Sørensen, and R. R. Ernst, *J. Magn. Reson.* **84,** 14–63 (1989).
33. D. Marion, P. C. Driscoll, L. E. Kay, P. T. Wingfield, A. Bax, A. M. Gronenborn, and G. M. Clore, *Biochemistry* **28,** 6150–6156 (1989).
34. D. Marion, L. E. Kay, S. W. Sparks, D. A. Torchia, and A. Bax, *J. Am. Chem. Soc.* **111,** 1515–1517 (1989).
35. E. R. P. Zuiderweg and S. W. Fesik, *Biochemistry* **28,** 2387–2391 (1989).
36. M. Ikura, L. E. Kay, R. Tschudin, and A. Bax, *J. Magn. Reson.* **86,** 204–209 (1990).
37. G. M. Clore, A. Bax, and A. M. Gronenborn, *J. Biomol. NMR* **1,** 13–22 (1991).
38. J. Cavanagh and M. Rance, *J. Magn. Reson.* **96,** 670–678 (1992).
39. S. W. Fesik and E. R. P. Zuiderweg, *J. Magn. Reson.* **78,** 588–593 (1988).
40. G. M. Clore and A. M. Gronenborn, *Annu. Rev. Biophys. Chem.* **20,** 29–63 (1991).
41. T. Frenkiel, C. Bauer, M. D. Carr, B. Birdsall, and J. Feeney, *J. Magn. Reson.* **90,** 420–425 (1990).
42. M. Ikura, A. Bax, G. M. Clore, and A. M. Gronenborn, *J. Am. Chem. Soc.* **112,** 9020–9022 (1990).
43. L. E. Kay, G. M. Clore, A. Bax, and A. M. Gronenborn, *Science* **249,** 411–414 (1990).
44. D. R. Muhandiram, G. Y. Xu, and L. E. Kay, *J. Biomol. NMR* **3,** 463–470 (1993).
45. G. M. Clore, L. E. Kay, A. Bax, and A. M. Gronenborn, *Biochemistry* **30,** 12–18 (1991).
46. A. Bax and S. S. Pochapsky, *J. Magn. Reson.* **99,** 683–643 (1992).
47. G. W. Vuister, G. M. Clore, A. M. Gronenborn, R. Powers, D. S. Garrett, R. Tschudin, and A. Bax, *J. Magn. Reson., Ser. B* **101,** 210–213 (1993).

48. J. Schleucher, M. Sattler, and C. Griesinger, *Angew. Chem. Int. Engl. Ed.* **32,** 1489–1491 (1993).

49. G. Zhu and A. Bax, *J. Magn. Reson.* **98,** 192–199 (1992).

50. L. E. Kay, M. Ikura, G. Zhu, and A. Bax, *J. Magn. Reson.* **91,** 422–428 (1991).

51. A. Bax, G. M. Clore, P. C. Driscoll, A. M. Gronenborn, M. Ikura, and L. E. Kay, *J. Magn. Reson.* **87,** 620–627 (1990).

52. L. E. Kay, M. Ikura, and A. Bax, *J. Am. Chem. Soc.* **112,** 888–889 (1990).

53. M. Ikura, L. E. Kay, and A. Bax, *J. Biomol. NMR* **1,** 299–304 (1991).

54. A. Bax, G. M. Clore, and A. M. Gronenborn, *J. Magn. Reson.* **88,** 425–431 (1990).

55. E. T. Olejniczak, R. X. Xu, and S. W. Fesik, *J. Biomol. NMR* **2,** 655–659 (1992).

56. L. E. Kay, M. Ikura, and A. Bax, *J. Magn. Reson.* **91,** 84–92 (1991).

57. A. J. Shaka, C. J. Lee, and A. Pines, *J. Magn. Reson.* **77,** 274–293 (1988).

58. S. P. Rucker and A. J. Shaka, *Mol. Phys.* **68,** 509–517 (1989).

59. M. Kadkhodaie, O. Rivas, M. Tan, A. Mohebbi, and A. J. Shaka, *J. Magn. Reson.* **91,** 437–443 (1991).

60. A. Mohebbi and A. J. Shaka, *Chem. Phys. Lett.* **178,** 374–378 (1991).

61. G. M. Clore, A. Bax, P. C. Driscoll, P. T. Wingfield, and A. M. Gronenborn, *Biochemistry* **29,** 8172–8184 (1990).

62. M. Ikura, L. E. Kay, and A. Bax, *Biochemistry* **29,** 4659–4667 (1990).

63. L. E. Kay, M. Ikura, R. Tschudin, and A. Bax, *J. Magn. Reson.* **89,** 496–514 (1990).

64. R. H. Fogh, D. Schipper, R. Boelens, and R. Kaptein, *J. Biomol. NMR* **4,** 123–128 (1994).

65. M. L. Remerowski, T. Domke, A. Groenewegen, H. A. M. Pepermans, C. W. Hilbers, and F. J. M. van de Ven, *J. Biomol. NMR* **4,** 257–278 (1994).

66. S. Grzesiek, H. Döbeli, R. Gentz, G. Garotta, A. M. Labhardt, and A. Bax, *Biochemistry* **31,** 8180–8190 (1992).

67. S. Grzesiek and A. Bax, *J. Magn. Reson.* **96,** 432–440 (1992).

68. F. Delaglio, D. A. Torchia, and A. Bax, *J. Biomol. NMR* **1,** 439–446 (1991).

69. B. T. Farmer II, L. D. Venters, L. D. Spicer, M. G. Wittekind, and L. Müller, *J. Biomol. NMR* **2,** 195–202 (1992).

70. A. Bax and M. Ikura, *J. Biomol. NMR* **1,** 99–104 (1991).

71. D. R. Muhandiram and L. E. Kay, *J. Magn. Reson., Ser. B* **103,** 203–216 (1994).

72. R. T. Clubb, V. Thanabal, and G. Wagner, *J. Magn. Reson.* **97,** 213–217 (1992).

73. R. Powers, A. M. Gronenborn, G. M. Clore, and A. Bax, *J. Magn. Reson.* **94,** 209–213 (1991).

74. A. G. Palmer, W. J. Fairbrother, J. Cavanagh, P. E. Wright, and M. Rance, *J. Biomol. NMR* **2,** 103–108 (1992).

75. S. Grzesiek and A. Bax, *J. Magn. Reson., Ser. B* **102,** 103–106 (1993).

76. S. Grzesiek and A. Bax, *J. Am. Chem. Soc.* **114,** 6291–6293 (1992).

77. S. Grzesiek and A. Bax, *J. Magn. Reson.* **99,** 201–207 (1992).

78. G. T. Montelione, B. A. Lyons, S. D. Emerson, and M. Tashiro, *J. Am. Chem. Soc.* **114,** 10974–10975 (1992).

79. R. T. Clowes, W. Boucher, C. H. Hardman, P. J. Domaille, and E. D. Laue, *J. Biomol. NMR* **3,** 349–354 (1993).

80. S. Grzesiek, J. Anglister, and A. Bax, *J. Magn. Reson., Ser. B* **101,** 114–119 (1993).

81. S. Grzesiek and A. Bax, *J. Biomol. NMR* **3,** 185–204 (1993).

82. T. M. Logan, E. T. Olejniczak, R. X. Xu, and S. W. Fesik, *J. Biomol. NMR* **3,** 225–231 (1993).

83. B. A. Lyons and G. T. Montelione, *J. Magn. Reson., Ser. B* **101,** 206–209 (1993).

84. B. A. Lyons, M. Tashiro, L. Cedergren, B. Nilsson, and G. T. Montelione, *Biochemistry* **32,** 7839–7845 (1993).

85. A. C. Wang, P. J. Lodi, J. Qin, G. W. Vuister, A. M. Gronenborn, and G. M. Clore, *J. Magn. Reson., Ser. B* **105,** 196–198 (1994).

86. L. E. Kay and A. Bax, *J. Magn. Reson.* **86,** 110–126 (1990).
87. M. Billeter, D. Neri, G. Otting, Y. Q. Qian, and K. Wüthrich, *J. Biomol. NMR* **2,** 257–274 (1992).
88. G. Montelione and G. Wagner, *J. Am. Chem. Soc.* **111,** 5474–5475 (1989).
89. G. T. Montelione, M. E. Winkler, P. Rauenbuehler, and G. Wagner, *J. Magn. Reson.* **82,** 198–204 (1989).
90. G. Wider, D. Neri, G. Otting, and K. Wüthrich, *J. Magn. Reson.* **85,** 426–431 (1989).
91. G. Montelione and G. Wagner, *J. Magn. Reson.* **87,** 183–188 (1990).
92. O. W. Sørensen, *J. Magn. Reson.* **90,** 433–438 (1990).
93. A. S. Edison, W. M. Westler, and J. L. Markley, *J. Magn. Reson.* **92,** 434–438 (1991).
94. M. Kurz, P. Schmieder, and H. Kessler, *Angew. Chem. Int. Ed. Engl.* **30,** 1329–1331 (1991).
95. P. Schmieder, V. Thanabal, L. P. McIntosh, F. W. Dahlquist, and G. Wagner, *J. Am. Chem. Soc.* **113,** 6323–6324 (1991).
96. G. Wagner, P. Schmieder, and V. Thanabal, *J. Magn. Reson.* **93,** 436–440 (1991).
97. E. R. P. Zuiderweg and S. W. Fesik, *J. Magn. Reson.* **93,** 653–658 (1991).
98. U. Eggenberger, Y. Karimi-Nejad, H. Thüring, H. Rüterjans, and C. Griesinger, *J. Biomol. NMR* **2,** 583–590 (1992).
99. S. D. Emerson and G. T. Montelione, *J. Am. Chem. Soc.* **114,** 354–356 (1992).
100. S. D. Emerson and G. T. Montelione, *J. Magn. Reson.* **99,** 413–420 (1992).
101. C. Griesinger and U. Eggenberger, *J. Magn. Reson.* **97,** 426–434 (1992).
102. M. Sattler, H. Schwalbe, and C. Griesinger, *J. Am. Chem. Soc.* **114,** 1126–1127 (1992).
103. S. Seip, J. Balbach, and H. Kessler, *Angew. Chem. Int. Ed. Engl.* **31,** 1609–1611 (1992).
104. G. W. Vuister and A. Bax, *J. Biomol. NMR* **2,** 401–405 (1992).
105. R. X. Xu, E. T. Olejniczak, and S. W. Fesik, *FEBS Lett.* **305,** 137–143 (1992).
106. J. C. Madsen, O. W. Sørensen, P. Sørensen, and F. M. Poulsen, *J. Biomol. NMR* **3,** 239–244 (1993).
107. M. D. Sørensen, S. M. Kristensen, J. J. Led, and O. W. Sørensen, *J. Magn. Reson.* **A103,** 364–368 (1993).
108. Y. Karimi-Nejad, J. M. Schmidt, H. Rüterjans, H. Schwalbe, and C. Griesinger, *Biochemistry* **33,** 5481–5492 (1994).
109. S. Seip, J. Balbach, and H. Kessler, *J. Magn. Reson., Ser. B* **104,** 172–179 (1994).
110. R. Weisemann, H. Rüterjans, H. Schwalbe, J. Schleucher, W. Bermal, and C. Greisinger, *J. Biomol. NMR* **4,** 231–240 (1994).
111. S. J. Archer, M. Ikura, D. A. Torchia, and A. Bax, *J. Magn. Reson.* **95,** 636–641 (1991).
112. A. Bax, D. Max, and D. Zax, *J. Am. Chem. Soc.* **114,** 6923–6925 (1992).
113. P. R. Blake, M. F. Summers, M. W. W. Adams, J.-B. Park, Z. H. Zhou, and A. Bax, *J. Biomol. NMR* **2,** 527–533 (1992).
114. S. Grzesiek, M. Ikura, G. M. Clore, A. M. Gronenborn, and A. Bax, *J. Magn. Reson.* **96,** 215–221 (1992).
115. S. Grzesiek, G. W. Vuister, and A. Bax, *J. Biomol. NMR* **3,** 487–493 (1993).
116. G. W. Vuister and A. Bax, *J. Am. Chem. Soc.* **115,** 7772–7777 (1993).
117. G. W. Vuister, A. C. Wang, and A. Bax, *J. Am. Chem. Soc.* **115,** 5334–5335 (1993).
118. G. W. Vuister, T. Yamazaki, D. A. Torchia, and A. Bax, *J. Biomol. NMR* **3,** 297–306 (1993).
119. A. Bax, G. W. Vuister, S. Grzesiek, F. Delaglio, A. C. Wang, R. Tschudin, and G. Zhu, *Meth. Enzymol.* **239,** 79–105 (1994).
120. A. S. Edison, F. Abildgaard, W. M. Westler, E. S. Mooberry, and J. L. Markley, *Meth. Enzymol.* **239,** 3–79 (1994).

8

SEQUENTIAL ASSIGNMENTS AND STRUCTURE CALCULATIONS

The first seven chapters of this text have focused on the theory of NMR spectroscopy and the experimental methods used for multidimensional NMR spectroscopy of proteins. Analysis of the information contained in the NMR spectra of proteins is the subjects of this final chapter. Exhaustive coverage of these topics would constitute an entire additional text; instead, this chapter has three objectives: (1) to indicate the types of analyses commonly performed using NMR spectra of proteins, (2) to rationalize usage of the experiments described in Chapters 6 and 7 for a thorough investigation of proteins by NMR spectroscopy, and (3) to provide a bibliography of the primary literature to augment discussions of objectives 1 and 2.

8.1 Resonance Assignment Strategies

In the initial stage of any investigation by NMR spectroscopy, each nuclear magnetic resonance must be associated with a specific nucleus in the molecule under investigation. Resonance assignments must be *sequence-specific*: each resonance must be assigned to a spin in a particular amino acid residue in the protein sequence. NMR spectroscopy provides three types of information useful for spectral assignments: through-bond interactions (via scalar couplings), through-space interactions (via dipolar coupling), and chemical environment (via the chemical shift). The strategies employed for resonance assignments depend on whether only homonuclear ^{1}H NMR spectra are available (unlabeled proteins) or whether ^{13}C and ^{15}N heteronuclear correlation spectra are available (isotopically labeled proteins).

8.1.1 ^{1}H RESONANCE ASSIGNMENTS

With few exceptions, correlations resulting from ^{1}H–^{1}H scalar couplings can be observed only between protons separated by two or three bonds in proteins. Cross-peaks in ^{1}H homonuclear correlation NMR spectra occur between protons within the same amino acid residue or *spin system*, but not between protons in different residues (because the interresidue $^4J_{H^N_{i+1}H^\alpha_i}$ coupling constant is negligible). This observation guides the procedures by which ^{1}H resonance assignments are made. Scalar correlation experiments, such as COSY (Section 6.2), MQF-COSY (Section 6.3), MQ (Section 6.4), and TOCSY (Section 6.5) experiments are used to identify resonance positions within each amino acid spin system, and the NOESY experiment (Section 6.6) is used to sequentially connect the amino acid spin systems.

Initially, the ^{1}H resonances are categorized on the basis of their chemical shifts. As indicated in Fig. 8.1, for the vast majority of residues in a protein, the backbone amide protons (^{1}H^N) resonate between 10.0 and 7.0 ppm, the backbone α-protons (^{1}H$^\alpha$) resonate between 6.0 and 3.5 ppm, aliphatic side-chain protons resonate between 3.5 and 1.0 ppm, and methyl protons resonate at chemical shifts less than 1.5 ppm. The dependence of ^{1}H chemical shifts on amino acid type and on protein secondary structure is discussed elsewhere (*1–6*). Figure 8.2 plots the distribution of ^{1}H chemical shifts (averaged over all observed conformations) for 13 proteins in a database compiled by Kraulis (*7*).

The backbone amide protons are usually the best resolved set of resonances within a protein ^{1}H NMR spectrum; thus, the ^{1}H$^\alpha$ and side-chain resonance positions are most readily determined in scalar correlation

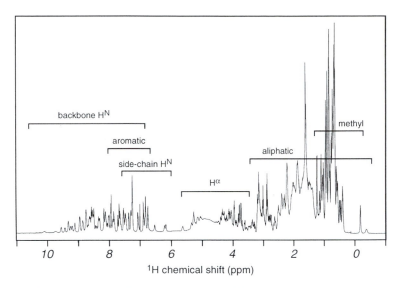

FIGURE 8.1 Chemical-shift ranges observed for the various types of ^1H resonances in ubiquitin. The spectrum was recorded in H_2O solution using presaturation and a Hahn echo (Section 3.6.4.2) following the acquisition pulse. The residual water resonance was removed from the FID by convolution (Section 3.5.4) prior to Fourier transformation.

experiments by the observation of direct and relayed cross-peaks to the backbone amide protons (8,9). In theory, most ^1H 2D spectra are symmetrical about the $F_2 = F_1$ diagonal (multiple-quantum spectra are the principal exceptions); hence a particular cross-peak involving an amide proton is observable above ($F_2 > F_1$) or below ($F_2 < F_1$) the diagonal. In real spectra, this symmetry often is broken because of experimental artifacts, the most obvious of which is usually a streak of noise at the F_2 frequency of the water resonance that arises from incomplete solvent suppression. In unfavorable circumstances, ridges emanating from the central stripe extend 1.0 ppm or more from the water resonance and obscure many correlations in this region. The distortions at the F_1 frequency of the water resonance are rarely comparable in width or intensity to the F_2 distortions. This asymmetry means that cross-peaks involving amide protons are most readily observed in the above-diagonal region. Concentrating on these cross-peaks also has the advantage of combining the high digital resolution of the F_2 dimension with the high chemical-shift dispersion of the amide resonances. For these reasons, the discussion of R-COSY, DR-COSY, and TOCSY experiments performed in H_2O in Chapter 6 focused on the side chain to amide proton correlation cross-peaks occurring above the diagonal.

The principal process of determining ^{1}H resonance assignments is known as the *sequential assignment strategy*, and has been developed by Wüthrich and coworkers (*10*). This strategy is summarized in Fig. 8.3. The first stage of analysis makes use of scalar couplings to establish sets of ^{1}H^N, ^{1}H$^\alpha$, and aliphatic side-chain resonances that belong to the same amino acid residue spin system. A protein of N residues has N distinct backbone-based spin systems. The N spin systems are assigned an amino acid type (or one of several possible types) based on the coupling topology and resonance chemical shifts, as described in Fig. 8.4.

Aromatic protons of tyrosine, phenylalanine, tryptophan, and histidine residues, and side-chain amide protons of glutamine and asparagine residues, are not scalar-coupled to the remainder of the side chain. Association of the side-chain and backbone resonances of these spin systems has to be made on the basis of intraresidue NOE correlations. For example, in the case of a tyrosine or phenylalanine residue, the most intense NOE correlations to ^{1}H$^\delta$ are generally from ^{1}H$^\beta$ because these protons are always less than 2.8 Å apart (*10*). As a second example, the side-chain amide protons *trans* to the side-chain carbonyl oxygen (^{1}H$^{N\delta2}$ in asparagine and ^{1}H$^{N\varepsilon2}$ in glutamine) are always closer to the other side-chain protons than are the *cis* protons (^{1}H$^{N\delta1}$ in asparagine and ^{1}H$^{N\varepsilon1}$ in glutamine). Therefore, the proton resonance with the more intense intraresidue NOEs is assigned to the *trans* position (*11*).

In the second stage of the assignment process, every spin system is assigned to a particular residue within the polypeptide chain by using through-space dipolar coupling (NOE) interactions to sequentially connect the spin systems identified from scalar correlations. Statistical analysis of the proton positions inferred from X-ray-crystal structures of proteins has shown that the majority of short interproton distances between ^{1}H^N, ^{1}H$^\alpha$, and ^{1}H$^\beta$ are between residues adjacent in the primary sequence (*12*). Thus, identification of intense NOEs from ^{1}H^N, ^{1}H$^\alpha$ and/or ^{1}H$^\beta$ of one spin system to ^{1}H^N of a second spin system suggests that the two spin systems are adjacent in the primary sequence with the first spin system nearer to the N-terminus of the protein (i.e., the spin systems correspond to residue i and residue $i + 1$, respectively). The sequential NOEs commonly are given the short-hand notation d_{NN}, $d_{\alpha N}$, and $d_{\beta N}$, respectively (*10,13*). Identification of a series of sequential NOE interactions places several spin systems in the order $i, i + 1, i + 2, ..., i + n$ (Fig. 8.5). As more spin systems are connected, the sequence of spin systems eventually will match a unique section of the primary amino acid sequence of the protein (which must be known a priori); at this point, the spin systems

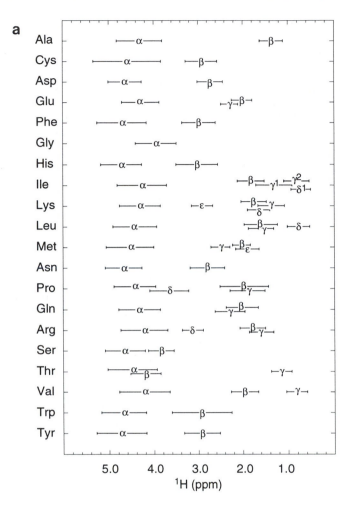

FIGURE 8.2 (a) Aliphatic and (b) aromatic and side chain amide ^{1}H chemical shifts. The mean and standard deviation for each proton type was obtained from the database of 13 proteins described by Kraulis (7). The bars extend one standard deviation in either direction from the mean value. The chemical shifts of backbone amide protons show very little dependence on amino acid type and have a mean of 8.3 ppm and a standard deviation of 0.7 ppm.

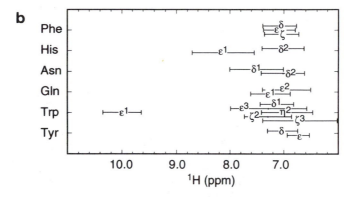

FIGURE 8.2—*Continued*

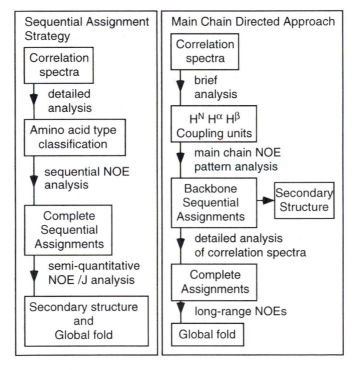

FIGURE 8.3 Comparison of the steps involved in the sequential assignment method (left) and the main-chain-directed method (right) of ¹H resonance assignment. Both schemes require a knowledge of the primary sequence and make use of the same experimental data, although this information is utilized at different stages of the process.

amino acid type	H$^\beta$ shifts only	H$^\beta$ H$^\gamma$ shifts	H$^\beta$ H$^\gamma$ H$^\delta$ shifts	H$^\beta$ H$^\gamma$ H$^\delta$ shifts and intraresidue NOEs
Glycine	GLY	GLY	GLY	GLY
Alanine	ALA	ALA	ALA	ALA
Threonine	THR	THR	THR	THR
Valine		VAL	VAL	VAL
Isoleucine		ILE	ILE	ILE
Leucine			LEU	LEU
Arginine	LONG	LONG	ARG	ARG
Lysine			LYS	LYS
Glutamine				GLN
Glutamate		5-SPIN	5-SPIN	GLU
Methionine				MET
Cysteine				CYS
Aspartate				ASP
Asparagine				ASN
Phenylalanine	3-SPIN	3-SPIN	3-SPIN	PHE
Tyrosine				TYR
Tryptophan				TRP
Histidine				HIS
Serine	SER	SER	SER	SER

FIGURE 8.4 Categorization of spin systems based on knowledge of chemical shifts and spin topology. Each box indicates a unique category containing the amino acid types shown on the left. The four columns of categories indicate how spin systems can be more finely subdivided as chemical shifts further along the side chain are included in the analysis. The shading indicates that assignment of threonine and serine spin systems may be ambiguous because of the partial overlap of the ^{1}H$^\beta$ chemical shifts of these residues with ^{1}H$^\beta$ of three-spin systems. The three columns on the left of the figure indicate categorizations possible using correlation spectra only; intraresidue NOEs are needed to perform the most complete categorization shown in the fourth column. Categorization in this fashion assumes that correlations to methyl groups can be differentiated from other interactions on the basis of chemical shift, intensity, or cross-peak fine structure.

are assigned *sequence-specifically*. If the spin system types are well characterized (i.e., the majority of side-chain resonance positions have been identified), then alignment of four or five spin systems usually is sufficient to achieve sequence specific assignment.

The observation of ^{1}H^N–^{1}H^N, ^{1}H$^\alpha$–^{1}H^N or ^{1}H$^\beta$–^{1}H^N NOEs is not limited to sequential interactions, and may also occur between nonsequential

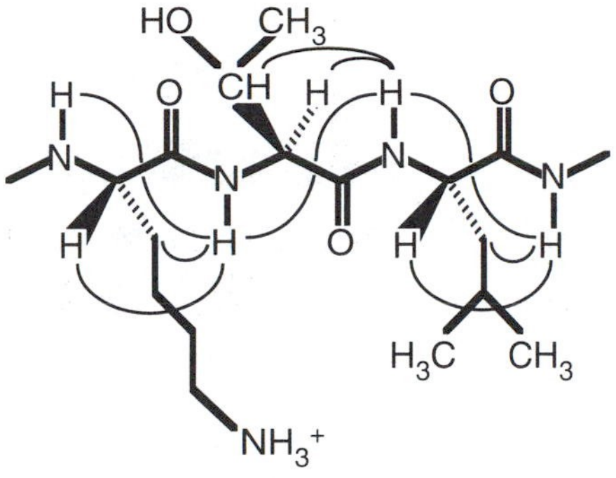

Lys6	Thr7	Leu8
Ile13	Thr14	Leu15
Arg54	Thr55	Leu56
Ser65	Thr66	Leu67

FIGURE 8.5 The use of backbone–backbone NOEs to obtain sequence-specific ^{1}H resonance assignments. In the top portion of the figure, the d_{NN}, $d_{\alpha N}$, and $d_{\beta N}$ connecting the three spin systems together are shown by the thin curved lines. Observation of a threonine prior to a leucine residue occurs four times in ubiquitin, hence such pairs of spin systems cannot be assigned sequence specifically. Once a lysine residue preceding the threonine is identified, the tripeptide segment can arise only from residues Lys6, Thr7, and Leu8. Because of the similarity of ^{1}H$^\beta$ and ^{1}H$^\gamma$ shifts of lysine, isoleucine, and arginine residues, ^{1}H$^\delta$ chemical shifts must be identified in the first residue of this tripeptide to distinguish between Lys6, Ile13 and Arg54, as described in Fig. 8.4.

residues as a result of secondary or tertiary structure in the protein (*12*). The ambiguity in the assignment process is reduced by the identification of sequential ^{1}H^N–^{1}H^N, ^{1}H$^\alpha$–^{1}H^N and ^{1}H$^\beta$–^{1}H^N NOEs. Additionally, the sequential ordering of spin systems must match the primary sequence; ^{1}H^N–^{1}H^N, ^{1}H$^\alpha$–^{1}H^N, or ^{1}H$^\beta$–^{1}H^N NOEs between spin systems that are never adjacent in the primary sequence result from longer-range contacts. In the limit, the assignments encompass *all* spin systems, and self-consistency is the best measure of the validity of the results. Tabulations of the relative intensities of the sequential NOEs are customarily presented graphically to demonstrate that complete, self-consistent sequential assignments have been achieved (Fig. 8.6).

An alternative strategy, known as the *main-chain-directed* approach (MCD), has been developed by Wand and coworkers (*14,15*). In the MCD approach, scalar coupling connectivities are used initially to identify ^{1}H^N–^{1}H$^\alpha$–^{1}H$^\beta$ units only. Assignment of the spin systems by amino acid

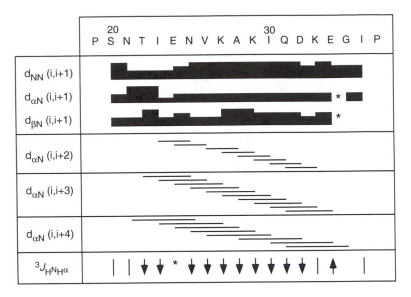

FIGURE 8.6 Summary of sequential NOEs observed for a portion of ubiquitin. The sequence is displayed along the top of the box, and the bars connect residues between which sequential NOEs are observed. The thickness of the bars indicates the intensity of the NOE. Commonly, medium-range NOEs and coupling constant data are included in such tabulations to help identify regions of regular secondary structure. Downward-pointing arrows, vertical lines, and upward-pointing arrows indicate $^3J_{H^NH^\alpha} < 6.0$ Hz, 6.0 Hz $< {}^3J_{H^NH^\alpha} < 8.0$ Hz, and $^3J_{H^NH^\alpha} > 8.0$ Hz, respectively. Asterisks indicate data that were not observed due to resonance overlap. This particular section of ubiquitin contains a helix from residues Thr22 to Glu34, as indicated by the intense d_{NN} NOEs, medium-range NOEs, and small coupling constants.

type is not attempted. Next, the $^1H^N$–$^1H^\alpha$–$^1H^\beta$ units are aligned sequentially by systematically searching the NOESY spectrum for patterns of sequential NOEs. Different elements of secondary structure give rise to specific patterns of NOEs (*10*), and a search is made for these motifs in the following order: helix, antiparallel sheet, parallel sheet, turns, and loops. The search process is amenable to computer automation (*16*). Once all the backbone coupling units have been aligned sequentially and categorized by secondary structural element, determination of the amino acid type of several side chains permits the defined elements of secondary structure to be aligned with the primary sequence. The sequential and MCD methods are compared schematically in Fig. 8.3.

The advantage of the MCD approach is that the sometimes troublesome identification of amino acid type from scalar coupling data is not attempted initially, so at least sequential backbone assignments can be made. Eventually, however, complete assignments of the side chains are needed because NOEs to all resonances must be assigned to determine precise three-dimensional structures. In practice, elements of both the sequential and MCD assignment approaches frequently are combined. The initial identification of spin systems in the sequential approach is rarely complete because of problems of resonance degeneracy, spectral artifacts, and absent cross-peaks (due to small coupling constants). However, even limited knowledge of amino acid types can assist in the MCD sequential assignment process: once several residues have been connected sequentially, identification of the next residue is trivial if only one of the candidates is the correct type of amino acid. Further, sequential assignment of the backbone resonances facilitates assignment of the side-chain resonances because the expected spin-system topologies (amino acid types) are known.

Although computer-automated analysis of ^{1}H 2D NMR spectra of proteins has been discussed for some years (*17,18*), satisfactory completely automated methods for resonance assignment do not exist. Attempts to semiautomate the assignment process have been more successful (*19–22*). Resonance degeneracy, cross-peak overlap, and incomplete correlations in the ^{1}H spectra are the main obstacles to complete automation. Homonuclear three-dimensional spectra potentially can circumvent resonance overlap (*23–26*), and novel search algorithms and assignment methods continue to be tested (*27*).

8.1.2 Heteronuclear Resonance Assignments

The conventional sequential or MCD assignment strategies also are used to obtain sequence specific assignments from 3D heteronuclear NMR spectra. The principal advantage of using 3D heteronuclear-edited NOESY and TOCSY experiments (Section 7.2) (rather than homonuclear 2D experiments) for resonance assignments is the significant reduction in cross-peak overlap. The 3D ^{1}H–^{15}N NOESY-HSQC experiment is used to identify sequential through-space $d_{\alpha N}$, $d_{\beta N}$, and d_{NN} connectivities. Amino-acid spin systems are identified by observation of direct and relayed through-bond connectivities between the ^{1}H^N spins and the ^{1}H$^\alpha$ and aliphatic side-chain protons using the 3D ^{1}H–^{15}N TOCSY-HSQC experiment (Section 7.2.2). Alternatively, complete side-chain assignments also can be obtained from HCCH-COSY and HCCH-TOCSY experiments (Section 7.3). In this case, the side-chain spin systems are connected with the backbone ^{1}H^N and ^{15}N resonances via ^{1}H^N–^{15}N–^{1}H$^\alpha$ and other correlations observed in a ^{1}H–^{15}N TOCSY-HSQC spectrum, or by using correlations observed in one or two of several triple-resonance experiments (Section

7.4) [e.g., HNCA, HN(CO)CA, HCA(CO)N, CBCANH, HBHA(CB-CA)NH, CBCA(CO)NH, or HBHA(CBCACO)NH]. The heteronuclear-edited NOE-based sequential assignment method has been successfully applied to a number of proteins, with molecular masses of up to ~20 kDa (28–34).

The triple-resonance experiments introduced in Section 7.4 offer an alternative to the NOE-based strategy for sequentially assigning $^1H^N$, ^{15}N, ^{13}CO, $^1H^\alpha$, $^{13}C^\alpha$, $^1H^\beta$, and $^{13}C^\beta$ resonances. Using these experiments, sequential correlations are established via the relatively uniform and well-resolved heteronuclear one-bond and two-bond couplings, without any prior knowledge of spin system types. Side-chain assignments are completed using the TOCSY-HSQC, HCCH-COSY, and HCCH-TOCSY experiments. Potential errors that arise from misassignment of sequential and long-range connectivities in the NOE-based procedures are avoided because assignments are based solely on predictable through-bond scalar correlations.

For example, the combined use of the HNCA (Section 7.4.1) experiment, which provides intraresidue (together with some sequential) correlations between $^1H^N$, ^{15}N, and $^{13}C^\alpha$ resonances; and the HN(CO)CA experiment (Section 7.4.2), which gives solely interresidue correlations between the $^1H^N$ and ^{15}N resonances of one residue and the $^{13}C^\alpha$ resonance of the preceding residue, provides an obvious route to sequential assignment. Each $^{13}C^\alpha$ resonance is linked to both its intraresidue and sequential $^1H^N$ and ^{15}N resonances. Ambiguities caused by chemical-shift degeneracy are solved by using additional experiments that provide alternative correlations. For instance, the HNCO (Section 7.4.4.1) and HN(CA)CO (Section 7.4.4.2) experiments correlate the $^1H^N$ and ^{15}N resonances with both the intraresidue and sequential ^{13}CO signals (rather than the $^{13}C^\alpha$ spins), whereas the H(CA)NH (Section 7.4.3), HCACO (Section 7.4.4.3), and HCA(CO)N (Section 7.4.4.4) experiments provide correlations analogous to those for $^1H^\alpha$ spins (rather than the $^1H^N$ spins). Combinations of these experiments give at least two and often more independent pathways to link adjacent amino acids, without prior knowledge of the spin system types. To "align" the backbone sequential assignments with the protein amino acid sequence, side-chain amino acid spin systems are identified from HCCH-COSY, HCCH-TOCSY, and ^1H–^{15}N TOCSY-HSQC experiments.

A different set of experiments, CBCA(CO)NH and CBCANH, together with the closely related HBHA(CBCACO)NH and HBHA(CB-CA)NH experiments (Section 7.4.5), offer an alternative sequential assignment strategy for proteins with $^{13}C^\alpha$ and $^1H^\alpha$ chemical-shift degeneracy. Using these experiments, the $^1H^N$ and ^{15}N resonances are correlated with the intraresidue and sequential $^{13}C^\alpha$ and $^{13}C^\beta$ (or $^1H^\alpha$ and $^1H^\beta$) resonances. Information regarding amino acid type can be obtained from the $^{13}C^\alpha$ and

$^{13}C^{\beta}$ chemical shifts (see discussion below). The HCC(CO)NH-TOCSY (35–39) and HCCNH-TOCSY (39,40) experiments, which are not discussed in Chapter 7, also provides sequential and spin-system assignment information. These experiments are similar in principle to the CBCA-(CO)NH and CBCANH experiments, except that ^{13}C isotropic mixing is used instead of COSY-type transfer to relay aliphatic side-chain magnetization to the $^{13}C^{\alpha}$ nucleus. Sequential and intraresidue correlations are obtained between backbone $^{1}H^{N}$ and ^{15}N resonances and the side-chain (either ^{13}C or ^{1}H) resonances.

Information on amino acid type can be obtained from ^{13}C chemical-shift data. The distributions of aliphatic and aromatic ^{13}C chemical shifts for different amino acid residues from a database of 13 proteins (7) are plotted in Fig. 8.7. The characteristic $^{13}C^{\alpha}$ and $^{13}C^{\beta}$ chemical shifts of alanine, threonine, serine, and glycine residues (Fig. 8.7) allow ready identification of these amino acid types from CBCA(CO)NH and CBCANH data. Clearly, knowledge of other aliphatic ^{13}C chemical shifts can also be used to assign a given spin system to a unique or limited number of possible amino acid types. This information, coupled with the alignment of sequential connected spin systems with the known amino acid sequence, leads to unambiguous sequential assignment.

The relative simplicity and predictability of triple-resonance 3D and 4D spectra used for protein assignment purposes makes these experiments particularly amenable to automated or semiautomated analysis. A number of protocols have been described recently that automate the resonance assignment process to varying degrees (41–45). Current efforts at automated assignment generally begin with automatic peak picking. The reduced resonance overlap in 3D and 4D triple-resonance spectra, relative to 2D homonuclear spectra, increases the reliability of this process; however, the final peak lists usually must be edited (by the spectroscopist) to distinguish "real" resonance peaks from spectral artifacts. The peak lists are searched automatically to find expected intraresidue and interresidue correlations, and the spin systems are identified and sequentially ordered according to these correlations. Information regarding spin system type, which may be obtained from $^{13}C^{\alpha}$ and $^{13}C^{\beta}$ chemical-shift values for instance, is incorporated and the ordered spin systems are aligned with the known amino acid sequence.

8.2 Three-Dimensional Solution Structures

The sequence specific assignment of NMR resonances typically requires many different experiments, days to weeks of spectrometer time,

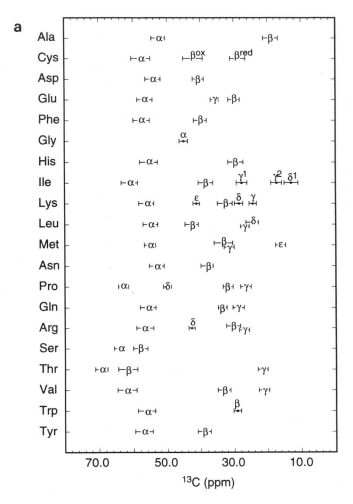

FIGURE 8.7 Aliphatic (a) and aromatic (b) ^{13}C chemical shifts. The mean and standard deviation in the mean for every carbon in each amino acid type was obtained from the database of 13 proteins described by Kraulis (7). The bars extend one standard deviation in either direction from the mean value. Two ranges are included for ^{13}C$^{\beta}$ of cysteine, due to the significant effect that disulfide bond formation has on these chemical shifts. The ^{13}CO chemical shifts range from $\sim$172 to $\sim$180 ppm with a mean random coil chemical shift of 176 ppm.

and a substantial effort from the spectroscopist. The mere achievement of resonance assignments, however, usually does not reveal much of interest about the structure and dynamics of the protein under investigation. The assignments are a necessary prerequisite to achieving these

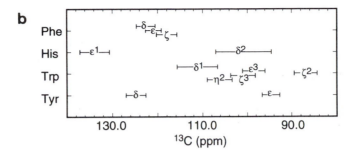

FIGURE 8.7—*Continued*

more biochemical goals. This section discusses procedures for determining the three-dimensional structures of proteins with the aim of highlighting the information that can be obtained from the NMR experiments described in Chapters 6 and 7.

8.2.1 NMR-DERIVED STRUCTURAL RESTRAINTS

Chemical shifts, scalar coupling interactions, and dipolar relaxation are sensitive to molecular conformation, and quantification of these parameters permits structural analysis by NMR spectroscopy. The scalar coupling and dipolar interactions form the basis of all protein structures calculated to date using NMR data, and are discussed in more detail below. Interpretation of chemical shifts in structural terms is more difficult because of the subtle interplay between internal dynamics and conformation, although research continues apace (*46,47*).

By far the most important NMR-observable parameter used in determining protein structure is the NOE. The dipolar cross-relaxation rate constant is proportional to the inverse sixth power of the distance between the two interacting protons (Section 5.5). In the initial rate approximation, NOE cross-peak intensities are proportional to the cross-relaxation rate constants. Thus, if one interproton distance, r_{ref}, is known (e.g., from covalent geometry), then another, unknown interproton distance, r_i, is determined by the relationship (ignoring differential internal mobility):

$$r_i = r_{\text{ref}} \left(\frac{S_{\text{ref}}}{S_i} \right)^{1/6}, \qquad [8.1]$$

in which S_{ref} and S_i are the integrated cross peak intensities. In practice, NOESY mixing times short enough to satisfy the initial-rate approximation are impractical because the cross-peak intensities have low signal-to-noise ratios. For longer mixing times, the intensities of NOESY cross-peaks are no longer directly proportional to the cross-relaxation rate constants between the interacting spins because magnetization is transferred between spin in multiple steps via "spin diffusion" (Section 6.6.1.2). Without recourse to complex calculations (48), precise ^{1}H–^{1}H separations cannot be determined from the NOE cross-peak intensities.

Instead, NOE cross-peaks typically are grouped on the basis of their intensities into several categories (strong, medium, and weak, etc.). Each category is associated with an upper-bound separation between the interacting spins. The cross-peak volume limits and the upper-bound distances for the categories are estimated from the NOE intensities observed for protons of known covalent geometry (e.g., geminal methylene protons, or vicinal protons in aromatic rings) or between protons in regions of regular secondary structure (e.g., sequential ^{1}H$^\alpha$–^{1}H^N, cross-strand ^{1}H$^\alpha$–^{1}H$^\alpha$ and cross-strand ^{1}H$^\alpha$–^{1}H^N NOEs in β-sheets, or ^{1}H$^\alpha$–^{1}H^N medium-range NOEs in helices). Conservative upper-bound distances compensate for cross-peaks affected by spin diffusion or partial overlap; otherwise, failure to allow for such artifacts results in structures that have large violations of the input restraints, high energies, and artificially high precisions (49).

As was first described by Karplus (50), the magnitude of a ^3J scalar coupling constant is a function of the dihedral angle, θ, formed by the three covalent bonds:

$$^3J = A \cos^2 \theta + B \cos \theta + C. \qquad [8.2]$$

The constants A, B, and C depend on the particular nuclei involved in the covalent bonds. Historically, dihedral angle restraints for ϕ and χ_1 dihedral angles have been derived only from $^3J_{H^N H^\alpha}$ and $^3J_{H^\alpha H^\beta}$ coupling constants, respectively (51–53). In addition, several experiments have recently been developed that allow measurement of ^{13}C–^{13}C, ^{13}C–^{15}N, ^{1}H–^{15}N, and ^{1}H–^{13}C three-bond coupling constants (Section 7.5) (54,55).

Accurate parameterization of the Karplus equation is necessary in order to relate measured values of ^3J coupling constants to dihedral angle ranges. Numerous attempts have been made to refine the original Karplus approach to account for the chemical nature of the atoms involved in the coupling (56). However, such theoretical approaches lead to complex relationships that are not generally applicable. Instead, semiempirical methods of obtaining values for the constants A, B, and C in [8.2] have been more successful. The most relevant values for the constants have been derived by correlation of observed ^3J values with the corresponding dihedral angles measured in protein structures determined by X-ray crystallography or NMR methods (52,57–61). Values of these constants for the two readily measured ^{1}H scalar couplings, $^3J_{H^N H^\alpha}$ and $^3J_{H^\alpha H^\beta}$, are given

TABLE 8.1
Parameterization of the Karplus Equation

Coupling constant	Dihedral	A	B	C	Ref.
$^3J_{H^N H^\alpha}$	$H^N-N-C^\alpha-H^\alpha$	6.4	−1.4	1.9	(52)
$^3J_{H^N H^\alpha}$	$H^N-N-C^\alpha-H^\alpha$	6.0	−1.4	2.4	(60)
$^3J_{H^N H^\alpha}$	$H^N-N-C^\alpha-H^\alpha$	6.7	−1.3	1.5	(59)
$^3J_{H^N H^\alpha}$	$H^N-N-C^\alpha-H^\alpha$	6.51	−1.76	1.60	(61)
$^3J_{H^\alpha H^\beta}$	$H^\alpha-C^\alpha-C^\beta-H^\beta$	9.5	−1.0	1.4	(57)
$^3J_{H^\alpha H^\beta}$	$H^\alpha-C^\alpha-C^\beta-H^\beta$	9.5	−1.6	1.8	(58)

in Table 8.1. Calibration of the Karplus equation and determination of dihedral angle restraints is complicated by relaxation-induced self-decoupling that reduces the apparent coupling constant in larger proteins (Section 5.4.2) (62).

Given the shape of the Karplus curve, shown in Fig. 8.8, as many as four different conformations can yield the same value of ^3J, although some of the possible orientations may be sterically unfavorable. Dihedral angle constraints enforced during structural calculations must include all of the valid solutions to the Karplus curve. Intramolecular motions give rise to

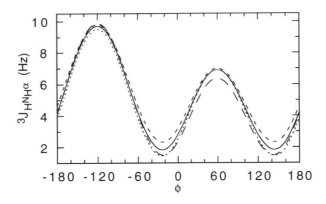

FIGURE 8.8 The Karplus curve describing the variation of $^3J_{H^N H^\alpha}$ with backbone dihedral angle ϕ. The dihedral angle between $^1H^N$ and $^1H^\alpha$ is given by $\theta = \phi - 60°$. The four curves shown were calculated using [8.2] with the constants A, B, and C, as given in Table 8.1; (——) $A = 6.4$, $B = -1.4$, $C = 1.9$; (----) $A = 6.0$, $B = -1.4$, $C = 2.4$; (·····) $A = 6.7$, $B = -1.3$, $C = 1.5$; (---) $A = 6.51$, $B = -1.76$, $C = 1.6$.

values of ^3J that are averaged over distributions of dihedral angles. Motional averaging often means that useful restraints sometimes cannot be derived from accurately measured coupling constants (*63*). For example, extensive backbone motion results in a $^3J_{H^N H^\alpha} \approx 7.0$ Hz in short, unstructured linear peptides. Thus, $^3J_{H^N H^\alpha}$ coupling constants observed in the range 6.0–8.0 Hz generally are not utilized as protein structural restraints, because they may reflect motional averaging of multiple conformations. The closer ^3J is to one of the extrema expected from the Karplus relationship, the less the likelihood of significant motional averaging. Dihedral angle restraints are defined conservatively (usually not restrained to ranges less than 60° or 120°) to allow for limited motional averaging of the observed coupling constant.

Slow rates of amide exchange (relative to the intrinsic rates for unstructured peptides; see Fig. 3.26) are associated with shielding of amide protons from solvent, and most commonly result from hydrogen-bonding interactions of the amide protons (*64*). Amide exchange rates are usually measured in one of two ways, depending on the rate of exchange. When the rate is comparable to, or faster than the spin–lattice relaxation rate ($k_{ex} > 0.1$ s^{-1}), the rate constant is most easily determined from a saturation transfer experiment (analogous to the transient NOE experiment discussed in Section 5.1.2) (*65,66*). For slower exchange ($k_{ex} < 0.01$ s^{-1}), exchange usually is measured by rapidly exchanging the protein from H$_2$O into D$_2$O solution, and repeatedly acquiring homonuclear TOCSY (Section 6.5.5) or ^{1}H–^{15}N HSQC (Section 7.1.1.2) spectra to observe the decrease in amide proton resonance intensities with time. Observation of a slow amide proton exchange rate implies that the ^{1}H^N atom may be involved in a hydrogen bond, but does not identify the atoms acting as hydrogen-bond acceptors. Hydrogen-bond restraints have a large impact on the nature and precision of the resulting structures and are usually enforced only in well-defined regions of regular secondary structure, in which only one possible hydrogen-bond acceptor is consistent with the NOE data.

8.2.2 STRUCTURE DETERMINATION

Details of the local backbone geometry can be obtained by an extension of the sequential assignment process; the relative intensities of d_{NN}, $d_{\alpha N}$, and $d_{\beta N}$ NOE cross-peaks and the measurement of the backbone $^3J_{H^N H^\alpha}$ are required. The observation of intense d_{NN} NOEs and small $^3J_{H^N H^\alpha}$ coupling constants (<6.0 Hz) are indicative of helical or turn sections of polypeptide (e.g., see Fig. 8.6); observation of intense $d_{\alpha N}$, weak d_{NN}, and $d_{\beta N}$ NOEs and large $^3J_{H^N H^\alpha}$ coupling constants (>8.0 Hz) are indicative of extended β-strands of polypeptide (*10*). The combination of

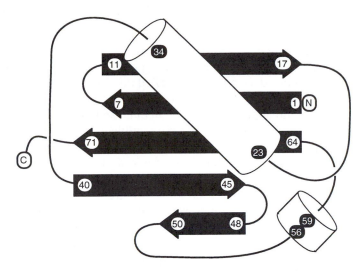

FIGURE 8.9 Definition of the global fold using long-range NOEs. Previously identified elements of regular secondary structure (see Fig. 8.6) may be arranged in a low-resolution global fold by the observation of a few key long-range NOEs involving protons at the termini of the strands and helices. This figure depicts the strands of β-sheet and sections of helix present in ubiquitin as black arrows and white cylinders, respectively.

sequential NOE and $^3J_{H^NH^\alpha}$ coupling constant data together with medium range and a few long range NOEs is capable of providing details of the regions of regular secondary structure within the protein. The elements of secondary structures can be connected together to give a crude view of the global fold by the identification of a few key long-range NOEs (Fig. 8.9). Thus, without recourse to extensive calculations, important structural results (albeit of low absolute resolution) can be obtained in a straightforward manner.

The majority of the restraints used to calculate the three-dimensional structure of a protein come from many hundreds of NOE cross-peaks, and are represented as upper bounds for the separation of ^{1}H pairs. Usually, the lower-bound separation for pairs of protons are set to the sum of the van der Waals radii. In addition, the distance constraints are supplemented with a smaller number of dihedral angle restraints, with upper and lower bounds determined from the observed scalar coupling constants, and hydrogen-bond distance restraints (two per restrained hydrogen bond). Often, restraints are specifically imposed to enforce proper chirality and the planarity of aromatic rings and peptide units. Finally, the covalent structure of the protein, including bond lengths, bond angles, and other elements of standard covalent geometry, are enforced.

A variety of methods have been developed to calculate protein structures using restraints derived from NMR data. These protocols aim to determine coordinates for the protein atoms that will satisfy the input distance and dihedral angle restraints in an unbiased fashion while exploring all regions of conformational space compatible with the observed NMR parameters. The NMR data do not uniquely define a 3D protein structure, because the restraints are included as ranges of allowed values, the data contain experimental uncertainties, and only a sparse subset of all possible distance, dihedral and hydrogen bonding constraints are observable. Instead, the structure calculation is repeated frequently to determine an ensemble of (low-energy) structures consistent with the input data. Thus, a "good" ensemble of structures minimizes violations of the input restraints and maximizes the root-mean-square deviation (RMSD) between members of the ensemble (*49,67*).

The two most common approaches for generation of structures are distance geometry (DG) and restrained molecular dynamics (rMD). Popular implementations of DG use either the metric matrix algorithm (*68,69*) or the variable target function approach (*70,71*). Distance geometry determines ensembles of 3D structures consistent with an incomplete set of distance constraints. The constraints are incomplete because not all distances can be characterized (because the NOE is limited to distances less than approximately 5 Å) and because the distance constraints are not known precisely. The metric matrix algorithms in particular tend to be very computationally expensive as the size of the protein increases.

Restrained molecular dynamics provides an alternative to the distance geometry approaches. Molecular dynamics force fields are supplemented by square-well pseudoenergy terms based on the NMR-derived restraints (*72,73*). These potentials drive the structure toward a conformation that will reduce the violation of the restraints during a forced heatup–cooldown annealing cycle. The most computationally efficient implementations of the rMD method use a simplified force field in which bond length, bond angle, and repulsive van der Waals terms are retained (electrostatic and attractive van der Waals terms are ignored), and are referred to as dynamical simulated annealing (SA) (*74*).

In many laboratories, a hybrid method is the chosen means of calculating structures (*75*). Initial structures are generated by DG; in order to ease the computational burden, only a subset of the atoms may be included for large proteins, with the remainder added by reference to standard amino acid templates. The resulting structures have the correct global fold but poor local geometry, and are refined (annealed) using rMD. The annealing process removes many of the local violations of NMR restraints and covalent inconsistencies present in the DG structures.

The precision to which a structure can be calculated is directly related to the number of experimental restraints used to generate it. Structures of low resolution may be obtained with as few as five restraints per residue, whereas the most precise structures obtained from ^{1}H data alone may have up to 15 restraints per residue. For the latter cases, the mean RMS deviation of backbone atoms from the mean structure may be as low as 0.5 Å for the well-defined regions of secondary structure. The use of ^{15}N or ^{13}C labeling usually allows many more ^{1}H–^{1}H NOEs to be unambiguously identified, increasing the number of restraints per residue to between 20 and 25, and leading to the highest precision structures (backbone RMSD in the 0.3–0.5-Å range). Thus, many hundreds of NOE cross-peaks must be assigned unambiguously for use in these calculations. The ability to stereospecifically assign prochiral groups (especially $C^\beta H_2$ groups, and $(CH_3)_2CH$ groups of valine and leucine residues) are critical to obtaining structures of high precision (76).

Once three-dimensional structures have been determined, interproton distances can be measured from the atomic coordinates, the dipolar relaxation rate matrix can be estimated from the interproton distances, and the theoretically expected NOE intensities can be calculated from the rate matrix using the results of Chapter 5. This process is known as complete relaxation matrix analysis (CORMA) (48) or back-calculation. After suitable scaling, the calculated NOE intensities are compared to the experimentally observed NOE intensities to judge how well the structures reproduce the experimental data. Usually, the agreement is judged by a figure of merit, or *R*-factor (77). The difference between observed and back-calculated NOE intensities also can be included as a pseudoenergy term during rMD calculations (78,79). Thus, the structures are driven during the calculation to minimize the *R*-factor and maximize the agreement between the structures and the experimental NOE intensities. This process requires a complete relaxation matrix analysis to be performed at every step of the annealing cycle and therefore is very computationally demanding.

For any of these methods to be valid, the accuracy of the experimental NOE intensities must be high. NOESY spectra must have very flat baselines (even slight rolls or tilts in the baseline can have dramatic effects on cross-peak volumes) and be otherwise free of spectral artifacts; such topics are discussed in Chapters 6 and 7. In addition, the dynamical model must be capable of reproducing the effects of molecular motion known to be present in proteins, such as fast rotation of methyl groups and 180°

flips of aromatic side chains (Section 5.3) (*80*) and systematic deficiencies in the experimental data (*81*). Back-calculation provides one of the few ways to independently assess the accuracy of protein structures derived from NMR data.

8.3 Conclusion: Beyond Structure

This text has two objectives: (1) presentation of the theory of NMR spectroscopy applicable to studies of biological macromolecules and (2) explication of the types of NMR experiments utilized for resonance assignments and three-dimensional structure determinations of proteins. Knowledge of the three-dimensional structure of a protein is the beginning, not the endpoint of structural biology and biophysics of protein function. Proteins are not rigid isolated molecules, and understanding protein function depends on a detailed knowledge of structure, dynamics, and interactions with other biomolecules. Fortunately, NMR spectroscopy is not limited in application to structure determination. Many NMR-derived parameters depend on both the time-average three-dimensional structures and the intramolecular dynamical properties (fluctuations about the time-average structures) of proteins. In addition, structural and kinetic aspects of intermolecular interactions can be studied by the techniques used for monomeric proteins (presented herein) and by experimental methods that have been designed specifically for studying such interactions. Figure 8.10 illustrates the parallel richness of both biophysics and NMR spectroscopy of protein function.

The experiments discussed in this text for resonance assignments and structure determination have a certain universality and are used in nearly all comprehensive studies of proteins by NMR spectroscopy (the topology of amino acid spin systems and secondary structural elements serve as unifying features). In contrast, NMR experiments for investigations of intramolecular dynamics and intermolecular interactions are more heterogeneous and must be matched to the timescale of the phenomenon under study. For example, protein–peptide interactions can be studied by transferred NOE experiments if the exchange rate is fast (*82,83*) or by ω_1-filtered NOESY experiments if the exchange rate is slow (*84,85*). In the interest of brevity, this text does not discuss the many types of NMR experiments used for investigations "beyond structure." Nonetheless, the theoretical and experimental principles presented herein are equally valid and important to the understanding and optimization of all experiments for high-resolution NMR spectroscopy of biomolecules in solution.

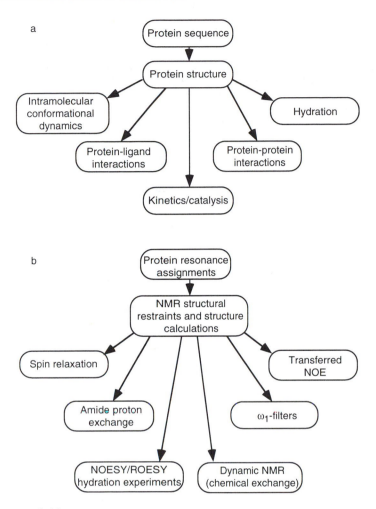

FIGURE 8.10 (a) Protein biophysics and (b) protein NMR spectroscopy.

Since the first observation of NMR absorption in condensed phase in 1946, nuclear magnetic resonance has developed into a major technique in biology, chemistry, and medicine. Yet, surveying the developments of the past decade suggests that the "Golden Age" of biological NMR spectroscopy is in its infancy. Continued advances in sample preparation, magnet, and rf technology, NMR spectroscopic techniques, and computational methods promise to make the next 50 years as exciting as the last.

References

1. D. C. Dalgarno, B. A. Levine, and R. J. P. Williams, *Biosci. Rep.* **3**, 443–452 (1983).
2. K.-H. Gross and H. R. Kalbitzer, *J. Magn. Reson.* **76**, 87–99 (1988).
3. A. Pastore and V. Saudek, *J. Magn. Reson.* **90**, 165–176 (1990).
4. M. P. Williamson, *Biopolymers* **29**, 1423–1431 (1990).
5. D. S. Wishart, B. D. Sykes, and F. M. Richards, *J. Mol. Biol.* **222**, 311–333 (1991).
6. D. S. Wishart and B. D. Sykes, *Meth. Enzymol.* **239**, 363–392 (1994).
7. P. J. Kraulis, *J. Mol. Biol.* **243**, 696–718 (1994).
8. W. J. Chazin and P. E. Wright, *Biopolymers* **26**, 973–977 (1987).
9. W. J. Chazin and P. E. Wright, *J. Mol. Biol.* **202**, 623–636 (1988).
10. K. Wüthrich, "NMR of Proteins and Nucleic Acids," pp. 1–292. Wiley, New York, 1986.
11. A. D. Kline, W. Braun, and K. Wüthrich, *J. Mol. Biol.* **204**, 675–724 (1988).
12. M. Billeter, W. Braun, and K. Wüthrich, *J. Mol. Biol.* **155**, 321–346 (1982).
13. K. Wüthrich, *Biopolymers* **22**, 131–138 (1982).
14. D. L. Di Stefano and A. J. Wand, *Biochemistry* **26**, 7272–7281 (1987).
15. S. W. Englander and A. J. Wand, *Biochemistry* **26**, 5953–5958 (1987).
16. S. J. Nelson, D. M. Schneider, and A. Wand, *Biophys. J.* **59**, 1113–22 (1991).
17. G. J. Kleywegt, R. M. N. J. Lamerichs, R. Boelens, and R. Kaptein, *J. Magn. Reson.* **85**, 186–197 (1989).
18. F. M. van de Ven, *J. Magn. Reson.* **86**, 633–644 (1990).
19. M. Billeter, V. J. Basus, and I. D. Kuntz, *J. Magn. Reson.* **76**, 400–415 (1988).
20. P. J. Kraulis, *J. Magn. Reson.* **84**, 627–633 (1989).
21. G. J. Kleywegt, R. Boelens, and R. Kaptein, *J. Magn. Reson.* **88**, 601–608 (1990).
22. C. Eccles, P. Güntert, M. Billeter, and K. Wüthrich, *J. Biomol. NMR* **1**, 111–130 (1991).
23. C. Cieslar, T. A. Holak, and H. Oschkinat, *J. Magn. Reson.* **89**, 184–190 (1990).
24. H. Oschkinat, T. A. Holak, and C. Ciesler, *Biopolymers* **31**, 699–712 (1991).
25. G. W. Vuister, R. Boelens, A. Padilla, and R. Kaptein, *J. Biomol. NMR* **1**, 421–438 (1991).
26. R. Bernstein, A. Ross, C. Cieslar, and T. A. Holak, *J. Magn. Reson., Ser. B* **101**, 185–188 (1993).
27. J. Xu, S. K. Strauss, B. C. Sanctuary, and L. Trimble, *J. Magn. Reson., Ser. B* **103**, 53–58 (1994).
28. G. M. Clore, A. Bax, P. C. Driscoll, P. T. Wingfield, and A. M. Gronenborn, *Biochemistry* **29**, 8172–8184 (1990).
29. P. C. Driscoll, G. M. Clore, D. Marion, P. T. Wingfield, and A. M. Gronenborn, *Biochemistry* **29**, 3542–3556 (1990).
30. M. D. Carr, B. Birdsall, T. A. Frenkiel, C. J. Bauer, J. Jimenez-Barbero, V. I. Polshakov, J. E. McCormick, G. C. K. Roberts, and J. Feeney, *Biochemistry* **30**, 6330-6341 (1991).
31. R.T. Clubb, V. Thanabal, C. Osborne, and G. Wagner, *Biochemistry* **30**, 7718–7730 (1991).
32. W. J. Fairbrother, J. Cavanagh, H. J. Dyson, A. G. Palmer, S. L. Sutrina, J. Reizer, M. H. Saier, and P. E. Wright, *Biochemistry* **30**, 6896–6907 (1991).
33. W. J. Fairbrother, A. G. Palmer, M. Rance, J. Reizer, M. H. Saier, and P. E. Wright, *Biochemistry* **31**, 4413–4425 (1992).
34. B. J. Stockman, N. R. Nirmala, G. Wagner, T. J. Delcamp, M. T. DeYarman, and J. H. Freisheim, *Biochemistry* **31**, 218–229 (1992).
35. G. T. Montelione, B. A. Lyons, S. D. Emerson, and M. Tashiro, *J. Am. Chem. Soc.* **114**, 10974–10975 (1992).

36. R. T. Clowes, W. Boucher, C. H. Hardman, P. J. Domaille, and E. D. Laue, *J. Biomol. NMR* **3**, 349–354 (1993).

37. S. Grzesiek, J. Anglister, and A. Bax, *J. Magn. Reson., Ser. B* **101**, 114–119 (1993).

38. T. M. Logan, E. T. Olejniczak, R. X. Xu, and S. W. Fesik, *J. Biomol. NMR* **3**, 225–231 (1993).

39. B. A. Lyons, M. Tashiro, L. Cedergren, B. Nilsson, and G. T. Montelione, *Biochemistry* **32**, 7839–7845 (1993).

40. B. A. Lyons and G. T. Montelione, *J. Magn. Reson., Ser. B* **101**, 206–209 (1993).

41. R. Powers, D. S. Garrett, C. J. March, E. A. Frieden, A. M. Gronenborn, and G. M. Clore, *Biochemistry* **31**, 4334–4346 (1992).

42. M. S. Friedrichs, L. Mueller, and M. Wittekind, *J. Biomol. NMR* **4**, 703–726 (1994).

43. R. P. Meadows, E. T. Olejniczak, and S. W. Fesik, *J. Biomol. NMR* **4**, 79–96 (1994).

44. J. B. Olson, Jr. and J. L. Markley, *J. Biomol. NMR* **4**, 385–410 (1994).

45. D. Zimmerman, C. Kulikowski, L. Wang, B. Lyons, and G. T. Montelione, *J. Biomol. NMR* **4**, 241–256 (1994).

46. K. Ösapay and D. A. Case, *J. Am. Chem. Soc.* **113**, 9436–9444 (1991).

47. K. Ösapay and D. A. Case, *J. Biomol. NMR* **4**, 215–230 (1994).

48. J. W. Keepers and T. L. James, *J. Magn. Reson.* **57**, 404–426 (1984).

49. T. Havel, in "Proteins: Structure, Dynamics, Design" (V. Renugopalakrishnan, P. R. Carey, I. C. P. Smith, S. Huans, and A. L. Storer, eds.), pp. 110–115. ESCOM Science Publishers, Leiden, Holland, 1991.

50. M. Karplus, *J. Phys. Chem.* **30**, 11–15 (1959).

51. V. F. Bystrov, *Progr. Nucl. Magn. Reson.* **10**, 41–82 (1976).

52. A. Pardi, M. Billeter, and K. Wüthrich, *J. Mol. Biol.* **180**, 741–751 (1984).

53. G. Wagner, W. Braun, T. F. Havel, T. Schaumann, N. Gō, and K. Wüthrich, *J. Mol. Biol.* **196**, 611–639 (1987).

54. G. T. Montelione, S. D. Emerson, and B. A. Lyons, *Biopolymers* **32**, 327–334 (1992).

55. G. W. Vuister, S. Grzesiek, F. Delaglio, A. C. Wang, R. Tschudin, G. Zhu, and A. Bax, *Meth. Enzymol.* **239**, 79–105 (1994).

56. M. J. Minch, *Concepts Magn. Reson.* **6**, 41–56 (1994).

57. C. M. Deber, D. A. Torchia, and E. R. Blout, *J. Am. Chem. Soc.* **93**, 4893–4897 (1971).

58. A. DeMarco, M. Llinás, and K. Wüthrich, *Biopolymers* **17**, 617–636 (1978).

59. S. Ludvigsen, K. V. Andersen, and F. M. Poulsen, *J. Mol. Biol.* **217**, 731–736 (1991).

60. L. J. Smith, M. J. Sutcliffe, C. Redfield, and C. M. Dobson, *Biochemistry* **30**, 986–996 (1991).

61. G. W. Vuister and A. Bax, *J. Am. Chem. Soc.* **115**, 7772–7777 (1993).

62. G. S. Harbison, *J. Am. Chem. Soc.* **115**, 3026–3027 (1993).

63. S. G. Hyberts, W. Märki, and G. Wagner, *Eur. J. Biochem.* **164**, 625–635 (1987).

64. G. Wagner, *Quart. Rev. Biophys.* **16**, 1–57 (1983).

65. S. Forsén and R. A. Hoffman, *Acta Chem. Scand.* **17**, 1787 (1963).

66. S. Forsén and R. A. Hoffman, *J, Chem Phys.* **39**, 2892 (1963).

67. S. Hyberts, M. S. Goldberg, T. F. Havel, and G. Wagner, *Protein Sci.* **1**, 736–751 (1992).

68. G. M. Crippen and T. F. Havel, "Distance Geometry and Molecular Conformation," Research Studies Press, Taunton, (UK), 1988.

69. T. F. Havel, *Prog. Biophys. Mol. Biol.* **56**, 43–78 (1991).

70. W. Braun and N. Gō, *J. Mol. Biol.* **186**, 611–626 (1985).

71. P. Güntert, W. Braun, and K. Wüthrich, *J. Mol. Biol.* **217**, 517–530 (1991).

72. A. T. Brünger, G. M. Clore, A. M. Gronenborn, and M. Karplus, *Proc. Natl. Acad. Sci.* (USA) **83**, 3801–3805 (1986).

73. G. M. Clore, A. T. Brünger, M. Karplus, and A. M. Gronenborn, *J. Mol. Biol.* **191,** 523–551 (1986).

74. M. Nilges, G. M. Clore, and A. M. Gronenborn, *FEBS Lett.* **239,** 129–136 (1988).

75. M. Nilges, G. M. Clore, and A. M. Gronenborn, *FEBS Lett.* **239,** 317–324 (1988).

76. P. Güntert, W. Braun, M. Billeter, and K. Wüthrich, *J. Am. Chem. Soc.* **111,** 3997–4004 (1989).

77. C. Gonzalez, J. A. C. Rullmann, A. M. J. J. Bonvin, R. Boelens, and R. Kaptein, *J. Magn. Reson.* **91,** 659–664 (1991).

78. P. Yip and D. A. Case, *J. Magn. Reson.* **83,** 643–648 (1989).

79. A. M. J. J. Bonvin, R. Boelens, and R. Kaptein, *J. Biomol. NMR* **1,** 305–309 (1991).

80. H. Liu, P. D. Thomas, and T. J. James, *J. Magn. Reson.* **98,** 163–175 (1992).

81. M. J. Dellwo, D. M. Schneider, and A. J. Wand, *J. Magn. Reson., Ser. B* **103,** 1–9 (1994).

82. R. Levy, O. Assulin, T. Scherf, M. Levitt, and J. Anglister, *Biochemistry* **28,** 7168–7175 (1989).

83. P. A. Campbell and B. D. Sykes, *J. Magn. Reson.* **93,** 77–92 (1991).

84. G. Otting and K. Wüthrich, *Quart. Rev. Biophys.* **23,** 39–96 (1990).

85. M. Ikura and A. Bax, *J. Am. Chem. Soc.* **114,** 2433–2440 (1992).

SUGGESTED READING

The literature concerning NMR spectroscopy contains a number of texts that amplify and complement the present text. A selection of these sources is listed below.

Biomolecular NMR Spectroscopy

Bertini, I., and Luchinat, C., "NMR of Paramagnetic Molecules in Biological Systems," pp. 1–319. Benjamin/Cummings, Menlo Park, CA, 1986.

Clore, G. M., and Gronenborn, A. M. (Eds.), "NMR of Proteins," pp. 1–307. CRC Press, Boca Raton, 1993.

James, T. L., and Oppenheimer, N. J. (Eds.), *Meth. Enzymol.* **239,** 1–813 (1994).

Roberts, G. C. K. (Ed.), "NMR of Macromolecules," pp. 1–399. IRL Press, Oxford, 1993.

Wüthrich, K., "NMR of Proteins and Nucleic Acids," pp. 1–292. Wiley, New York, 1986.

NMR Spectroscopy

Abragam, A., "Principles of Nuclear Magnetism," pp. 1–599. Clarendon Press, Oxford, 1961.

Corio, P. L., "Structure of High-Resolution NMR Spectra," pp. 1–548. Academic Press, New York, 1967.

Ernst, R. R., Bodenhausen, G., and Wokaun, A., "Principles of Nuclear Magnetic Resonance in One and Two Dimensions," pp. 1–610. Clarendon Press, Oxford, 1987.

Farrar, T. C., and Harriman, J. E., "Density Matrix Theory and Its Applications in NMR Spectroscopy," pp. 1–211. Farragut Press, Madison, WI, 1992.

Goldman, M., "Quantum Description of High-Resolution NMR in Liquids," pp. 1–268. Clarendon Press, New York, 1988.

Harris, R. K., "Nuclear Magnetic Resonance Spectroscopy," pp. 1–260. Longman Scientific and Technical, Harlow, 1986.

Munowitz, M., "Coherence and NMR," pp. 1–289. Wiley, New York, 1988.

Slichter, C. P., "Principles of Magnetic Resonance," pp. 1–655. Springer-Verlag, Berlin, 1990.

Quantum Mechanics

Blum, K., "Density Matrix Theory and Applications," pp. 1–217. Plenum, New York, 1981.

Levine, I. N., "Quantum Chemistry," pp. 1–566. Allyn and Bacon, Boston, 1983.

McQuarrie, D. A., "Quantum Chemistry," pp. 1–517. University Science Books, Mill Valley, CA, 1983.

Merzbacher, E., "Quantum Mechanics," pp. 1–621. Wiley, New York, 1970.

INDEX

The words "nuclear," "spectroscopy," and "spin" are not indexed unless necessary for clarity. For example, "nuclear magnetic moment" is indexed as "magnetic moment," but "spin diffusion" is indexed in full as "spin diffusion." Nuclear isotopes are indexed under the atomic symbol, rather than the atomic number. For example, ^{1}H is indexed under the letter "H."

559